WITHDRAWN

TRUE BUGS OF THE WORLD

(Hemiptera: Heteroptera)

Copyright © 1995 by Cornell University

All rights reserved. Except for brief quotations in a review, this book, or parts thereof, must not be reproduced in any form without permission in writing from the publisher. For information, address Cornell University Press, Sage House, 512 East State Street, Ithaca, New York 14850.

First published 1995 by Cornell University Press.

Library of Congress Cataloging-in-Publication Data
Schuh, Randall T.
    True bugs of the world (Hemiptera:Heteroptera) : classification
and natural history / Randall T. Schuh, James A. Slater.
       p. cm.
    Includes bibliographical references and index.
    ISBN 0-8014-2066-0 (alk. paper)
    1. Hemiptera. 2. Hemiptera—Classification. I. Slater, James
Alexander. II. Title. III. Title: Heteroptera.
QL521.S38 1995
595.7\54—dc20   94-32643

Printed in the United States of America.
Frontispiece printed in Hong Kong.

∞ The paper in this book meets the minimum requirements
of the American National Standard for Information Sciences—
Permanence of Paper for Printed Library Materials, ANSI Z39.48-1984.

Pedro W. Wygodzinsky, 1916–1987

René H. Cobben, 1925–1987

Occasionally, remarkable individuals emerge in a given field of study, leaving for posterity writings transcendent in their scope and organization. We dedicate this volume to two such individuals, our late colleagues René Cobben and Pedro Wygodzinsky. Although they died prematurely with much important work unfinished, both left a legacy of knowledge on the heteropterans of their special interest as well as a profound influence on our conception of the broader aspects of heteropteran morphology and classification. As authors of the present volume, we have been inspired by the activities and influenced by the opinions of these two great heteropterists, a fact reflected by our frequent citation of their works. We regret only that they could not be with us now to influence the next generation with their unbounded enthusiasm for the study of bugs.

CONTENTS

Preface xi

Chapter 1.	A History of the Study of the Heteroptera	1
Chapter 2.	Major Workers on the Heteroptera	6
Chapter 3.	Sources of Information	14
Chapter 4.	Collecting, Preserving, and Preparing Heteroptera	17
Chapter 5.	Habitats and Feeding Types	20
Chapter 6.	Wing Polymorphism	23
Chapter 7.	Mimicry and Protective Coloration and Shape	27
Chapter 8.	Heteroptera of Economic Importance	32
Chapter 9.	Historical Biogeography	38
Chapter 10.	General Adult Morphology and Key to Infraorders of Heteroptera	41

HETEROPTERA

Chapter 11.	Enicocephalomorpha	*by Pavel Štys*	67
Chapter 12.	Aenictopecheidae	*by Pavel Štys*	68
Chapter 13.	Enicocephalidae	*by Pavel Štys*	70

EUHETEROPTERA

Chapter 14.	Dipsocoromorpha	*by Pavel Štys*	74
Chapter 15.	Ceratocombidae	*by Pavel Štys*	75
Chapter 16.	Dipsocoridae	*by Pavel Štys*	78
Chapter 17.	Hypsipterygidae	*by Pavel Štys*	80
Chapter 18.	Schizopteridae	*by Pavel Štys*	80
Chapter 19.	Stemmocryptidae	*by Pavel Štys*	82

NEOHETEROPTERA

Chapter 20.	Gerromorpha	84

Mesoveloidea

Chapter 21.	Mesoveliidae	88

Hebroidea

Chapter 22.	Hebridae	90

Chapter 23.	Paraphrynoveliidae	92
Chapter 24.	Macroveliidae	93

Hydrometroidea

Chapter 25.	Hydrometridae	95

Gerroidea

Chapter 26.	Hermatobatidae	97
Chapter 27.	Veliidae	98
Chapter 28.	Gerridae	102

PANHETEROPTERA

Chapter 29.	Nepomorpha	107

Nepoidea

Chapter 30.	Belostomatidae	111
Chapter 31.	Nepidae	114

Ochteroidea

Chapter 32.	Gelastocoridae	116
Chapter 33.	Ochteridae	118

Corixoidea

Chapter 34.	Corixidae	119

Naucoroidea

Chapter 35.	Potamocoridae	122
Chapter 36.	Naucoridae	124
Chapter 37.	Aphelocheiridae	126

Notonectoidea

Chapter 38.	Notonectidae	127
Chapter 39.	Pleidae	129
Chapter 40.	Helotrephidae	130
Chapter 41.	Leptopodomorpha	134

Saldoidea

Chapter 42.	Aepophilidae	136
Chapter 43.	Saldidae	137

Leptopodoidea

Chapter 44.	Omaniidae	141
Chapter 45.	Leptopodidae	142
Chapter 46.	Cimicomorpha	146

Reduvioidea

Chapter 47.	Pachynomidae	148
Chapter 48.	Reduviidae	150

Velocipedoidea

Chapter 49.	Velocipedidae	161

Microphysoidea

Chapter 50.	Microphysidae	161

Joppeicoidea
    Chapter 51.   Joppeicidae   164

Miroidea
    Chapter 52.   Thaumastocoridae   165
    Chapter 53.   Miridae   169
    Chapter 54.   Tingidae   180

Naboidea
    Chapter 55.   Medocostidae   184
    Chapter 56.   Nabidae   186

Cimicoidea
    Chapter 57.   Lasiochilidae   190
    Chapter 58.   Plokiophilidae   190
    Chapter 59.   Lyctocoridae   194
    Chapter 60.   Anthocoridae   195
    Chapter 61.   Cimicidae   199
    Chapter 62.   Polyctenidae   202

Chapter 63.   Pentatomomorpha   205

Aradoidea
    Chapter 64.   Aradidae   208
    Chapter 65.   Termitaphididae   214

Pentatomoidea
    Chapter 66.   Acanthosomatidae   215
    Chapter 67.   Aphylidae   218
    Chapter 68.   Canopidae   219
    Chapter 69.   Cydnidae   220
    Chapter 70.   Dinidoridae   225
    Chapter 71.   Lestoniidae   227
    Chapter 72.   Megarididae   228
    Chapter 73.   Pentatomidae   229
    Chapter 74.   Phloeidae   234
    Chapter 75.   Plataspidae   236
    Chapter 76.   Scutelleridae   238
    Chapter 77.   Tessaratomidae   241
    Chapter 78.   Thaumastellidae   243
    Chapter 79.   Urostylidae   245

Lygaeoidea
    Chapter 80.   Berytidae   246
    Chapter 81.   Colobathristidae   249
    Chapter 82.   Idiostolidae   251
    Chapter 83.   Lygaeidae   251
    Chapter 84.   Malcidae   264
    Chapter 85.   Piesmatidae   266

Pyrrhocoroidea
    Chapter 86.   Largidae   268
    Chapter 87.   Pyrrhocoridae   270

Coreoidea

Chapter 88.	Alydidae	271
Chapter 89.	Coreidae	274
Chapter 90.	Hyocephalidae	279
Chapter 91.	Rhopalidae	281
Chapter 92.	Stenocephalidae	283

Literature Cited   285

Glossary   317

Index   323

About the Authors   337

PREFACE

The Heteroptera, or true bugs, are the largest and most diverse group of insects with incomplete metamorphosis. They are generally treated as a suborder of the Hemiptera, and a majority of the 75 families occur on all continents (except Antarctica) and on many islands. Their great age and apparent adaptability have resulted, over evolutionary time, in extreme structural and biological diversity.

The earliest volume to deal solely with the Hemiptera (including Heteroptera) was Fabricius's *Systema Rhyngotorum* (1803). It was followed 40 years later by the much more comprehensive and influential *Histoire naturelle des insectes hémiptères* of Amyot and Serville (1843). In subsequent years many treatments have appeared, most of them dealing with a single family or a specific region. More general treatments have usually appeared in textbooks, wherein individual families are given spare coverage and less familiar families are often excluded completely because of their rarity or a lack of space or both. The number of taxa known by Fabricius was comparatively small, and thus he was able to treat the world fauna at the species level. Today, with over 38,000 known species of Heteroptera, such an undertaking would take several lifetimes and many volumes.

Until now, the most recent attempt to treat all families of Heteroptera was *The Biology of the Heteroptera* (Miller, 1956a, 1971). This work concentrated heavily on the Reduviidae, the family of Miller's primary interest. His treatments of most of the other families were generally too brief to be adequately informative and in some cases contained numerous factual errors.

The present volume was conceived as a way of providing a general summary of what is currently known about the Heteroptera. We first offer a nonsystematic introduction to the group in chapters that cover the history of its study; a review of the major workers, techniques, and sources of specimens; attributes of general biological interest; selected taxa of economic importance; and basic morphology. Second, we present a current classification of the Heteroptera, synthesizing to the subfamily and sometimes tribal level the enormous and scattered literature and supplying diagnoses, keys, figures, general natural history information, a summary of distributions, and a listing of important faunistic works. Third, we list references for over 1350 published works dealing with Heteroptera. Finally, we provide a glossary as an aid to organizing and interpreting the welter of terms that have appeared over the years and that often differ for the same structure from family to family.

Restating the view of Cobben (1968:360): "Authors of textbooks and faunistic works are . . . faced with a bewildering array of 'higher classification[s]' and proper selection is extremely difficult. Confusion and error [are] bound to occur." Although a great number of classifications have been proposed for the Heteroptera, many are inadequately documented. We have therefore attempted to provide a classification for the group which is best supported by existing information—down to the subfamily level. This classification, we hope, will serve all biologists studying the Heteroptera as a framework for the presentation of other comparative information.

Taxa are grouped by infraorder, superfamily, family, and subfamily. Although we provide diagnoses for infraorders, families, and subfamilies, we do not provide them for superfamilies because superfamily concepts, particularly in the Pentatomomorpha, are in flux, with little agreement in the literature among the various authors. The sequence of family presentation is usually alphabetic in those groups for which no credible phylogenetic evidence has been published. The lack of such evidence is most glaringly obvious in the Pentatomomorpha.

The structure of information presented in the family treatments is parallel for the most part. For a few of the larger families, however, we have included information on natural history and distribution and faunistics under each subfamily, because we found it impossible to produce useful generalizations at the family level, in contrast to the approach taken for most of the smaller families.

Pavel Štys, Charles University, Prague, deserves special thanks for his substantial contribution to this work. He wrote the sections on Enicocephalomorpha and Dipsocoromorpha and prepared the keys for the Nepomorpha chapters. Dr. Štys provided critical commentary on several of the introductory chapters as well as all chapters dealing with the Cimicomorpha. Last, but not least, he made many helpful suggestions regarding the general organization and contents of the volume.

The following colleagues assisted in the preparation of this work. For reading and commenting on the entire manuscript we thank W. R. Dolling, I. M. Kerzhner, G. M. Stonedahl, and M. H. Sweet. For reading and commenting on portions of the manuscript we thank N. M. Andersen, J. Grazia, T. J. Henry, D. A. Polhemus, J. T. Polhemus, C. W. Schaefer, and M. D. Schwartz. For assistance in the library and with securing references we thank S. O. Fischl, C. Chaboo Michalski, R. Packauskus, J. T. Polhemus, M. D. Schwartz, G. M. Stonedahl, the staff of the American Museum of Natural History library, and the many colleagues who have faithfully sent us papers over the years. For the loan of original figures we thank R. C. Froeschner, J. D. Lattin, A. S. Menke, C. W. Schaefer, Kathleen Schmidt, G. M. Stonedahl, and the Division of Entomology, CSIRO, Canberra, Australia. For preparation of original figures we thank Kathleen Schmidt. For the loan of specimens used in preparation of the scanning micrographs of Nepomorpha we thank J. T. Polhemus.

Donna Englund and Beatrice Brewster assisted with editing and printing various drafts of the manuscript. Peling Fong Melville and William Barnett, Interdepartmental Laboratory, American Museum of Natural History, assisted in the preparation of the scanning electron micrographs. M. D. Schwartz assisted in scanning and assembling many of the black-and-white line illustrations, with help from J. M. Carpenter, G. Sandlant, and S. Stock. Lee Herman facilitated preparation of the index.

We especially thank our wives, Brenda Massie and Elizabeth Slater, for attentively listening to progress reports and other tedious details concerning the preparation of this volume.

Robb Reavill, Cornell University Press, worked patiently with us and offered many encouraging words during the lengthy preparation of the manuscript. Helene Maddux and Margo Quinto helped bring the text into its final form.

All previously published figures are used with permission. We gratefully acknowledge the generous cooperation of publishers, authors, and artists for allowing us to reproduce their work.

The College of Agriculture and Life Sciences Fund, Cornell University, helped support preparation of the artwork for this volume.

RANDALL T. SCHUH
JAMES A. SLATER

New York, New York
Storrs, Connecticut

TRUE BUGS OF THE WORLD

(Hemiptera: Heteroptera)

1
A History of the Study of the Heteroptera

Classification of the Heteroptera has reached its present state through a long evolutionary process beginning, insofar as modern systematics is concerned, with the work of Linnaeus. Taxonomic studies did not take place in an intellectual vacuum, but rather were the result of forces arising from both the general scientific community and society at large. Reviews of this historical development have been published by Štys and Kerzhner (1975) and Göllner-Scheiding (1991).

Early Attempts at Higher Classification

The first recognized higher group to include the true bugs was the Hemiptera of Linnaeus, a group that also included thrips, aphids, scale insects, and cicadas. Although, as the name indicates, Linnaeus based his group on the structure of the wings, he also recognized the distinctive nature of the hemipteran rostrum, subdividing the group into those insects with the "rostrum inflexum" (true bugs, cicadas, and other Auchenorrhyncha) and "rostrum pectorale" (scales and some other Sternorrhyncha). The true bugs were divided by Linnaeus in the tenth edition (1758) of the *Systema naturae* into three genera: *Notonecta, Nepa,* and *Cimex*. These are all familiar modern-day generic names, but the concepts attached to them have become more restricted over time, particularly for *Cimex.*

Fabricius, a student of Linnaeus, placed those insects with distinctive sucking mouthparts in his group Rhyngota (Rhynchota of later authors) and was the first to prepare a "monograph" of the group, the *Systema Rhyngotorum* (1803), in which he recognized 29 genera. Fabricius's greatest misconception, from the view of modern classifications, was the inclusion of *Pulex* (Siphonaptera), the fleas, in the Rhyngota.

The French naturalist Latreille used the Linnaean term Hemiptera to refer to the Rhyngota of Fabricius (but excluded the fleas). He formally named the subgroups Homoptera and Heteroptera (Latreille, 1810) and later divided the Heteroptera into the Geocorisae and Hydrocorisae, groupings based on the structure of the antennae.

Dufour (1833) subsequently divided the Geocorisae of Latreille, recognizing the Amphibicorisae (modern-day Gerromorpha). Fieber (1861) introduced the redundant descriptive terms Gymnocerata and Cryptocerata for the Geocorisae and Hydrocorisae, respectively.

A decade after the appearance of Dufour's (1833) monograph, the first comprehensive family-group classification of the Heteroptera was published by Amyot and Serville, their *Histoire naturelle des insectes Hémiptères* (1843). Many of the included names between the level of order and modern families have fallen into disuse because they were descriptive, rather than following the modern convention of being based on generic names. Nonetheless, the Amyot and Serville classification was a fundamental advance, and it has had a lasting impact.

Not all subsequent authors followed the nomenclature of Latreille, causing confusion as to which names should be applied to the subgroups of insects with their distinctive sucking mouthparts. The term Hemiptera has been applied by many North Americans only to the true bugs (the Heteroptera of Latreille) as an order coordinate with the Homoptera, rather than treating the two as subgroups of the more inclusive Hemiptera as done by most Europeans. Modern textbook authors such as Borror, Triplehorn, and Johnson (1989) have argued that the Heteroptera should be called Hemiptera and treated at the ordinal level because they are sufficiently morphologically distinct from the Homoptera.

We recognize what appear to be monophyletic groups and apply to them names of longest standing, irrespective of categorical rank. Thus, whereas the terms Hemiptera *sensu lato,* Coleorrhyncha, and Heteroptera identify monophyletic groups, it now seems clear that the Homoptera are not a natural group (e.g., CSIRO, 1991), and would better be referred to as Sternorrhyncha and Auchenorrhyncha (see Fig. 1.1).

Descriptive Foundations of Heteropteran Classification

The establishment of specific entities and the means for their recognition form the building blocks upon which all meaningful phylogenetic, biological, ecological, and physiological studies are based. Early work, including that of Linnaeus and Fabricius, in addition to establishing what has come to be recognized as a monophyletic assemblage, began the long tradition of describing, in a formal system, genera and species from all parts of the world. Their work was strongly influenced by the many exotic insects that were brought to Europe by explorers and early scientific expeditions. The impetus given to basic taxonomic work by these hitherto unknown species persists to the present day.

For each family of any size there exist descriptive developments, sometimes of great complexity, which mirror developments in higher classification. We have attempted to summarize these in discussions of individual families, and discuss only more general trends in the following paragraphs.

Slater (1974), in a discussion of the South African Heteroptera fauna, recognized four periods, which can also be applied to the development of basic knowledge of Heteroptera throughout the world: Classical Period, Period of European Specialists, Intermediate Period, and Contemporary Period.

The classical period, beginning with Linnaeus, extends roughly to 1870. Most of the workers during this time studied the Heteroptera in general. Although many published works were important syntheses encompassing the first major higher classifications, nearly all contained large amounts of basic descriptive work. Most works up to this point, with some notable exceptions, were devoid of illustrations, even up to and through the magnificently detailed and influential works of Carl Stål.

A series of specialized works began to appear primarily in Europe in the 1870s. Knowledge of some families had reached a point where what we today call revisional studies became imperative, while at the same time the opening of many previously inaccessible parts of the world brought rich collections into European centers and stimulated a great deal of descriptive work. Some of this work was faunistic, as for example the great *Biologia Centrali Americana* and *Fauna of British India*. It was a period of dominance for systematics, and some of the best scientific minds were involved. During this time Lethierry and Severin (1893–1896) produced the first and only world catalog of the Heteroptera. Their catalog was followed by comprehensive catalogs for the Palearctic (Oshanin, 1906–1909, 1912) and Nearctic (Van Duzee, 1916, 1917) faunas. This period of descriptive intensity, which continued until about 1920, was marked by the publication of impressive faunal compendia, and the inclusion of illustrations became much more common.

European dominance of systematics began to decline after World War I, and heteropteran studies greatly increased in the United States and Japan. Many important works were published during this period, including scores of papers by prolific authors such as C. J. Drake, H. H. Knight, H. B. Hungerford, and T. Esaki. Novel approaches were seen in works of authors such as H. H. Knight, who consistently applied the use of male genitalic characters in species recognition.

The modern period, beginning in about 1950, is remarkable in the history of heteropteran classification. It embraces the completion of many fundamental revisional studies, some covering the entire world, such as those of Usinger and Matsuda (1958) on the Aradidae, Usinger (1966) on the Cimicidae, and Andersen (1982a) on the Gerromorpha. World catalogs of several families also appeared, for example, Miridae (Carvalho, 1957, 1958a, b, 1959, 1960), Lygaeidae (Slater, 1964b), Tingidae (Drake and Ruhoff, 1965), and Reduviidae (Putchkov and Putchkov, 1985, 1986–1989; Maldonado, 1990), as well as the only up-to-date regional catalog, that of Henry and Froeschner (1988) on the North American fauna.

The First World Specialists

Many concepts of heteropteran classification up through the time of Fieber and his *Die europäischen Hemiptera* (1861) were based largely on the European fauna. With the appearance of the works of Carl Stål in the late 1850s, things began to change. In the course of two decades Stål monographed in a series of papers of increasing geographic scope—the best known and most sweeping being the *Enumeratio Hemipterorum* (1870–1876)—the Reduviidae, Lygaeidae, Coreidae, Pentatomidae, and several other families. In some groups such as the Coreidae, no subsequent equivalent work has appeared, and Stål's keys still serve as important aids to identification.

Stål's successor as dean of heteropteran taxonomy was O. M. Reuter. Whereas Stål had concentrated primarily on the Pentatomomorpha and Reduviidae, Reuter devoted most of his taxonomic efforts to the Miridae, Saldidae, and Cimicoidea. He authored the definitive works on the Palearctic fauna for the first two groups (Reuter, 1878–1896, 1912b) and produced a world classification of the Miridae (Reuter, 1910).

Reuter's contemporaries such as C. Berg, E. Bergroth, G. Breddin, G. C. Champion, W. L. Distant, G. Horvath, V. E. Jakovlev, G. W. Kirkaldy, J.-B. A. Puton, P. R. Uhler, and many others were involved mostly in the production of descriptions of new taxa in faunistic works, tremendously broadening knowledge of the world fauna.

Comparative Morphology in the Study of the Heteroptera

Beginning with the work of Tullgren (1918) on abdominal trichobothria, followed by the works of Singh-Pruthi (1925) on male genitalia, Poisson (1924) and Ekblom

(1926, 1930) on a variety of structural systems, and Spooner (1938) on the head capsule, comparative morphological studies in the Heteroptera began to develop in a way not encountered since the work of Dufour, and they began to be integrated into heteropteran classification. Although the impact of our understanding of morphological variation on classifications is in most cases discussed below under the relevant section dealing with morphology, some areas merit special mention.

Genitalia. Genitalia were first used in heteropteran classification by Verhoeff (1893) for the females and Sharp (1890) for the males. Although Poisson (1924) made an important contribution to detailing the structure of male genitalia in the Gerromorpha and Nepomorpha, Singh-Pruthi (1925) was the first to examine comparatively the detailed structure of the phallus for nearly all major heteropteran groups. His study has some limitations from a modern perspective, because he did not examine all major groups—notably the Enicocephalidae—and examined only gross structure. He nevertheless demonstrated that there are distinctive genitalic types within the Heteroptera, and he proposed a scheme of interrelationships based on those characters. Notable was his recognition of the pentatomoid genitalic type, although his reduvioid genitalic type almost certainly forms a group based on symplesiomorphy.

With the exception of the works of Ekblom (1926, 1930), the conclusions of Singh-Pruthi were not reexamined or extended until the comparative studies of Kullenberg (1947) on the Miridae and Nabidae. Further broad-based studies on genitalia (Pendergrast, 1957; Scudder, 1959), as well as those for individual families (e.g., Slater, 1950: female Miridae; Ashlock, 1957: male Lygaeidae; Kelton, 1959: male Miridae; Carayon: many papers on male and female Naboidea and Cimicoidea), produced results that greatly advanced our understanding of heteropteran relationships at all levels.

Internal anatomy. It was not until about 1940 that significant additional comparative documentation of anatomical details beyond that acquired by Dufour began to appear. Notable works include those of Baptist (1941) on the scent-gland system and Miyamoto (1961a) and Goodchild (1963) on the alimentary canal. Whereas many anatomical studies have not produced information on variation that proved to be of value in higher-level classification, existing knowledge makes it clear that the gut type found in the Pentatomomorpha is clearly distinctive and, along with many other characters, argues for the monophyly of that group. Other internal structures, such as the dorsal vessel and nervous system, are relatively homogeneous across all Heteroptera.

Ovariole numbers were first investigated on a comparative basis by Woodward (1950) and Carayon (1950a) and later by Miyamoto (1957, 1959) and Balduf (1964). Testis follicle numbers were initially studied by Woodward (1950) and later for the Miridae by Leston (1961). Although both sets of structures suggest certain patterns of relationships, they are nonetheless relatively simple and possess substantial variability. It is only in the Cimicoidea, a group first investigated through studies of the human bed bug and later broadened by Carayon (e.g., 1977) to include all cimicoids, that internal reproductive anatomy has offered a wealth of detail pertinent to classification of the group.

Methodological Issues in Heteropteran Classification

The earliest heteropteran classifications were based largely on single character systems, such as the structure of the wings or mouthparts. Additional discriminating characters—such as color, size, pronotal shape, antennal length—were largely those used to differentiate species.

Some classifications, such as that of Schiødte (1869, 1870) founded on coxal types, were based not only on single character systems, but also on ones that showed little concordance with other available information; consequently they were adopted by only a few subsequent authors. Others, as pointed out by Štys and Kerzhner (1975), were totally undocumented, introducing new taxonomic concepts in conjunction with the publication of checklists or catalogs.

Reuter (1905, 1910, 1912a) stated in his treatises on heteropteran phylogeny that the early classifications of the Heteroptera were linear in character, and it was therefore not appropriate to make deductions about phylogenetic relationships. As Reuter (1910:31) noted, when Fieber (1861) was "compelled to select characters for his key, which permitted him to put related families together in a series, it is certain that the selection of the structure of the tarsi, as a consequence of which he put the families Phymatidae, Aradidae, Tingididae, and Microphysae [sic] after each other, was a mistake" (our translation). Reuter further rejected the idea of deriving one group from another, as in the classifications of Kirkaldy.

Reuter's phylogenetic work incorporated several important innovations: first, he reviewed all of the previously proposed classifications of the group; second, he summarized and discussed the characters on which those classifications were based; and third, he made a list of characters he believed to be diagnostic for the groups he proposed. His effort was very nearly the preparation of a synapomorphy scheme. For example, with reference to the structure of the antennae in the Nepomorpha (Cryptocerata) he said (Reuter, 1910:26): "The water bugs have short, concealed antennae. This . . . type is doubtless . . .

an adaptation to life in the water that was acquired later [from long exposed antennae] and it might not be at all excluded that it represents a heterophyletic homomorphism unless other conditions were at hand which made it probable that at least most of the so-called Cryptocerata were homophyletic" (our translation). Reuter was also consistent in his reference to the composition of groups, using the terms homophyletic and heterophyletic, to refer to what in modern parlance would be monophyletic and polyphyletic.

Character polarity. Unfortunately, as has been the case with the work of many other authors old and new, Reuter's method for determining character polarity was probably the weakest aspect of his work. In some cases—such as number of tarsal segments—he used the "common equals primitive" principle. In other cases he treated some groups as more ancient than others because they appeared to bear a greater number of primitive characters. In this vein, Reuter's thinking was pervaded by the idea that the Heteroptera are strictly diagnosable on the structure of the "hemelytra." He stated on more than one occasion that it is clear that the corium, clavus, and membrane were present in the primitive forms. Thus, in his view the undifferentiated mesothoracic wings in the Enicocephalomorpha, Dipsocoromorpha, and Gerromorpha were explicitly derived from the differentiated type (Reuter, 1910). This point of view contrasts markedly with recent studies involving the search for congruence among more characters, which suggest that the relatively undifferentiated wings represent the primitive condition, the true "hemelytron" being derived.

Cobben (1978:5) commented that some authors who had worked on feeding behavior and mouthpart structure in phytophagous Hemiptera "failed to consider information previously published in a wide variety of papers on comparative morphology and systematics . . . or [believed] . . . that the Homoptera are more generalized (or more symplesiomorphous) than the Heteroptera." Although Cobben was at times inconsistent in his arguments—as for example, in treating the Gerromorpha as the basal heteropteran group—his criticism can be applied to the works of China and Myers (1929) and China (1933), who were of the opinion that the primitive Heteroptera were phytophagous and were homopteroid in character, and to Miles (1972) and Sweet (1979), who asserted that, because many pentatomomorphs produce a salivary sheath or feeding tube similar to that found in the Sternorrhyncha and Auchenorrhyncha, they must possess the primitive heteropteran feeding type.

Homology problems. Some authors arrived at erroneous conclusions concerning homology of structures. For example, China (1933) noted the "absence of arolia" in *Leotichius* Distant, which meant that there were no fleshy pads between the claws. Such remarks, which are widespread in the literature, overlook the fact that nearly all Heteroptera possess similarly placed—although sometimes differently formed—structures between the claws. Fleshy structures were arolia by definition, but setiform structures (parempodia) of similar position were not treated as homologous. Tullgren (1918) attempted to resolve this problem, but as indicated by the labeling in his figure 11, his concept of an arolium was based solely on a definition without respect to structural homology, and because he reasoned from a false premise he arrived at a false conclusion.

Adaptationist arguments. Character analysis in much of the earlier literature was permeated by adaptationist arguments. For example, China and Myers (1929), China (1933), and Kullenberg (1947) argued that Reuter had placed an "exaggerated importance" on the arolia and claws as a guide to relationships in the Heteroptera—and particularly the Miridae—because "these organs are far too plastic to serve as a fundamental group character" (China, 1933:192). Such arguments overlooked the fact that if characters did not vary they would be of no use in forming groups and that only by examining this variability for congruence with other characters does one arrive at a hierarchic classification.

Autapomorphy. Several authors have been impressed with the number of autapomorphous characters possessed by certain heteropteran taxa and have selectively used novelty of appearance as a measure for determining classificatory rank. Possibly the earliest of these was Linnaeus, in his recognition of *Nepa* and *Notonecta* as distinctive from his omnibus *Cimex*, the last clearly a group devoid of defining characters other than those of the true bugs as a whole. Distant (1904) originally (and in our view correctly) placed *Leotichius* in his Leptopinae (Leptopodidae). China (1933) elevated the group to family status, concluding that the taxon could not possibly belong to the Leptopodidae, not because it did not have many of the attributes of other members of that group, but rather because it had several distinctive features of its own including, among others, only 3 veins in the membrane with no closed cells, and a 1-2-2 tarsal formula. Along similar lines Froeschner and Kormilev (1989) and Maldonado (1990) argued that the Phymatinae are so distinctive as to merit family status, whereas all the features that diagnose the Reduviidae are also found in the Phymatinae, although sometimes in a more highly modified form.

The advent of cladistics and its more rigorous approach to character analysis is beginning to have an impact on heteropteran classification (Schuh, 1986c). As noted by Štys and Kerzhner (1975), the dismemberment of the classic Geocorisae comes from the realization that the group was based on only a few symplesiomorphies and that groups such as the Enicocephalomorpha and Dipsocoromorpha are not closely related to the two major groups of land bugs, the Cimicomorpha and Pentatomomorpha.

Modern Higher Classification of the Heteroptera

The classic subdivisions of Latreille and Dufour were used in heteropteran classifications well into the twentieth century. Although Reuter and others proposed subordinal names, most were never consistently adopted. Leston, Pendergrast, and Southwood (1954) introduced the terms Cimicomorpha and Pentatomomorpha in the first formal attempt to recognize natural groups within the polyphyletic Geocorisae. Their conclusions were based on accumulated evidence from comparative studies of internal anatomy and external morphology of the Heteroptera. The influence of their work was widely felt; other authors adopted these groups and applied the "morpha" nomenclature to additional higher groups in the Heteroptera, although not without some variant spellings along the way (see, e.g., Popov, 1971). More important, the work of Leston et al. (1954) spurred the attempt to document the monophyly of higher groups within the Heteroptera, with the eventual recognition of seven such groups—termed infraorders—all with typified names, as outlined by Štys and Kerzhner (1975). We present below diagnoses and references for each of these infraordinal groupings. The evidence for the relationships among them is discussed in the following paragraphs.

The first documented higher-level scheme for the seven heteropteran infraorders was that of Schuh (1979). The character data were drawn mostly from information assembled by Cobben (1978). Cobben (1981a, b) criticized Schuh's scheme because all the characters used to support it showed some within-group variability, even though he had argued that many of those same characters were possibly diagnostic for certain groups. Other authors found certain portions of the scheme of interest, as for example Andersen (1982a) who agreed on the placement of the Gerromorpha among the more primitive heteropteran infraorders. Slater's (1982) arrangement of the seven infraorders was in line with that of Schuh's scheme. Štys (1985a) provided a set of names for the basal inclusive groups on the cladogram (Fig. 1.1). More recently the authors of the *Insects of Australia* (CSIRO, 1991) have portrayed the Coleorrhyncha (Peloridiidae) as the sister

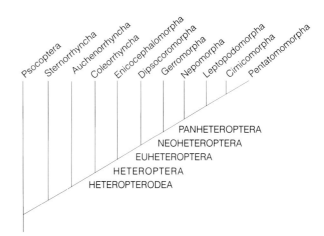

Fig. 1.1. Phylogenetic relationships of heteropteran infraorders (after Wheeler et al., 1993).

group of the Heteroptera (Heteropterodea; see Schlee, 1969), and the Auchenorrhyncha as the sister group of those two combined (e.g., Emel'yanov, 1987), thereby treating the classic Homoptera as paraphyletic. These outgroup relationships were not explicitly specified by Schuh (1979), who did not provide arguments for the polarity of characters used in his scheme.

Little new evidence for higher-group relationships within the Heteroptera was adduced since the work of Schuh (1979), until the publication of 18s nuclear rDNA sequences by Wheeler et al. (1993) for 29 hemipteran taxa, representing all infraorders and six outgroup taxa, including the Psocoptera, Sternorrhyncha, Auchenorrhyncha, and Coleorrhyncha. Their scheme is shown in Fig. 1.1, indicating substantial congruence between the molecular data and most of the morphological data used by Schuh (1979).

Certain things about this scheme are at variance with traditional ideas concerning phylogenetic relationships within the Heteroptera. First, the Heteroptera are primitively predaceous, contrary to the beliefs of China and Myers (1929), Miles (1972), Sweet (1979), and others. Second, the Enicocephalomorpha are the basal heteropteran lineage, rather than the Nepomorpha (Hydrocorisae) as proposed by Reuter (1910) and China (1933), or the Gerromorpha as proposed by Cobben (1978), or the Pentatomomorpha as strongly implied by Sweet (1979). Third, the "hemelytron" is not a ground-plan character for the Heteroptera, but rather a synapomorphy for the Panheteroptera. We now await additional evidence to test this scheme, around which much of the remainder of the present work is based.

2
Major Workers on the Heteroptera

The following section provides brief descriptions of now deceased influential or controversial workers on the Heteroptera from the time of Linnaeus. Where possible we provide dates and places of birth and death, a short biographical sketch, citations of particularly influential contributions to the field, and references to published bibliographies. Most of the information was located by consulting the works of Derksen and Göllner-Scheiding (1963–1975) and Gilbert (1977). We have not covered all workers because of limitations on space and available information. Names that could be considered of equal significance to many of those cited might include: G. C. Champion, A. Costa, E. F. Germar, F. E. Guérin-Méneville, R. F. Hussey, C. L. Kirschbaum, F. L. Laporte, I. LaRivers, O. Larsén, O. Lundblad, S. Matsumura, G. Mayr, W. L. McAtee, L. R. Meyer-Dür, P. Montrouzier, V. Motschulsky, E. Mulsant, E. C. Reed, C. P. Thunberg, F. Walker, J. O. Westwood, and A. Wroblewski. Living workers are dealt with by reference to their works at appropriate places in the text.

AMYOT, CHARLES JEAN-BAPTISTE. b. Vandeuvre, Aisne, France, September 23, 1799; d. Paris, October 13, 1866. Orphaned at early age, taken in by wealthy businessman Pavet with home in Paris opposite that of Audinet-Serville. Lawyer and influential person in French society. Coauthor with Serville of *Histoire naturelle des insectes Hémiptères* (1843). Had no personal collection; worked with that of Serville and Muséum National d'Histoire Naturelle, Paris. Bibliography published by Signoret (1866).

ASHLOCK, PETER D. b. San Francisco, California, USA, August 22, 1929; d. Lawrence, Kansas, January 26, 1989. Ph.D., University of California, Berkeley. Curator, Bishop Museum, Honolulu (1964–1967); professor, University of Kansas (1968–1989). Specialist on Orsillinae (Lygaeidae) and author of major classification of group (1967); also strong interest in systematic methodology. Collection deposited University of Kansas. Bibliography and list of names proposed published by Slater and Polhemus (1990).

BARBER, HARRY G. b. Hiram, Ohio, USA, April 20, 1871; d. Washington, D.C., January 27, 1960. Secondary school teacher in New York City. After retirement a specialist in Heteroptera, U.S. Department of Agriculture, Washington, D.C. Expert on Lygaeidae, particularly North American and Caribbean faunas. Collection deposited National Museum of Natural History, Washington, D.C. Bibliography of over 100 papers and list of names proposed published by Ashlock (1960).

BERG, CARLOS. b. Tuckum (Tukums), Curlandia, Latvia, April 2, 1843; d. Buenos Aires, Argentina, January 19, 1902. First employed as biologist at museum in Riga, Latvia, later in Baltic Polytechnic, Riga. Moved to Argentina June, 1873, to fill position in museum in Buenos Aires, of which H. Burmeister was then director. Lectured in local colleges and universities, succeeding Burmeister as director of museum in 1892. Published widely in biology; best known to heteropterists for *Hemiptera Argentina* (1879, 1884). Majority of types deposited in Museo de la Plata, a small number in Museo Nacional de Historia Natural, Buenos Aires. Bibliography published by Gallardo (1902).

BERGEVIN, ERNEST DE. b. 1859; d. 1933. Author of many alphataxonomic, faunistic, and bionomic papers on fauna of North Africa. Collection deposited in Muséum National d'Histoire Naturelle, Paris.

BERGROTH, ERNST EVALD. b. Jakobstad (Pietarsaari), Finland, April 1, 1857; d. Ekenäs, Finland, November 22, 1925. Swedish-speaking Finn, trained as a medical doctor. Lived in United States 1905–1911 (Alaska, Minnesota, Massachusetts), remainder of life in Finland. Known for linguistic ability, mastery of many heteropteran families (notably Aradidae), and biting critiques of work of others. Did little field work but mostly identified collections of others; numerous types deposited in University Zoological Museum, Helsinki, and elsewhere. Bibliography of 317 papers published by Lindberg (1928).

BLATCHLEY, W. S. b. North Madison, Connecticut, USA, October 6, 1859; d. Indianapolis, Indiana, May 28, 1940. Master's degree from Indiana University with thesis on butterflies of Indiana. State geologist of Indiana 1894–

1910. Individualist, all-around naturalist, entomological generalist, prolific publisher, and sometime antagonist of several heteropterist contemporaries (e.g., Knight, 1927; Blatchley, 1928). His *Heteroptera of Eastern North America* (1926), a singular and still widely respected work. Collection deposited at Purdue University, Lafayette, Indiana. Bibliography and list of new taxa published by Blatchley (1930, 1939).

BLIVEN, B. P. Resident of Eureka, California, USA; d. Eureka, ca. 1980. Between 1954 and 1973 described numerous species of Auchenorrhyncha and Heteroptera from western North America in his own journal, *Occidental Entomologist*. R. L. Usinger and others argued that the names of the reclusive Bliven should be suppressed because contemporaries were unable to examine specimens on which his taxa were based. Collection now available in California Academy of Sciences.

BREDDIN, GUSTAV. b. Magdeburg, Germany, February 25, 1864; d. Oschersleben, Germany, February 25, 1909. Director of a secondary school. Author of about 70 papers, mainly on Pentatomomorpha and Reduviidae from Orient, South America, and Africa. Collection deposited in Deutsches Entomologisches Institut, Berlin-Dahlem; some material in Senkenberg Museum, Frankfurt.

BURMEISTER, HERMANN CARL CONRAD. b. Stralsund, Germany, January 15, 1807; d. Buenos Aires, Argentina, May 2, 1892. Professor of zoology, University of Halle (1837–1861). Traveled and collected extensively in Brazil (1850–1852) and later Argentina, Peru, Panama, and Cuba (1857–1860). Left wife and family, moved to Buenos Aires, Argentina, in 1861 to begin new life as director of fledgling Museo Publico de Historia Natural, where he studied primarily fossil vertebrates. Prodigious worker, best known to entomologists and heteropterists for *Handbuch der Entomologie* (1832–1839; Heteroptera in vol. 2, 1835). Collection in Zoological Museum, Martin Luther University, Halle-Wittenburg, Germany.

BUTLER, EDWARD A. b. Alton, Hants, England, March 17, 1845; d. Clapham, England, November 20, 1925. Schoolmaster who studied natural history of Heteroptera, summarizing his life's observations in *A Biology of the British Hemiptera-Heteroptera* (1923). Collections deposited in The Natural History Museum, London.

CHINA, WILLIAM EDWARD. b. London, England, December 7, 1895; d. Mousehole, Cornwall, England, September 17, 1979. University education Trinity College, Cambridge University, D.Sc. 1948. Curator of Heteroptera (and Keeper of Entomology, 1955–1961), The Natural History Museum, London. General heteropterist: described new species, genera, and suprageneric taxa; elucidated genitalic and other morphology; established relationships among poorly known taxa; and resolved nomenclatorial problems. Known especially for assisting other heteropterists and for keys to world families and subfamilies of Heteroptera (China and Miller, 1959). Bibliography of 260 papers published by W. J. Knight (1980).

COBBEN, RENÉ H. b. Netherlands, 1925; d. Rhenen, Netherlands, 1987. Gifted morphologist and illustrator; specialist in taxonomy and morphology of Leptopodomorpha; professor of entomology, Netherlands Agricultural University, Wageningen, from 1954 until death. Author of monumental monographic treatises on egg structure and embryogenesis (1968a) and feeding structures (1978), each with voluminous additional observations and evolutionary syntheses; other important papers detailing classification and description of Leptopodomorpha. Collections from Caribbean, Africa, and Europe deposited in Agricultural University, Wageningen, Netherlands. Bibliography of 64 papers published by de Vrijer (1988).

DALLAS, WILLIAM SWEETLAND. b. London, England, January 31, 1824; d. London, May 28, 1890. Worked for a time in The Natural History Museum, London (around 1850), producing list of Hemiptera in its collections (1851–1852). Published additional books and translations and single-handedly prepared first five volumes of Insecta for *Zoological Record*. Collection in The Natural History Museum, London.

DE GEER, CARL. b. Finspang, Sweden, February 10, 1720; d. Stockholm, March 8, 1778. Wealthy Swedish nobleman of Dutch ancestry. Author of now rare work *Mémoires pour servir a l'histoire naturelle des insectes* (1752–1771). Collection in Swedish Museum of Natural History, Stockholm.

DISTANT, WILLIAM LUCAS. b. Rotherhithe, England, November 12, 1845; d. Wanstead, Essex, England, February 4, 1922. Businessman in hide-tanning industry. Traveled to Malaya and South Africa. From 1899 to 1920 part-time assistant, The Natural History Museum, London. Prolific species describer, ardent student of Cicadidae and Heteroptera. Author of *Rhopalocera Malayana* and *A Naturalist in the Transvaal;* best known to heteropterists for sections of *Biologia Centrali Americana* (1880–1893) and *Fauna of British India* (1902–1918). Criticized by Bergroth, Kirkaldy, Horvath, and Reuter (e.g., 1905) for failing to appreciate their works on heteropteran classification. Many types, deposited in The Natural History

Museum, London; additional types in Genoa, Italy. Bibliography of several hundred papers published by Dolling (1991a).

DOUGLAS, JOHN WILLIAM. b. Putney, England, November 15, 1814; d. Harlesden, England, August 28, 1905. Worked briefly at Kew Gardens, and then for over 50 years as a customshouse officer. Author of numerous papers on British Heteroptera, including *The British Hemiptera* (1865), coauthored with John Scott, first comprehensive work on subject. Types in Hope Museum, Oxford.

DRAKE, CARL J. b. Eaglesville, Ohio, USA, July 28, 1885; d. Washington, D.C., October 2, 1965. Ph.D. from Ohio State University under direction of Herbert Osborn. Longtime head of the Department of Zoology and Entomology, Iowa State College. World authority on Tingidae, and prolific publisher on Saldidae, Gerromorpha, and aspects of applied entomology. Primary interest in description of new species; included many excellent habitus illustrations of Saldidae and Tingidae in later papers. Collection and considerable estate willed to National Museum of Natural History, Washington, D.C. Bibliography of 519 papers and list of names proposed published by Ruhoff (1968).

DUFOUR, LEON. b. St. Sever, France, April 11, 1780; d. St. Sever, April 18, 1865. Published on a great diversity of insect groups. Author of first detailed anatomical work on Heteroptera, *Recherches anatomiques et physiologiques sur les Hémiptères* (1833), a magnificently illustrated study of salivary glands, alimentary tract, and reproductive and respiratory systems. Bibliography of more than 230 papers published by Laboulbéne (1865). Collection in Muséum National d'Histoire Naturelle, Paris.

ESAKI, TEISO. b. Tokyo, Japan, July 15, 1899; d. December 14, 1957. Professor of entomology, Kyushu University, Kyoto, Japan. Traveled widely in Europe, working with G. Horvath in Budapest and at The Natural History Museum, London. Best known for extensive work on aquatic Heteroptera, and particularly for discovery of Helotrephidae. Bibliography published by Hasegawa (1967).

FABRICIUS, JOHANN CHRISTIAN. b. Tonder, Denmark, January 7, 1745; d. Kiel, Germany, March 3, 1808. Trained as physician, close associate of Linnaeus, from early age interested in entomology. Described nearly 10,000 species. Appointed professor of natural history at University of Kiel at early age by patron. Author of *Systema Rhyngotorum* (1803), first monograph of Heteroptera. Status of Fabrician types and other information summarized in Zimsen (1964). Most material deposited in Zoological Museum, Copenhagen.

FALLÉN, CARL FRIEDRICH. b. 1764; d. 1830. Professor of Natural History, Lund University, Lund, Sweden. Provided first detailed treatments of Heteroptera of Scandinavia in *Monographia Cimicum Sueciae* (1807) and subsequent publications. Types in Swedish Museum of Natural History, Stockholm.

FIEBER, FRANZ XAVIER. b. March 1, 1807; d. Chrudim, Bohemia, Czech Republic, February 22, 1872. Director of Royal Imperial District Court at Chrudim. Author of earliest comprehensive publications on European fauna, culminating in *Die europäischen Hemiptera* (1861). Established modern generic classification of Miridae based on European fauna. Most types in Muséum National d'Histoire Naturelle, Paris and Naturhistorisches Museum, Vienna.

GMELIN, JOHANN FRIEDRICH. b. Tübingen, Germany, August 8, 1748; d. Göttingen, Germany, November 1, 1804. Professor of chemistry in Göttingen. Compiler of final (13th) edition of *Systema naturae* of Linnaeus.

HAHN, CARL WILHELM. b. December 16, 1786; d. 1836. German naturalist who worked in Nuremberg, on spiders as well as Heteroptera. Author of volumes 1–3 of *Die wanzenartigen Insecten* (1831–1835), in which were described as new large numbers of European and extra-european genera and species. Work completed by G. A. W. Herrich-Schaeffer.

HANDLIRSCH, ANTON. b. Vienna, Austria, January 20, 1865; d. Vienna, August 28, 1935. Longtime employee of Naturhistorisches Museum, Vienna. Specialist in taxonomy of Sphecidae and later Heteroptera (particularly Phymatinae), insect paleontology, and higher-level classification. Bibliography published by Beier (1935).

HEIDEMANN, OTTO. b. Magdeburg, Germany, September 1, 1842; d. Washington, D.C., USA, November 17, 1916. Emigrated to United States 1876, employed as an illustrator and engraver and later by U.S. Bureau of Entomology (USDA), Washington, D.C. Student of North American Heteroptera, mostly Aradidae, Miridae, and Tingidae. Collection deposited in Cornell University Insect Collection, Ithaca, New York. Bibliography of 34 papers published by Howard et al. (1916).

HERRICH-SCHAEFFER, GOTTLIEB A. W. b. Regensburg, Germany, December 17, 1799; d. Regensburg,

April 14, 1874. As did father, served as physician in courts in Regensburg, from 1833 until retirement in 1856. Author, among other works on Heteroptera, Lepidoptera, and so on of volumes 4–9 of *Die wanzenartigen Insecten* (1839–1853), completing work of C. W. Hahn. Types destroyed during World War II.

HORVATH, GEZA. b. Csecs, northern Hungary, November 23, 1847; d. Budapest, September 8, 1937. Trained as M.D., University of Vienna. First worked in control of pest insects; instrumental in elimination of *Phylloxera* from European vineyards; later appointed head, Department of Zoology, National Museum, Budapest. Traveled widely, active in international scientific affairs and influential in careers of other well-known heteropterists. Published on systematics of Heteroptera, Homoptera, and other subjects. Like A. N. Kiritshenko, reputedly made all observations with a hand lens. Bibliography of 467 papers published by Csiki (1944). Most types in Budapest.

HSIAO, TSAI-YU. b. Shandong (Shantung) Province, China, July 25, 1903; d. Tianjin, China, June 27, 1978. Basic education in China, graduate work in United States at University of Illinois, Oregon State College (M.S.), and Iowa State College (Ph.D.) under H. H. Knight. Returned to China in 1946 after 10 years in United States and became dean and professor of biology at Nankai University, Tianjin, and head of Tianjin Natural History Museum. Suffered during Cultural Revolution of late 1960s and early 1970s. Published on systematics of Heteroptera, most comprehensive being two volumes of *A Handbook for the Determination of Chinese Hemiptera-Heteroptera* (Hsiao, 1977, 1981). Bibliography published by Schaefer and Sailer (1980).

HUNGERFORD, HERBERT BARKER. b. Mahaska, Kansas, USA, August 30, 1885; d. Lawrence, Kansas, May 13, 1963. Ph.D., Cornell University. Professor of entomology, University of Kansas, Lawrence; mentor of many students in Gerromorpha and Nepomorpha. Author of *The Biology and Ecology of the Aquatic and Semiaquatic Hemiptera* (1919) and comprehensive works on *Notonecta* of the world (1933) and Corixidae of the Western Hemisphere (1948), as well as numerous other papers. Outstanding collection of Gerromorpha and Nepomorpha deposited University of Kansas. Bibliography published by Woodruff (1956; 1963).

JACZEWSKI, TADEUSZ L. b. St. Petersburg, Russia, February 1, 1899; d. February 25, 1974. Studied University of St. Petersburg; moved to Warsaw University, Poland, 1920. Employed Warsaw Zoological Museum and Warsaw University. Organized several Polish faunal series. Primary systematic interests in Nepomorpha, particularly Corixidae, often in cooperation with O. Lundblad. Prepared sections on aquatic and semiaquatic bugs for *Keys to the Insects of the European USSR* (Kerzhner and Jaczewski, 1964). Bibliography published by Wroblewski (1974).

JAKOVLEV, VASILIY E. b. Tsaritsyn (now Volgograd), Russia, February 9, 1839; d. Eupatoriya, Ukraine, August 15, 1908. A government official living in different locations, for 20 years in Astrakhan, for 12 years in Irkutsk; upon retirement in 1898 lived briefly in St. Petersburg and then in Eupatoriya (Crimea) until death. Conducted some vertebrate studies; best known to heteropterists for work on Russian fauna. Collections deposited in Zoological Institute, St. Petersburg. Bibliography and list of names proposed published by Semenov-Tian-Shanski (1910).

JEANNEL, RENÉ. b. 1879; d. Paris, February 20, 1965. One-time director of Muséum National d'Histoire Naturelle, Paris; ardent collector, coleopterist, biogeographer, and monographer of Enicocephalidae (1941). Bibliography published by Delamare Deboutteville and Paulian (1966). Collections in Muséum National d'Histoire Naturelle, Paris.

KIRITSHENKO, ALEXANDR NIKOLAYEVICH. b. Berdyansk, Zaporozh'ye Province, Russia (now Ukraine), September 9, 1884; d. St. Petersburg, January 23, 1971. Longtime curator of Heteroptera, Zoological Institute, St. Petersburg; worked actively to time of death. Author of numerous papers on Palearctic fauna. Built extensive collection of Palearctic Heteroptera at Zoological Institute, St. Petersburg. Bibliography published by Kerzhner and Stackelberg (1971).

KIRKALDY, GEORGE WILLIS. b. Clapham, England, July 26, 1873; d. San Francisco, California, USA, February 2, 1910. Educated in England, moved to Honolulu, Hawaii, 1903. Worked initially for USDA Board of Agriculture and Forestry, later for Hawaiian Sugar Planters' Experiment Station. Author of numerous works on Auchenorrhyncha and Heteroptera, mainly on taxonomy, nomenclature, bibliography, and natural history, possibly best known among them a world catalog of Pentatomoidea (1909). Bibliography published by Dolling (1991a). Collections deposited The Natural History Museum, London, Bishop Museum, Honolulu; Snow Entomological Museum, University of Kansas, Lawrence; and National Museum of Natural History, Washington, D.C.

KNIGHT, HARRY HAZELTON. b. Koshkonong, Missouri, USA, May 13, 1889; d. Ames, Iowa, September 6, 1976. Ph.D., Cornell University. Faculty member, University of Minnesota 1919–1924, then Iowa State University until retirement, mid-1950s. Nearly single-handedly detailed genera and species of North American Miridae, describing in excess of 1300 species in 182 papers, notable among them Miridae of Connecticut (1923), Miridae of Illinois (1941), and Miridae of Nevada Test Site (1968). Prolific collector in early years. First American worker to use male genitalia for species recognition. Collection, including nearly all types, deposited in National Museum of Natural History, Washington, D.C., with limited material also at Texas A&M University and Biosystematics Research Centre, Ottawa.

LATREILLE, PIERRE-ANDRÉ. b. Brive, Corréze, France, November 29, 1762; d. Paris, February 6, 1833. Primarily an entomologist, having worked in Muséum National d'Histoire Naturelle, Paris, and as a professor of zoology. Particularly worthy of mention in this volume for his proposal of term Heteroptera and classificatory work on the Heteroptera.

LESTON, DENNIS. b. ca. 1919, England; d. 1979, Florida, USA. An iconoclastic commoner of flashing intellect. Author of numerous papers on Pentatomoidea, particularly of Africa, and Cimicoidea, testes follicle numbers, stridulatory mechanisms, and general classification of Heteroptera. Coauthor with T. R. E. Southwood of *Land and Water Bugs of the British Isles* (1959). Tropical ecologist and collector stationed in Ghana and Brazil. Collections deposited in The Natural History Museum, London, with limited material in American Museum of Natural History, New York.

LETHIERRY, LUCIEN. b. Lille, France, 1830; d. Lille, April 4, 1894. Entomologist of diverse interests, including Heteroptera, best known for his joint publication with M. Severin of *Catalogue général des Hémiptères* (3 volumes, 1893–1896), only nearly complete world catalog of group ever published, but not including Miridae. Types in Muséum National d'Histoire Naturelle, Paris.

LINDBERG, HAKAN. b. Joroinen, Finland, May 24, 1898; d. Helsinki, August 6, 1966. Swedish-speaking Finn educated at and longtime professor in Department of Zoology, University of Helsinki. Best known for work on Heteroptera fauna of Canary (1953) and Cape Verde (1958) islands. Collection deposited in University Zoological Museum, Helsinki.

LINNAEUS, CARL. b. Rashult, Swaland, Sweden, May 24, 1707; d. Uppsala, January 10, 1778. Professor, University of Uppsala. Father of systematic biology, author of earliest systematic concepts in Heteroptera.

MATSUDA, RYUICHI. b. Kajiki, Kagoshima Prefecture, Japan, July 8, 1920; d. Ottawa, Canada, June 19, 1986. Ph.D., Stanford University, under direction of G. F. Ferris. Held positions at University of Kansas, University of Michigan, and Biosystematic Research Centre, Ottawa. Published extensively with H. B. Hungerford on Gerridae, R. L. Usinger on Aradidae (1959), as well as on general insect morphology. Bibliography published by Ando (1988).

MILLER, NORMAN CECIL EGERTON. b. Ramsgate, Kent, England, July 13, 1893; d. Sturminster, Newton, England, May 26, 1980. Without formal training in entomology, worked as entomologist for Department of Agriculture, Kuala Lumpur, Malaysia, 1928–1947; retired to Zimbabwe, 1947–1949; joined staff of Commonwealth Institute of Entomology in London, 1949–1958. Primary interest in Reduviidae (and Acrididae); probably best known for his book *The Biology of the Heteroptera* (1956a, 1971), a work heavily weighted to his anecdotal observations on Reduviidae, and "Notes on the Biology of the Reduviidae of Southern Rhodesia" (1953). Collections deposited in The Natural History Museum, London. Bibliography published by Dolling (1987b).

MONTANDON, ARNOLD LUCIEN. b. Besançon, France, 1852; d. Iasi (Jassy), Romania, March 1, 1922. Author of over 120 descriptive taxonomic papers on many groups of Heteroptera, especially Nepomorpha, Reduviidae, Lygaeidae (especially Geocorinae), Coreidae, and Plataspidae. Collection sold in parts, most in The Natural History Museum, London, and Natural History Museum, Bucharest.

OSHANIN, VASILIY F. b. Politovo, Lipetsk Province, Russia, December 21, 1844; d. St. Petersburg, January 22, 1917. Educated Moscow University and later in western Europe. Worked briefly in Moscow and from 1872 to 1906 in Tashkent as director of a silk mill, and later as a teacher and director of Tashkent Girls' School. Participated in cultural life in Tashkent, mostly with political deportees, among them G. A. Lopatin, a friend of Marx and Engels. Upon retirement in 1906 worked as an associate in Zoological Institute, St. Petersburg. Best known to heteropterists for his *Verzeichnis der paläarktischen Hemipteren* (1906–1909), *Katalog der paläarktischen Hemipteren* (1912), and *"Vade mecum"* (1916), a guide to literature on identification of Heteroptera; also translated rules of zoological and botanical nomenclature into Russian. Many Central Asia specimens destroyed by dermestids; some specimens in Zoological Institute, St. Petersburg, others in Zoological Museum, Moscow University. Bibliography published by Kiritshenko (1940).

PARSHLEY, HOWARD MADISON. b. Hallowell, Maine, USA, August 7, 1884; d. May 19, 1953. Ph.D., Harvard University; professor and administrator, Smith College, Amherst, Massachusetts; accomplished musician. Early student and author on Aradidae and other groups in eastern North America, but possibly best known for *A Bibliography of North American Hemiptera-Heteroptera* (1925).

POISSON, RAYMOND A. b. Briouze (Orne), France, February 18, 1895; d. Lucon, France, November 28, 1973. Professor of zoology, Faculty of Sciences, Rennes. Notable among more than 300 publications, those on Gerromorpha and Nepomorpha of Palearctic and Ethiopian regions as well as papers on morphology, anatomy, and biology of Nepomorpha. Author of Heteroptera chapter in *Traité de zoologie* (Poisson, 1951). Collection purchased by Zoological Museum, Copenhagen (Veliidae) and National Museum of Natural History, Washington, D.C. (other groups). Bibliography published by Grassé (1974).

POPPIUS, ROBERT BERTIL. b. Kyrkslatt (Kirkkonommi), Finland, July 28, 1876; d. Copenhagen, Denmark, November 27, 1916. Swedish-speaking Finn, educated at University of Helsinki; worked University Zoological Museum, Helsinki, 1900–1914. Important revisionary work on Miridae, particularly Ethiopian region (1912, 1914) and cimicomorphan fauna of Old World tropics; also an accomplished student of Coleoptera. Most types in University Zoological Museum, Helsinki, although many specimens originally from other museums.

PUTON, JEAN-BAPTISTE AUGUSTE. b. Remiremont, France, 1834; d. Remiremont, April 8, 1913. M.D.; student of Palearctic and particularly French Heteroptera. Published over 150 papers, including taxonomy, faunistics, and biology of western European fauna, including several catalogs. Described numerous new species. Collection in Muséum National d'Histoire Naturelle, Paris.

REUTER, ODO MORANAL. b. April 28, 1850, Turku, Finland; d. Turku, September 2, 1913. One of most prolific and influential heteropterists of all time. Ph.D. from and longtime professor at Imperial Alexander University (University of Helsinki); student of Miridae, Anthocoridae, Saldidae, and many other families of Heteroptera and other orders. Notable works include lavishly illustrated monograph of European Miridae entitled *Hemiptera Gymnocerata Europae* (1878–1896), first character-based phylogeny of Heteroptera (1910), and first world classification of Miridae (1905, 1910). A gifted poet, who wrote of his own late-in-life blindness. Many types in Zoological Museum, Helsinki, and other institutions. Bibliography of more than 500 papers published by Palmén (1914).

RUCKES, HERBERT. b. New York, New York, USA, February 1, 1895; d. New York City, December 23, 1965. Ph.D., Columbia University; professor of biology, City College of New York; Research Associate, American Museum of Natural History. Student of Pentatomidae and osteology of Chelonia. Author of approximately 50 papers, most published in *Journal of the New York Entomological Society* and *Bulletin* and *Novitates* of American Museum.

SAY, THOMAS. b. Philadelphia, Pennsylvania, USA, June 27, 1787; d. New Harmony, Indiana, October 10, 1834. Father of entomology in North America. Spent short time as druggist; at 25 devoted himself to natural history, working at Academy of Natural Sciences, Philadelphia. Made two major expeditions to western U.S. territories. Left Philadelphia permanently in 1825 to live in New Harmony, an ill-fated utopian community in southern Indiana. Most of Say's writings rare in original; reprinted completely by LeConte (1859). Described numerous heteropterans in a variety of families, mostly in his *Descriptions of New Species of Heteropterous Hemiptera of North America* (1831). Most of collections destroyed by dermestids.

SCHOUTEDEN, HENRI. b. Brussels, Belgium, May 3, 1881; d. Brussels, November 15, 1972. D.Sc., Free University, Brussels, 1905. Soon joined staff of Musée Royal du l'Afrique Centrale; appointed head of section 1919; director 1927–1946; worked actively to time of death. Student of Pentatomoidea, Coreidae, Tingidae, and Reduviidae. Authored several lavishly illustrated volumes in Wytsman's *Genera insectorum* (for example, Schouteden, 1905b, 1907, 1913). Possibly best known for work on birds of the former Belgian Congo (Zaire).

SCOTT, JOHN. b. Morpeth, England, September 21, 1823; d. Morpeth, August 30, 1888. Civil engineer; author of papers on Tineidae, Psyllidae and other Auchenorrhyncha, and Heteroptera (frequently with J. W. Douglas), particularly of Great Britain. To heteropterists, best known for contribution on Miridae to *The British Hemip-*

tera, Vol. 1: *Hemiptera-Heteroptera* (1865), coauthored with J. W. Douglas. Collection deposited in The Natural History Museum, London.

SEIDENSTÜCKER, GUSTAV. b. Nuremburg, Germany, June 1, 1912; d. November 18, 1989. Worked as health insurance official in Bavaria. Published 99 papers, many reporting results of his travels in Mediterranean and Turkey, particularly on Miridae and Lygaeidae, including concise descriptions, keys, and useful illustrations. Collection deposited in Zoologische Staatssammulung, Munich. Bibliography published by Heiss (1990).

SERVILLE, JEAN-GUILLAUME AUDINET. b. Paris, France, November 11, 1775; d. Gerte sous-Touarre, March 27, 1858. Born into wealthy French family; father secretary to a prince. Became acquainted with wealthy Madame Tigny, who entertained influential biologists of the time, including Latreille. Among many works, most important contribution on Heteroptera *Histoire naturelle des insectes Hémiptères* (1843) coauthored with C. J.-B. Amyot. Possessed one of finest insect collections of his time, later sold to a number of individuals.

SIGNORET, VICTOR. b. Paris, France, April 6, 1816; d. Paris, April 3, 1889. Pharmacist and medical doctor. Among most important works "Révision du groupe des Cydnides" (1881–1884). Collection of Heteroptera deposited in Natural History Museum, Vienna. Bibliography published by Fairmaire (1889).

SPINOLA, MAXMILLIAN. b. Toulouse, France, July 1, 1780; d. Tassarolo, Italy, November 12, 1857. Author of 54 papers and books, mainly on Coleoptera, Hymenoptera, and Heteroptera, including *Essai sur les genres . . . des Hétéroptères . . .* (1837). Some of later works confusing for inclusion of genera without species and other features not in conformity with modern practice. Collection deposited in Museo Regionale di Scienze Naturali, Turin, Italy (see Vidano and Arzone, 1976).

STÅL, CARL. b. Castle of Carlberg, Stockholm, Sweden, March 21, 1833; d. Stockholm, June 13, 1878. One of most respected and influential heteropterists of all time. Primarily a student of Pentatomomorpha and Reduviidae, but also worked extensively in other groups. Studied under C. H. Boheman at Uppsala University and later received Ph.D. from Jena. Appointed assistant in entomology, National Zoological Museum, Stockholm, 1859; head of section and professor, 1867. Traveled widely in Sweden, continental Europe, and England, studying collections and conferring with colleagues. According to O. M. Reuter, Stål had a well-nigh inspired eye for "essential characters significant of natural affinity." Published numerous works on Heteroptera, notable among them compendious *Enumeratio Hemipterorum* (1870–1876), and significant works on Orthoptera and Chrysomelidae. Bibliography published by Spångberg (1879). Collections in Swedish Natural History Museum, Stockholm.

TORRE-BUENO, JOSÉ R. DE LA. b. Lima, Peru, October 6, 1871; d. Tucson, Arizona, USA, May 3, 1948. Graduated from School of Mines, Columbia University, New York City, 1894, later worked for General Chemical Company, New York. Longtime editor of *Bulletin of the Brooklyn Entomological Society*, author of *A Glossary of Entomology* (1937), "Synopsis of North American Heteroptera" (1939, 1941), and numerous smaller papers and notes. Collection deposited Snow Entomological Museum, University of Kansas, Lawrence.

UHLER, PHILLIP REESE. b. Baltimore, Maryland, USA, June 3, 1835; d. Baltimore, October 21, 1913. Graduate of Harvard University; student of Louis Agassiz. Librarian at Peabody Institute, Baltimore, and associate professor at the newly formed Johns Hopkins University. Father of North American heteropterology. Monographer of North American Cydnidae and Saldidae; describer of numerous species from newly explored western territories of United States. Collection deposited National Museum of Natural History, Washington, D.C. Bibliography published by Schwarz et al. (1914).

USINGER, ROBERT L. b. Fort Bragg, California, USA, October 24, 1912; d. Berkeley, California, October 1, 1968. Student (B.S., Ph.D.) and longtime professor of entomology, University of California, Berkeley. Gifted and internationally recognized entomological organizer and administrator. Author or coauthor of many papers and several books, including, *Methods and Principles of Systematic Zoology* (Mayr et al., 1953), *Aquatic Insects of California* (1956), *Monograph of Cimicidae* (1966), and, with R. Matsuda, *Classification of the Aradidae* (1959). Life and personality chronicled in autobiography (1972). Bibliography and list of names proposed published by Ashlock (1969).

VAN DUZEE, EDWARD PAYSON. b. New York, New York, USA, April 6, 1861; d. June 2, 1940. Librarian at Grosvenor Library, Buffalo, New York, from 1885 to 1912; later assistant librarian and curator of entomology, California Academy of Sciences, San Francisco. Ardent collector, bibliographer, and author of over 250 publications. Probably best known for *Catalogue of North American Hemiptera* (1917); author of many new taxa, particularly from California. Extensive collections from western United States form core of extensive Hemiptera collections of California Academy of Sciences.

VILLIERS, ANDRÉ. b. 1915; d. June 8, 1983. Worked in laboratories of Jeannel and T. Monod; traveled extensively in Africa. Became subdirector of Laboratory of Entomology, Muséum National d'Histoire Naturelle, Paris, 1956, with responsibility for Coleoptera. Published 661 papers, majority on Cerambycidae; also produced immense body of work on Enicocephalidae and Reduviidae of Africa and Madagascar. Bibliography published by Quentin (1983). Collection in Muséum National d'Histoire Naturelle, Paris.

WAGNER, EDUARD. b. Hamburg, Germany, June 20, 1896; d. September 11, 1978. Educated and worked as a schoolteacher. Student of western European and Mediterranean fauna, particularly of Miridae, important works including *Die Tierwelt Deutschlands* (1952, 1966, 1967) and "Die Miridae des Mittelmeerraumes" (1971–1978). First to consistently illustrate detailed structure of aedeagus in Miridae. Collection deposited University of Hamburg Zoological Museum. Bibliography of 553 papers and list of taxa proposed published by H. H. Weber (1976).

WOODWARD, THOMAS EMMANUEL. b. Auckland, New Zealand, June 8, 1918; d. Brisbane, Australia, November 22, 1985. Ph.D., Imperial College, London, under direction of O. W. Richards. Worked briefly in New Zealand, then as professor, Department of Entomology, University of Queensland, Brisbane. Primarily a student of Australian and New Zealand Lygaeidae. Bibliography of 64 papers published by Monteith (1986).

WYGODZINSKY, PETR (PEDRO) WOLFGANG. b. Bonn, Germany, October 5, 1916; d. Middletown, New York, USA, January 27, 1987. Ph.D., University of Basel, Switzerland; student of Eduard Handschin. Lifelong student of Microcoryphia and Zygentoma. Emigrated to Brazil in 1940, began study of Heteroptera; moved to Argentina in 1948, began study of Simuliidae at Instituto Miguel Lillo, Tucuman; moved to American Museum of Natural History, New York City in 1962, publishing on all groups of interest. Specialist on Enicocephalomorpha, Dipsocoromorpha, and Reduviidae; published scores of papers (many heavily illustrated), including "Monograph of Emesinae" (1966) and "Revision of the Triatominae" with Herman Lent (1979). First to examine detailed structure in Dipsocoromorpha, revealing wealth of morphological detail (e.g., Wygodzinsky, 1947b, 1948a). Collection deposited American Museum of Natural History, New York. Bibliography published by Schuh and Herman (1988).

ZETTERSTEDT, J. W. b. Göteborg, Sweden, May 24, 1785; d. December 23, 1874. Professor, Lund University, Sweden. Studied for doctorate under Retzius, a contemporary of Linnaeus. Student of Scandinavian fauna; produced some of earliest faunistic treatments of area, including *Fauna insectorum Lapponica* (1828).

3
Sources of Information

Literature

The present volume is designed to provide a window on the literature dealing with heteropteran biology and classification. All text citations are organized into a list of references at the back of the book, and most of those are derived from the family treatments. The following introduces the most general literature on the Heteroptera, primarily that which is not mentioned elsewhere in this volume.

General sources. Presented with only the name of a taxon, such as a family, the simplest approach to finding literature is through the *Zoological Record* (ZR). First published in 1864, the ZR annually indexes the literature on systematics and other subject areas. Beginning in 1972 the Hemiptera are treated in a separate volume. Possessed by most university and specialty libraries, the ZR offers relatively easy access to the literature if one has the time to search it systematically. Beginning with the 1970 volume it is available for on-line computer searches. Earlier literature can be located using bibliographies published by Horn and Schenkling (1928–1929), Derksen and Göllner-Scheiding (1963–1975), and Gaedike and Smetana (1978–1984).

Biological Abstracts first appeared in 1926, reviewing the literature on a monthly basis. It is less strongly oriented toward systematics than is the ZR and covers all of biology. Therefore it is more cumbersome to use than ZR, albeit more current. *Biological Abstracts* is available in most university libraries and beginning in the early 1970s is available for on-line computer searches.

Catalogs. Entomologists, and particularly systematists, have long relied on systematic catalogs to aid them in understanding classifications and for gaining access to the massive literature on insects. The most basic catalogs are in the form of checklists, some are synoptic and a bit more complete, and the most thoroughgoing include substantial numbers of references from the general biological as well as systematic literature.

The only world catalog ever completed for the Heteroptera was that of Lethierry and Severin (1893–1896). This once valuable work is now badly outdated and presents antiquated classifications in most groups. All other general heteropteran catalogs are regional in coverage. World catalogs at the family level are mentioned elsewhere in the text.

One of the earliest regional catalogs was that of Oshanin (1906–1909), also published as a checklist (1912), for the Palearctic fauna. The only other Palearctic work of similar coverage is the checklist of Stichel (1956–1962), a work that is sparely documented, although remarkably comprehensive as a listing of available names.

Two North American catalogs exist—Van Duzee, 1916 as a checklist and Van Duzee, 1917 as a comprehensive catalog, works that are badly out of date. Henry and Froeschner, 1988 carries on the Van Duzee tradition, providing synoptic coverage of the North American fauna north of Mexico.

Australia, the Orient, Africa, and Central and South America are without regional catalogs, and workers must therefore refer to family catalogs for groups or go to the primary literature.

Bibliographies. Two historical bibliographies exist for the Heteroptera. The *Vade mecum* (Oshanin, 1916) and the *Bibliography of the North American Hemiptera-Heteroptera* (Parshley, 1925). Both are still valuable for accessing earlier literature. The recent paper by Stonedahl and Dolling (1991), "Heteroptera Identification: A Reference Guide, with Special Emphasis on Economic Groups," lists many papers in addition to those dealing strictly with groups of economic importance, and as such serves as a valuable general reference.

General biological and morphological treatments. Few general treatments have appeared on bugs. Weber's (1930) *Biologie der Hemipteren* treats general biological phenomena and morphology in the Hemiptera in a way not seen in any other book. The *Handbuch der Zoologie* (Beier, 1938) and the Heteroptera chapter in the *Traité de zoologie* (Poisson, 1951) also offer excellent general treatments. *The Biology of the Heteroptera* (Miller, 1956a, 1971) offers some general observations and summary information, with emphasis on the Reduviidae.

China and Miller (1959) provided keys to the heteropteran families and subfamilies on a world basis. Few texts deal with the group comprehensively, nearly all being regional, as for example Borror et al. (1989) with its emphasis on the North American fauna and *The Insects of Australia* (Carver et al., 1991). The *Classification of Living Organisms* (Slater, 1982) provided diagnoses for all families, a limited number of references, and some helpful habitus illustrations.

Monographs. Only the earliest works on Heteroptera pretended to be monographic on a world basis, as for example Amyot and Serville (1843). Thus, there is no up-to-date volume that treats all Heteroptera.

Faunistic studies. Faunistic studies are concentrated on the Northern Hemisphere fauna. Notable in its completeness is *The Land and Water Bugs of the British Isles* (Southwood and Leston, 1959), with its species-by-species treatment, keys, and habitus illustrations. More recent is Dolling's (1991b) *The Hemiptera,* which deals with the Heteroptera and other hemipterans of the British Isles at the family level. The western Palearctic is covered in the *Die Tierwelt Deutschlands* (Wagner, 1952, 1966, 1967), *Illustrierte Bestimmungstabellen der Wanzen* (Stichel, 1956–1962), and *Keys to the Insects of the European USSR* (Kerzhner and Jaczewski, 1964, 1967), whereas the *Heteroptera of Yakutia* (Vinokurov, 1979, 1988) and *Insects of the Far Eastern USSR* (Vinokurov et al., 1988) treat the fauna of the eastern Palearctic. The serious student will wish to inquire into the several additional available treatments.

The North American fauna was most recently covered by Slater and Baranowski (1978). The classic *Heteroptera of Eastern North America* (Blatchley, 1926) dealt with the eastern fauna in detail.

Three historical studies are deserving of mention here. The *Enumeratio Hemipterorum* (Stål, 1870–1876) was a singular work in its time, and it is still useful in the identification of some groups that have not been the subject of subsequent faunistic studies, such as the Coreidae. The *Biologia Centrali Americana* (Distant, 1880–1893; Champion, 1897–1901) and the *Fauna of British India* (Distant, 1902–1918) dealt with the fauna of their respective regions in a way that has never again been seen for a tropical area. Although now out of date in terms of classification and nomenclature, both are extensively illustrated and are still useful for identification of members of these faunas.

Collections

The museums of the world are repositories for collections made from the time of Linnaeus until the present day. We describe below the holdings of some of those institutions that have particularly significant collections of Heteroptera. Information is available in Horn and Kahle (1935–1937), Sachtleben (1961), and Horn et al. (1990). It might be noted that in the United States most collections segregate holotypes from the remaining material, whereas many of the major European collections store holotypes and other material together.

Many universities and other organizations maintain important regional collections and should be contacted by anyone interested in revisionary or faunistic studies. The following are notable in the United States and Canada: Cornell University, Ithaca, New York; Oregon State University, Corvallis; University of California, Berkeley; Texas A&M University, College Station; the J. A. Slater Collection, University of Connecticut, Storrs; University of Michigan Zoological Museum, Ann Arbor; Florida State Collection of Arthropods, Gainesville, Carnegie Museum, Pittsburgh; and University of British Columbia, Vancouver. A number of European museums not listed below also contain valuable material, including those in Berlin, Geneva, Leiden, Munich, and Oxford. Finally, any study of a regional nature will almost certainly benefit from consulting the collections maintained by local institutions.

North America

American Museum of Natural History, New York. Rich in relatively modern collections, but with limited historical material. Incorporates collections amassed by Pedro Wygodzinsky, Randall Schuh, much material collected or acquired by J. A. Slater, and the collection of Rauno Linnavuori (including most types), as well as the Heteroptera formerly deposited in the Museum of Comparative Zoology, Harvard University, including Meyer-Dür material. Strongest in the New World, but also with substantial material from New Guinea and Africa as well as other areas.

Biosystematics Research Centre, Agriculture Canada, Ottawa. A collection with strong holdings from North America, including Mexico, with particular emphasis on the Miridae of Canada. Also occasional series of exotic material.

Bishop Museum, Honolulu. The single most important repository for material from the western Pacific, including Hawaii. Contains unparalleled collections from New Guinea, Borneo, and the Philippines. Much unworked material. Strongest in groups that can be collected through the use of lights.

California Academy of Sciences, San Francisco. Contains the collections of E. P. Van Duzee from the western United States (including nearly all of his types), I. LaRivers, B. P. Bliven, and the types of R. L. Usinger. Particularly rich in the Miridae.

National Museum of Natural History, Smithsonian Institution, Washington, D.C. Extensive holdings in nearly all major groups, with special strength in the North American fauna. Repository for the collections of P. R. Uhler, C. J. Drake, H. H. Knight, R. A. Poisson (part), J. C. M. Carvalho (part), and J. T. Polhemus (although still maintained by him). Rich in types. Collections originally begun by the U.S. Department of Agriculture and still maintained in close association with that organization.

Snow Entomological Museum, University of Kansas, Lawrence. Possibly the single most important collection of aquatic and semiaquatic Heteroptera, including numerous types, amassed by H. B. Hungerford. Also contains significant material in other groups, including the collections of J. R. de la Torre-Bueno and P. D. Ashlock.

Europe

Lund University Museum of Zoology and Entomology, Lund, Sweden. Extensive collections from Central America, West Africa, South Africa, and Sri Lanka.

Musée Royal de l'Afrique Centrale, Tervuren, Belgium. Repository for collections from the former Belgian Congo, especially those involving the work of Henri Schouteden. Important material also present in the Institut Royal de Sciences Naturelles de Belgique, Brusselles.

Muséum National d'Histoire Naturelle, Paris. Important collections for continental Europe and the former and present French colonies. Alone among most world class collections in maintaining the physical integrity of collections of early workers, including, for example, those of Bergevin, Dufour, and Puton, with many types.

National Museum, Prague. Collection assembled largely through efforts of Ludvik Hoberlandt, with extensive holdings from Czechoslovakia, the Balkans, Turkey, and Iran.

Natural History Museum, Budapest. One of the most important European repositories, containing much material worked by G. Horvath.

The Natural History Museum, London. Formerly British Museum (Natural History), containing collections on which *Fauna of British India* and *Biologia Centrali Americana* were based. Extremely rich in types, numbering in the many thousands. Most species represented by short series or by types alone. Modern collections of Heteroptera limited, but with significant material collected by Dennis Leston in Ghana.

Naturhistorisches Museum, Vienna. A historical collection, containing comparatively large numbers of types, including those of Signoret and some of Stål.

Swedish Museum of Natural History, Stockholm. Repository of the collections studied by Carl Stål. Extremely rich in type material, but with limited recent acquisitions.

University Zoological Museum, Helsinki. One of the three or four most important historical collections of Heteroptera, containing many types and other material of O. M. Reuter, B. Poppius, and J. R. Sahlberg. Particularly rich in Miridae.

Zoological Institute, St. Petersburg. The national insect collection of the former USSR. One of the truly great collections of Heteroptera, developed by A. N. Kiritshenko, containing material of most of the important Russian heteropterists and long series of many species.

Zoological Museum, Copenhagen. Repository for collection of J. C. Fabricius. Contains much material in the Gerromorpha resulting from the work of N. M. Andersen.

Zoological Museum, University of Hamburg, Hamburg. Repository for the collection of Eduard Wagner.

Central and South America

Departamento de Entomología, Museo de La Plata, La Plata, Argentina. Repository for most Carlos Berg types. A general Heteroptera collection of limited scope.

Instituto de Biología, Universidad Autónomo de México, Mexico, D.F. A recent collection amassed primarily through the efforts of Harry Brailovsky. Contains a broad representation of the extremely diverse Mexican fauna, including much unworked material.

Museo Argentino de Ciencias Naturales "Bernardino Rivadavia," Buenos Aires. Repository for the Nepomorpha collection of José de Carlo, with a limited amount of other material, including a small number of Carlos Berg types.

National Museum of Natural History, Rio de Janeiro. Repository for many of the types and other material of José Carvalho.

Africa

National Collection of Insects, Plant Protection Research Institute, Pretoria. Probably the most important collection of Heteroptera on the African continent. Contains much material, although often unworked. Other South African institutions of significance include the South African Museum (Cape Town), Natal Museum (Pietermaritzburg), and Transvaal Museum (Pretoria).

The Orient

Department of Biology, Nankai University, Tianjin, People's Republic of China. The most extensive and well-organized collection of Heteroptera in China.

Entomological Laboratory, Kyushu University, Kyoto, Japan. Repository of types and other material studied by T. Esaki, S. Miyamoto, and co-workers.

Institute of Zoology, Academia Sinica, Beijing. A relatively large but mostly unworked collection of Heteroptera from China.

National Science Museum (Natural History), Tokyo. Repository of Collections of M. Tomokuni and others.

Australia

Australian National Insect Collection, Canberra. One of three collections in Australia containing significant holdings of Heteroptera (others are South Australian Museum, Adelaide, and Queensland Museum, Brisbane).

4
Collecting, Preserving, and Preparing Heteroptera

Heteroptera occur in a wide variety of habitats and therefore are most effectively collected through the application of a diversity of methods. The following section briefly describes collecting equipment, techniques, and methods for mounting and preservation. Explanations of techniques and instructions on how to construct equipment can be found in most general entomology texts.

Collecting Equipment and Techniques

Aquatic nets. As with any group of aquatic insects, water bugs are best collected with a net designed specifically for the task. A net with a relatively long sturdy handle, a heavy hoop, and a heavy, coarse-mesh bag will usually work best, especially in larger bodies of water. For collecting in shallow water, or in confined areas such as rock pools, and among aquatic vegetation, a metal or plastic strainer is often most effective. Many stream-dwelling species, particularly in the tropics, can be found only by disturbing the marginal ground and vegetation and then collecting the drifting debris and bugs in a net or strainer.

Aspirator. The aspirator is probably the single most effective tool available for capturing small Heteroptera. It can be employed when collecting from a light trap sheet, beating net, sweeping net, when sorting litter by hand or collecting directly in the litter, and in many other situations. It should not be used when collecting around carrion, dung, bat guano, and similar substrates.

Beating. For groups that occur on woody vegetation, sweeping is a relatively ineffective method of collection, although often used. A preferable method employs a beating sheet or net. Beating sheets usually consist of a piece of canvas stretched on a lightweight frame; an umbrella with a short handle and a light-colored covering can also be used. They offer the advantage of considerable surface area, but for groups such as the Miridae, which often fly soon after they are dislodged from the foliage or other plant parts, many specimens will escape before they can be captured.

An alternative, compact, and often more effective design is a shallow oval net made of light or dark material and fitted with a short handle. When rapidly flying species are knocked into the net, they generally fly and alight on the sides of the net first and can be more readily captured than from a large flat sheet.

Berlese funnels. Funnels provide one of the best methods of extracting Heteroptera from forest litter. Although litter (or litter concentrated by a sifter) can be collected and examined by hand on a white sheet or in a yellow pan (or similar substrate), many heteropterans will be difficult to see (e.g., Dipsocoromorpha, small tingids) or capture (fast running or flying species). Funnels are used most effectively by first concentrating the litter through the use of a concentrating sifter and then placing the residue in the funnel to be processed for a day or two, depending on the amount of moisture in the soil-litter layer. Funnels may be metal and stationary or built of fabric and easily transportable.

Canopy fogging. Use of pyrethrin mist blowers to dislodge insects from the forest canopy has become a standard technique among workers in the tropics. Large numbers of heteropterans have been discovered using this approach.

Flight intercept traps. These are of two types, Malaise and pan traps. They are not the most effective methods for collecting large numbers of Heteroptera, but are more commonly and effectively used for Diptera and Hymenoptera. Nonetheless, certain groups will be difficult to find by other means, and some taxa or morphs, such as winged Vianaidinae (Tingidae), have been collected only by this method.

Killing bottles. Many groups of Heteroptera can be collected directly into 70% ethanol, probably the simplest and safest approach. Others, particularly the Miridae and some of the Pentatomoidea with coloration derived from plant pigments, are better killed by other means. Two obvious choices are ethyl acetate and potassium or sodium cyanide. The former works well for nearly all heteropterans, has the advantage of being relatively safe to use, and keeps the specimens moist and relaxed during the time they are in the killing jar. Ethyl acetate bottles always "sweat" to a certain extent, and therefore, especially for specimens of Miridae and other tiny cimicomorphans, such as Microphysidae and Cimicoidea, cyanide is the preferred killing agent. One must use great care with cyanide because it is a lethal poison; also, if pigmented specimens are left in cyanide for long periods they will change color. Specimens can be carefully layered in "cellucotton" packing material or pinned directly from the killing bottle.

Lights. Many winged insects—and the Heteroptera are no exception—can be collected most easily through the use of lights or light traps, particularly sources containing a certain amount of ultraviolet. Self-starting mercury vapor lamps are especially effective. Groups such as the Dipsocoromorpha, which are difficult to see under the best of conditions, can often be easily collected at lights; the Cydninae (Cydnidae) frequently come to lights in large numbers in the tropics but may be virtually impossible to locate by other means. Many Heteroptera are crepuscular, so the light must be set up and running just at sundown. However, some species, particularly of Reduviidae, often appear later in the evening.

Collecting aquatic heteropterans at lights requires the use of a trap or a sheet arranged in such a way that when the insects fly to the light but are unable to cling to the vertical surface, they are not missed when they fall to the ground. Many interesting taxa can be easily overlooked because of their inability to grasp the surface of the sheet or light trap.

Pitfall traps. Burying a plastic container so that the top is flush with the soil level and placing a centimeter or two of ethylene glycol in the bottom is an effective way to capture many ground-dwelling species. In areas of high rainfall, it is imperative that the traps be changed frequently or that a "roof" be provided to keep the trap from filling with water.

Sweep nets. Many groups of Heteroptera occur on the foliage of herbaceous plants or grasses and can be effectively collected using a typical insect net with a sturdy bag.

Mounting and Preservation Techniques

Pin mounting. Many larger heteropterans can be pinned directly. In tropical environments only stainless steel pins should be used. Members of the Pentatomoidea (and others with a large scutellum) are usually pinned through the right side of the scutellum. In taxa with a small scutellum (such as many Reduviidae), the pin many be run through the right hemelytron anteriorly or placed near the posterior margin of the pronotum.

Card mounting. Gluing specimens on small rectangular cardboard mounts is the traditional method of preparation among European workers. This approach has the advantage of protecting the specimens from breakage and allows for attractive arrangement of the appendages. It has the disadvantage of obscuring the ventral surface of the bug, including the labium and bucculae, and makes study of the pretarsus, and at times the genitalia, difficult. The use of water-soluble adhesive makes removal of specimens relatively easy.

Point mounting. Most American workers use "card points" (small elongate triangular mounts), for which punches are available in a variety of sizes and shapes for preparation of smaller specimens. This method does not protect the specimens directly (as card mounting does) although the locality and identification labels generally do. It also does not obscure the ventral body surface or the pretarsi of the specimen. As with card mounting, the use of water-soluble (or at least softenable) adhesive simplifies removal of specimens, should the need arise. Most workers follow the convention of placing the point on the right side of the specimen. If there is a question about whether a specimen should be point mounted or pinned directly, we recommend point mounting whenever possible.

Ethanol preservation. Ethyl alcohol (70%) is an excellent short- and long-term preservative for nearly all groups of Heteroptera, with the obvious exception of the Miridae (which will shed their legs) and species whose coloration is derived from plant pigments. For best preservation, alcohol should be changed a few hours after collecting. Members of the Enicocephalomorpha and Dipsocoromorpha should be collected directly into alcohol, and are best stored that way, until special preparation such as slide mounting or for use on the scanning electron microscope (SEM) is required.

Ultrasonic cleaning. Some heteropterans, including the Gerromorpha and Leptopodomorpha, have a dense vestiture and often become covered with soil particles or other debris during the collecting process. These and other dirty specimens often benefit greatly from use of the ultrasonic cleaner before mounting. Specimens can be placed in some dilute household ammonia or laboratory detergent for removal of debris and then returned to alcohol after removal from the ultrasonic cleaner to remove water and facilitate drying. The ultrasonic cleaner can also be used with great success to clean most specimens for SEM observation.

Critical point drying. This process is beneficial in the preparation of delicate specimens from alcohol (most Enicocephalomorpha, Dipsocoromorpha, Microphysidae, Cimicoidea, and many nymphs), in that body shape is retained, whereas when such specimens are simply air dried they collapse. The process requires specialized equipment, a carbon dioxide tank, and anhydrous alcohol.

Slide mounting. Certain groups of Heteroptera can be studied most efficiently when slide mounted whole or in part, for example, small Enicocephalomorpha, most Dipsocoromorpha, and some Cimicomorpha (notably Cimicidae and Polyctenidae). All of these groups are best collected directly into alcohol, if possible, in anticipation that slide mounting might be required.

Specialized Preparation and Observation Techniques

Scanning electron microscopy. Since the mid-1970s the use of the scanning electron microscope has become commonplace in the study of insects. The instrument is perfectly suited for the observation of details of the exoskeleton. Air-dried specimens may be mounted whole or dismembered. Material stored in alcohol may benefit from critical point drying in that specimens should be as dry as possible.

Dissection and preparation of genitalia. Techniques for the study of male and female genitalia, including inflation of the phallus in some taxa, have been described in the literature by many authors (e.g., Slater, 1950; Kelton, 1959; Ashlock, 1967). The abdomen of dried specimens must first be softened with dilute alcohol or other suitable relaxing fluid so that the pygophore can be teased from the abdomen or the abdomen removed entirely. The muscle tissue can then be removed by hydrolysis in potassium hydroxide (5–10% solution). The preparation should then be washed in dilute acetic acid and distilled water, dehydrated in alcohol, and set up for observation. Minute structures will require slide mounting; temporary mounts in lactophenol, glycerine, or glycerine jelly are often effective and allow for easy reorientation of complex structures, after which the specimens can be permanently stored in glycerine. Permanent slide mounts are optically better and allow for observation at very high magnification, but preclude reorientation.

Carayon (1969) described a technique for staining with chlorazol black, an approach that allows visualization of delicate chitinous structures that would otherwise be virtually impossible to see. Great care should be taken when using this chemical because of its recognized toxicity.

5
Habitats and Feeding Types

No other major group of insects successfully utilizes such an enormous array of different habitats as do the Heteroptera. They live as parasites of birds and mammals, feed on all parts of seed plants—and a very few ferns—from roots to pollen grains, feed on the mycelia of fungi, prey on other arthropods, live in spider and embiopteran webs and in the water and on its surface; a few species even occupy the open ocean. Only in their limited ability to burrow into woody tissues or to internally parasitize other organisms are heteropterans more limited than insects in some other orders. This diversity is especially remarkable when one realizes that in numbers of species the Heteroptera constitute one of the smallest of the "major" orders or suborders of insects. Their numbers dwindle into insignificance as compared with the 300,000 or so species of beetles. Yet this relatively small monophyletic group of approximately 38,000 described species must be considered one of the most successful of the Exopterygota, and their habitat diversity suggests a very long evolutionary history.

Heteropterans are phytophagous, predatory, or hematophagous. Significant numbers of representatives of each type are known, and it is this diversity that has led many authors to comment on the success with which the Heteroptera have exploited the range of possible food sources—as well as habitats. The individual feeding habits of the various groups are discussed in the family and subfamily treatments.

Feeding Types

Phytophagy. Plant-feeding species make up the majority of Heteroptera. As indicated by the results of phylogenetic studies discussed elsewhere in this volume, the phytophagous habit has been acquired independently at least twice from predatory precursors over the long millennia in which heteropterans have evolved.

Phytophagous feeding types may also be categorized in a functional manner, these being referred to as stylet-sheath feeders and lacerate-flush feeders (see discussion in Cobben, 1978). Miles (1972) believed incorrectly that the first type was restricted to the Pentatomomorpha, based on the fact that the salivary sheath—widespread in the Sternorrhyncha and Auchenorrhyncha—was thought to be secreted in the Heteroptera only by members of that group, although it is known to occur in the Reduviidae (Friend and Smith, 1971) and probably many other families. In this type, the salivary glands produce a feeding cone that attaches the apex of the labium to the feeding substrate, and sometimes they also produce a salivary sheath that lines the puncture through which the stylets are inserted. It has been observed that a given species of lygaeine lygaeid may secrete a feeding cone only when feeding on seeds and using the lacerate-flush method, whereas when feeding on plant sap will produce a stylet-sheath. This observation probably applies to many other pentatomomorphans, including true predators such as the geocorine Lygaeidae and asopine Pentatomidae.

Lacerate-flush feeders use the barbed apical portion of the mandibles to macerate tissues within the host (plant or animal), which are then mixed with saliva (usually containing digestive enzymes) and sucked up through the food canal.

In the Pentatomomorpha plant feeding appears to be the ancestral condition, and the great majority of species possess it, albeit feeding occurs on many parts of the plant. Most Cydnidae live in the ground—sometimes at considerable depths—feeding upon roots. Several of the chinch bugs occur about the roots. Stem, leaf, and bark feeding also occur, and members of at least three families—Aradidae, Termitaphididae, and Canopidae—feed on fungi.

Many species whose habits have been studied feed on portions of the plants where the nutritional return for a given amount of feeding is high. Thus the great majority of Lygaeoidea, Pyrrhocoroidea, and Rhopalidae feed on mature seeds and are what Cobben (1978) referred to as "lacerate flush" feeders. The pentatomoids, remaining coreoids, and Piesmatidae may feed on developing fruits or on the flowers, but most appear to gain their nourishment by extracting plant sap directly from the vascular system, particularly the phloem vessels (e.g. Maschwitz et al., 1987). These bugs produce from salivary secretions a feeding cone formed on the plant surface and in the feeding lesion.

Phytophagy is by no means limited to the pentatomomorphan evolutionary line. Within the Cimicomorpha the largest family is the Miridae. While some mirids are predaceous, probably the majority are phytophagous. These species tend to concentrate on the growing portions of the plant, chiefly the flowers, buds, and new foliage, but some have specialized on pollen. In this great family we also find species that feed on both plant and animal material. Likewise the Tingidae are entirely phytophagous, feeding more frequently on leaf tissue than do the Miridae, and usually on mature foliage. These two families (and the Thaumastocoridae) contain a vast number of species in an otherwise predaceous evolutionary line. The method of feeding in phytophagous cimicomorphans is quite different from that in many pentatomomorphans. They slash through the tissue with their stylets and in a sense whip a liquid from the cells in a fashion similar to that of their predatory relatives (Cobben, 1978).

The Aradoidea have succeeded in utilizing the mycelia of various fungi. This ability has been accomplished by a remarkable elongation of the stylets, which lie in a vast coil within the head of the insect when it is not feeding and extend for a long distance into the slender threads of the fungus when the insect is feeding.

Predation. Despite the numerous phytophagous heteropterans, the majority of families are predaceous upon insects and other arthropods. Among these there are exquisite specializations for success. There are not only predators searching along the ground and in vegetation for prey, but also species that live in webs of spiders and Embiidina, feeding upon the small insects trapped there. Others conceal themselves in flowers, acting much as small praying mantids, seizing the bee, wasp, fly or butterfly that comes within reach. Some are very general predators, feeding upon almost anything small enough to be overwhelmed, others are highly specialized to the point of feeding only upon a single prey species.

The carnivorous habit has arisen secondarily more than once in the Pentatomomorpha from phytophagous ancestors. The pentatomid subfamily Asopinae is an obvious example, most species of which feed upon soft-bodied caterpillars. Similarly, in the Lygaeidae, the majority of whose species are seed feeders, carnivory has become at least the most common feeding habit in the subfamily Geocorinae.

It has been shown that at least some predaceous taxa, such as members of the Reduviidae and Nabidae, will at times feed on plant substances, probably to secure moisture or to tide them over in times of shortage of suitable prey. They are not able to complete development on such a diet and die in a relatively short time (e.g., Stoner et al., 1975).

Hematophagy. Perhaps the most remarkable development has been the ability of heteropterans to utilize the blood of vertebrates. This feeding type has arisen at least three and possibly four times independently. As in the bed bug family, Cimicidae, all of whose members feed on bats and birds, most blood-feeding Heteroptera live in the nests of their hosts. In the otherwise predominantly arthropod-feeding assassin bugs of the family Reduviidae, blood feeding is dramatically demonstrated in the Triatominae, many species of which carry the dread Chagas' disease, which is of great importance to health in South America. In the small rhyparochromine lygaeid tribe Cleradini, blood feeding has evolved in an otherwise seed-feeding lineage. This habit reaches its greatest degree of specialization in the Polyctenidae (bat bugs), all species being permanent ectoparasites on bats.

Habitat Types

Inquilinous or commensal. Species in a number of families are associated with ants and ant nests. Most of these do not prey on the ants. Included are species of Enicocephalidae, Tingidae (Vianaidinae), Anthocoridae, Cydnidae, and Lygaeidae.

All species of Termitaphididae live in obligate association with termites.

Members of at least four families have perfected the ability to live in webs. Plokiophiline Plokiophilidae, phyline mirids of the genus *Ranzovius* Distant, nabids in the genus *Arachnocoris* Scott, and some emesine Reduviidae live in the webs of spiders, usually feeding on insects ensnared there. They coexist with the spiders, and in the case of a few emesines, apparently feed on them (Snoddy et al., 1976). Embiophiline Plokiophilidae, on the other hand, inhabit the webs of the Embiidina, feeding on eggs and weak and dead individuals.

Water surface-dwelling. Heteroptera of the infraorder Gerromorpha compete for the water surface only with the beetle family Gyrinidae. Members of the group have modifications, including specialized pretarsi, unwettable body surfaces, and novel communication mechanisms, that enable them to thrive in this habitat.

Aquatic. Only the Heteroptera and Coleoptera have successfully adapted to a true aquatic existence in the adult stage. In the Nepidae and Belostomatidae air is breathed directly from the atmosphere through a "siphon," through the use of an air bubble captured on the venter in the Notonectidae, Pleidae, Helotrephidae, and some Naucoridae (as in the Hydradephaga), and through the use of a plastron in certain Naucoridae and all Aphelocheiridae, a condition found elsewhere only in dryopoid beetles.

Intertidal. A number of species of Leptopodomorpha live only in the intertidal zone, the most specialized among them being *Aepophilus bonnairei* Signoret with its plastron respiration and greatly reduced compound eyes and hemelytra.

Competition

Much has been written concerning competition, or means of avoiding it, between closely related species. In the Heteroptera such subtle interactions have not been studied for most species, but there are some striking examples. Blakley (1980) studied the relationships of two species of the lygaeid genus *Oncopeltus* Stål in the West Indies. He found that *Oncopeltus fasciatus* (Dallas) is able to survive only upon plants of the milkweed *Asclepias curassavica* when seeds and pods are present, whereas *O. cingulifer* Stål is able to pass its entire life cycle on the vegetative tissues of the plants. Thus while *O. fasciatus* has greater flight ability, it is more restricted in local distribution because of the limitations placed upon its available breeding sites.

A striking case of two species not at all related taxonomically but feeding upon some of the same host plants also involves a species of *Oncopeltus*. Blakley and Dingle (1978) reported that on the West Indian island of Barbados the monarch butterfly has completely eradicated the milkweed bug *O. sandarachatus* (Say). What appears to have happened in this case was that the monarch built up large populations on the introduced milkweed *Calotropis procera*, a plant not suitable for *O. sandarachatus*. Since the monarch also feeds on *A. curassavica*, the large populations eliminated the latter plant, upon which the milkweed bug is dependent.

Hamid (1971b) studied three *Cymus* spp. that live on the identical sedges and rushes in the same habitats. Their life cycles are largely nonoverlapping so that when one species is adult the others are either in the egg stage or early nymphal stage and thus utilize the hosts in seasonal succession.

It is known that in the extensive fauna of the ground-living Lygaeidae that feeds on fallen figs there is also a succession of species: from those that appear as soon as the fruits fall to the ground, through a changing fauna, to finally a group of minute species that appear after most of the seeds have been "eaten" and search out seeds that have fallen into cracks and crevices and thus, in a sense, "scavenge" for the leftovers.

Taylor (1968) and Streams and Newfield (1972) studied the habits of backswimmers (Notonectidae) that occur in the same ponds and discovered that they divide the available niches both in time and space.

Stonedahl (1988a) found that in the mirid genus *Phytocoris* Fallén a number of species occur simultaneously on the same host plant. In most such cases resource partitioning occurs, wherein different species occupy different parts of the host plant such as the cones, foliage, branches, boles, and bark. These preferences may be in part due to the different degrees of phytophagy and predation in different species. Cooper (1981) found that the bark-inhabiting species *Phytocoris neglectus* Knight and *P. nobilis* Stonedahl appeared to be occupying the same part of the host plant, *Abies procera,* at the same time in Oregon. However, they are actually temporally isolated and occur together for only a brief period in mid-August.

Dispersal

It is apparent that to succeed, populations of insects must be able to move from an unsuitable habitat to a suitable one. For the Heteroptera this ability is a remarkable phenomenon and involves evolutionary time adaptations as well as contemporary ecological ones. The evolutionary consequences of dispersal or the lack of need for it are discussed in Chapter 6. Here we will be concerned with the ecological adaptations.

Dispersal and *dispersion* are terms that are used loosely in the literature. Perhaps it is proper to restrict the term *dispersal* to the extension of the range of a species across a barrier that is unsuitable for it to live in or on. If this is done, then the movement of individuals within a population can be considered dispersion and does not imply an increase in range. One must recognize of course that the actual range of a given species is not constant and that there is continual increase and decrease along the margins of a range depending upon better or worse conditions for the population at any given time. Nevertheless, species do have ranges that can be stated rather accurately in a general sense.

One has only to go to a light at night in summer or in the warm tropics to see that heteropterans are dispersing almost constantly. Indeed all we know of many species that live in very specialized or inaccessible habitats is that they actually exist, because they have been collected at lights and thus far their actual habitats are unknown.

Monteith (1982b) reported a remarkable aggregation of insects in northeastern Australia. Here the dry season is very harsh, and insects migrate to the pockets of monsoon forest. He noted four species of Heteroptera (two alydids, a coreid, and a plataspid), some of which occur in dense clusters comprising large numbers of individuals. Monteith believed that these bugs do not feed and that the aggregating behavior is an example of group reinforcement of natural chemical defenses, in the case of the Heteroptera of course the scent-gland secretions. He reported that when sudden massed flights were elicited by disturbance of aggregations of *Leptocorisa acuta* (Thunberg) and *Coptosoma lyncea* Stål flight was accompanied not only by discharge of the glands but also by a sudden loud buzzing similar to that of a disturbed paper-wasp nest. This combination yielded considerable fright reaction in humans coming suddenly on an aggregation.

6
Wing Polymorphism

Biologists usually equate the success of insects with their ability to fly. Yet, when flight is no longer advantageous, insects frequently lose this ability and often the wings themselves. Nowhere in the insect world is this more evident than in the Heteroptera. In family after family the ability to fly has been lost. The usual result is the reduction, modification, or loss of the wings. In some cases flight loss is less evident; the wings appear normal, but flight muscles become reduced after a migration period, as in some species of *Dysdercus* Guérin-Ménéville (F. J. Edwards, 1969; Dingle and Arora, 1973).

There are many reasons why heteropterans have lost the ability to fly. In the ectoparasites (Cimicidae, bed bugs; Polyctenidae, bat bugs), termitophiles (Termitaphididae, some Lygaeidae, and others), and myrmecophiles, flightlessness is obviously associated with the adaptation to living on the bodies of warm-blooded vertebrates or in the nests of termites or ants.

Darwin (1859) considered the problem of why species living on the ground, under bark, or on the water surface should develop flightless morphs. He interpreted the phenomenon in what today's jargon calls the "energy budget"; that is, an individual, or a population, has a given amount of energy and if that energy is not needed for "making" wings, it can be better used for other purposes such as increased reproduction. Thus, if it is not an advantage to be able to fly, individuals in the population that do not utilize energy in developing wings capable of flight will be favored, and eventually the entire population will become flightless.

One sex may become flightless in some species while the other will remain fully capable of flight. This phenomenon is not well understood experimentally, but in some myrmecomorphic Miridae, where only the female is strongly mimetic (and flightless), fewer eggs need to be produced because of the reduction in mortality (McIver and Stonedahl, 1993). In many semiaquatic heteropterans, populations will be flightless during most of the year but have a winged generation during one season.

Heteropterans may be entirely wingless or the wings may appear normally developed but flight muscles are reduced. In a number of ground-living species the forewings become shell-like and closely resemble the elytra of beetles, an adaptation that reduces water loss in arid habitats (Slater, 1975).

The different degrees of wing modification may be categorized as follows. These descriptions apply most easily to the Panheteroptera, which have a distinct clavus, corium, and membrane in the forewing (see Slater, 1975).

Aptery: Fore- and hind wings completely absent.

Microptery: Forewings reduced to minute leathery pads covering, at most, only the base of the abdomen. Hind wings either minute flaps or absent.

Staphylinoidy: Forewings covering no more than basal half of abdomen; clavus and corium indistinguishably fused; membrane, if present, only a narrow marginal rim; distal ends truncate. Hind wings reduced to flaps or absent.

Brachyptery: Forewings reduced, not covering abdominal terga 6 and 7. Clavus and corium fused or not, but elements recognizable. Hind wings reduced but usually not flaplike.

Coleoptery: Forewings shell-like, with clavus and corium elongated, usually fused and meeting evenly along midline; membrane of forewing reduced. Hind wing absent or reduced to small flaps. Anterior abdominal terga often membranous.

Submacroptery: Clavus and corium distinct and usually not reduced. Membrane of forewing reduced leaving posterior abdominal segments exposed. Hind wings either slightly reduced or elongate.

Macroptery: Clavus and corium distinct, membrane well developed. Hind wing elongate.

Caducous (= deciduous): Wings fully developed, but broken off or eliminated at some time during adult life. A condition commonly found in Enicocephalomorpha, Mesoveliidae, Veliidae, and occurring in some Aradidae.

There is relatively little experimental evidence for the mechanisms producing flightless morphs. The one feature that seems to be common to most of the flightless species is relative permanency of habitat (Brown, 1951; Southwood, 1962; Sweet, 1964; Fujita, 1977; Slater, 1977; Vepsäläinen, 1978). Many islands also contain a high proportion of flightless morphs (see Darlington, 1971, 1973 for general discussion).

Wing reduction in the Nepomorpha was long thought not to exist except where there was an adaptation for underwater respiration. Several species of the naucorid genus *Cryphocricos* Signoret display wing reduction (Sites, 1990). In *C. hungerfordi* Usinger in Texas, a species that lives in permanent streams, Sites found that in a sample of 790 adults only six were macropterous and one submacropterous. The others (98%) were strongly brachypterous, the wings being what Slater (1975) termed staphylinoid, reaching only over the basal third of the abdomen and truncate or subtruncate along the distal margins with the membranes absent. J. T. Polhemus (1991a) described the Neotropical naucorid *Procryphocricos perplexus* Polhemus as having the wings reduced to small pads that reach only the second abdominal tergum. Both of these naucorid genera must have plastron respiration, allowing them to stay permanently submerged.

The literature shows, however, that several genera of Nepidae also have reduced hind wings; in one case these wings are leathery. These nepids breathe atmospheric air by means of the long terminal appendages. Brown (1951) observed that, in Britain, Corixidae that were brachypterous were confined to permanent habitats. He also pointed out that species of water boatmen normally found in temporary habitats migrated a great deal, whereas those in permanent habitats did not, and he implied that brachyptery was always associated with permanent habitats. Thus, it seems that adaptation to underwater respiration is not necessary for flightlessness to evolve in the Nepomorpha.

Slater (1977) noted that permanency of habitat may be thought of in an evolutionary-time sense as well as in a shorter ecological-time sense. He observed that in old stable areas, such as the southern Cape region of South Africa and southwestern Australia, the majority of ground-living Lygaeidae have a flightless morph, and that in such areas the degree of wing modification is greater than in less stable areas. Similarly, Malipatil (1977) noted that the New Zealand ground-living lygaeid fauna is composed almost entirely of members of the tribe Targaremini and that 95% of the species are flightless, with coleoptery being the predominant condition. He observed high proportions of flightlessness in several other orders of New Zealand insects.

The genetic or environmental factors involved in species with wing polymorphism are still understood for only a very few species. Solbreck (1986) showed that the ground-living lygaeid *Horvathiolus gibbicollis* (Costa) from the Mediterranean region possessed both macropterous and submacropterous forms. In the latter the forewings are about two-thirds the length of those in macropterous forms and the hind wings less than one-third the length of those of macropterous forms. He found that each morph bred true to wing length when reared under variable density, food, and temperature conditions for several generations. All F1 offspring between crosses of the two morphs were submacropterous. In the F2 generation approximately 25% were macropterous and 75% submacropterous, implying a monogenic control of the wing form. His study also demonstrated that short-winged morphs reach adulthood more rapidly than do macropterous forms and have a distinctly shorter adult preoviposition period. Solbreck's study suggested that a simple gene modification may be involved and offered a possible explanation for why macropterous forms will persist in populations for a long period of time. The ability to develop flightless populations, but also to retain a proportion of fully winged individuals, gives such populations the ability to make maximal use of a limited food supply for which they may be competing and still retain the ability in the population to colonize new food sources at some distance from the original population.

Klausner et al. (1981) studied a population of brachypterous *Oncopeltus fasciatus* (Dallas) on Guadeloupe. They found brachyptery to be caused by a single recessive Mendelian unit and that neither temperature nor photoperiod had any effect on the expression of the brachypterous trait in the homozygous short-winged genotype.

Southwood (1961) believed that wing polymorphism in Heteroptera was under hormonal control. He felt that the short-winged condition resulted from an excess of juvenile hormone and thus wing reduction was essentially a neotenic condition. Southwood noted that more species living at high altitudes were flightless than were species of the same groups at low altitudes. He suggested that at high altitudes low temperatures acted on the hormone balance or that a longer time was spent in the nymphal stage and therefore the insect was under the influence of the juvenile hormone for a longer period of time. Although this theory may account for the occasional production of short-winged morphs, it cannot account for most species with wing reductions. First, while low temperature—as shown by Brinkhurst (1961)—certainly influences wing development in some taxa, it is not clear that this is due to excess juvenile hormone. Second, the short-winged adults are not neotenous, as the short wing is not the wing pad of the nymph but morphologically an adult wing. Third, the presence of reduced wings in many terrestrial insects living at high altitudes may be the result of living in areas of long-term habitat stability. Such insects are basically in stable environments, for when conditions change in montane habitats the insect populations need move only a relatively short distance up or down to find themselves in the same habitat they were in before. By contrast, were the same variation in conditions to take place on a level surface, it would require movement of an insect population hundreds of miles to find ecologically similar conditions.

An appreciable number of flightless heteropterans live

in extremely hot situations such as deserts, tropical savannahs, and beaches. The last may not at first glance appear to be a stable habitat, but it actually is, in the same sense as mountains are, in that insects living in beach habitats must move only a short distance to remain in the same habitat as the beach expands or retreats over a period of time. The most important consideration, we believe, is to understand the genetic system at work and to realize that what is a stable habitat for one group may not be for another.

Heteropterans living upon the surface of the water, especially in temperate habitats, face a somewhat different ecological challenge. Brinkhurst (1961) and Vepsäläinen (1978) showed that gerrids and veliids living in rivers and streams (relatively permanent habitats) consist of populations that are univoltine and predominantly apterous. Univoltine populations living in ponds, however, are usually long-winged. In bivoltine species the situation is much more complicated. Most species of gerrids studied by Brinkhurst are polymorphic. The overwintering generation is chiefly long-winged, and summer generations of short-winged individuals are the offspring of long-winged overwintering mothers. Brinkhurst found that wing development was under the control of a single gene and that short-winged homozygotes were lethal but that short-winged heterozygotes were under environmental influence so that expression of a "switch gene" came into play. For example, when he crossed short-winged with short-winged forms he obtained either all long-winged forms or 2.5:1 ratio of short-winged to long-winged forms depending upon the temperature. Vepsäläinen (1978), however, indicated that the most important factor in determining wing length was photoperiod.

Not all semiaquatic bugs behave under the model described above. For example, Muraji et al. (1989) showed that in the veliid *Microvelia douglasi* Scott, which lives in relatively permanent habitats as well as ephemeral ones, the percentage of macropterous forms declines dramatically in autumn, and the populations are then almost entirely flightless. Temporary habitats such as rice paddies are apparently recolonized each year, and the flightless fall generation hibernates only near permanent water. These authors found that the strongest determinant for an increase in the proportion of macropterous forms is density, but that this factor is modified by photoperiod, temperature, and abundance of food. Muraji and Nakasuji (1988) showed in a comparative study of three species of *Microvelia* Westwood, that the species that lived in the most permanent habitats developed a macropterous morph only when populations reached high densities, and it was the slowest of the three species to respond by production of macropterous forms as density increased. Zera and Tiebel (1991) studied the effects of photoperiod on wing length in the pterygopolymorphic gerrid *Limnoporus canaliculatus* (Say). They found that reduction in photoperiod had a strong positive effect on the production of the long-winged overwintering morph.

Brinkhurst's belief that wing polymorphism is under the control of a single gene was challenged by Andersen (1973), who discussed the condition in two species in Denmark, noting that in bivoltine species there is a loss of wing muscles in the overwintering generation after a flight period in the spring. Vepsäläinen (e.g., 1971a, b) demonstrated that, at least in temperate-latitude species, the development of short- and long-winged forms was due to changes in photoperiod. Andersen (1973) reviewed the evidence in detail and presented experimental data on changes in fifth-instar nymphs in several water striders that are correlated with pigmentation, aspects of reproductive dimorphism, and diapause.

Andersen (1993) used cladistic techniques to correlate wing polymorphism and habitat type in temperate-latitude Gerridae. He found that species with the ancestral pattern of wing polymorphism occupy the most stable habitats. He therefore hypothesized that the ancestral type of habitat was relatively permanent (streams, permanent ponds, lakes) and the ability of water striders to colonize less stable or temporary water bodies had evolved more recently. Andersen (1993) concluded that wing dimorphic or short-winged morphs are the ancestral states and that where only long-winged morphs exist it is because the short-winged morph has been lost. He further reasoned that two distinct types of short-winged gerrids exist, one emerging from fifth-instar nymphs with distinct wing pads, the other from nymphs with reduced wing pads. He suggested that the two kinds of short-winged conditions are associated with different mechanisms of morph determination involving a polygenic system and environmental switches adapted to features of the habitat.

Waloff (1983) pointed out that in arboreal species, habitat architecture may militate against the development of flightless morphs. She noted that in the British Heteroptera 75 of the species associated with grasses or forbs show wing polymorphism, whereas only 11 of the 111 tree-living species have short-winged forms. Moreover 10 of these 11 species occur outside of the niches occupied by typically phytophagous insects (9 are predaceous and 1 lives under bark), the 11th species lives on dwarf willow and heather. She believed this may be true only of temperate populations and cited the high incidence of flightless arboreal tropical Acridoidea. There is no evidence at present, however, that tropical arboreal Heteroptera tend to become flightless to any greater degree than do temperate ones. Waloff's paper would be even more compelling if she had distinguished between those forb-inhabiting species that live above the ground (arboreal in the sense of Slater [1977]) and those that are geophilous for the most part.

Rolston (1982a) described a strongly staphylinoid pentatomid from an elevation of about 1800 meters in Haiti. He assumed it was arboreal because all other members of the tribe are. If Rolston is correct, this is an example of wing reduction in a group where the phenomenon is rare.

7
Mimicry and Protective Coloration and Shape

The questions of mimicry, aposematic coloration, and protective resemblance have created much controversy and a large literature over the years. The Heteroptera contain striking examples of each. Mimicry has traditionally been grouped into distinct types, although there are cases that are somewhat intermediate or involve more than one of the classical categories.

Mimicry

Batesian mimicry. An insect that is not protected comes to look like a distasteful or dangerous insect such as a wasp, bee, or ant. This phenomenon is widespread in the Heteroptera; ant mimicry alone occurs in no fewer than seven families.

Müllerian mimicry. Several distasteful or dangerous species come to look alike in color or structure or both to reduce predation as the potential predator learns to avoid a given color or form pattern. This phenomenon also is widespread in the Heteroptera. Documented cases are difficult to establish experimentally, and most evidence is based upon taxa not closely related, coming to look like other species in sympatry but unlike each other in areas of allopatry.

Mertensian mimicry. The model is moderately distasteful or dangerous, whereas at least some of its mimics are lethal. This type of mimicry has not been established in the Heteroptera; but it probably occurs.

Wasmannian mimicry. Mimicry in which an insect attempts to "fool" the ant with which it is associated is not yet documented in the Heteroptera, although it is common in the Staphylinidae and related beetles.

Aggressive mimicry. A predator looks like a prey species to "fool" the latter and enable predation to be enhanced. McIver and Stonedahl (1993) note this type may occur in some species that feed on ants. This phenomenon should be studied carefully as the reduviid-pyrrhocorid mimetic situation discussed below might appear to be such a situation, but is probably a case of Batesian mimicry.

Aposematism and Mimicry

Among the Miridae, *Lopidea* Uhler, *Hadronema* Uhler, and most Resthenini exhibit aposematic coloration. McIver and Lattin (1990) studied *Lopidea nigridia* Uhler in detail, noting that it possessed a clumped distribution, was relatively sedentary, and fed on a plant species (*Lupinus caudatus*) known to produce several alkaloid compounds. They experimentally tested the palatability of *L. nigridia*, using spiders as predators, and found it to have limited acceptance as a prey item when compared with a nonaposematically colored mirid, *Coquillettia insignis* Uhler.

Many species of lygaeids in the genera *Lygaeus* Fabricius and *Oncopeltus* feed on milkweed plants that contain cardenolides. Such species have been shown to be extremely distasteful to vertebrate predators. It has long been believed that brightly colored insects living in similar habitats would be protected by resembling these showy red and black, and presumably aposematically colored, milkweed-feeding insects.

The problem with this concept has always been to account for the great similarity of many of the presumably protected species. For example, in Africa some species of *Spilostethus* Stål (a genus closely related to *Lygaeus*) are almost carbon copies of one another in color pattern. Similarly several species of *Oncopeltus* are also almost identical in color pattern, and indeed Zrzavý (1990b) marshaled evidence from study of the west Palearctic fauna to indicate that similar color patterns have evolved many times. Such close resemblance suggests that greater protection from predators is obtained the closer the resemblance of one group of species is to another (i.e., less learning needed by the potential predator) and that, although bright color may afford some protection, it is only partially effective. Yet, when species from a number of families resemble one another, several will have well-developed scent glands (supposedly rendering them distasteful), and determination of who is the model and who is the mimic awaits experimental work.

McLain (1984) showed that the red and black mirid *Lopidea instabile* (Reuter) and the similarly colored lygaeid *Neacoryphus bicrucis* (Say) are both unpalatable to *Anolis* lizards when fed on the alkaloid-containing composite *Senecio smallii*. When fed on sunflower seeds the bugs are palatable to anoles. However, the lizards rarely attack either bug species, once they have attempted to eat a bug that had fed on an alkaloid-containing plant, thus suggesting a Mullerian mimicry complex.

Sillen-Tullberg et al. (1982) demonstrated that in the case of the Palearctic lygaeid *Lygaeus equestris* (Linnaeus), two aposematic forms must be very similar to gain full mutual protection. This study provided evidence why not only Batesian mimics, but also Müllerian mimics, and their aposematic models develop such extremely close resemblances.

A striking example of what appears to be a combination of Batesian and Müllerian mimicry occurs in the milkweed and related bugs of the family Lygaeidae and similar-appearing species in the Rhopalidae, Largidae, Pyrrhocoridae, and Reduviidae.

Van Doesburg (1968) noted that many of the *Dysdercus* spp. (Pyrrhocoridae) in the Western Hemisphere appear to form mimicry "rings." He pointed out that in a given geographic area several species have a similar color pattern. In the case of *Dysdercus obscuratus* Distant five subspecies are recognized, four of these having adult color patterns very different from each other and each resembling another species of the genus that occurs in the same region. This case appears to be one of classic Müllerian mimicry, but it is complicated by the resemblance of some of these insects to species of Lygaeidae (*Oncopeltus*) and to species of Coreidae and Miridae. Complex combinations of Müllerian and Batesian mimicry may be involved. Involvement of such combinations is further suggested by conflicting evidence in the literature concerning the palatability of members of the genus and whether the scent glands produce distasteful liquids. The relationship of the insects to potential predators has not been investigated. The nymphs of most species are bright red, conspicuous, almost certainly aposematic, and do not vary in coloration geographically, suggesting that the adult coloration has evolved through terminal addition in ontogeny.

Further developments of this type occur in a number of African pyrrhocorids and some of their assassin bug predators. These cotton stainers often occur on or near malvaceous host plants in aggregations of hundreds of individuals. Stride (1956) showed that for several different species of Pyrrhocoridae an assassin bug of the genus *Phonoctonus* Stål lives in the aggregations and feeds upon the cotton stainers. Each of these assassin bugs closely resembles the particular cotton stainer species with which it lives. Stride studied three species of *Phonoctonus* in West Africa, each of which was found most frequently feeding on the pyrrhocorid that it most closely resembled. However, the assassin bugs also occurred regularly both as nymphs and adults, laid eggs with, and fed upon species of cotton stainers with which they showed lesser degrees of resemblance. Stride further found that one *Phonoctonus* sp. was extremely variable in color, one color morph resembling a brown species of pyrrhocorid and the other a red and white species. When he fed nymphs of this reduviid species upon both red and white and brown pyrrhocorids the resulting adults were always red and white. What this mimetic situation almost certainly suggests is that the predator is protected from vertebrate predation by developing a close resemblance to its aposematically colored prey. Other examples include an African assassin bug of the genus *Rhynocoris* Hahn, which bears red and black stripes and closely resembles the lygaeid species of *Spilostethus* Stål that live in the same area. A Neotropical assassin bug of the genus *Zelus* Fabricius is strikingly like the aposematically colored species of *Oncopeltus*.

Although these cases have not been supported by experimental work, it is difficult to believe that assassin bugs whose relatives are dull brown or gray would develop such exacting patterns of black and red coloration except as a mimetic resemblance. There is no evidence, as had previously been hypothesized, that these assassin bugs are exhibiting aggressive mimicry of the phytophagous heteropterans with which they are associated.

Poppius and Bergroth (1921) suggested that African species of *Poeantius* Stål, ground-living myrmecomorphic lygaeids, were predatory on ants and that this was a case of aggressive mimicry. They believed that the lygaeids lived in ant nests. However, the lygaeid bugs are not predators but seed feeders, and it is probable that the mimicry is simple Batesian. Many studies suggest or demonstrate that vertebrate predators, especially birds and lizards, are the principal agents involved in the development of mimicry. However, Berenbaum and Miliczky (1984) reported that the Chinese mantid *Tenodera ardifolia sinensis* learns to avoid the toxic *Oncopeltus fasciatus* after several trials. The mantid regurgitates and shows signs of poisoning when cardenolides are present and will not attack the milkweed bugs when they are reared without cardenolides. McIver and Stonedahl (1993) stated that the reduviid *Sinea diadema* (Fabricius) will accept nonmimetic species of *Lopidea* much more often than they will the myrmecomorphic *Coquillettia insignis* (both Miridae). Tyshchenko (1961) believed that spiders and predatory insects were at least as important as vertebrates in mimicry development.

Ant Mimicry, Myrmecomorphy, and Other Insect Resemblances

One of the most striking mimetic conditions found in the Heteroptera is the development of myrmecomorphy. This phenomenon has developed independently not only in more than seven different families (McIver and Stonedahl, 1993), but also in different phylogenetic lines of the same family. Furthermore, the degree of mimetic resemblance is extremely diverse, ranging from cases where even the experienced investigator in the field cannot always differentiate the hemipteran from the ant, to situations where in museum specimens there would appear to be no mimicry at all. Yet the movements of the insects in the field are distinctly antlike, apparently sufficiently so to provide the potential prey species with a degree of protection when disturbed by a vertebrate (or other) predator.

The Neotropical lygaeid *Neopamera bilobata* (Say) does not appear to be ant mimetic in collections, but in the field when a large population under a food plant is disturbed, the insects rush out in all directions with jerky irregular motions that make them difficult to distinguish from ant species that live in similar habitats. It is our belief that such "action mimicry" will often cause a predator to hesitate for a period long enough to allow most of the lygaeids to escape.

Frequently myrmecomorphy involves the loss of flight (and possibly the reverse). The most common condition is a shortening of the wings, an enlargement of the abdomen with a concomitant constriction at the base, and a modification of the shape of the head. This condition may be enhanced by various color patterns, especially but not exclusively, in the Miridae by streaks of silvery hairs that break up the body shape into a three-lobed ant form. In macropterous forms myrmecomorphy is often achieved by patches of white color on a dark insect, usually involving the forewings—but sometimes also the pronotum—which have patches of color placed so as to give the illusion of a "wasp waist," but which still allow the insect to fly. In some of the most striking mimics the head may be modified to possess ridges along the lateral margins that closely resemble the mandibles of ants.

Ant mimicry is widespread in the Lygaeidae and has developed many times in different phyletic lines. *Bledionotus systellonotoides* Reuter, a ground-living species from the Near East that occurs under plants growing on rocks, is said to sometimes occur with, and closely resemble, the mimetic mirid *Mimocoris rugicollis* (Costa). The Pamphantini, which are often brightly colored with orange and black, are speciose in Cuba and Hispaniola, where they form mimicry complexes with beetles and spiders, all of which mimic species of *Solenopsis* ants (Myers and Salt, 1926). The Pamphantini, in contrast to *Bledionotus,* frequently occur on trees; Brailovsky (1989) described several species from the Brazilian rainforest canopy. Many Rhyparochrominae are myrmecomorphic and, because of the large incrassate forefemora, were long thought to be predaceous, at least in part on ants, but all evidence indicates strict seed-feeding habits. Some Heterogastrinae are also strikingly myrmecomorphic.

Ant mimicry appears to take two basic evolutionary pathways. The first is to live in a habitat frequented by many species of ants. The mimicry often is not to any particular species but to a general antlike habitus. The second is probably less common but has the most exquisite adaptations, in which a specific ant appears to be the model. Sometimes more than one species of insect or spider will mimic the same ant species. Myers and Salt (1926) noted that each of several species of *Solenopsis* ants in Cuba was mimicked by a spider, a beetle, and at least one pamphantine lygaeid. They commented that the mimicry was so striking that even after a period of study, when they attempted to pick a lygaeid bug mimic from a branch, they were astonished to see it drop by a thread from the branch and realized that it was a spider. Some of these *Solenopsis* ants are strikingly colored orange and black. In addition to the pamphantines there are two lygaeid bugs belonging to different genera that develop this striking orange and black color on Cuba, but elsewhere in the Neotropics other species of these genera are dull brown and black. Parenthetically it may be added that such information can be useful in interpreting questions of vicariance versus dispersal for given taxa (Slater, 1988). In the case above, the lygaeid *Heraeus triguttatus* (Guérin-Méneville) is the only orange and black lygaeid occurring in extreme southern Florida as well as on Cuba. Since the ant model is not present in Florida the obvious conclusion would seem to be that this species has reached Florida by overwater dispersal from Cuba subsequent to its evolution of mimetic coloration.

McIver and Stonedahl (1987a, b) conducted experimental studies on myrmecomorphic species of *Coquillettia* Uhler and *Orectoderes* Uhler and showed that their arthropod predators were capable of learning the antlike habitus and associating it with unpleasant experiences. In some cases it is clear that myrmecomorphic mirids are associated with ants, and with a particular ant species, whereas in other cases, it is not obvious that such an association exists (see Fig. 53.2C, D).

Heteropteran nymphs often mimic different species of ants as they grow (McIver and Stonedahl, 1993). Species of the mirid genera *Coquillettia* and *Paradacerla* Carvalho and Usinger—and the lygaeid genus *Slaterobius* Harrington—have two color morphs that resemble two different ant species. In many mirids only the female is strongly mimetic, whereas in alydids and lygaeids both sexes are mimetic. McIver and Stonedahl (1993) believed that within given phyletic lines of orthotyline and phyline mirids the most derived taxa are myrmecomorphic and the most plesiomorphic are nonmimetic.

Many genera of Alydidae contain species that are striking ant mimics as nymphs, but most species as adults have behavioral and color patterns of wasps. This phenomenon may best be illustrated by an experience that we had in South Africa some years ago. We were attracted to the behavior of a species of alydid that was alighting on a sandy area in bright sunlight. When the insects landed they moved their bodies in a jerky fashion and opened the forewings to display their bright orange dorsal abdominal coloration. The movements and color pattern closely resembled the actions of sand wasps working in the same area. Interestingly, adjacent to this site was a stand of *Acacia* trees under which was a large population of alydid nymphs that were, as usual, strongly ant mimetic. The irony of this whole association was that upon returning to the laboratory with a sample of nymphs we discovered that among them were several adult specimens of a tiny lygaeid that so closely resembled the ant-mimetic, early-instar alydid nymphs that we had not differentiated them in the field.

Possible beetle mimicry is less well understood. Many ground-living and some arboreal hemipterans develop coriaceous front wings that meet down the middle of the back in a coleopteroid fashion, but it has not been shown that these bugs are beetle mimics. Some of these cases are associated with flightlessness, and it appears to be an adaptation to prevent water loss in xeric habitats. Nevertheless, there are peculiar longitudinal markings on the shell-like covering of the scutellum of some Scutelleridae that greatly enhance the beetle-like appearance of the insects and do not seem to have any other protective value.

Protective Coloration and Shape

Protective coloration is present to some extent in almost every family of Heteroptera. Frequently this is a simple matching of the color of the insect to that of its environment. In many Miridae that feed on flowers, the nymphs in particular are the same color as the flowers on which they develop. The occurrence of green coloration in phytophages and brown or gray coloration in bark-dwelling predators is widespread in the family.

More striking is the modification of body shape to "mimic" the substrate on which the insect lives. In the South American pentatomoids of the family Phloeidae, which live on tree bark, the sides of the head, thorax, and abdomen are expanded into great flattened lobes that make the insect almost indistinguishable from the background on which it lives. A convergent condition occurs on Borneo, where the pentatomid genus *Serbana* Distant, whose habits appear to be unknown, has developed similar flattened, expanded body lobes and presumably also will prove to live on bark.

Insects of several families, such as the stenodemine mirids, monocot-living alydids, tingids, lygaeids, predatory reduviids, and even pentatomids living among grasses and sedges develop elongate slender bodies that resemble closely the elongate slender grass and sedge stems and leaves among which they live.

The lygaeid bugs of the subfamilies Cyminae and Pachygronthinae live chiefly in the seed heads of sedges. Both nymphs and adults closely resemble the shape and color of the seeds on which they feed.

Some heteropterans have the ability to change color with the seasons or the places where they live to increase their protective (or hiding) ability. Species of the mirid genus *Stenodema* Laporte are green during the early part of the host growing season, but pale brown later as the grasses on which they feed mature and become brown. This color change can also be seen in such predators as the ambush bug *Phymata pennsylvanica* Handlirsch. It is green and black early in the season, but in the autumn, when it lives most frequently in the heads of yellow goldenrod (*Solidago* spp.), it becomes a bright yellow and black. This latter change may be considered protective coloration but it also is concealing coloration for these "sit and wait" predators, making them much less conspicuous to potential prey insects. Some tropical assassin bugs exhibit this concealing pattern in color and sometimes in structural modifications that make them virtually indistinguishable from the flowers on or in which they sit.

Even more striking color modifications occur in toad bugs of the genera *Gelastocoris* Kirkaldy and *Nerthra* Say. We have observed a species of the former living along the banks of the Platte River in Nebraska, where in the same area individuals were black when on a dark substrate, tan when on a brown substrate, and speckled when on sand of white and dark grains. In Western Australia a species of *Nerthra* that lives in dry habitats among rock chips becomes reddish, black, or dull white depending on the color of the rocks among which it lives. To our knowledge there has been as yet no study to determine whether such striking color variations are genetically controlled or whether individuals are able to change color to match the substrate on which they live.

Protective coloration also may take the form of deflective patterns which are perhaps best illustrated by the white banding on the terminal segments of many ground-living heteropterans and by the occurrence of white patches near the end of the abdomen. There has been little experimental work done on this phenomenon, but to the field collector it becomes obvious that these are deflective markings that cause the potential vertebrate predator either to miss the rapidly moving heteropteran by striking at the white marking at the posterior end, or in the case of the antennal marking striking at the area anterior to the insect's body and at best obtaining the terminal antennal segment. These color patterns occur in family after family of geophilous heteropterans and merit serious experimental work.

The leaflike broadened and flattened antennal segments and hind tibiae of several Coreidae may be either a type of camouflage to break up the insectlike profile or another example of deflective patterning that causes the predator to focus on an appendage rather than on the body of the prey species.

Bright coloration, especially iridescence, might seem at first glance to be aposematic. However, it occurs most frequently in tropical forest species and is probably protective in habitats where light and dark contrast strongly.

Protective coloration may also involve the use of foreign substances. The masked bedbug hunter, *Reduvius personatus* (Linnaeus), a black shining insect when adult, is pale and soft-bodied as a nymph. These nymphs, which live in houses, habitually cover themselves with a thick layer of dust particles so that they look exactly like a small piece of dust moving about in corners and under furniture. They are often called dust bugs for this reason. The origin of this habit presumably is not due to their association with humans, because a similar phenomenon also occurs in some species in dry dusty habitats, where the nymphs cover themselves with small particles of sticks and other debris and become almost invisible among the detritus in which they live (see Chapter 48, Reduviidae).

8
Heteroptera of Economic Importance

Although it is probably true that no single species of Heteroptera is as economically important as some species in other orders, it is also true that the variety of ways in which heteropterans affect humans and their environment is certainly as great as, if not greater than, that of any other major insect group.

Economic importance traditionally has been viewed in terms of the direct damage that insects do to the crops upon which humans depend for existence and of the diseases that are transmitted to humans and their crops and domestic animals by insect vectors. As ecological sophistication has increased in recent decades so has our understanding of the exquisite interrelationships that exist between all of the organisms in an area, with ecologists now speaking of the impact that species and communities have on the ecosystem. It is thus more difficult to speak of heteropterans in traditional economic terms, as the removal of what today may be thought of as a species of relatively little direct consequence to human welfare may in fact set in motion a chain of events leading to irreversible damage to the habitat in which we live.

It is also true that our knowledge of the biology of most Heteroptera is still fragmentary or nonexistent. There is a great opportunity here for the field biologist (Gross, 1975–1976).

In this discussion we shall limit ourselves chiefly to the more traditional concepts of economic importance—that is, the immediate damage or benefit caused by heteropterans to humans and to their associated crops and animals—and we will confine the discussion to a review of only a few of the heteropterans involved. Otten (1956) should be consulted for a detailed discussion of economically important plant-feeding Heteroptera.

Because the majority of heteropteran species are phytophagous, it is obvious that among these plant-feeding species some will feed on the crops that humans utilize for food, medicines, and esthetic pleasure, and some will feed on plants that are essential for the continuation of the habitat. Plant damage by Heteroptera is, in most cases, direct damage. Although some plant diseases are carried by Heteroptera, the various species that transmit diseases do so to a much lesser extent than do the Auchenorrhyncha and Sternorrhyncha.

The economic importance of various Heteroptera also involves many species that are beneficial in that they feed on economically destructive insects. Some species are ectoparasitic on humans and domestic animals, and a few carry serious diseases of humans. The following discussion is organized alphabetically by family.

Alydidae. *Camptopus lateralis* Germar is an alfalfa seed pest of considerable importance in southern Eurasia and North Africa (Mukhamedov, 1962).

The elongate slender *Leptocorisa acuta* (Thunberg) is often a major pest of rice in the Orient. In India it is known as the Gandhi bug. Other species of *Leptocorisa* Latreille such as the rice ear bug, *L. oratorius* (Fabricius) (Rothschild, 1970), and *L. chinensis* Dallas sometimes cause serious injury in Asian paddies.

Various species of *Riptortus* Stål, the "pod-sucking bugs," are serious pests of beans (*Phaseolus* and other genera) from West Africa to Japan. *Mirperus jaculus* (Thunberg) is a pest of beans throughout tropical Africa. Alydids feed on seeds of rice and beans at the "milky grain" stage, when they are fully formed but unripe.

Anthocoridae. Various species of *Orius* Wolff are important predators on Thysanoptera, mites, and the eggs of Lepidoptera, in both greenhouse and field crop situations. *Montandoniola maraquesi* (Puton) has been widely introduced as a biocontrol agent of Thysanoptera on figs and olives.

Berytidae. Wheeler and Henry (1981) summarized information on *Jalysus wickhami* Van Duzee (the tomato stilt bug) and *J. spinosus* (Say) in North America. The former feeds on a great variety of dicotyledonous plants and sometimes is destructive to tomatoes. *Jalysus spinosus* feeds chiefly on monocotyledonous plants. Elsey and Stinner (1971) considered *J. wickhami* to be an important predator of aphids and the eggs of Lepidoptera on tobacco. They stated that development could not be completed without utilization of some animal food.

Cimicidae. All species feed on vertebrate blood, and among them are the common temperate and tropical bed bugs, *Cimex lectularius* Linnaeus and *C. hemipterus* (Fabricius). It has not been established that either of these species is a disease carrier, although both may cause nutritional deficiencies and allergies (Ryckman et al., 1981). Human bed bugs as well as such species as the poultry bug, *Haematosiphon inodorus* (Duges), are often serious pests in poultry houses.

Bed bug bites have been one of the minor plagues of human societies for untold centuries. It is surprising how recently society has been able to free itself to a considerable extent from these pests through the use of insecticides. (One of us remembers vividly as a young student going on a "field trip" to a house, only three blocks from his large campus, where bed bugs were so abundant that when wallpaper was stripped back there was an almost solid mass of swollen red bugs.)

Oeciacus vicarius Horvath is a vector of Fort Morgan virus found in Cliff Swallows and House Sparrows and related to western equine encephalitis.

Colobathristidae. Several species of *Phaenacantha* Horvath are important pests of sugar cane in the Orient (Kormilev, 1949).

Coreidae. A number of species are of considerable economic importance. For example, the North American squash bug *Anasa tristis* (De Geer) often is a serious pest of cucurbits. Other *Anasa* spp. become destructive to cucurbits in localized areas.

In East Africa *Pseudotheraptus wayi* Brown causes "gummosis," or early fall, of coconuts (Brown, 1955a), and in the Solomon Islands the same type of damage is caused by *Amblypelta cocophaga* China (Brown, 1955).

Leptoglossus gonagra (Fabricius) is an important pest of garden and truck crops throughout the Pacific and in southern Africa (see Slater and Baranowski, 1990). *Leptoglossus phyllopus* (Linnaeus) is a pest on a variety of cultivated plants in the southern United States, and *L. clypealis* (Heidemann) causes a high percentage of loss to the pistachio crop in California. *Leptoglossus corculus* (Say) and *L. occidentalis* Heidemann are often destructive to the seeds of various species of *Pinus* in North America, at times destroying over 50% of the seed crop (Krugman and Korber, 1969).

Anoplocnemis curvipes (Fabricius) is a polyphagous pest of field crops in tropical Africa. *Corecoris* spp. attack sweet potato and other crops in the tropics and subtropics of the Western Hemisphere. *Phthia picta* (Drury) is a pest of tomatoes in the Caribbean.

In the Orient *Clavigralla gibbosa* Spinola, the tur pod bug, has been reported in an abundance of 10 adults per plant on pigeon pea (*Cajanus cajan*), where it can cause almost complete destruction of the crop. *Clavigralla elongata* Signoret is a pest of certain legumes in East Africa (most of the economic literature is under the name *Acanthomia horrida*), and *C. tomentosicollis* Stål is common throughout most of subsaharan Africa and is particularly injurious to beans.

In Australia the crusader bug, *Mictis profana* (Fabricius), is often a serious pest in warm areas.

Several species of the genus *Chelinidea* Uhler were introduced into Australia in an attempt to control the prickly pear cactus but with minor success.

Cydnidae. *Aethus indicus* (Westwood) is sometimes a pest of cereal crops in tropical Africa and Asia. *Scaptocoris castaneus* Perty is a pest of sugar cane and has been reported feeding on the roots of cotton, bananas, tomatoes, and pimentos. A much more complex relationship is that of *S. divergens* Froeschner to the production of bananas. This insect injures banana plants by feeding on the roots. In many parts of Central America the banana crop has been destroyed by the presence of *Fusarium* wilt. Roth (1961) summarized what is known concerning the relationship of *Fusarium* wilt to populations of *S. divergens*. This cydnid is found primarily in sandy soils. Where the cydnids were abundant, feeding on the roots of bananas, the plants were healthy, whereas nearby plants with few cydnids were dying from *Fusarium* wilt. Roth noted records of populations as high as 76 individuals per square foot, with 532 insects feeding around the roots of a single healthy plant. The scent-gland secretions of nymphal and adult cydnids are toxic to *Fusarium*. This is an excellent example of how complex the relationships of insects to humans can be, for, while in many areas this cydnid is still considered harmful, in other areas it is possible to grow bananas only when the bugs are present.

Dinidoridae. *Coridius janus* (Fabricis) is a serious pest of cucurbits in the Orient, where it attacks both pumpkins and gourds of the genera *Cucumis*, *Luffa*, and *Lagenaria* (Rastogi and Kumari, 1962). *Cyclopelta obscura* (Lepeletier and Serville) is a pest of legumes in Southeast Asia.

Lygaeidae. Historically the most destructive plant-feeding North American heteropteran was *Blissus leucopterus* (Say). As early as the mid-1700s, this small black and white insect was the scourge of corn fields and did considerable damage to other grain crops. Unlike many major insect pests it is a native species, and presumably before the advent of vast monoculture techniques it fed on native grasses. In the course of adapting to human's crops it apparently shifted from being predominantly flightless to always being capable of flight and migrating to and from the fields each year. The ravages of this insect in the midwestern United States in the 1930s enlisted most of the farming community in burning, trenching, and spraying. The chinch bug is still an important pest, but its abundance seems to be tied to drought conditions, for in wet summers its numbers are held in check by a disease organism. At present an eastern subspecies (*B. leucopterus hirtus* Montandon) and a southern relative (*B. insularis* Barber) are of great importance because they attack lawn grasses. In Florida whole fleets of trucks with their "chinch bug control" logos may be seen, from which professional workers busily attempt to rid householders of this lawn destroyer.

In South Africa the chinch bug *Macchiademus diplopterus* (Distant) is a serious pest of wheat. It appears to have adapted to this plant from a native grass in a manner similar to that of the North American chinch bug. The Oriental chinch bug *Cavelerius sweeti* Slater and Miyamoto is a serious pest of sugar cane.

In Africa, the Orient, and Australia several species of *Oxycarenus* Fieber, especially *O. hyalinipennis* (Costa), are important pests of cotton, where they damage the crop not only by feeding but also by staining the bolls. *Oxycarenus hyalipennis* has been introduced into South America.

In Australia the rutherglen bug, *Nysius vinitor* Bergroth, is a major pest as are *Nysius ericae* Schilling (= *niger* Van Duzee) and *N. raphanus* Howard in the United States. These oligophagous plant feeders damage cultivated crops most when native plants in arid regions dry up. At times the numbers become extremely high and the insects even become nuisances as they swarm into human habitations. In southern Africa a similar phenomenon occurs.

Species of *Spilostethus*, especially *S. pandurus* (Scopoli), damage many different crops in Europe, Africa, and Asia. *Elasmolomus sordidus* (Fabricius) is a pest of various seed crops, especially peanuts (groundnuts) both in the field and in storage situations in Africa and Asia.

Some species of Geocorinae are considered of value in biological control of crop pests. A substantial literature on North American species of *Geocoris* Fallén has developed in recent decades.

Malcidae. Several species of *Chauliops* Scott have been reported as destructive to cultivated legumes in the Orient (see Slater, 1964b).

Miridae. This large family contains many species that damage crops. Problems range from occasional outbreaks or injury to crops of minor importance, to serious injury to crops of great economic value.

At least 40 species of the subfamily Bryocorinae are known to feed on cacao. Members of the subtribes Odoniellina and Monaloniina are particularly destructive. In Africa the principal species are *Sahlbergella singularis* Haglund and *Distantiella theobroma* Distant. At least 11 species of *Monalonion* Herrich-Schaeffer are serious pests in South and Central America; in southeast Asia and Africa species of *Helopeltis* Signoret may be equally or more destructive; and in Madagascar the principal damage is done by *Boxiopsis madagascariensis* Lavabre (Lavabre, 1977).

Several species of the genus *Helopeltis* in the Orient are serious pests of crops other than cacao. The primary destruction is to tea, cashew, and cinchona, but other economically important plants including allspice, black pepper, apples, grapes, and guava are also attacked. Before the use of modern insecticides crop losses on tea plantations often reached 100% (Stonedahl, 1991).

Calocoris norvegicus (Gmelin) is a serious potato pest in many temperate areas. *Adelphocoris lineolatus* (Goeze) was accidentally introduced into North America from Europe, spread rapidly, and is a very serious pest in areas where alfalfa (*Medicago*) is grown for seed (Wheeler, 1974). It is also a serious pest in the Palearctic.

The tarnished plant bug, *Lygus lineolaris* (Palisot de Beauvois), and closely related species are very general feeders, attacking a wide variety of plants. When feeding on fruits such as peaches and pears they cause a blemish called "cat facing," that seriously damages the fruit. There is an extensive literature dealing with this group as pests of many garden and crop plants (see Scott, 1977, 1981; Graham et al., 1984; O. P. Young, 1986). *Lygus hesperus* Knight and *L. elisus* Van Duzee are also of major importance, the latter causing serious damage to cotton crops in the western United States. Other cultivated plants damaged by these species include beans, strawberries, peaches, and many seed crops, the most important being alfalfa and rape seed grown for oil. Stern (1976) stated that in California alone in 1974 *L. elysus* and *L. hesperus* caused an estimated loss of $56 million, more than 10% of the total loss caused by all insects and mites to all California crops for that year. The history of control of these two pests offers a striking example of the importance of knowledge of the ecological relationships of insect pests. Stern (1976) found that *L. hesperus* actually does not attack cotton if other suitable plants are available. When alfalfa was interplanted as a "trap crop," cotton was raised without any insecticide application. *Lygus rugulipennis* Poppius and *L. pratensis* (Linnaeus) are destructive to many cultivated plants in the Palearctic.

The predatory nature of several *Lygus* spp. as well as other mirids further complicates the picture. Wheeler (1976) concluded that many Miridae usually considered to be phytophagous feed to some extent on small arthropods.

Taylorilygus pallidulus (Blanchard) is a polyphagous pantropical pest. *Creontiades pallidus* (Rambur) injures sorghum, *Solanum* spp., and other crops in southern Europe, Africa, and Asia. *Cyrtopeltis* (*Nesidiocoris*) *tenuis* Reuter is a pantropical pest of tomato and other solanaceous crops.

Several mirids do significant damage to rangeland grasses and small grain crops. Among these are species of *Irbisia* Reuter, *Labops* Burmeister, *Leptopterna* Fieber, and *Trigonotylus* Fieber.

Tytthus mundulus (Breddin) is well known among predatory mirids, because of its successful use in Hawaii as a biological control agent of the sugar cane leafhopper *Perkinsiella saccharicida* Kirkaldy (Zimmerman, 1948).

Members of the deraeocorine genus *Stethoconus* Flor, are either primary or exclusive predators of Tingidae. *Stethoconus frappai* Carayon (1960) has been recorded as an important population control agent of *Dulinius unicolor* (Signoret) on coffee in Madagascar and *S. japonicus* Schumacher of the *Rhododendron* lace bug *Stephanitis rhododendri* Horvath in Japan and the United States (Henry et al., 1986).

Pentatomidae. Most stink bugs are plant feeders, and several of them are of great economic importance. *Nezara viridula* (Linnaeus) is a large green species found in the warmer parts of the Old and New Worlds. It attacks over 90 species of plants, including many vegetables, and is often a serious pest of beans and tomatoes.

Murgantia histrionica (Hahn), the brightly colored harlequin cabbage bug, is a serious pest of some crucifers in the United States and Mexico (Paddock, 1918). This species is native to Mexico but has spread northward within the past hundred years.

Stink bugs severely damage cacao in the Old and New Worlds, not only by direct feeding but, more important, by introducing pathogens. Among the more destructive species are *Mecistorhinus tripterus* (Fabricius) (Sepúlveda, 1955) and *Bathycoelia thalassina* (Herrich-Schaeffer). The latter species has become increasingly important, apparently because of the destruction of its natural enemies by the use of insecticides, the destruction of its wild host plants, and the introduction of new and more susceptible varieties of cacao (Owusu-Manu, 1977).

Pentatomids have also been implicated in the transmission of pathogens to palms (Dolling, 1984).

Other destructive pentatomids include the following: *Scotinophara* spp. are pests of cereals, particularly *S. lurida* (Burmeister) on rice (Kiritani et al., 1963), and grasses in Africa and Asia; *Eysarcoris ventralis* (Westwood) is sometimes a pest of rice in Southeast Asia; *Oebalus pugnax* (Fabricius) is a rice pest in the Western Hemisphere (Sailer, 1944); *Piezoderus hybneri* (Fabricius) feeds on various legumes in Africa and Asia; *Bagrada cruciferarum* Kirkaldy feeds on crucifers in the Old World; *Biprorulus bibax* Breddin is a pest of citrus (McDonald and Grigg, 1981); *Antestiopsis lineaticollis* (Stål) is a widespread coffee pest (Greathead, 1966); various species of *Euschistus* Dallas are desructive to soybeans in North America (Boas et al., 1980); *Agonoscelis versicolor* (Fabricius) attacks millet, cotton, and cacao; and *Aelia rostrata* (De Geer) is a serious wheat pest in the Near East. The last species, like several scutellerids and some other pentatomids has an unusual life cycle, in that the insects leave the wheat fields in late summer and migrate as much as 50 km to overwinter in the mountains at elevations of 1500–1800 meters in *Quercus* scrub. The migration is not a continuous one, but occurs in stages over several days (Brown, 1965). In Morocco the related *Aelia germari* Kuster is known to migrate several hundred kilometers. Similar migrations in the Near East have been observed in species of *Dolycoris* Mulsant and Rey, *Carpocoris* Kolenati, *Codophila* Mulsant and Rey, and *Eurydema* Laporte.

The Asopinae are predaceous pentatomids, some of which are significant biological control agents. Particularly important among these is *Perillus bioculatus* (Fabricius), which feeds almost exclusively on the Colorado potato beetle. The beetle, which apparently originally lived on wild solanaceous weeds in the western United States, adapted early to potatoes and spread rapidly eastward across the country with the predatory stink bug following along. Unfortunately the beetle was introduced into Europe, where it has become even more destructive than it is in North America. *Perillus* Stål has been introduced into Europe as a control measure, but apparently it has been a failure.

Several species of *Podisus* Herrich-Schaeffer are considered valuable predators, but since most of them are rather general feeders they are for the most part not capable of major control.

Eocanthocona furcellata (Wolff) preys on larvae of Limacodidae on palms in Southeast Asia. In Australia the pentatomid *Agonoscelis rutila* (Fabricius) is widely known for its control of horehound.

Piesmatidae. *Piesma cinereum* (Say) transmits the virus known as "savoy" that causes serious damage to sugar beets. *Piesma capitatum* (Wolff) is a pest of sugar beets in Russia, but does not transmit any disease. *Piesma quadratum* (Fieber) transmits a leaf-curl disease of sugar and fodder beets in Germany.

Pyrrhocoridae. Several species of the genus *Dysdercus* Guérin-Méneville are serious pests of cotton in tropical areas (Fuseini and Kumar, 1975). They frequently form large aggregations both on cotton and on wild hosts such as *Hibiscus* and *Ceiba*. The colonial nature of these aposematically colored bugs aids to warn predators away from the distasteful clusters. Van Doesburg (1968) reported that in the New World several species are sometimes present in feeding clusters that are confined almost entirely to ripening fruits and seeds. Also, it has been demonstrated that when several insects feed on a seed, an enzyme pool is secreted that makes feeding more advantageous for the entire group. These aggregations are formed by visual, tactile, and olfactory cues.

Reduviidae. Perhaps the most destructive heteropteran is not a plant feeder but an assassin bug that feeds upon the blood of humans and other mammals. Throughout the warmer parts of the Western Hemisphere, but especially in South America, various species of *Triatoma* Laporte and related genera transmit the trypanosome that causes the debilitating Chagas' disease in humans. These insects are thought to be one of the most important reasons for the impoverishment of many areas of South America. The insects often live in thatched houses, and they transmit the trypanosome parasite to the inhabitants by subsequent defecation in the wound that they make when feeding on the sleeping victim. Whole communities may be infected and, as with African sleeping sickness, infected people become lethargic, are unable to function efficiently, and are susceptible to a variety of other diseases. It is interesting to note that Charles Darwin wrote in his journal of being bitten by the "great black bug of the pampas" when he lived in Argentina with the gauchos. For years there has been controversy concerning the lifelong illness of Darwin, with theories ranging from psychosomatic to genetic, and including the possibility that Darwin suffered from Chagas' disease.

Some Reduviidae are beneficial. *Tinna wagneri* Villiers preys on mosquitoes in dwellings in East Africa; *Phonoctonus* spp. feed on *Dysdercus* spp.; *Sycanus collaris* (Fabricius) preys on Limacodidae larvae in Southeast Asia; and *Amphibolus venator* (Klug) and *Peregrinator biannulipes* (Montrouzier and Signoret) prey on stored-products pests.

Rhopalidae. The box elder bug, *Boisea trivittata* (Say), is often abundant on box elder trees (*Acer negundo*) in North America, but it is chiefly a nuisance insect for its habit of entering houses in large numbers for hibernation.

Liorhyssus hyalinus (Fabricius) is a cosmopolitan pest of many low-growing crops, especially Asteraceae. *Leptocoris hexophthalmus* (Thunberg) often destroys coffee and cotton in Africa.

Scutelleridae. The most destructive species are members of the genus *Eurygaster* Laporte. In the Near East they destroy large quantities of wheat and related grain crops. Most infamous is the soun bug, *Eurygaster integriceps* Puton. In the southern part of the former USSR, Turkey, Iraq, and Iran this species is a major pest of wheat, and there is an enormous economic literature. The Russian literature is summarized in Fedotov 1947–1960. Studies by Brown (1962a, b, 1963, 1965) are especially important. In the spring these insects attack wheat and other cereal crops, sucking the green shoots and causing the terminal buds to die. Later they feed on the seeds. The crop may be damaged so that it is not worth harvesting, and even when the damage is of a lesser magnitude the feeding affects palatability and lowers baking quality. These insects have a remarkable life cycle. In late summer they fly en masse from the wheat fields to mountainous areas, where they overwinter. The flights may be as long as 200 km, but usually they average 20–30 km. The migration from the lowlands is to areas as much as 1000 or even 2000 meters higher in elevation. The insects remain there for nine months, massed under cushionlike or dome-shaped ground-hugging plants such as species of *Astragalus* and *Artemisia*. Brown (1962b) reported populations of up to 900 bugs per plant and up to 1000 bugs per square meter. Overwintering consists of two stages. The first is actually an aestivation while the weather is relatively warm. It terminates, at least in Turkey and Iran, in a second movement, to an elevation as much as 100 meters lower. In the late spring the population migrates back to the wheat fields. Thus, a given individual lives for an entire year.

Although *E. integriceps* is by far the most important species economically, it is only part of a more widespread phenomenon in the Near East that also includes pentatomids, coreids, alydids, and rhopalids. At least three species of *Odontotarsus* Laporte and *Ventocoris fischeri* (Herrich-Schaeffer) also migrate from the mountains to cultivated fields, but they feed to a large extent on weeds rather than on wheat and other cereals. Brown (1965) showed that migration is influenced by wind conditions and to some extent by the direction of the sun.

In the case of *E. integriceps* there are periods of enormous abundance and periods of relatively low populations. During outbreak periods there is increased homogeneity in the populations and increased resistance to environmental conditions. These changes in the population structure are passed on for several generations, a situation rather similar to that found in migratory locusts. It seems likely that the original populations of *E. integriceps* and the other migratory scutellerids lived on mountain grasses at middle elevations and moved to abundant food sources in the lowlands as agriculture became more concentrated.

Tessaratomidae. *Musgraveia sulciventris* (Stål), known as the bronze orange bug in Australia, is a pest of citrus in the northeastern part of the continent. Its native host plants are also members of the Rutaceae. It is a very large insect (over 20 mm) with flattened ovoid nymphs. In China *Tessaratoma papillosa* Drury is destructive to the fruit of the litchi (Yang, 1935).

Tingidae. Lace bugs of the genus *Stephanitis* Stål are frequently serious pests of ornamental azaleas, rhododendrons, and andromedas. *Stephanitis pyri* (Fabricius) and *Monosteira unicostata* (Mulsant and Rey) are both pests of rosaceous orchard trees, especially pear, in Europe. *Corythuca ciliata* (Say), the sycamore lace bug, damages *Platanus* spp. in the eastern United States and is now causing concern in southern Europe, where it was recently introduced. *Urentius hystricellus* (Richter) damages aubergine in the Old World tropics. *Corythuca gossypii* (Fabricius) is a pest of cotton, beans, citrus, and solanaceous crops in the Caribbean and Central and

South America. *Gargaphia torresi* Costa Lima attacks cotton in Brazil. *Corythaica cyathicollis* (Costa) is a pest of Solanaceae in the Caribbean and South America, and *Corythaica monacha* (Stål) is destructive to cotton and beans in South America.

Several phytophagous heteropterans have been used as biological control agents against invasive weeds. None of these has been spectacularly successful. Among the best known is the tingid *Teleonemia scrupulosa* (Stål), which has been widely introduced in an attempt to control *Lantana*. For additional information see Drake and Ruhoff (1965).

Veliidae. Some species of *Microvelia* Westwood have been employed as predators in biological control schemes; for example, *Microvelia pulchella* Westwood has been used to control mosquitoes (Miura and Takahashi, 1987, 1988). *Microvelia* spp. are effective predators of rice planthoppers (Reissig et al., 1985).

9
Historical Biogeography

Our understanding of the meaning of heteropteran distributions has undergone revolutionary change since the 1960s. Two factors have been of particular importance. First, most working scientists have accepted the concept of continental drift, despite continuing controversy over details. Our knowledge of the heteropteran fossil record—no matter how fragmentary—makes it evident that most major lineages diverged early in the Mesozoic, from Pangaean, Laurasian, or Gondwanan patterns, and therefore understanding the former relationships of continents is crucial to understanding many present-day distributions. Second, the rise of cladistic methodology has provided a more rigorous means of recognizing monophyletic groups and the relationships among them.

This acceptance of continental movement and cladistic methods shifted biogeographic emphasis from centers of origin, classification of subregions, and methods and abilities of dispersal, to attempts to establish areas of endemism and degrees of distributional concordance between different taxa and the areas they occupy (e.g., see Nelson and Platnick, 1981).

It must be kept in mind that historical biogeography can be no better than the facts upon which interpretations are made. There is a great deal that we do not know about heteropteran distributions both factually and causally. Not only are there still large numbers of undescribed taxa in every large family, but knowledge of the distributions of most species and even of some higher groups is fragmentary at best. Thus, there is a continuing need for careful faunistic studies of local areas. Although most groups of Heteroptera are reasonably well known in Europe and North America, most of the rest of the world is in desperate need of additional study, in both temperate and tropical areas. This is true because of the danger of irrecoverable loss of information due to extinction, particularly at the hand of humans.

Because most families are almost worldwide in distribution, it is at the level of the subfamily, tribe, and below that we ordinarily begin to perceive and attempt to understand patterns of endemism. In the chapters that follow dealing with the various families of Heteroptera, the distributions of families, subfamilies, and sometimes tribes are summarized.

Because the meaning of distributions is subject to interpretation, it seemed to us important to first enumerate some working principles, especially today, when so much controversy exists as to the most productive approaches to the study of historical biogeography. We then provide some interpretive remarks regarding apparent general patterns of distribution within the Heteroptera.

A first principle is that whether or not one wishes to recognize paraphyletic groups, only monophyletic groups can be the subject of biogeographic analysis.

A second principle is that in interpreting disjunct distributions, a vicariance model should be considered first, and only when this model proves to be unsatisfactory should dispersalist hypotheses be considered.

A third principle is that, even though only positive occurrences provide information, distributions of lineages of many taxa, including recent representatives, changed dramatically during the Tertiary and before, as is clearly shown by the amber-fossil record, for example. Thus, extinction should never be discounted as a possible explanation for seemingly anomalous patterns.

A fourth principle in interpreting any distributional pattern is to ask what geologic or climatic event might best explain the observed pattern.

Perhaps a final principle of any meaningful distributional study should be a clear statement of the methodological basis of the study.

The classical separation of the terrestrial faunas by Sclater (1858) into Nearctic, Palearctic, Neotropical, Ethiopian, Oriental, and Australian still has merit, but it tends to mask relationships across continental lines. We give here only a few examples from the Heteroptera which more or less conform to Sclater's system but which also help to point out its defects.

Neotropical Region. This area has long been treated as including all of Central and South America and the West Indies. This well-known pattern has been documented in the Miridae by Schuh and Schwartz (1985, 1988). The far southern part of South America, however, has a heteropteran fauna that is either depauperate or more closely allied to that of Australia and New Zealand, with some elements extending far northward at higher elevations in the Andes. The latter situation is well demonstrated by the Saldidae. Many elements of the Nearctic fauna extend southward along the cordilleras through Mexico deep into Central America and in some cases Colombia. The West Indies are for the most part composed of a mixture of South American elements in the Lesser Antilles and Central American elements in the Greater Antilles, with only a small part of the fauna being derived from temperate North America. Slater (1988) believed that much of the West Indian lygaeid fauna is the result of over-water dispersal despite the complex geologic history of the Greater Antilles and the strong evidence for the influence of plate movements in their formation.

Ethiopian Region. Like that of South America, the African fauna is composite. Southern Africa has a degree of relationship with Australia and New Zealand, as shown by the Plinthisini and Stygnocorini (Lygaeidae). Tropical Africa has a high proportion of Paleotropical elements found in family after family. There is also close relationship with the Palearctic, particularly north of the Sahara, but also in the east African savannah fauna. There is evidence, however, of an older relationship between the Macchia plant community of South Africa and the Magreb of North Africa. Africa does not have the mountain chains so characteristic of most other continents. Nonetheless, the isolated volcanic peaks of East Africa have a fauna showing relationships to the Palearctic, to the Drakensberg range of South Africa, and to Madagascar.

Madagascar shows a high degree of endemicity, which increases as the fauna becomes better known. This endemicity, however, is overlaid with a recent invasive African savannah fauna. The vaunted Oriental relationship is not yet well documented in the Heteroptera, although it is known to occur in the Ptilomerinae (Gerridae).

Australian Region. Australia contains many Paleotropical elements in the north, often shared with New Guinea. The temperate Bassian subregion shows a high degree of endemicity. In some groups such as the udeocorine Lygaeidae a marked radiation has occurred in this area (Slater, 1975). Small families such as the Lestoniidae and Aphylidae are endemic here. As in the case of South America and Africa, the Australian Heteroptera fauna is to a great degree a composite, despite the relatively homogeneous mammal fauna on which most classical biogeographic scenarios have been based.

Holarctic. The faunas of the northern Hemisphere are much better understood taxonomically than those of the Southern Hemisphere. There is a marked Holarctic element in the Miridae, Saldidae, Gerridae, and other families (see Wheeler and Henry, 1992). Genera in many families show considerable endemicity (Holarctic). Nonetheless, in many families the faunas of both the Nearctic and Palearctic comprise northern extensions of the rich tropical faunas to the south.

Other types of intercontinental patterns are also readily found in heteropteran distributions.

Gondwana patterns. The term *Gondwanan* has often been used in reference to taxa found in Australia, New Zealand, southern South America, and southern Africa. In fact, many far southern distributions do not include Africa, apparently because the northward movement of Africa has eliminated cold-adapted elements from the southern part of the continent. There is no known heteropteran distribution similar to the classic Gondwana pattern proposed by Brundin (1966) for the Chironomidae. Slater and Sweet (1970) documented a similar pattern for stygnocorine lygaeids, but the group is not known from South America (perhaps because of a lack of nymphs, which are critical for its recognition). But there are several groups with distributions that are difficult to interpret as other than Gondwanan.

The Enicocephalomorpha, the apparent sister group of all other Heteroptera (Fig. 1.1), have their greatest diversity in New Zealand and the southwest Pacific, with the fauna of the remainder of the world being relatively monotonous and composed almost entirely of members of the Enicocephalidae. Similarly, the Aradidae are diverse in New Zealand, with all eight subfamilies occurring there, while the Northern Hemisphere fauna is composed of a limited number of phyletic lines. A similar pattern occurs in the Acanthosomatidae.

Monteith (1980) documented an apparently ancient pattern in the Chinamyersiinae (Aradidae), a group that occurs in the mountains of eastern Australia, New Caledonia, and New Zealand. A similar pattern was found in some cylapine Miridae, which also occur on Lord Howe Island, although not in New Zealand (Schuh, 1986a).

For such taxa as the Idiostolidae, considered by several authors to be the most primitive of the lygaeoid families, and the aradid subfamily Isoderminae, the situation is less clear. Some would consider such groups to represent a classic East Gondwanaland pattern including southern Australia, Tasmania, New Zealand, and southern Chile. But, because there is no cladistic analysis of these groups, it is impossible to exclude the possibility of a more recent relationship through Antarctica long after the breakup of East Gondwanaland.

There is, however, strong evidence in the Heteroptera of West Gondwana relationships—that is, between tropical and subtropical Africa and South America. Examples are the Pachynomidae, with the Aphelonotinae restricted to Africa and South America, and the Pachynominae, with one South American genus and two Old World genera (these extending from Africa east into India also). Similar patterns can be found in the Cetherinae and Holoptilinae (Reduviidae), Plokiophilinae (Plokiophilidae), and Madeoveliinae (Mesoveliidae).

The family Pyrrhocoridae demonstrates a possible origin in the Eastern Hemisphere with subsequent migration to and diversification of one taxon in the New World. Van Doesburg (1968) noted that of the 30 recognized genera, all Western Hemisphere species belong to one subdivision of *Dysdercus* that otherwise occurs only in Africa. If van Doesburg is correct in believing that the family reached the New World by over-water migration, then it must have been sufficiently long ago for the elaborate mimicry complexes to have evolved. To believe that this is a vicariance pattern involving West Gondwanaland means that the other African pyrrhocorids were not present or subsequently have become extinct in the New World.

Paleotropical patterns. Distributions encompassing the Old World tropics and subtropics are extremely common in the Heteroptera. Such patterns frequently include tropical northern Australia and the western Pacific Islands—sometimes as far east as the Marquesas. Some taxa are absent from Africa, and others, while present there, are restricted to West Africa or also the Congo basin. Schuh (1984), Schuh and Stonedahl (1986), and Stonedahl (1988b) have termed such patterns in the Miridae "Indo-Pacific." At the family level only a few groups show such a pattern (Plataspidae, Malcidae), but there are untold numbers of examples at the subfamily, tribal, and generic levels.

Transpacific patterns. Several groups of Heteroptera occur exclusively or primarily in the Southern Hemisphere but do not occur in Africa. This striking distribution is well illustrated by the Colobathristidae (Štys, 1966a, b) and triatomine Reduviidae (Lent and Wygodzinsky, 1979); both show substantial diversity in the New World tropics and range in the Old World from India to northern Australia. In the Triatominae this distribution may even be present at the specific level, as a single species is tropicopolitan, but may have been spread by humans as may also be the case with the only known Eastern Hemisphere species of the coreid genus *Leptoglossus* Guérin-Méneville. Several other reduviid distributions appear to be transpacific (Peiratinae, Physoderinae, and Vesciinae). The Thaumastocoridae may show such a pattern, although of a more relictual nature. There is insufficient information in all these cases to identify the next area of relationship. Nevertheless, in the eccritotarsine Miridae, Africa appears to contain the sister group of a clade whose distribution is otherwise transpacific. Lygaeid clades such as the Antillocorini and Ozophorini also appear to have such a pattern, although the higher-level cladistic relationships are not yet sufficiently well understood. This transpacific pattern is widespread in flowering plants.

Relict patterns. When we look at present-day distributions, it is evident that monotypic groups (such as the Joppeicidae and Medocostidae) and small families with limited distributions (Canopidae, Megarididae, Phloeidae, and Lestoniidae) provide little information of general interest. On the other hand, small families such as the Paraphrynoveliidae, with only two known species, one in the Cape region of South Africa and the other in Lesotho, show a distribution similar to that of many other insect and plant taxa, suggesting a localized vicariance pattern, perhaps due to aridity as recently as the Pleistocene, but conceivably much older.

Oceanic islands. Traditionally the faunas of oceanic islands have been thought to be solely the result of long-distance dispersal over water. This explanation fails to appreciate the great age of many lineages within the Heteroptera and discounts movements of land masses as a causal factor in explaining distributions. Consequently, the origins of island faunas must be reexamined.

As an example, the Hawaiian fauna had been thought to have arrived within the last 5 million years. Yet the recognition of the Emperor Sea Mount chain, which contains the remnants of formerly larger islands that passed over the Hawaiian "hot spot," suggests that the fauna may well be much older. Nevertheless, these islands also possess an imbalanced fauna, suggesting that over-water dispersal may have been important in populating them. This is well illustrated by the mirid fauna which Zimmerman (1948) placed in several subfamilies, whereas Schuh (1974) demonstrated that all groups that have diversified belong to a single tribe in the Orthotylinae. The diverse lygaeid fauna belongs almost entirely to the subfamily Orsillinae, a group notorious for its ability to colonize remote islands. The Nabidae also radiate remarkably in these islands, but within only one or a few lineages.

The history of other islands such as New Caledonia, the New Hebrides, the Galapagos, and the Juan Fernandez remains enigmatic because the relationships of most groups occurring there are still ambiguous. New Zealand may be an exception. It shows ancient Gondwana relationships, more recent ones through a warm Antarctica, and evidence of recent over-water dispersal. For example, the previously mentioned diversity of the aradid and enicocephalid faunas is in striking contrast to the imbalanced lygaeid and mirid faunas and the complete absence of Coreidae and Reduviidae (other than Emesinae).

The Heteroptera will be important to any understanding of historical biogeography on a world scale. Further progress toward such understanding will come only with an improved knowledge of distributions and cladistic relationships.

10

General Adult Morphology and Key to Infraorders of Heteroptera

Head

The heteropteran head ranges from elongate both anterior and posterior to the large compound eyes, as in the Enicocephalomorpha (Fig. 13.1A), to barely longer than the compound eyes, as in many Miridae (Fig. 10.1). The *compound eyes* are generally well developed, and in many groups such as the Nepomorpha (Fig. 33.1) and Leptopodomorpha (Fig. 45.1A), they are large in comparison with the total size of the head. In other groups such as the Gerromorpha (Fig. 20.2C) and many members of the Pentatomomorpha (Fig. 64.1A) they are relatively much smaller. In some groups such as the Vianaidinae (Tingidae) (Fig. 54.1) they are reduced, whereas in some Enicocephalomorpha and all Polyctenidae and Termitaphididae they are completely absent (Fig. 65.1A, B). In first-instar nymphs the number of ommatidia ranges from 4 or 5 to many, and 1 or 2 setae arise from the center and/or periphery of these facets (Fig. 10.10E). Primitively 2 *ocelli* are situated between or sometimes slightly behind the compound eyes in adults (Fig. 13.1A) and may be close to one another or widely separated; these are lost in many Nepomorpha, consistently in some families of other infraorders, often in forms with reduced wings, and are always absent in nymphs.

Posteriorly the head forms a *neck* that inserts into the pronotum. Anteriorly there is a more or less prominent *clypeus* (or tylus), which is subdivided into an *anteclypeus* (tylus) and *postclypeus* in some taxa, such as most Leptopodomorpha. Adjacent—and in most taxa posterior—to the clypeus, lie dorsally the *mandibular plates* (or juga) and ventrally the *maxillary plates* (or lora) (Fig. 10.1), which internally form the attachments for the mandibular and maxillary levers and their associated stylets (Spooner, 1938; Cobben, 1978).

The *labrum*, triangular in form, is generally situated below the clypeus. The *mandibles* and *maxillae* are modified—as in all Hemiptera—into tubular concentric stylets (Fig. 10.2B), the former surrounding the latter. The mandibular stylets serve a cutting and lacerating function and are often adorned apically with minute barbs or teeth. The maxillary stylets—like the mandibular—are held together by interlocking keyways and contain the *salivary canal* and *food canal* (Fig. 10.2B). The *labium*, which is tubular, either straight or curving, and in many taxa capable of substantial flexion during feeding, possesses 3 or 4 segments (Fig. 10.2A). The 3-segmented condition is always the result of reduction of the basal segment. The labium varies from relatively short and barely reaching the posterior margin of the head—as in many Nepomorpha and Reduviidae—to relatively long and reaching to near the apex of the abdomen—as in some Pentatomomorpha; distally it bears a field of specialized sensors (Fig. 10.7B) (Cobben, 1978). Maxillary and labial palpi are completely absent. The *epipharyngeal sense organ* lies between the base of the labrum and the labium. In the family chapters we refer to the ultimate labial segment as 4 and count backward, so that the basal segment in the case of a 3-segmented labium will be segment 2.

Ventrally the head possesses a distinct *gula*. The *bucculae* are usually well developed, lying along each side of the labium and ordinarily extending to near the posterior margin of the head. Posteriorly the gula may be open and conjoin the neck, or it may be closed by a distinctively sclerotized *buccular bridge*.

The heteropteran *antenna* consists of 4 basic segments: the *scape* (1), *pedicel* (2), *basiflagellum* (3), and *distiflagellum* (4). During postembryonic development additional sclerites may be formed or subdivisions may take place. Most obvious among these are the presence of a *prepedicellite* between segments 1 and 2, which forms the fifth segment in the Prostemmatinae (Nabidae) (Fig. 56.1), and the subdivision of the pedicel to form 5-segmented antennae in the Pachynomidae (Fig. 47.1) and many Pentatomoidea (Fig. 66.1). A few taxa, including some Nepomorpha and Phloeidae show segmental loss or fusion (Figs. 40.1C, 74.2A). Antennal structures, including other less obvious intersegmental sclerites were described in detail by Zrzavý (1990c).

Thorax

The *prothorax* may possess anteriorly a distinct rounded or flattened *collar* (or collum) or may take the form of a finely reflexed margin. Dorsally there is usually a transverse impression delimiting the *anterior lobe* and

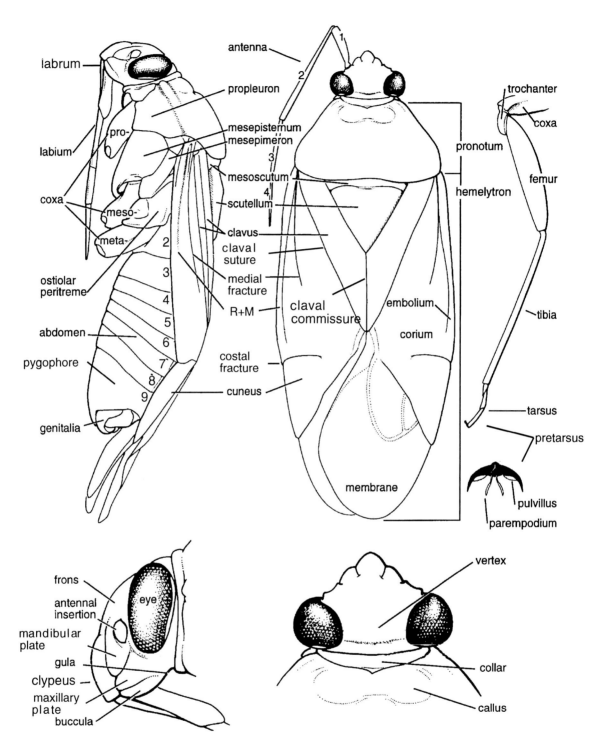

Fig. 10.1. General heteropteran morphology. *Lygus* sp. (Miridae) (modified from Schwartz and Foottit, 1992).

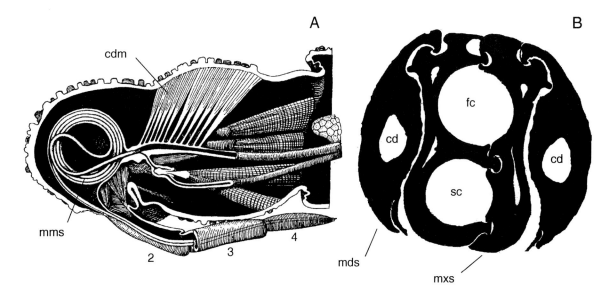

Fig. 10.2. Mouthpart structures. **A.** Longitudinal section of head of Aradidae (from Weber, 1930). **B.** Cross section of mandibles and maxillae, *Oncopeltus fasciatus* (modified from Forbes, 1976). Abbreviations: cd, central duct; cdm, cibarial dilator muscles; fc, food canal; mds, mandibular stylets; mms, mandibular and maxillary stylets; mxs, maxillary stylets; sc, salivary canal.

posterior lobe, the former usually possessing a pair of somewhat protuberant *calli*, which are grossly developed in some Saldidae and Miridae into conical structures, the external expression of foreleg muscle attachments. The *humeral* (posterolateral) angles of the pronotum may be rounded or projecting and sometimes spinelike as in many Reduviidae, Coreidae, and Pentatomidae. Laterally the prothorax may be either rounded or carinate, explanate, or reflexed.

The dorsum of the mesothorax is usually visible in the form of the triangular *scutellum* (mesoscutellum), which is bordered anteriorly by the sometimes visible *mesoscutum;* both may be occasionally completely obscured by the posterior lobe of the pronotum; in some Pentatomoidea the scutellum is greatly enlarged, entirely covering the wings and abdomen (Fig. 68.1A).

The thoracic pleuron is structurally comparatively monotonous and assumes the form similar to that found in most other pterygote insects. The *metepisternum* is distinguished by the presence of the *peritreme*, the actual point of release onto the body surface of fluid from the metathoracic scent glands, and the associated *evaporatory area* (evaporatorium) found in most Cimicomorpha and Pentatomomorpha (see also Exocrine Glands).

The thoracic sternum may be relatively broad or almost completely obscured by closely spaced coxae. The *prosternum* sometimes is produced ventrally in the form of a *xyphus*, or it may be longitudinally grooved for reception of the labium. The *mesosternum* and *metasternum* may bear longitudinal spines in some Pentatomoidea.

Spiracles are located dorsally on the pleuron on the meso- and metathorax, the anterior pair usually being obscured.

Wings

Wing venation in the Heteroptera has been investigated by Tanaka (1926), Hoke (1926), Davis (1961), Leston (1962), and Wootton and Betts (1986), among others. There is no absolute agreement on the homologies of all veins.

The wings are folded flat over the abdomen. The *forewings* may be either of uniform texture (tegminal) (Fig. 10.3A)—as in the phylogenetically more primitive groups—or in the form of *hemelytra*, that is, divided into a distinctly coriaceous anterior portion and a membranous posterior portion (Fig. 10.3C) (with sometimes reduced venation), as in the Panheteroptera. The anterior margin of the wing, which lies laterally in repose, is bordered by the *subcosta* (Sc) or *costa* plus subcosta (C+Sc), and often reflexed and thickened ventrally into a *hypocostal ridge* or *lamina*. Although venation in all Heteroptera is somewhat reduced, it is most well developed in the Enicocephalomorpha, with a distinct *radius* (R), *media* (M), and *cubitus* (Cu), all attaining the posterior margin of the wing. R and M are often fused and accompanied by a fracture line known as the *medial furrow*, which may be continuous with the *costal fracture*, a break in C+Sc, and distal to which may be formed a distinct triangular *cuneus*. The *clavus*, containing the *1st* and *2nd anal veins*, lies adjacent to the scutellum, and in the Panheteroptera

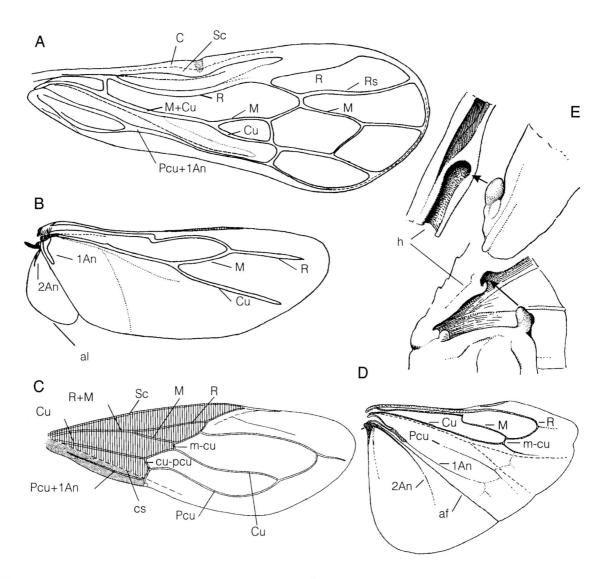

Fig. 10.3. Wings and wing coupling. **A.** Forewing, *Australostolus monteithi* Štys (Aenictopecheidae) (from Štys, 1980). **B.** Hind wing, *A. monteithi* (from Štys, 1980). **C.** Forewing, *Triatoma rubrofasciata* (De Geer) (from Lent and Wygodzinsky, 1979). **D.** Hind wing, *Notonecta undulata* (Say) (from Davis, 1961). **E.** Wing-to-body coupling in Panheteroptera (from Weber, 1930). Abbreviations: af, anal furrow; al, anal lobe; An, anal vein; C, costa; cs, claval suture; Cu, cubitus; cu-pcu, cross vein; h, hemelytron; M, media; m-cu, cross vein; Pcu, postcubitus; R, radius; Rs, radial sector; Sc, subcosta.

(see Fig. 10.1) is interlocked with the latter and usually to the opposite wing along the *claval commissure* by the *frenum*. The clavus is separated from the *corium* by a line of weakness, the *claval furrow* (claval suture), which runs obliquely from the basal articulation of the wing toward the posterodistal margin between the cubitus and the anal veins. The outer portion of the corium (morphologically anterior to R+M) is often expanded and reflexed and referred to as the *embolium* (or exocorium). *Membrane* venation may be totally lacking, consist of a few veins, 1 to 5 closed cells, or in some cases—such as some Nepomorpha and many Pentatomomorpha—consist of numerous anastomosing veins; homology of membrane veins in most Panheteroptera is the subject of substantial disagreement. The corium plus the membrane is sometimes referred to as the *remigium*.

Hind wing (Fig. 10.3B, D) venation is similar to that of the forewing, but often shows even greater reduction, and the wing is largely membranous except for the veins. In those taxa with the scutellum greatly enlarged the wings are often rather elaborately folded. The hind wings are frequently subject to substantial reduction in size or complete loss.

Wing-to-body coupling mechanisms are not elaborately developed in the Enicocephalomorpha, Dipsocoromorpha, and Gerromorpha. The forewings in the Panheteroptera are coupled to the body anterolaterally by the Druckknopfsystem (Cobben, 1957) (Fig. 10.3E) and posteriorly (mesially) by the above-mentioned frenum. Wing-to-wing coupling mechanisms consist of a holding structure on the posterior margin of the forewing (Fig. 12.1C), which provides a sliding attachment for the recurved anterior margin of the hind wing (Teodoro, 1924; Schneider and Schill, 1978; Wygodzinsky and Schmidt, 1991).

Variation in wing development is treated in Chapter 6.

Legs

Heteropteran legs show substantial structural variation within and among groups. Much of the diversity is correlated with life habits (see review by Putchkova, 1979), although certain easily observed structures, such as trichobothria in the Miridae, serve functions that are as yet unknown. We treat leg structure in substantial detail, because in addition to the above-mentioned points, it is of great importance in the classification of true bugs.

Cursorial. The legs of many terrestrial bugs are of the obviously cursorial type, with femora and tibiae elongate and relatively slender. They serve the functions of walking and running.

Saltatorial. Some bugs are capable of jumping. For example, species of *Halticus* Hahn and *Coridromius* Signoret (Miridae: Halticini) have the hind femora conspicuously enlarged and capable of propelling the bug a substantial distance. In the Saldidae, Gelastocoridae, and Ochteridae the legs are not obviously enlarged, but nonetheless provide substantial propulsive force.

Raptorial. The term *raptorial* has been applied to the forelegs of many heteropteran groups, most particularly the obvious predators such as the Phymatinae (Reduviidae), Prostemmatinae (Nabidae), and the true aquatic bugs (Nepomorpha) (Fig. 10.4A). The forefemur is weakly to strongly enlarged and often armed with spines (Fig. 10.4B); the foretibia is also often enlarged or otherwise modified so as to be addressed against the forefemur in repose. In Phymatinae the foretarsus is nonfunctional in the prey-grasping process, either lying in a scrobe (Fig. 10.4G, H) or being completely absent. The femur in species of *Carcinocoris* Handlirsch (Carcinocorini) projects beyond the tibiofemoral articulation to form a chela in conjunction with the opposable tibia.

Modifications of the forelegs in the Enicocephalomorpha are restricted to the distal end of the tibia and the tarsus; these structures form an opposable grasping apparatus (Figs. 12.1D, 13.1C, D). As in many of the truly aquatic bugs, the claws are either reduced to only one or are apposed in such a way as to function as a single claw.

In many phytophagous Heteroptera, notably the Blissinae and Rhyparochrominae (Lygaeidae), the forelegs have the basic structure of the raptorial type. Rather than being used to capture and hold animal prey, they manipulate seeds or other vegetable foodstuffs in the latter group. Their function in the former group is not easily explained by prevailing theories.

Fossorial. Only a limited number of bugs spend any part of their lives underground. Most obvious are members of the Cydninae (Cydnidae) (Fig. 69.1C, D). In this group the forelegs are armed with heavy spines, and in some cases the tarsus is inserted proximal to the tibial apex. The tarsus may be absent from the propulsive hind legs, as it is in some species of *Scaptocoris* Perty.

Rowing. All members of the Gerromorpha are capable of walking on the water surface film, but rowing legs are restricted to some members of the Veliidae (Rhagoveliinae: Fig. 20.3A; some Veliinae) and all Gerridae (Fig. 20.3B–D). The middle legs are longer than the hind legs and serve the function of propulsion on the water surface; also, the claws are usually inserted before the apex of the tibia.

Natatorial. Modifications for swimming are of several types. In the Belostomatidae, the middle and hind femora and tibiae are usually flattened (Fig. 30.1) and fringed with hairs (except in the African snail predator *Limnogeton* Mayr), and both pairs of legs participate in the swimming function. The same is true of most Naucoridae. The legs of Nepidae show little modification for true swimming, but appear to be better suited to assist the animals in "crawling" through the water (Fig. 31.1). The same can be said of the Pleidae and Helotrephidae.

In the Notonectidae (Fig. 38.1) and Corixidae (Fig. 34.1), the hind legs are flattened and fringed with setae and function like oars.

Coxae. Schiødte (1869, 1870), followed by Kirkaldy (1906), classified bug coxae as rotatory and cardinate (trochalopodous and pagiopodous, respectively). Most modern authors (e.g., Cobben, 1978) have abandoned this system, observing that the elongate and globose types are not discrete, but are actually the extremes of a continuum, and they certainly do not correspond to phyletic lines. The coxae of some Veliidae and all Gerridae are distinctive by virtue of their rotation into the horizontal plane.

Trochanters. The heteropteran trochanter is usually simple. In the Miridae it appears to be consistently divided and forms the point at which the legs can be autotomized (see Dolling, 1991).

Femora. The heteropteran forefemur is often swollen and fitted with spines or tubercles associated with grasping. The hind femur may be ornamented with large spines (Coreidae) (Fig. 89.1) or enlarged to serve in jumping.

Tibiae. The heteropteran tibia is usually straight and cylindrical, although particularly the hind tibia may be foliaceous in some Coreidae (e.g., species of *Diactor* Stål) or flattened and bladelike in some Miridae (e.g., species of *Pilophorus* Hahn, *Diocoris* Kirkaldy). The tibial shaft is usually adorned with scattered or regularly arranged spines, as long as or longer than the tibial diameter. At least in most Miridae, it is also adorned with several, longitudinal, parallel rows of tiny spicules. Distally on one or more tibiae there may be a grooming comb (Fig. 10.4C, D).

The Heteroptera appear to be unique in their possession of not only a well-sclerotized tibial flexor sclerite at the base of the tibia, but also a tibial extensor pendant sclerite (Furth and Suzuki, 1990).

Tarsi. The heteropteran tarsus shows substantial variability (Cobben, 1978). The number of segments is usually the same for all legs, although some taxa show variation in this regard. First-instar nymphs of the Enicocephalomorpha, Dipsocoromorpha, Gerromorpha, and many Nepomorpha have 1 tarsal segment, whereas in the Leptopodomorpha, Cimicomorpha, and Pentatomomorpha the tarsi are generally 2-segmented. Some groups add segments during nymphal development. Adults most commonly have 3 tarsal segments, although there is substantial variation. In a few cases (*Scaptocoris* spp., some Macrocephalinae, and other Reduviidae) the tarsi may be lost; in others (e.g., Polyctenidae), a fourth (supernumerary) segment is present (Fig. 62.1A). In many true bugs the ventral surface of the first tarsal segment bears a long slender seta (Figs. 10.4D, 24.1D); the dorsal surface may bear a campaniform sensillum (Fig. 24.1D).

Pretarsi. The greatly variable heteropteran pretarsus has long been utilized in classifications, although certain structural details are still poorly understood or misunderstood. Terminology was reviewed by Crampton (1923) and Dashman (1953), but was not consistently applied within the Heteroptera until about 1970. The most thorough comparative review of these structures was offered by Cobben (1978).

The basic units of the pretarsus are two: the unguitractor plate, which is positioned internally at the apex of the tibia (Figs. 29.2D; 53.5C, F) with a proximally attached retractor tendon, and the claws (ungues), whose bases have a flexible attachment to the unguitractor plate and the distal end of the tibia (Fig. 53.5E). These structures are common to all Heteroptera (where a tarsus is present), although in some groups such as the Enicocephalomorpha and Nepomorpha there is often only a single claw. Several other structures may or may not be present.

Pulvilli. Pulvilli are padlike structures arising from the claws (Figs. 10.5G–I, 53.5E); with light microscopy they usually appear white in contrast to the claws themselves, whereas with scanning electron microscopy they are often difficult to distinguish from the claws except for subtle differences in surface texture. Pulvilli are present in nearly all pentatomomorphans (Fig. 10.5G–I), their structure being rather monotonous (Goel and Schaefer, 1970). They are also present in many Miridae (Fig. 53.5E) (where they were referred to as *pseudarolia* by Knight), Oriini (Anthocoridae), and in Xylastodorinae (Thaumastocoridae) (Fig. 52.3E) (but see discussion under Accessory Parempodia).

Pentatomomorphan pulvilli often possess a distinctly lamellar, grooved, or striate surface. It has been suggested that this grooved surface produces and serves to conduct lipids across the pulvillus (Ghazi-Bayat and Hassenfuss, 1979, 1980a, b), providing the pads with an adhesive function. The evidence for the origin and nature of these secretions is very limited.

Parempodia. Parempodia are usually a symmetrical pair of setiform structures arising from the distal surface of the unguitractor plate between the claws (Figs. 10.5A, 53.5F); some Nepomorpha have either 3 parempodia (Fig. 29.2B) or 2 pairs (Fig. 29.2E). Parempodia are part of the ground plan of the Heteroptera and are observable in some form in nearly all families, although they are occasionally completely absent (e.g., some Plokiophilidae) (Fig. 10.5E), are modified into low elevations or protuberances (adult Saldidae), or exist in the form of fleshy pads in many Miridae (Fig. 53.5C).

The term *arolium* has been used to refer simply to a "fleshy pad." In the writings of most European authors, the statement "arolia absent" meant that no fleshy parempodia or accessory parempodia were present (see, e.g., discussion of the construction of the claws in Reuter, 1910), even though homologous setiform parempodia are often present; in some cases, such as in the Tingidae and Thaumastocorinae (Thaumastocoridae), the failure to observe the parempodia appears to have been the result of their small size. In the North American literature (e.g., Knight, 1941) dealing with the Miridae, the term *arolium* was applied to a parempodium of either setiform or fleshy structure, recognizing that the differently formed structures were homologous, but using the term *arolium* in a way not applied to most other insect groups.

The function of the parempodia is not well understood. The setiform type suggests simple mechanoreceptors associated with the placement of the tarsi and claws. Parempodia in many Miridae are fleshy and have surface structural details similar to those of pentatomomorphan pulvilli and accessory parempodia of other Miridae and some other heteropterans. If the adhesive function theory is correct, then one might postulate a similar function for fleshy parempodia.

Arolia. These unpaired medial structures arise between the bases of the claws, dorsal to the unguitractor plate (contra Goel and Schaefer, 1970; Goel, 1972) and

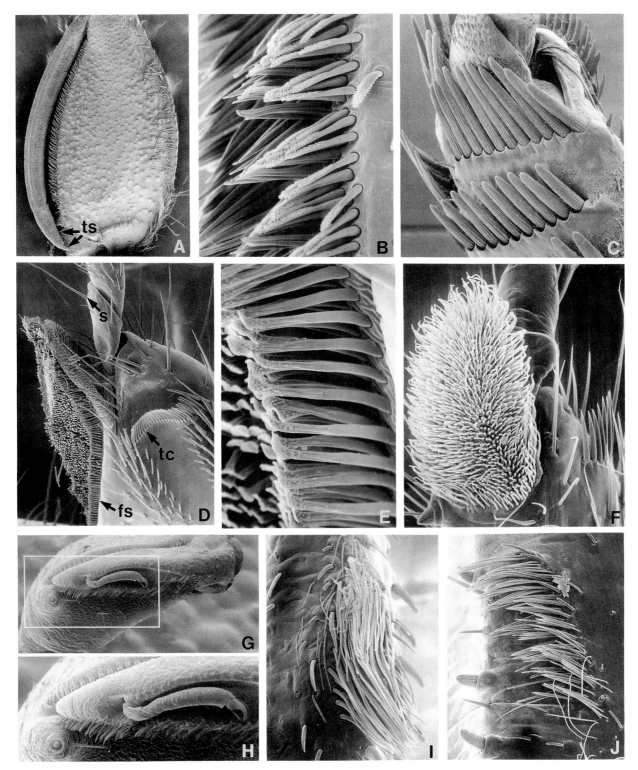

Fig. 10.4. Leg structures. **A.** Right foreleg, *Cryphocricos latus* Usinger (Naucoridae). **B.** Grasping setae on forefemur, *Ranatra* sp. (Nepidae). **C.** Tibial combs, apex of middle tibia, *C. latus* (Naucoridae). **D.** Apex of tibia and base of tarsus of foreleg, *Prostemma* sp.; fossula spongiosa, tibial comb, and elongate seta ventrally on tarsal segment 1. **E.** Detail of fossula spongiosa, *Prostemma* sp. (Nabidae). **F.** Fossula spongiosa and tibial comb, foretibia, *Nabis* sp. (Nabidae). **G.** Foreleg, *Phymata* sp. (Reduviidae). **H.** Same as G, detail. **I, J.** Foreleg antennal cleaner (I, tibiae; J, femur), *Gelastocoris oculatus* (Fabricius) (Gelastocoridae). Abbreviations: fs, fossula spongiosa; s, seta; tc, tibial comb; ts, tarsus.

General Adult Morphology and Key to Infraorders

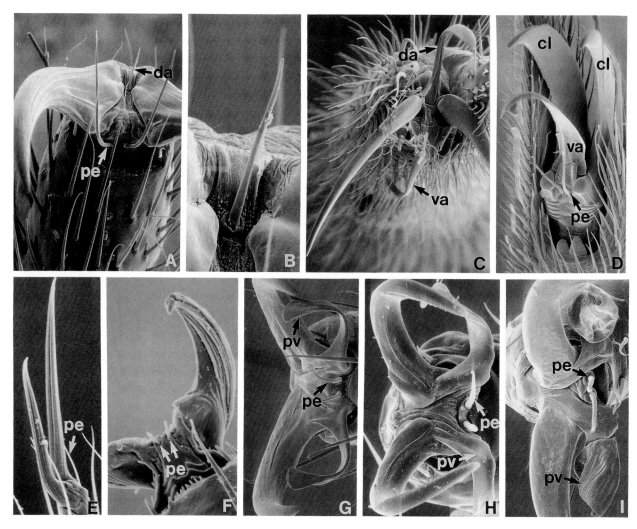

Fig. 10.5. Pretarsal structures. **A.** Adult middle leg pretarsus, *Oncylocotis curculio* (Kirschbaum) (Enicocephalidae). **B.** Detail of dorsal arolium, adult middle leg, *O. curculio* (Enicocephalidae). **C.** Pretarsus, adult *Paravelia rescens* (Drake and Harris), showing dorsal and ventral arolia (Veliidae). **D.** Hind leg pretarsus, adult *Rhagovelia* sp. (Veliidae). **E.** Nymphal pretarsus showing asymmetrical claws and greatly reduced parempodia, *Lipokophila* sp. (Plokiophilidae). **F.** Adult pretarsus, *Zetekella minuscula* (Barber) (Tingidae) (from Schuh, 1976). **G.** Adult pretarsus, *Mezira* sp. (Aradidae). **H.** Adult pretarsus, *Termitaradus guianae* (Morrison) (Termitaphididae). **I.** Adult pretarsus, *Edessa* sp. (Pentatomidae). Abbreviations: cl, claw; da, dorsal arolium; pe, parempodium; pv, pulvillus; va, ventral arolium.

are usually present in conjunction with parempodia. The *dorsal arolium* is found in nymphs and adults of Enicocephalomorpha (Fig. 10.5A, B), Gerromorpha (Fig. 10.5D), some Dipsocoromorpha and Nepomorpha (Fig. 29.2F), and nymphs and adults of most Leptopodomorpha (Fig. 41.2A, C). The *ventral arolium* is present in some Dipsocoromorpha, all Gerromorpha (Fig. 10.5C), and probably in all nymphal Nepomorpha (Fig. 29.2D, F, G). The occurrence of arolia may vary between sexes in the Dipsocoromorpha. Furthermore, arolia may not be present on all pairs of legs in Dipsocoromorpha and Nepomorpha. Arolia are absent in all life stages of all members of the Cimicomorpha and Pentatomomorpha (Cobben, 1978).

Accessory parempodia (pseudopulvilli). Some heteropterans possess paired fleshy structures at the base of the claws. These structures appear to be connected, at least in part, to the unguitractor plate. They were observed by Schuh (1976) and Cobben (1978) in the Bryocorini (Fig. 53.5E, F), Dicyphina (Fig. 53.5F), Monaloniina, and Odoniellina (Miridae). Schuh referred to them as *pseudopulvilli*, and Cobben as *accessory parempodia*. Cobben (1978) also believed that the "pulvilli" found in Xylastodorinae (Thaumastocoridae) were accessory parempodia, a term he also used to describe pretarsal structures found in the Rhagoveliinae (Veliidae).

Fossula spongiosa (spongy fossa). This pad of specialized setae is found in many Reduvioidea, Naboidea, and

Cimicoidea. The structure is located apicoventrally on the tibia (usually foretibia, and less commonly on the middle and hind tibiae) (Fig. 10.4D–F). It presumably serves to assist members of these predatory groups in prey capture.

Tibial appendix. This structure arises from the distal end of all tibiae in the Thaumastocorinae (Thaumastocoridae) (Fig. 52.3C, D) (incorrectly stated as foretibia only in Schuh and Štys, 1991). It was previously referred to as a fossula spongiosa, although clearly the two are not homologous. Drake and Slater (1957) used the term "lobate sensory appendage." The tibial appendix apparently functions in grasping the substrate.

Trichobothria. Femoral trichobothria are known in two groups of Heteroptera. In the Gerridae and some Veliidae a seta arises ventrodistally from the trochanter and one or more setae arise from the femur. They may function in communication between conspecifics on the water surface and in prey location. In the Miridae, 3 to 8 (rarely more) trichobothria are located on the lateral and ventral surfaces of the meso- and metafemora (Figs. 10.8I, 53.3E, F) (Schuh, 1976).

Viscid setae. Sticky setae are known to occur on at least the foretibiae of some harpactorine Reduviidae. Miller (1956a) described how the bugs smeared resins from trees on their forelegs, the resulting sticky setae aiding in prey capture. It appears that other taxa actually produce viscid secretions that serve to accomplish the same purpose (Cobben, 1978).

Abdomen

In its simplest form the pregenital abdomen in the Heteroptera consists of terga 1–7 and sterna 2–7 in females and terga 1–8 and sterna 2–8 in males; these are commonly divided into *mediotergites* and *dorsal laterotergites* and sometimes *ventral laterotergites* (paratergites) and/or *laterosternites*, collectively forming the *connexivum*. More rarely they also are divided into *inner laterotergites* lying between the dorsal laterotergites and the mediotergite (Sweet, 1981). Primitively there are 8 abdominal *spiracles*, spiracle 1 always located on tergum 1, spiracles 2–8 ordinarily located on the respective sternum or laterosternite; spiracle 1 may be nonfunctional or absent (e.g., Gerromorpha and many Cimicomorpha) as is spiracle 8 in many groups, or the total number of functional spiracles may be further reduced to as few as 3 in some Schizopteridae. One or more of spiracles 2–8 may be located dorsally, usually on the laterotergites, as in most Leptopodoidea and many Lygaeidae. Segments 8 and 9 in females and segment 9 in males are incorporated into the genitalia. Segment 10 forms the *proctiger,* containing the *anus,* which also bears the remnants of segment 11.

Male genitalia. Sternum 9 is developed in most taxa into a distinct *genital capsule* (pygophore, pygopher) (Fig. 10.1) containing the *phallus*. The phallus was first studied comparatively by Singh-Pruthi (1926) and subsequently has been examined at a more restricted taxonomic level by Galliard (1935; Reduviidae), Larsén (1938; Nepomorpha), Kullenberg (1947; Miridae, Nabidae), Bonhag and Wick (1953; Lygaeidae), Ashlock (1957; Lygaeidae), Kelton (1959; Miridae), and Davis (1966; Reduviidae). The literature and terminology were reviewed by Dupuis (1970). A welter of terms has been proposed for its various parts. Although showing substantial variability, the phallus is usually composed of a *basal articulatory apparatus* (incorporating the basal plates) to which is attached the *aedeagus* (intromittent organ) and the *parameres* (claspers or harpagones). The *ductus ejaculatorius* passes through the *basal foramen* into the lumen of the aedeagus. The intromittent organ is formed proximally in most taxa of an at least partially sclerotized *phallotheca* (theca or phallosoma) into which much of the membranous distal portion of the aedeagus, the *endosoma,* is often drawn in repose. The endosoma is sometimes formed of a proximal *conjunctiva* and distal (often tubular) *vesica,* the former of which may bear lobes, spines, and other ornamentation when inflated. Upon passage into the phallotheca, the ductus ejaculatorius forms the *ductus seminis,* which proximally may be modified into an *ejaculatory reservoir*. The position of the *secondary gonopore* determines the termination of the ductus seminis and may be in the conjunctiva (e.g., Pentatomoidea), tubular vesica (e.g., Lygaeoidea), or membranous vesica (e.g., most Miridae).

Portions of the phallus may be asymmetrical, as also are the parameres in most Nepomorpha, some Cimicomorpha, and occasionally in some other groups; one or both parameres may be lost. Asymmetries of pregenital abdominal segments occur in Dipsocoromorpha and Nepomorpha.

There are normally 2–7 testis follicles.

Female genitalia. The heteropteran *ovipositor* is composed of the *first* (anterior) and *second* (posterior) *valvulae* (gonapophyses, gonostyli), usually fused with the *first* and *second valvifers* (gonocoxae, gonocoxopodites), respectively (Scudder, 1959; Dupuis, 1970) (Fig. 10.6). The valifers are derived from abdominal segments 8 (first valifer) and 9 (second). Segment 9 may also bear a pair of *third valvulae* (gonoplacs, styoids), which form an ovipositor sheath, although these frequently are absent in the Nepomorpha and always are absent in the Pentatomomorpha and a few other taxa. In many groups the valvulae are elongate and laterally compressed to form a *laciniate ovipositor* suited for depositing eggs in plant tis-

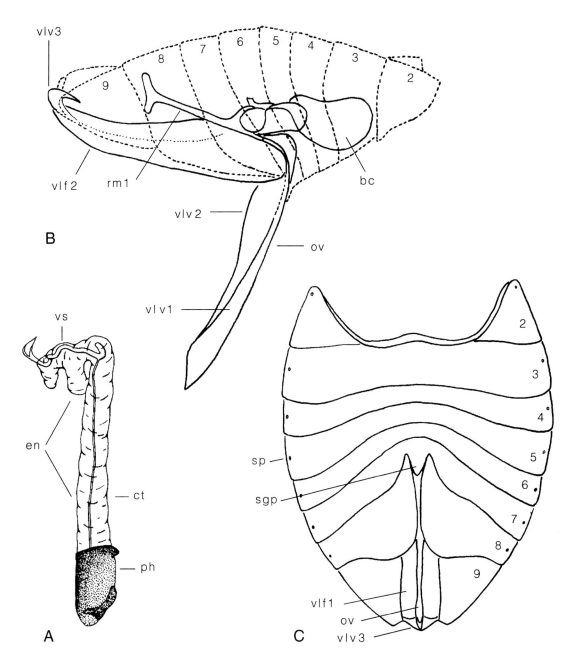

Fig. 10.6. Male and female genitalia. **A.** Aedeagus, Lygaeidae (from Ashlock, 1957). **B.** Female abdomen and external genitalia, lateral view, Miridae (modified from Davis, 1955). **C.** Ibid, ventral view. Abbreviations: bc, bursa copulatrix; ct, conjunctiva; en, endosoma; ov, ovipositor; ph, phallotheca (phallosoma); rm 1, first ramus; sgp, subgenital plate; sp, spiracle; vlf1, first valifer; vlf2, second valifer; vlv1, first valvula; vlv2, second valvula; vlv3, third valvula; vs, vesica.

sue, whereas in others the valvulae are greatly reduced or absent and the eggs are laid free or cemented to surfaces, as on plants. Although the ovipositor is morphologically associated only with abdominal segments 8 and 9, it may encroach on sternum 7 and partially or completely divide it.

The portion of the internal genitalia associated with copulation, sperm reception, and sperm storage is sometimes referred to as the *gynatrial complex*. The *spermatheca* (Figs. 12.1G, 16.1B, 40.1J, 78.2E), or sperm-storage organ, is primitively present in the Heteroptera and shows distinctive structural variation from group to group (Pendergrast, 1957). It is modified into a *vermiform gland* (spermathecal gland), lost, or nonfunctional in all Cimicomorpha, being replaced by alternative sperm-storage organs, such as paired *pseudospermathecae* in the Pachynomidae, Reduviidae (Fig. 48.3L), and Tingidae, and the *bursa copulatrix* in the Miridae (Fig. 53.3G, I). Heteropteran ovariole numbers vary from at least 2 to as many as 17. Insemination and fertilization are unremarkable, except in the Cimicoidea, in which sperm are usually introduced into the abdominal cavity and migrate to the oviduct. Fertilization in cimicoids usually takes place before completion of chorionic development, as does some development (Larsén, 1938; Bonhag and Wick, 1953; Davis, 1955, 1966; Scudder, 1959; Dupuis, 1970; Dolling, 1991b).

Sound Production and Reception Organs

The first systematic survey of sound-producing devices in the Heteroptera was that of Handlirsch (1900a, b). Many individual notices and several summary papers have appeared since.

The function of sounds produced by Heteroptera is less well understood than are the structures that produce them. Sound production in some cases clearly is associated with courtship by males and the readiness of females to mate (Corixidae), agonistic behavior (Corixidae) (Finke, 1968; Jansson, 1976), a defensive repertoire (Reduviidae), and aggregative behavior.

Stridulatory Structures

Observations of sound production exist for taxa other than those for which putative stridulatory devices are listed below, but no physical mechanism has been found and certainly many additional structures remain to be discovered or are not yet documented in the literature. Most of the following information is derived from the works of Torre-Bueno (1903, 1905), Kormilev (1949), Usinger (1954a), Leston (1957b), Miller (1958), Ashlock and Lattin (1963), Štys (1961a, 1964a), Jansson (1972), Schuh (1974, 1984), Wilcox (1975), Akingbohungbe (1979), Schaefer (1980b), Andersen (1981a), Schaefer and Pupedis (1981), Gogala (1984), J. T. Polhemus (1985), Péricart and Polhemus (1990), and J. T. Polhemus (1994).

Stridulatory structures in the Heteroptera are usually composed of a movable and a stationary portion. The former was early on termed the *plectrum* (Fig. 10.7H), and that term has been widely adopted among heteropterists. The terminology referring to the stationary portion is more complicated; Ashlock and Lattin (1963) proposed using *stridulitrum* (Fig. 10.7F, G, I) in place of the many terms that had gone before, such as "pars stridens," and stridulitrum is now widely used.

Forewing edge–hind femur mechanism. This is the most widely distributed and easily observed type of stridulatory apparatus in the Heteroptera. It is known to occur in some members of the Dipsocoromorpha, Leptopodomorpha, Cimicomorpha (Miridae), and Pentatomomorpha (Lygaeidae: Orsillinae, Rhyparochrominae; Largidae; Alydidae). Ordinarily the corial margin is either finely transversely striate or finely tuberculate, and the mesal surface of the metafemur is longitudinally ridged or finely tuberculate. Whereas in most known occurrences the stridulitrum is located at the extreme margin of the hemelytron, in the Saldidae it may be located either on the hypocostal or secondary hypocostal ridge, and the peg field of the plectrum may be either proximal or distal on the femur.

Connexival margin–hind femur mechanism. Males and females of the veliid genera *Angilovelia* Andersen and *Stridulivelia* Hungerford have the stridulitrum located on the connexival margin and the plectrum on the inner surface of the hind femur.

Abdominal sternum–hind tibia or femur mechanism. Within a substantial number of Lygaeidae (Cleradini, Myodochini, Ozophorini), Pentatomidae (Macideini), Scutelleridae (Tetyrinae), and some Aradidae (Fig. 64.2L, M) the stridulitrum is composed of striae or minute tubercles located lateroventrally on one or more abdominal sterna, and the plectrum is positioned on the mesal surface of the metafemur or in some Aradidae on the metatibia. A single example in the lygaeid subfamily Blissinae (*Heteroblissus anomilis* Barber) has a stridulitrum consisting of a series of coarse elongate ridges on the abdomen, while the plectrum appears to consist of a series of small spines on the mesal surface of the metatibia.

Propleuron–forefemur mechanism. In nymphs and adults of the North American rhyparochromine lygaeid *Pseudocnemodus canadensis* (Provancher) the stridulitrum consists of an elongate cross-striate area beginning over the foreacetabulum and extending forward to the anterior prothoracic margin; the plectrum is located on the basal third of the forefemur.

Metapleuron–middle femur mechanism. This type of stridulatory apparatus is found in some calisiine Aradidae.

Head–forefemur mechanism. A lunate stridulitrum occurs laterally on the head in several genera of the Colobathristidae (e.g., *Peruda* Distant, *Trichocentrus* Horvath), the plectrum consisting of a small field of scattered tubercles on the mesal surface of the forefemur.

Forecoxa–forecoxal cavity mechanism. In *Ranatra* spp. (Nepidae) a roughened elevated area on the outer surface of the forecoxa is rubbed against the striate inner surface of the foreacetabulum.

Tibial comb–labial prong mechanism. Males of most species of *Buenoa* Kirkaldy and *Anisops* Spinola (Noto-

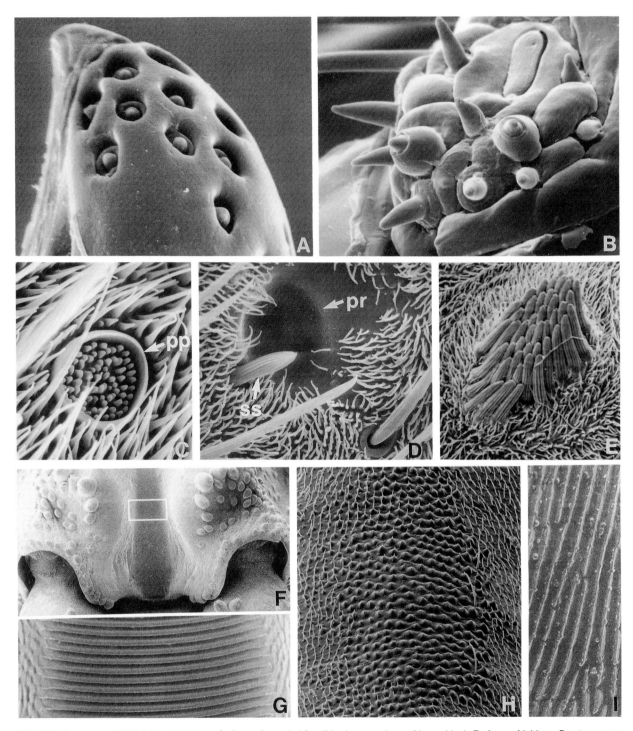

Fig. 10.7. Sensors and stridulatory structures. **A.** Apex of nymphal foretibia, *Laccocoris* sp. (Naucoridae). **B.** Apex of labium, *Prostemma* sp. (Nabidae). **C.** Peg plate, *Paravelia rescens* (Veliidae). **D.** Pit receptors with associated specialized setae, on thorax, *Rhagovelia* sp. (Veliidae). **E.** Pregenital ventral abdominal organ, *Hydrometra* sp. (Hydrometridae). **F.** Prosternal stridulatory sulcus, *Phymata* sp. (Reduviidae). **G.** Same as F, detail. **H.** Metafemoral stridulatory plectrum, *Hallodapus albofasciatus* (Motschulsky) (Miridae) (from Schuh, 1984). **I.** Ventral abdominal stridulitrum, *Ligyrocoris diffusus* (Uhler) (Lygaeidae). Abbreviations: pp, pegplate; pr, pit receptor (pit organ); ss, specialized seta.

nectidae) have an elongate elevated area proximally on the inner surface of the foretibia, on which lies a comb of flattened setae. This comb is apparently rubbed against prongs located on either side of labial segment 3.

Femoral ridge–coxal peg mechanism. Males of most *Buenoa* spp. (Notonectidae) have on the inner surface of the forefemur a series of approximately parallel ridges. These apparently are rubbed against a short heavy peg located on the lateral face of the forecoxa.

Base of labium–femoral apex mechanism. Males of most *Buenoa* spp. (Notonectidae) possess several stout setae distally on the inner surface of the forefemur. These may be plucked against the base of the labium.

Maxillary plate stridulitrum. Males and females of most Corixinae have the stridulitrum formed by the acute edge of the maxillary plate. The plectrum consists of a field of specialized pegs on the basomesal surface of the forefemur. The device is usually better developed in males, which are capable of making sounds louder than those made by the females.

Metathoracic wing vein stridulitrum. A variety of pentatomomorphans, and some other bugs, possess stridulatory structures combining a striate vein or accessory vein of the hind wing with a striate area on the thoracic or abdominal dorsum (Leston, 1954a). All species of *Kleidocerys* Stephens (Lygaeidae: Ischnorhynchinae) have an arcuate veinlike ridge located on the underside of the hind wing. This ridge contacts a transversely striate longitudinal ridge located near the lateral margin of the metanotum. In *Piesma quadrata* Fieber (Piesmatidae) the underside of the cubitus is transversely striate basally, making contact with a striate area located anterolaterally of abdominal tergum 1. In the Tessaratomidae, Cydnidae, some Scutelleridae, Thaumastellidae, and the Leptopodidae, the postcubitus, vannus, or other area of the hind wing has a striate area that interacts with a striate area on abdominal tergum 1.

Hypocostal lamina (or articulatory sclerite) stridulitrum. Males and females of some Coreidae have a striate ridge located posterolaterally on the undersurface of the prothoracic foramen. It functions in conjunction with a stridulitrum located either on the hypocostal lamina or on the articulatory sclerites of the forewing.

Underside of clavus stridulitrum. *Rhytocoris* spp. (Coreidae) have the stridulitrum located on the underside of the clavus; the plectrum is situated on the articulatory region of the hind wing.

Posterior margin of pygophore stridulitrum. In the Scutellerinae (Sphaerocorini) the posterodorsal margin of the pygophore is fitted with six rows of short stout setae, which may produce sound in conjunction with the use of vesical conjunctival appendages as a plectrum.

Prosternal groove stridulitrum. A transversely striate prosternal stridulitrum exists as a groove in nearly all members of the Reduviidae, the apex of the rostrum functioning as the plectrum.

Other Sound-Producing Organs

Tymbals. The dorsal surface of the abdomen may function as a tymbal (or drum), whereby terga 1 and 2 are usually fused together and vibrate at low frequencies over a hollow chamber within the abdomen. This feature occurs widely in those Pentatomidae and Acanthosomatidae that lack stridulatory structures associated with the abdominal dorsum, as well as in the Cydnidae, which have a stridulitrum anteriorly on the abdominal dorsum. Similar tymbals also appear to exist in at least some members of the Lygaeidae, Coreidae, and possibly Reduviidae.

Sound Reception

Scolopophorous organs. Scolopophorous organs are known to occur in many groups of insects. Those located on the meso- and metathoracic pleuron and on abdominal terga 1 and 2 of most nepomorphan families (Larsén, 1957; Mahner, 1993), including the Gelastocoridae (Parsons, 1962), appear to represent a set of structures distinctive to the infraorder. They are composed of sensory sensilla (scolopidia), which may be single or united into a single strand, whose distal end attaches to a membrane on the body wall or to the body wall itself. The proximal end is innervated and sometimes associated with the tracheal system. These organs are thought by some to be auditory in function, although Larsén believed that they might be involved in orientation in the aquatic Nepomorpha.

The scolopophorous (scolopidial) organ, located on the tarsus of all 3 pairs of legs in Notonectidae, is sensitive to wave frequencies of 20–50 Hz, a range typical of that produced by prey of some *Notonecta* spp. (Wiese, 1972).

Tympanal organs and physical gills. Tympanal organs exist on either side of the body in the mesothorax under the hind wing articulation in the Corixidae (Hagemann, 1910) and are in contact with the physical gill air bubble (Prager and Streng, 1982). They are innervated by two scolopidia, which are stimulated by airborne sound, those on opposite sides of the body differing in their responses (Prager, 1973, 1976). One of the scolopidia in each tympanal organ is sensitive to stridulation frequencies produced by conspecific bugs. Thus, there appears to be a credible description for a hearing method in the Corixidae. Equally plausible hearing methods are not available for most other families that are known to stridulate.

Mechanoreceptors

The Heteroptera possess a wide variety of mechanoreceptors, most of which are of a type found in nearly

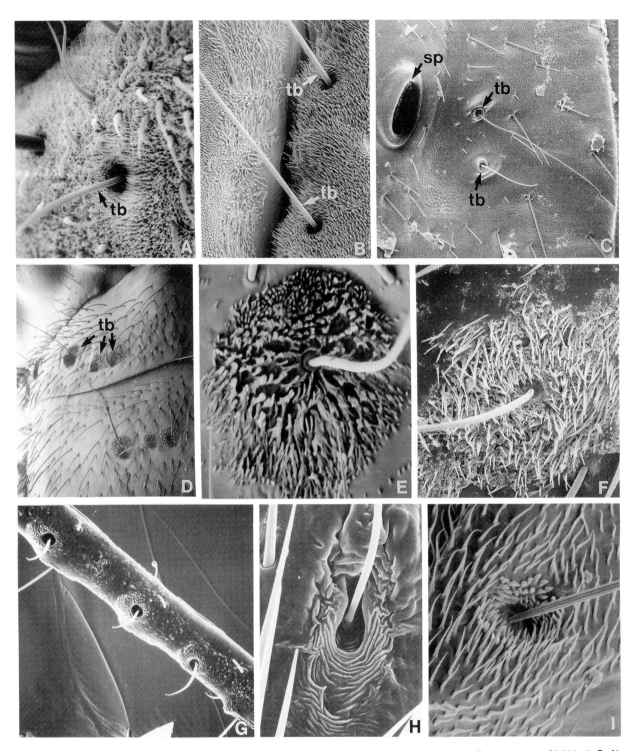

Fig. 10.8. Trichobothria. **A.** Cephalic trichobothrium, *Rhagovelia* sp. (Veliidae). **B.** Scutellar trichobothria, *Prostemma* sp. (Nabidae). **C.** Abdominal trichobothria, *Edessa* sp. (Pentatomidae). **D.** Medial abdominal trichobothria, *Paragonatas divergens* (Distant) (Lygaeidae). **E.** Detail of base of lateral abdominal trichobothrium, *P. divergens* (Lygaeidae) (from Schuh, 1975a). **F.** Trichobothrium, ventrally on posterolateral margin of laterotergite of abdominal segment 7, *Nabis* sp. (Nabidae). **G.** Antennal segment 2 showing trichobothria, *Barce* sp. (Reduviidae, Emesinae) (from Wygodzinsky and Lodhi, 1989). **H.** Base of antennal trichobothrium, *Aphelonotus* sp. (Pachynomidae) (from Wygodzinsky and Lodhi, 1989). **I.** Base of femoral trichobothrium, *Sthenaridea australis* (Schuh) (Miridae) (from Schuh, 1975a). Abbreviations: sp, spiracle; tb, trichobothrium.

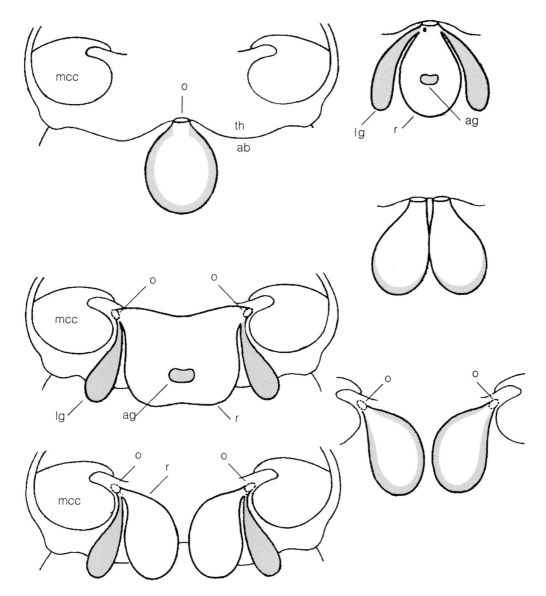

Fig. 10.9. Metathoracic scent-gland configurations (after Carayon, 1971a). Abbreviations: ab, abdomen; ag, accessory gland; lg, lateral gland; mcc, metacoxal cavity; o, orifice; r, reservoir; th, thorax.

all other groups of insects. The function of many of these is poorly understood or unknown. Notable among the mechanoreceptors are trichobothria (described earlier), specialized setae that have been implicated in sound reception, but for which function there is no solid evidence. They are distributed on many body regions in a great number of heteropteran families. Other mechanoreceptors worth mention are the static sense organs found on the abdominal sternum of Nepidae and Aphelocheiridae. Most such structures are documented and discussed in the family treatments. Although no comprehensive review of mechanoreceptors exists specifically for the Heteroptera, Grassé (1975) provided a good general introduction to the subject.

Exocrine Glands

The monophyly of the Heteroptera is based in part on their possession of scent glands, located on the abdominal dorsum in nymphs and in the metathorax of adults (Fig. 10.9). Like most insects, the Heteroptera communicate through the use of chemicals, and additional glandular structures continue to be discovered in the group. Most of the following summary is derived from information contained in the excellent review articles of Carayon (1971a), Cobben (1978), Staddon (1979), and Aldrich (1988). Additional references are cited below in the text.

Scent reception is much less well understood than are glandular structure and glandular products. Calla-

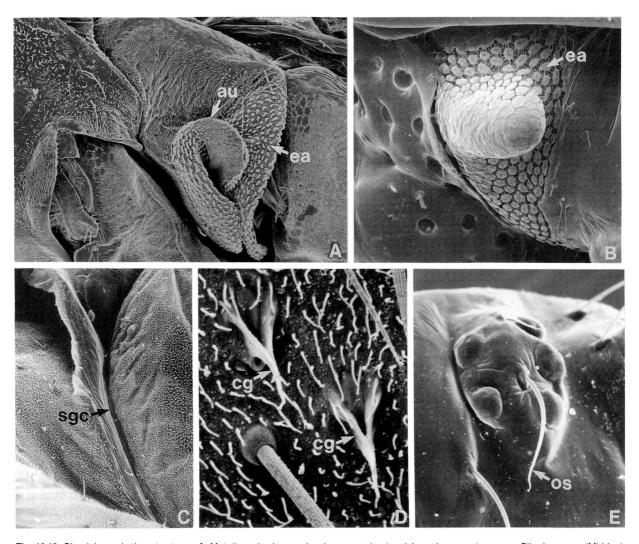

Fig. 10.10. Glandular and other structures. **A.** Metathoracic pleuron showing scent-gland auricle and evaporatory area, *Pilophorus* sp. (Miridae) (from Schuh, 1984). **B.** Metathoracic scent-gland evaporatory area, *Pronotacantha* sp. (Berytidae). **C.** Scent-gland channel on metathoracic pleuron, *Termitaradus guianae* (Termitaphididae). **D.** External manifestation of corial glands, *Lipokophila eberhardi* Schuh (Plokiophilidae). **E.** Ocular seta of nymphal instar 1, *L. eberhardi* (Plokiophilidae). Abbreviations: au, auricle; cg, corial gland; ea, evaporatory area; os, ocular seta; sgc, scent-gland channel.

han (1975) reviewed available evidence and proposed the theory that specialized setal receptors on the antennae of insects serve as "antennae" to receive specific frequency emissions of various semiochemicals. Grassé (1975) provided a general review of chemoreceptive organs in insects, although nothing exists specifically for the Heteroptera.

Nymphal dorsal abdominal scent glands. The nymphs of nearly all Heteroptera have scent glands located on the abdominal dorsum. These structures appear to have first been investigated by Gulde (1902); Cobben (1978) summarized their number and distribution in heteropteran families. They vary in number from 1 to 4 and are located at the anterior margin of terga 4–7. The most widespread condition is to have 3 functional glands, with their orifices located on the anterior margins of abdominal terga 4, 5, and 6 (3/4, 4/5, 5/6), this arrangement being found commonly in the Pentatomomorpha, but also occurring in many cimicomorphan families (except Miridae, 3/4, and Tingidae, 3/4, 4/5; Reduviidae show substantial variation). The Gerromorpha, Leptopodomorpha, and Nepomorpha, have either a single functional gland (3/4) or no glands at all, with the exception of the Corixidae, which typically have 3 functional glands (3/4, 4/5, 5/6). Ordinarily, functioning glands are lost from posterior to anterior, with the exception of the Coreoidea, where only glands 4/5 and 5/6 are functional. No taxa have 4 functional glands, and only Joppeicidae and some Schizopteridae have functional glands between terga 6/7.

The scent glands are multicellular, representing invaginations of the cuticle. They may be single or divided and may open through a single ostiole or a pair of ostioles. Muscles open the ostiole as well as compress the gland. Some taxa are capable of ejecting secretions from the glands as a spray.

Chemical components produced by these glands vary, but typically they include compounds such as 2-hexenal, 4-oxo-2-hexenal (or the corresponding octenal), and 2-hexenyl-acetate; the list of other substances is large and constantly growing. The glands appear for the most part to serve defensive or alarm functions (e.g., Calam and Youdeowei, 1968; Ishiwatari, 1974), although they may also be involved in the aggregation behavior seen in the nymphs of some taxa or serve a fungistatic role, as in the subterranean cydnid *Scaptocoris divergens* (Roth, 1961).

Dorsal abdominal glands in adults are usually absent or represented only by a scar, or they are nonfunctional and become smaller over time. In a few taxa they are well developed and functional. For example, in males of species of the predatory asopine pentatomid *Podisus* the 3/4 dorsal abdominal glands are enormous, producing a mixture of chemicals that serve as a powerful sex pheromone. These substances also function as kairomones, attracting scelionid egg parasitoids (Aldrich, 1988). The dorsal abdominal glands also remain active in adult Serinethinae (Rhopalidae) of the genera *Boisea* Kirkaldy and *Jadera* Stål (Aldrich et al., 1990). The two glands produce different compounds in adult *Jadera* spp., but the composition remains the same throughout nymphal and adult life and is the same for males and females, suggesting a pheromonal role, possibly in aggregation, in these aposematically colored bugs. Sexual attraction by functional abdominal scent glands in adults has been suggested for some Miridae, but Aldrich et al. (1988) were unable to experimentally substantiate that thesis with *Lygus* spp. and suggested that other as yet unrecognized glands were performing this function.

Adult metathoracic scent glands. Most adult Heteroptera have active scent glands located ventrally in the metathorax. The glands are integumentary invaginations. There may be a single gland with a single opening, a single gland opening into a pair of ostioles, or paired glands opening into closely to widely spaced ostioles. The glands themselves may consist of a single cuticle-lined corpus that produces and stores the scent fluid. More frequently, however, there is a reservoir with glands emptying into it. Sometimes there is also an accessory gland attached to the reservoir; this gland apparently serves to deliver enzymatic reaction products important in production of the final scent-gland chemicals. Dimorphism may exist between the sexes. For example, in several groups of Lygaeidae (Carayon, 1948b) and in the Lethocerinae (Belostomatidae) the glandular structures of the males are more highly developed than are those in the females. In some Enicocephalomorpha the glands are apparently present only in the males (Carayon, 1962).

The gland ostioles are always located ventrally on the metathorax and have associated with them a valve and a valve-opener muscle. The orifice is connected to a groove or canal that runs laterally onto the metathoracic pleuron, ordinarily terminating in an area termed the *peritreme*, that area usually surrounded by the *evaporatory area*, which is particularly well developed and of distinctive mushroomlike microsculpture (termed "flake cuticle" by Johansson [1957a]) in the Cimicomorpha and Pentatomomorpha (Fig. 10.10A, B) (Filshie and Waterhouse, 1969; Johansson and Braten, 1970; Carayon, 1971a). The distinctive structure of the evaporatorium is sometimes found on areas of the body such as the region around the thoracic spiracles (Remold, 1962, 1963; Schuh, 1984; Carver, 1990).

Two theories have been proposed for the function of the evaporatory area. One suggests that the modified cuticle offers a greater surface area from which the scent fluid can evaporate, thus increasing the effectiveness of the glands in defense. The other contends that the specialized evaporatory surface serves in restricting spread of the scent fluids to a circumscribed area of the body because the scent fluids are known to be toxic to the bugs that secrete them as well as to potential predators (Remold, 1962, 1963; Johansson, 1957a; Johansson and Braten, 1970; Carayon, 1971a; Carver, 1990). The observations of Falkenstein (1931) on *Tessaratoma papillosa* Drury indicate clearly that although the scent-gland secretions of both nymphs and adults serve a defensive function, they are also extremely toxic to both life stages when coming into contact with the cuticle of the individual that secretes them.

The main function of the adult scent glands appears to be defensive, as is the case in nymphs, although sexual, alarm, and aggregation functions may also exist. The composition of the glandular products is complex, but they often comprise substances similar to those listed above for the nymphs.

Several authors have shown that the shape of the peritreme and the shape and distribution of the evaporatorium have systematic value, as for example in the Cydnidae (Froeschner, 1960; Dethier, 1974) and Miridae (Cassis, 1984).

Brindley's glands. This pair of simple glands is located in many adult reduviids and pachynomids dorsolaterally in the anterior region of the abdomen; a gland of similar structure and location is found in the Tingidae. Brindley's gland opens onto the metepisternum, where in *Rhodnius prolixus* Stål a channel with a spongy surface runs some distance from the orifice of the gland. The main component of the glandular secretion in *Rhodnius* appears to be isobutyric acid, although paraffins may also be present. The secretions may serve to alert other individuals of danger in the case of aggregations of *Rhodnius;* they are nontoxic to adults, but toxic when administered to the cuticle of nymphs (Kalin and Barrett, 1975). Brindley's glands are not present in all reduviids, although when present the metathoracic scent glands are often inactive.

Ventral abdominal glands (paragenital glands; uradenies) of the Pentatomomorpha. "Ventral abdominal glands" were observed in female *Pyrrhocoris apterus* by Dufour (1833), who concluded that they were part of the reproductive system. They are now known to exist in the females of some members of the Lygaeidae (Lygaeinae, Orsillinae, Rhyparochrominae), Largidae, Pyrrhocoridae, Stenocephalidae, Coreidae, Alydidae, and Rhopalidae and in males of at least the Lygaeidae, Coreoidea, and Stenocephalidae, although apparently not in the Pyrrhocoroidea. These large multicellular glands have a single ventral opening in the intersegmental membrane between sterna 7 and 8, 8 and 9, or more rarely 9 and 10, depending on the species and the sex (Thouvenin, 1965). Gland secretions have been identified in the males of at least 10 species of Coreidae and a few Rhopalidae. Males produce volatile compounds, whereas females apparently do not. Aromatic compounds with the odors of cherries, roses, and other familiar substances are common in *Leptoglossus* spp., but aliphatic alcohols and esters are also produced by the coreids (Aldrich, 1988). The secretions of the ventral abdominal glands play a role in mate recognition, at least in the Rhopalidae (Aldrich et al., 1990).

Ventral abdominal glands of male Anthocoridae (Fig. 60.2E). Some male Anthocoridae of the tribe Scolopini possess internally paired glands with a single or double orifice opening onto abdominal sternum 4 or 5. At present nothing is known of the chemicals they produce or their function, although the latter is presumed to be sexual. At the orifice or between the orifices there is a brush of setae. Carayon considered these structures to be serially homologous with the ventral abdominal glands (*uradenies*) occurring in certain female Pentatomomorpha. He also suggested that they might be derived from the secretory integumentary cells found in some members of the Lasiochilidae and Lyctocoridae (Carayon, 1954, 1972a; Peet, 1979).

Eversible glands in abdomen of Saldidae. In both sexes of all members of the family Saldidae there exists a pair of glands that arise ventrolaterally in the membrane between abdominal segments 7 and 8. First observed by Cobben (1961), they can be everted by gently squeezing the abdomen of live or alcohol-preserved specimens. To date no function has been ascribed to them, and the composition of their secretions has not been determined.

Ventral (sternal or Carayon's) glands in Reduviidae. These paired, simple, saclike glands, known from the Elasmodeminae, Holoptilinae, and Phymatinae, are located in a position similar to the metathoracic scent glands, but each individual gland opens ventrally into the membrane between the thorax and abdomen. Despite the similarity of position to the metathoracic scent glands, the location of the glandular exit, structure of glands, and existence of both ventral glands and metathoracic glands in adults of *Phymata* Latreille (although both glands are not present in all taxa possessing ventral glands) attest to their distinctness. The secretions of these glands are described as oily, although their specific composition and function are unknown (Carayon et al., 1958; Staddon, 1979).

Subrectal gland in the Harpactorinae. Many female Harpactorinae (Reduviidae) have large bilobed saclike glands opening into the membrane between the styloids and the anus (Barth, 1961). When the bugs are disturbed, the bright orange glands are everted like a caterpillar osmeterium, producing a nauseatingly sweet odor (Barth, 1961; Davis, 1969; Aldrich, 1988).

Dermal glands. So-called type B dermal glands were first discovered and described in *Rhodnius* (Reduviidae: Triatominae) (Wigglesworth, 1933, 1948). They are composed of four cells (Lai-Fook, 1970), one of them assuming the secretory function, and they show their greatest activity at the time of molting. They have therefore been thought by most authors to play a role in the process of ecdysis or of cuticle formation (Wigglesworth, 1933, 1948; Baldwin and Salthouse, 1959; Lai-Fook, 1970). Such glands appear to be widespread over the cuticle of *Rhodnius,* and similar glands occur in other orders of insects, lending credence to the ecdysial-function theory.

"Floral glands" and "socket glands" have been described in *Dysdercus fasciatus* Signoret by Lawrence and Staddon (1975). Floral glands are multicellular, whereas the socket glands are unicellular. Both types are most common on the abdominal venter, and in *Dysdercus* apparently are confined to the adult male, strongly suggesting a pheromonal function.

Males and some females of most groups of Pentatomoidea (e.g., Pentatomidae, Plataspidae, Acanthosomatidae, and Scutelleridae) possess pheromone-producing unicellular tegumentary glands laterally on 3 or more abdominal sterna (Carayon, 1981). A special type of such a gland is found in males of several genera of Scutelleridae (e.g., *Tectocoris* Hahn, *Psacasta* Germar, *Odontoscelis* Laporte, and *Irochrotus* Amyot and Serville) and in at least some Oxycareninae (Lygaeidae). Referred to as an androconium, each gland consists of a single blind cell set in an alveolus. The contained aphrodisiacal substance, which

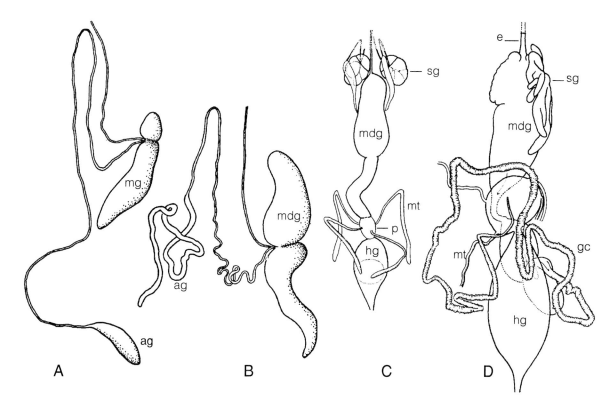

Fig. 10.11. Salivary glands and alimentary canal. **A.** Salivary gland, *Dictyonota strichnocera* Fieber (Tingidae). **B.** Salivary gland, *Stollia fabricii* Kirkaldy (Pentatomidae) (A, B from Southwood, 1955). **C.** Alimentary canal, *Hypselosoma hirashimai* Esaki and Miyamoto (Schizopteridae). **D.** Alimentary canal, *Megymenum gracilicorne* Dallas (Dinidoridae) (C, D Miyamoto, 1961a). Abbreviations: ag, accessory gland; e, esophagus (foregut); gc, gastric caeca; hg, hind gut; mdg, midgut; mg, main gland; mt, Malpighian tubules; p, pylorus; sg, salivary gland.

is produced by an endocuticular cell, can be released only if the androconium is ruptured in a fragile zone (Carayon, 1984).

Corial glands (Figs. 10.10D, 58.2B). These unicellular glands cover the corium, clavus, and cuneus in the Plokiophilidae. They were first observed by China (1953), who termed them "tubercular sense organs," their glandular function having been determined by Carayon (1974). They may play a role in maintaining a commensal relationship with the bugs in the webs of spiders and embiids.

Salivary Glands and Alimentary Canal

The classical work on the salivary glands is that of Baptist (1941) with additional useful comparative and structural information in the works of Barth (1954), Kumar (1967), Nuorteva (1956), and Southwood (1955), among others. The broad outlines of gut structure in the Heteroptera were presented by Dufour (1833). Details of the anatomy and function of the alimentary canal can be found in the works of Goodchild (1963) and Miyamoto (1961a).

Salivary Glands

The salivary glands (Fig. 10.11A, B) in the Heteroptera are positioned anteriorly in the thorax and usually are adpressed to the alimentary canal. The general structure consists of a *main gland* with 1–several lobes, most commonly 2–4, connected by an elongate slender duct to a tubular or vesicular *accessory gland*. The ducts of the main glands unite to form a common salivary duct, which leads to the *salivary pump* and hence to the salivary canal formed by the maxillary stylets. The main glands of the Heteroptera, as opposed to those of the Sternorrhyncha and Auchenorrhyncha, possess a distinct lumen that holds the secretory products produced in the gland walls.

The anterior portion of the main gland appears to produce sheath material that fixes the labium to the substrate at the point of insertion, although the glands may also produce substances that serve to lubricate the stylets. Such sheaths, or at least attachment flanges, appear to be best developed in the Heteroptera among the Pentatomomorpha, but they are also known in other groups, for example, predators such as the Reduviidae (Friend and Smith, 1971). The anterior lobes, which appear to produce zootoxic secretions important in immobilizing prey in the Reduviidae, are often greatly reduced in hematophagous species (Haridass and Ananthakrishnan, 1981). The posterior lobe (or lobes) of the main gland apparently produce hydrolyzing enzymes, which are proteolytic in the case of obligate predators. In the case of hematophagous species such as the Triatominae (Reduviidae), the saliva does not cause paralysis, but serves as an anticoagulant, the bite of the bug being undetected by the vertebrate host.

The glands of taxa that feed primarily in the plant vascular system (most Sternorrhyncha and Auchenorrhyncha, many Pentatomoidea and Coreidae) generally contain no hydrolyzing enzymes (at least when feeding in the phloem vessels), whereas those of taxa that feed on plant cell contents and seeds (e.g., Miridae, Tingidae, many Lygaeoidea, some Coreoidea) generally produce such enzymes. Substances that initiate gall formation are also produced in the salivary glands, such as in certain Tingidae (see Chapter 54).

The accessory glands may function as excretory organs, usually dispensing a dilute liquid, which apparently is extracted from the hemolymph. The function of this substance is not well understood, although it may serve to suspend materials being imbibed by so-called lacerate-flush feeders, in which case the accessory glands are usually better developed than in the stylet-sheath feeders; it might alternatively serve to dilute the concentrated secretions of the main glands.

Maxillary glands. These glands lie anteriorly in the head in a cavity formed by the proximal end of the maxilla and mesad of the maxillary protractor muscles. The glands consist of large peripheral cells and numerous smaller central cells. Chitinous ducts lead from the peripheral cells to a collecting reservoir on the floor of the preoral cavity which opens into the space occupied by the mandibular and maxillary stylets (Linder, 1956).

Nymphal thoracic glands. All nymphal instars of all Heteroptera possess thoracic glands, elongate bodies of heavily nucleated cytoplasm lying in close association with the salivary glands and attached to the accessory gland or the main gland. They are derived from an embryonic invagination of the second maxillary segment. During each nymphal instar the glands go through a phase of swelling, glandular discharge, and shrinkage. These cyclic changes appear to be mediated by secretions from the brain, the glands functioning in the hormonal control of molting. They atrophy and disappear within 72 hours of the imaginal molt (Wells, 1954).

Gut

The heteropteran alimentary canal (Fig. 10.11C, D) is formed of 3 principal divisions: foregut, midgut, and hindgut. The structure in nymphs and adults is essentially similar. The *foregut* consists of an elongate, chitin-lined *esophagus* connected to the food canal (Fig. 10.2A) of the maxillary stylets via the *cibarial pump* located dorsally in the head. The esophagus proper varies greatly in length, but it is never differentiated into a true crop as found in many insects.

The *midgut,* which is endodermal in origin, is variously subdivided, depending on the group. It comprises 2 or 3 sections in most groups, but in the Pentatomomorpha it often is divided into 4 or rarely 5 sections. The most anterior portion of the midgut forms a capacious chamber, or *stomach,* in all bugs. The remainder of the midgut may take on various forms, the penultimate section bearing 2–4 rows of *gastric caeca* in most Trichophora, except in the predatory Asopinae (Pentatomidae) and many seed feeders. The midgut of the Phyllocephalinae (Pentatomidae) forms a *filter chamber* similar to that found in many Auchenorrhyncha, apparently uniquely among the Heteroptera.

Some pentatomomorphans have a discontinuous midgut. This situation was investigated in detail by Goodchild (1963), who prepared histological sections of the guts of a relatively large number of taxa. The midgut in many sap-feeding taxa is occluded at a point just anterior to the section that contains the gastric caeca. There is a ligamentous connection between the anterior and posterior sections, but Goodchild found no lumen. This distinctive gut morphology exists in both nymphal and adult forms, but how it functions is not understood.

The *pylorus* (Miyamoto, 1961a), or *ileum* (Goodchild, 1963), is not lined with chitin, and therefore appears to represent part of the true midgut, rather than a portion of the hind gut as asserted by some authors. It may be well developed, particularly in the Pentatomomorpha, to virtually absent. The *Malpighian tubules,* which vary greatly in length and general conformation, arise from the pylorus, as an anterior and posterior pair in the Pentatomomorpha and as a dorsal and ventral pair in most other Heteroptera. Between the pylorus and the rectum there is generally a conspicuous valve, the *pyloric valve* (Myiamoto, 1961a:201), or *ileorectal valve* (Goodchild, 1963), composed of a cone of small columnar cells and surrounded by many strong circular muscle fibers.

The chitin-lined true *hind gut* consists primarily of the *rectum,* which is often pear-shaped. At least anterodorsally, if not to a much greater extent, it is modified into a *rectal gland,* or *rectal pad,* a relatively extensive area of specialized cells, which are thought to play an important role in extraction of water from the hemolymph.

Symbiotic Microrganisms

Gut symbionts. The gastric caeca in the Pentatomomorpha are filled with bacteria, which serve an undetermined function. Bacteria from a sample of bugs were described as species of *Pseudomonas* by Steinhaus et al. (1956). These symbionts are generally transmitted by the females' depositing symbiont-laden material in capsules among eggs or on freshly laid eggs (Carayon, 1951; Büchner, 1965). In the ovoviviparous rhyparochromine (Lygaeidae) *Stilbocoris,* however, the female deposits this material on the head of the freshly laid nymphs. The nymphs subsequently use the rostrum to imbibe the symbionts from the droplet (Carayon, 1963).

Mycetomes. Although gastric caeca are present only in most Pentatomomorpha, symbiotic microorganisms nonetheless appear to be present in most Heteroptera. They are harbored in the mycetomes (Büchner, 1921), which occur midlaterally in the abdomen in Cimicidae. In the male they are attached to the vas deferens, and in the female they lie unattached (Usinger, 1966). Transmission is apparently transovarial, and the identity of the symbionts is not agreed upon by all authors.

Intracellular symbionts. Intracellular symbionts are also known from a few Heteroptera, although, as for those described above, their function is not well understood.

Nervous System

The gross structure of the heteropteran nervous system was first studied comparatively by Brandt (1878) and later in greater detail by Pflugfelder (1937). The following account has been prepared from the works of Johansson (1957b), Parsons (1960a), and Livingston (1968). Other basic studies include those of Hamilton (1931), Rawat (1939), and Guthrie (1961).

Central nervous system. Like other phylogenetically more advanced insects, the Heteroptera show a great amount of ganglionic fusion. The so-called *brain,* or *supraesophageal ganglion,* is vaguely differentiated into a *protocerebrum, deutocerebrum,* and *tritocerebrum.* The first of these includes the *optic lobes,* which innervate the compound eyes and ocelli. From the second emanate both sensory and motor nerves for the antennae. The tritocerebrum and subesophageal ganglion show a strong degree of fusion in most Heteroptera. It is from this region that the mouthparts, salivary glands, and associated musculature are innervated.

The *first thoracic ganglion* is more or less distinct from the subesophageal ganglion, depending on the taxon. From it emanates foreleg innervation. The "posterior ganglionic mass," or *central ganglion,* represents the fusion of the mesothoracic ganglion, metathoracic ganglion, and all abdominal ganglia, supplying nerves to the wings, posterior 2 pairs of legs, and to the abdomen. There are fewer obvious pairs of nerves than abdominal segments, the largest pair being that which innervates the genitalia.

Stomodaeal nervous system (stomogastric or sympathetic nervous system) (Fig. 10.12B). The *frontal ganglion,* which has a nervous connection with the tritocerebrum, innervates the pharyngeal dilator muscles via the *procurrent nerve.* It is connected via the *recurrent nerve* with the *hypocerebral ganglion,* which may innervate the esophagus, and in some taxa is reported to have a nervous connection with the *corpus cardiacum.*

Endocrine glands (neurosecretory system). This specialized portion of the nervous system consists of the *corpora cardiaca* and the *corpora allata;* in some groups each exists as a single structure rather than in the paired condition. These structures are connected to the central or stomodaeal nervous system or both. During the first four nymphal stadia the corpora allata secrete juvenile hormone, which causes the insect to retain its immature form. The corpora cardiaca produce prothoracotropic hormone, which acts on the prothoracic gland to produce ecdysone, which causes metamorphosis into the adult form (Wigglesworth, 1984).

Circulatory System

Information on the insect dorsal vessel is taken from the following references: Hamilton, 1931; Malouf, 1933; Wooley, 1951; Hinks, 1966; and Miyamoto, 1981.

The insect circulatory system shows great similarity across a wide range of taxa. In the Heteroptera it consists of the *dorsal vessel* (Fig. 10.13A, B), which lies close against the abdominal and thoracic dorsum. This structure may be divided into a posterior *heart* and an anterior *aorta.* Hemolymph enters the heart through 2–7 pairs of *ostia,* which may lie in abdominal segments 1–7. The pumping action is peristaltic, being mediated by the alary and intrinsic musculature. Most terrestrial Heteroptera have 3 pairs of *alary muscles,* whereas the Nepomorpha usually have more, with a maximum of 8. The *intrinsic musculature* may be arranged in a circular or helical fashion. The Hemiptera and some other Acercaria are unique among the Insecta in generally having the heart restricted to the posterior portion of the abdomen. The anterior aorta is devoid of ostia and extends forward into the head capsule, often lying close along the dorsal surface of the alimentary canal and emptying over the supraesophageal ganglion of the nervous system.

The dorsal vessel is generally lined with specialized cells known as *nephrocytes;* these often are several layers thick and more prominent in the posterior region. They apparently function in the removal of waste substances from the hemolymph. Innervation of the heart wall and the alary muscles was reported by Hinks (1966), although the source of the nerves was not mentioned. Tracheation of the dorsal vessel is restricted to the heart region and may be largely dorsal or ventral, depending on the taxon. The Malpighian tubules are usually intertwined with the alary muscles in close association with the dorsal vessel.

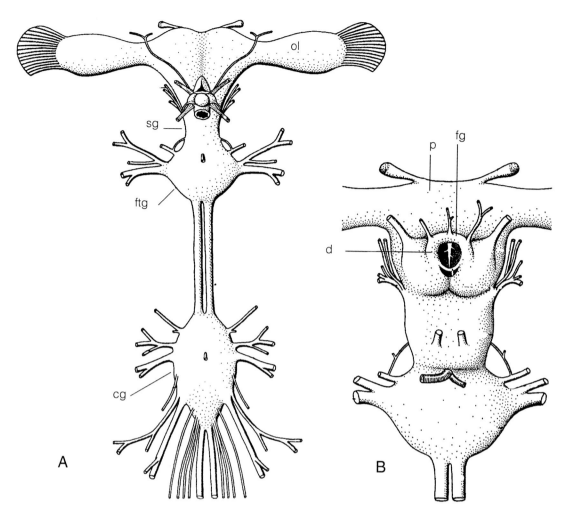

Fig. 10.12. Nervous system. *Gelastocoris oculatus* (Fabricius) (from Parsons, 1960a). **A.** Central nervous system. **B.** Detailed ventral view of cerebral area. Abbreviations: cg, central ganglion; d, deutocerebrum; fg, frontal ganglion; ftg, first thoracic ganglion; ol, optic lobe; p, protocerebrum; sg, subesophageal ganglion.

Eggs

Heteropteran eggs have been known since the time of Leuckhart (1855) to show substantial morphological variability, and particularly those of the pentatomoids have attracted attention because of their bright coloration and deposition on plant surfaces. Major modern reviews, from which most of the following material is taken, have been prepared by Southwood (1956) for terrestrial Heteroptera and Cobben (1968a) for all groups (see also works of Putchkova referenced in Cobben). The work of Ren (1992) contains many superb scanning micrographs of eggs, showing the details of their surface structure.

The eggs may be elongate (Fig. 10.14C), cylindrical and sometimes weakly curving (Fig. 10.14A, B), or barrel-shaped (Fig. 10.14D). Eggs are inserted into plant tissue, cemented to a surface, or laid free.

The *shell*, or *chorion*, is laid down by follicle cells in the telotrophic ovaries, after the yolk has completely developed, and except in the case of the Cimicoidea, before fertilization. The chorionic surface may range from shallowly pitted to a well-formed honeycomb, or it may even be rugose.

There may or may not be a distinct *operculum*, or *egg cap*, present at the cephalic pole, a feature strongly correlated with the type of eclosion mechanism. A true operculum, with a line of weakness referred to as a sealing bar, is present in the Cimicomorpha, all of which lack a true *egg burster;* eclosion takes place via pressure from within the egg and has been described by Cobben (1968a) as either embryonic or serosal. Most other groups have an egg burster present on the head of the ecloding embryos, and according to Cobben these may be generally classified as clypeal or frontal. Although the eggs of some Pentatomomorpha have a true operculum, eclosion in many members of the group takes place by an irregular rupture of the egg, rather than by lifting off a cap.

The cephalic pole of the egg of most bugs bears from

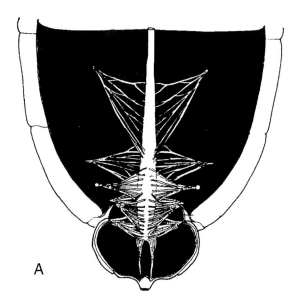
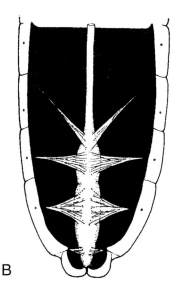

Fig. 10.13. Circulatory system (from Hinks, 1966). **A.** Dorsal vessel and alary muscles, *Phonoctonus fasciatus* (de Beauvois) (Reduviidae). **B.** Dorsal vessel and alary muscles, *Aquarius lacustris* (Linnaeus) (Gerridae).

1 or 2 to many *micropyles* (Fig. 10.14F, G), and in the case of most Cimicomorpha also bears what have been referred to as *aeropyles,* which apparently serve the function of gaseous exchange. In the case of the Pentatomomorpha, including the Aradoidea, the cephalic pole of the egg is also ornamented with *micropylar processes* (Fig. 10.14D, E), at which point the *micropylar canals* exit the egg shell. In all bugs except the Cimicoidea, sperm enter the egg during fertilization through the micropyles.

Sperm and Fertilization

Heteropteran sperm, like those of most other insects, are long and slender, with the nucleus present in the head (Phillips, 1970). On the basis of at least one representative of all infraorders except Enicocephalomorpha and Dipsocoromorpha, they appear to have three characteristics that do not occur in most other insects—additional evidence for the monophyly of the group. First, there are 2 or 3 crystalline bodies in the mitochondrial derivatives, rather than 1, as found in other Pterygota; second, 2 bridges join the mitochondrial derivatives to the axoneme at the level of doublets 1 and 5; and third, the axoneme lacks the longitudinal accessory bodies found in many closely related groups of insects (Dallai and Afzelius, 1980; Afzelius et al., 1985; Jamieson, 1987).

Some Pentatominae (Pentatomidae) produce sperm with abnormal chromosomal complements. These always come from one lobe of the testes, referred to as the harlequin lobe, and they do not participate in reproduction. The functional significance of their existence has not been satisfactorily explained.

Most male Heteroptera apparently deposit sperm in the female as a *spermatophore* (Ambrose and Vennison, 1990). Sperm are stored by the female in the spermatheca, or its functional equivalent in the Cimicomorpha. Fertilization and oviposition are ordinarily simultaneous in the Heteroptera. In most Cimicoidea, however, sperm are deposited in the abdominal cavity by injection through a specialized area of the body wall known as the *ectospermalege* (Carayon, 1966, 1977). From that point the sperm migrate to the base of the ovaries, and in many taxa embryogeny precedes oviposition.

Chromosomes and Karyotypes

Heteroptera, in common with other Hemiptera, have holocentric (holokinetic) chromosomes, in which the centromere is distributed along the entire length of the chromosome. Diploid chromosome complements range from 4 to 48 or 49, with some of the lowest and highest numbers occurring in the Nepomorpha. The total amount of genetic material is about the same in members of the same family with very different chromosome numbers, suggesting fragmentation rather than polyploidy as the ordinary method of increase in chromosome numbers. The Pentatomomorpha generally have diploid complements ranging from 12 to 18, whereas all other major groups (except Enicocephalomorpha, in which they are unknown) have diploid complements of 20 or above, but with substantial variation.

Sex chromosome systems may be either XY or XO. The XO system is found in Dipsocoromorpha, Gerromorpha, a few Nepomorpha, a few Lygaeidae, and all Lepto-

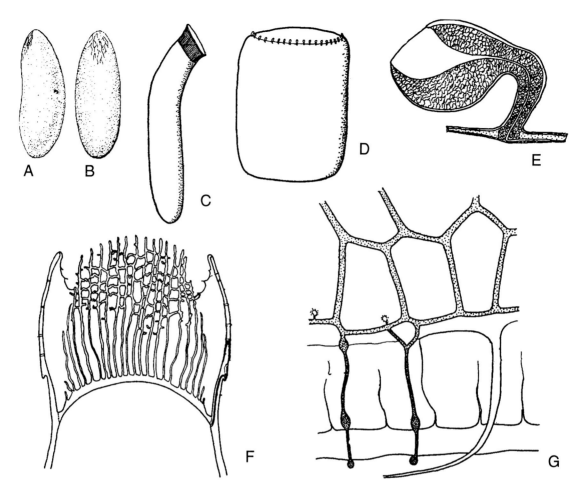

Fig. 10.14. Eggs and egg structures. **A.** Egg, lateral view *Saldula fucicola* (Sahlberg) (Saldidae). **B.** Egg, *S. fucicola* (A, B from Cobben, 1968a). **C.** Egg, *Nabis limbatus* Dahlbom (Nabidae). **D.** Egg, showing micropylar processes at cephalic pole. *Nezara viridula* (Linnaeus) (Pentatomidae). **E.** Micropylar process, *Oncopeltus fasciatus* Dallas (Lygaeidae). **F.** Operculum and micropylar region, *Rhinocoris* sp. (Reduviidae). **G.** Base of micropylar region, *Rhinocoris* sp. (C–G from Southwood, 1956).

podomorpha studied. Nearly all known Cimicomorpha and Pentatomomorpha have XY sex-determination systems, suggesting that XY is the derived condition. Many groups have multiple X chromosomes, with a maximum of 5 in some Reduviidae, and indeed, the greatest variation in chromosome complement is often the result of variation in the number of X chromosomes.

One or more supernumerary "m" chromosomes occur in many members of the Nepomorpha and in lygaeoid, pyrrhocoroid, and coreoid Pentatomomorpha (Ueshima, 1979).

Nymphs

Heteropteran nymphs typically resemble adults and live in similar environments. There are ordinarily five instars, although some variation does occur—as for example the consistent occurrence of four instars in *Mesovelia furcata* Mulsant and Rey (Zimmermann, 1984)—but without any apparent systematic pattern (Štys and Davidova, 1989). Especially in the Pentatomomorpha, the first-instar nymphs may be red and black, whereas in most other groups later instars are usually colored in a manner similar to the adults. Aside from the lack of ocelli, wings, and genitalia, nymphs are distinguished primarily by the presence of one less tarsal segment than found in adults, and in the case of the Pentatomoidea the subdivision of the pedicel to form the 5-segmented antennal condition usually appears at the imaginal molt. Yonke (1991) gave an excellent general treatment of heteropteran nymphs, including keys to all families, diagnoses, representative illustrations, and citations for additional sources. Cobben (1978) provided the best survey of nymphal characteristics of phylogenetic value, with illustrations, descriptions, and tabular summaries. Keys to families occurring in North America were published by DeCoursey (1971) and Herring and Ashlock (1971).

Key to Infraorders of Heteroptera

Diagnosis. Mouthparts typically hemipteran with mandibular stylets concentric and surrounding maxillary stylets; labium inserted anteriorly on head, a distinct gular area always present, often closed behind to form a buccular bridge; scent-gland structures, often paired, located and exiting ventrally in metathorax of adults, channels moving fluid to variously elaborated metathoracic pleuron; nymphs with paired scent glands located at junction of one or more of the following abdominal terga: 3/4, 4/5, 5/6, 6/7.

1. Compound eyes usually present, sometimes reduced, very rarely absent (a few Enicocephalomorpha); wings occasionally reduced or absent; mostly free-living 2
– Compound eyes always absent; apterous; obligate parasites of bats or termitophilous inquilines 14
2. Head constricted transversely, divided into 2 distinct lobes, ocelli (when present) placed on posterior lobe posterior to compound eyes (Figs. 12.1A, 13.1A); foretibiae flattened, with distinct distal spines (Fig. 13.1A, C, D); foretarsus usually 1-segmented, sometimes 2-segmented, always opposable to apex of foretibia (Figs. 12.1D, 13.1C, D); forewings, when present, of uniform texture, not differentiated into distinct corium and membrane, lacking claval commissure (Fig. 13.1A); compound eyes sometimes greatly reduced or absent .... Enicocephalomorpha
– Head generally not constricted and divided into lobes; if divided, prosternum with a distinct sulcus for reception of apex of labium (Fig. 10.7F) and foretibia and foretarsus never as above; forewings, if fully developed, of variable structure, divided into a distinct corium and membrane or not; compound eyes only very rarely reduced or absent (Figs. 42.1, 54.1) 3
3. Head with 3 or 4 pairs of conspicuous trichobothria placed near inner margin of compound eyes (Figs. 10.8A, 20.2C), inserted in distinct pits, bothrium obscured in pit, hair pile obviously developed around pit margin; forewings, when present, lacking claval commissure and not divided into a well demarcated corium-clavus and membrane (Figs. 20.2A, B, 20.3B); part or most of body always covered with a distinct pile of microsetae and spicules .... Gerromorpha
– Head without trichobothria, or if with one or more pairs of trichobothria, these never placed in deep pits; forewings, when developed, with or without claval commissure; venter infrequently with pile of microsetae and spicules (Nepomorpha: Aphelocheiridae; some Naucoridae; some Leptopodomorpha) 4
4. Antennae shorter than head and folded beneath it, usually concealed (Fig. 29.1A) in grooves (except Ochteridae lacking groove; Fig. 29.1D), at most apex of antenna slightly exposed beyond margin of large compound eyes (Fig. 29.1D); compound eyes present, usually large (Figs. 30.1, 33.1); prosternal sulcus never present; usually macropterous, rarely staphylinoid (some Naucoridae, Aphelocheiridae; Fig. 37.1B), claval commissure always present in macropterous forms (Fig. 30.1) .... Nepomorpha
– Antennae usually longer than head and never concealed in grooves below compound eyes, even when shorter than length of head; if antennae shorter than head, prosternal sulcus present; forewings variously developed, claval commissure present or absent 5
5. Abdominal sterna 3–7 usually with 2 or 3 (or rarely 1 or more than 3) trichobothria placed sublaterally (Pentatomoidea) (Fig. 10.8C) or submedially on sterna 3 and 4 and sublaterally on sterna 5–7 (other groups; Fig. 10.8D), occasionally present sublaterally only on sterna 5–7, or very rarely completely absent (some Piesmatidae); elongate pulvillus always attached near base of each claw (Fig. 10.5I) .... Pentatomomorpha (less Aradoidea)
– Abdomen at most with a single trichobothrium-like seta on either side of midline of one or more sterna and never arranged as described above, or with a single trichobothrium on ventral laterotergite 7 (some Nabidae; Fig. 10.8F); claws sometimes with pulvilli (most Aradoidea: Fig. 10.5G, H; some Miridae: Fig. 53.5E; and some Anthocoridae) 6
6. Antennal segments 1 and 2 short, subequal in length, segments 3 and 4 very long and slender and clothed with erect setae of length much greater than segmental diameter (Fig. 17.1); length usually under 2.5 mm; hemelytra, when present, tegminal or coleopteroid (Figs. 15.1A, 17.1, 18.1A) .... Dipsocoromorpha (less Stemmocryptidae)
– Antennal segment 2 usually longer than segment 1, if not (most Tingidae; Fig. 54.2B), then segments 3 and 4 lacking erect setae longer than segmental diameter and pronotum and hemelytra

covered with areoles; hemelytra, when developed, never tegminal and undifferentiated 7

7. Claws usually with distinct elongate pulvilli free from claw except at base (Fig. 10.5G); tarsi 2-segmented (Fig. 64.1A); body conspicuously flattened, sometimes encrustate (Fig. 64.1B); wings, when present, covering only discal area of abdomen, connexivum exposed (Fig. 64.1A); eyes small relative to size of head, pebblelike, but not with reduced numbers of ommatidia (Fig. 64.1A, B) .... Aradidae

– Claws sometimes with pulvilli (some Miridae: Fig. 53.5E; some Anthocoridae), but if present pulvilli rarely free from claw over most of length (some phyline Miridae) and then tarsi always 3-segmented; tarsi usually 3-segmented, less frequently 2-segmented; wings, when present, usually covering connexivum (except some Reduviidae); eyes often comparatively large relative to size of head (Fig. 53.2B), rarely pebblelike or with reduced numbers of ommatidia 8

8. Membrane of forewing present, in at least a reduced form, 3, 4, or 5 closed cells usually visible (Figs. 43.1A, B, 45.1A, B), never with veins emanating from posterior margin of cells .... Leptopodomorpha (part)

– Forewing membrane usually present, often with closed cells, usually 2 (Fig. 47.1), if more than 2 then always with veins emanating from their posterior margins 9

9. Forewings present in the form of hemelytra with a differentiated corium-clavus and membrane; compound eyes well developed .... Cimicomorpha (part)

– Forewings either tegminal and undifferentiated (Stemmocryptidae: Fig. 19.1A, B), coleopteroid (Omaniidae: Fig. 44.1A; Tingidae: Vianaidinae, Fig. 54.1), or greatly reduced (Aepophilidae: Fig. 42.1; some Cimicomorpha) 10

10. Forewings coleopteroid 11
– Forewings tegminal or greatly reduced 12

11. Eyes very large, occupying nearly entire sides of head; body length approximately 1 mm (Omaniidae; Fig. 44.1) .... Leptopodomorpha (part)

– Eyes consisting of only a few ommatidia; body length greater than 2 mm (Vianaidinae; Fig. 54.1) .... Tingidae (part)

12. Forewings developed, largely membranous with a few relatively strong veins (Stemmocryptidae; Fig. 19.1A) .... Dipsocoromorpha (part)

– Forewings completely absent or in the form of small pads 13

13. Most of body covered with a short hair pile; padlike forewings as in Fig. 42.1 (Aepophilidae) .... Leptopodomorpha (part)

– Body not covered with a hair pile; forewings greatly reduced or absent, but not exactly as in Fig. 42.1 .... Cimicomorpha (part)

14. Body very strongly flattened, with marginal laminae (Fig. 65.1A, B) (Termitaphididae) .... Pentatomomorpha (part)

– Body not so flattened, and without marginal laminae (Fig. 62.1A) (Polyctenidae) .... Cimicomorpha (part)

HETEROPTERA

11
Enicocephalomorpha

by Pavel Štys

General. The unique-headed bugs are also called four-winged flies for their habit of forming mating swarms and their similarity in appearance to the Nematocera when flying. These delicate, elongate bugs range in length from 2 to 15 mm and, although seldom collected, are widely distributed, even in temperate areas.

Diagnosis. Head elongate, porrect, subdivided into anterior and postocular lobes by a usually conspicuous postocular constriction (Fig. 13.1A); ocelli, if present, on posterior lobe, well removed from eyes (Fig. 13.1A); gula long; labium 4-segmented, short, straight to arcuate, never exceeding length of head; antennae flagelliform to terete, moderately long (Fig. 13.1A); forewings always completely membranous (tegminal) (Figs. 12.1A, 13.1A); ambient vein in remigium marginal or slightly submarginal, but if venation reduced, then vein represented at least by a continuous row of macrotrichia; forewings sometimes reduced or absent, occasionally caducous; medial fracture situated in front of R on forewing, in common only with some Dipsocoromorpha; base of forewing with bracelike cross vein connecting marginal veins with R and M+Cu; foreleg usually raptorial, tibia distoventrally produced, usually dilated, with 1 or 2 clusters of spiniform setae, opposed to ventral face of 1- or 2-segmented foretarsus, usually also with spines (Figs. 12.1D, 13.1C, D); male genitalia always symmetrical, with paired genital plates as in Auchenorrhyncha, greatly reduced to racquet-shaped guide in Enicocephalidae (Fig. 13.1E); ovipositor present (Fig. 12.1F) to absent; subgenital plate formed by sternum 8 rather than 7 as in other Heteroptera; spermatheca present (Fig. 12.1G).

Discussion. The Enicocephalomorpha, whose unique structure and phylogenetic position in the Heteroptera have been recognized by most modern authors (Štys and Kerzhner, 1975), is currently divided into two families—Aenictopecheidae and Enicocephalidae (Štys, 1989)—rather than the single family Enicocephalidae (e.g., Štys, 1970b). Enicocephalomorphans were at one time placed in the Reduvioidea e.g., Usinger, 1943), but they were combined with dipsocoromorphans by Miyamoto (1961a) and Popov (1971).

Grimaldi et al. (1993) illustrated and described aspects of morphology of a Lower Cretaceous amber fossil enicocephalomorph from Lebanon, helping to establish the great antiquity of this group. They also described in two Recent genera species of Oligocene-Miocene amber fossil Enicocephalidae from the Dominican Republic, noting that these taxa are virtually identical to living forms.

Major classical students of the family were Bergroth, Eckerlein, Breddin, and Usinger. Jeannel (1941) monographed the group, and Usinger (1945) provided a higher classification. Villiers (1958) monographed the fauna of Madagascar and later that of the Afrotropical and Malagasy regions (1969a) Usinger and Wygodzinsky (1960) revised the fauna of Micronesia. Štys (1969) revised the extinct Enicocephalomorpha and in many subsequent papers described new higher taxa and provided morphological information, listed the world supraspecific taxa (Štys, 1978), and established the current higher classification of the group. Wygodzinsky and Schmidt (1991) monographed the New World fauna (excluding *Systelloderes* Blanchard), adding much new structural data based on SEM observations.

Key to Families of Enicocephalomorpha

1. Males with a distinct inflatable or noninflatable phallus protruding out of the pygophore (Fig. 12.1E); parameres distinct, movable, articulated with phallus; ovipositor usually present, rarely strongly reduced; forewing in macropterous forms usually with a short costal fracture (Fig. 12.1A) (except Murphyanellinae) .... Aenictopecheidae
- Males with noninflatable intromittent organ never resembling typical heteropteran phallus (Fig. 13.1E); parameres distinct, but always immobile (Fig. 13.1E), hooklike, tuberculiform, plate-shaped to reduced to paired lateral parameral sclerites associated with supradistal plate or its vestige; ovipositor vestigial or wanting; forewing in macropterous forms without costal fracture (Fig. 13.1A) .... Enicocephalidae

12 Aenictopecheidae

by Pavel Štys

General. The aenictopecheids are inconspicuous, little known, and rarely encountered bugs. Some species have retained plesiomorphic characters not encountered in other enicocephalomorphans, or among Heteroptera in general. They range in length from approximately 3 to 10 mm.

Diagnosis. Posterior lobe of pronotum often abbreviated, poorly defined, not demarcated by sinuate lateral margin (Fig. 12.1A); Rs in forewing branching in Maoristolinae, costal fracture short (Figs. 12.1A, B); wing coupling mechanism as in Fig. 12.1C; pygophore never subdivided into tergum, laterotergites, and sternum; phallus typically heteropteran (Fig. 12.1E), inflatable or not, with movable parameres; ovipositor usually fully developed (Fig. 12.1F); nymphs with normally developed wing pads, not contiguous along midline.

Classification. The present conception of this family is on the basis of plesiomorphic characters. Nonetheless, none of its subfamilies possesses any apomorphies in common with the Enicocephalidae. This group is clearly relict, and many taxa badly need more detailed morphological study.

Four subfamilies are currently recognized comprising 10 genera and 20 species.

Key to Subfamilies of Aenictopecheidae

1. Eyes reduced to a single ommatidium; all legs fossorial; spiniform setae (markedly thicker than other macrotrichia) situated on ventral face of middle and hind femora, along the length of all tibiae, and on ventral apex of all tarsi, but not markedly clustered at the ventral apex of foretibia; micropterous, transverse scale-shaped forewings not extending onto abdomen; tarsal formula 1-1-1; posterior lobe of pronotum extremely short and not demarcated in lateral view; New Zealand and Tasmania .... Nymphocorinae
   - Eyes normally developed; forelegs raptorial, middle and hind legs cursorial; spiniform setae situated only on apices of tibiae in most taxa and ventrally on foretarsi (foretarsal spines absent in Murphyanellinae), those of ventral apex of foretibia concentrated into 1 or 2 clusters; macropterous, or if micropterous, then forewings not transverse, with remnants of venation, and extending onto abdomen 2
2. Tarsal formula 1-2-2; pronotum seemingly of 2 lobes only, posterior lobe usually reduced in size and/or not distinctly delimited in lateral view; collar rarely vaguely delimited and then pronotum seemingly not subdivided into lobes; macropterous forms with a short costal fracture, just interrupting forewing margin; foretarsus with ventral spiniform setae 3
   - Tarsal formula 1-1-1; pronotum with 3 distinct lobes; macropterous forms (only macropterous males known) without costal fracture; foretarsi without ventral spiniform setae; minute, less than 2 mm long; Singapore .... Murphyanellinae
3. Forewing with R branching into R_1 and Rs; foretibial ventral apical armature formed by clustered simple spiniform setae situated on dilated, but not projecting, tibial apex; male with a pair of inflatable pygophoral "vesicles"; ovipositor almost completely absent, no external traces of first valvulae; New Zealand .... Maoristolinae
   - Forewing with simple R; at least part of the tibial ventral apical armature formed by modified (broad, flattened, abbreviated, apically often rounded or bilobate) spiniform setae, situated on projecting tibial apex and subdivided into 2 clusters by a notch; no inflatable pygophoral "vesicles"; ovipositor retained, at least as internal remnants of valvulae .... Aenictopecheinae

AENICTOPECHEINAE (FIG. 12.1A). Small to medium-sized, sometimes robust (*Gamostolus* Berg); macropterous to strongly brachypterous, or sometimes dimorphic (*Boreostolus* Wygodzinsky and Štys); posterior lobe of pronotum reduced, usually not distinguishable in lateral outline (Fig. 12.1A); apical armature of foretibia subdivided into 2 clusters of abbreviated, thickened, often rounded or bilobate, peglike spiniform setae (Fig. 12.1D); tarsal formula 1-2-2; phallus noninflatable, permanently erected, and always strikingly protruding from lower part

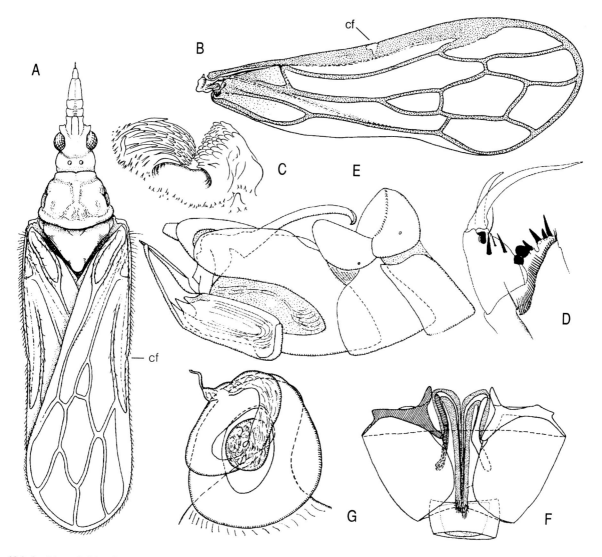

Fig. 12.1. Aenictopecheidae. **A.** *Tornocrusus penai* Wygodzinsky and Schmidt (from Wygodzinsky and Schmidt, 1991). **B.** Forewing, *Gamostolus subantarcticus* (Berg) (from Wygodzinsky and Schmidt, 1991). **C.** Wing-to-wing coupling mechanism, *G. subantarcticus* (from Wygodzinsky and Schmidt, 1991). **D.** Foreleg, distal portion, generalized Aenictopecheidae (A–D from Wygodzinsky and Schmidt, 1991). **E.** Male genitalia, *Monteithostolus genitalis* Štys (from Štys, 1981a). **F.** Female genitalia, ventral view apex abdomen, *Australostolus monteithi*. **G.** Spermatheca, *A. monteithi* (F, G from Štys, 1980). Abbreviation: cf, costal fracture.

of posterior foramen of pygophore (Fig. 12.1E); ovipositor either well developed with first valvifer fused with broad sternum 8 forming triangular submedial processes, or strongly reduced.

Six genera are grouped in two tribes. The Aenictopecheini includes *Aenictopechys necopinatus* Breddin from Indonesia and *Lomagostus jeanneli* Villiers from Madagascar. These are small species with complex patterns of vein fusion in the cubitoclaval area.

Members of the Gamostolini are usually larger and with more typical venation (Fig. 12.1B). *Gamostolus* Berg includes a single species, *G. subantarcticus* (Berg), from southern Argentina and Chile, but unidentified nymphs of the genus are known (Wygodzinsky and Schmidt, 1991) from the Venezuelan and Colombian cordilleras. *Tornocrusus* Kritsky includes several species from Central and South America. *Boreostolus* Wygodzinsky and Štys (1970) includes two boreal species. *Australostolus monteithi* Štys (1980) is Australian.

MAORISTOLINAE. Slender, medium-sized, macropterous to moderately brachypterous; foretibia with simple spiniform apical armature; tarsal formula 1-2-2; posterior lobe of pronotum strongly reduced, pronotum apparently formed of 2 lobes; forewings with short costal fracture and branching Rs; phallus inflatable; pygophore with a pair of large eversible vesicle-like structures; ovipositor strongly reduced.

This group comprises a single genus, *Maoristolus* Woodward (1956), with two species from New Zealand.

MURPHYANELLINAE. Pronotum formed of 3 distinct lobes; only macropterous males known; forewings lacking costal fracture; phallus distinct, inflatable in *Murphyanella aliquantula* Wygodzinsky and Štys, beaklike in *Timahocoris paululus* Wygodzinsky and Štys; parameres uniquely associated with base of phallus by pair of Y-shaped connectives resembling those in Auchenorrhyncha, tubercle-shaped and probably immobile in *Timahocoris* Wygodzinsky and Štys; *Murphyanella* Wygodzinsky and Štys with spiracle of abdominal segment 3 enlarged to about 11 times diameter of other abdominal spiracles.

This enigmatic group includes two tiny species described by Wygodzinsky and Štys (1982) from Singapore.

NYMPHOCORINAE. Small, yellowish, micropterous; eyes with a single ommatidium; femora and tibiae with many spiniform setae; foretibia with apical cluster of spiniform setae spreading along tibia; forelegs appearing fossorial; middle and hind legs cursorial; tarsal formula 1-1-1; phallus complex, inflatable, with lateral appendages; proctigeral region complicated; ovipositor complete.

This subfamily includes only *Nymphocoris* Woodward, with two species, *N. maoricus* Woodward (1956) from New Zealand and *N. hilli* Štys (1988) from Tasmania.

Specialized morphology. The Aenictopecheidae are most obvious for their lacking characters distinctive to the Enicocephalidae as well as having genitalia that are typically heteropteran in form (Fig. 12.1E, F) as opposed to the greatly reduced and modified structures found in most Enicocephalidae.

Natural history. Few details are known of the life habits of the Aenictopecheidae. *Gamostolus* has been found swarming, under stones, and in forest litter. *Australostolus* Štys is periodically attracted to lights in semidesert areas. *Boreostolus* spp. live under big stones on gravel and sand substrates along mountain streams in habitats characteristic of *Cryptostemma* spp. (Dipsocoridae). Maoristolines are found in litter, under bark, and in mosses. The Nymphocorinae are known to live in soil, litter, among tussocks of grass, and are found through the use of Berlese funnels and pitfall traps.

Distribution and faunistics. *Boreostolus* has a typical boreal amphipacific distribution: *B. americanus* Wygodzinsky and Štys in North America ranging from Washington and Oregon to Colorado, and *B. sikhotalinensis* Wygodzinsky and Štys occurring in eastern Russia from Vladivostok to Magadan, the latter locality being the northernmost record for any enicocephalomorphan.

The New World representatives are included in Wygodzinsky and Schmidt (1991). Old World taxa are treated in the papers referred to above.

13
Enicocephalidae

by Pavel Štys

General. The Enicocephalidae include about 95% of the species and 99.5% of known specimens of Enicocephalomorpha. The general facies of most species resembles small to medium-sized reduviids. As in the Aenictopecheidae, most species are dull-colored, usually uniformly yellow, brown, or blackish, less frequently with contrasting color patterns, sometimes of bright red. Length ranges from about 2 to 15 mm.

Diagnosis. Pronotum usually subdivided into 3 distinct lobes (except *Megenicocephalus* and *Alienates* Barber); wing polymorphism common, males sometimes macropterous and females brachypterous to apterous; costal fracture absent (weakly indicated in Megenicocephalinae); nymphal wing pads of later instars large, contiguous, and sometimes slightly overlapping along midline; foretibia and tarsus as in Fig. 13.1C, D; phallus in primitive taxa unlike that of other Heteroptera, formed from structures including paired elements homologous to the genital plates of Auchenorrhyncha, often transformed from condition found in *Monteithostolus genitalis* Štys to arcuate or racquet-shaped, distally perforated "guide" associated with the ventral margin of posterior foramen of pygophore (Fig. 13.1E); parameres always immobile, fused at bases or reduced to flat sclerites; external female genitalia absent or retained as remnants in Systelloderini and Megenicocephalinae; female genital opening covered by extensive subgenital plate formed by sternum 8.

Classification. Five subfamilies and several additional tribes are recognized (Štys, 1970c, 1989; Wygodzinsky and Schmidt, 1991). At present 33 genera and approximately 400 species are known. Because of the difficulty in diagnosing some taxa, distribution and wing development are emphasized in the key.

Key to Subfamilies of Enicocephalidae

1. Pronotum either with collar and middle lobe, or middle lobe and posterior lobe, if of 3 lobes then middle and posterior lobes distinguished by sinuation of lateral margin; males always with extremely simplified genitalia formed by "guide" and parameral sclerotizations only; forelegs variable; tarsal formula 1-1-1 or 1-2-2 2
   - Pronotum of all wing morphs usually subdivided in 3 well-delimited lobes (Fig. 13.1A); if lobes not distinct, then apterous/micropterous taxa from SW Pacific (including New Zealand) and subantarctic islands; male genitalia complex to simplified; forelegs always raptorial, with dilated ventral apicotibial projection and its clustered spiniform armature; tarsal formula 1-2-2 3
2. Large, length 9–17 mm, robust; coloration reddish; pronotum formed of 2 lobes, collar undifferentiated; eyes with many ommatidia; forelegs incrassate (particularly femora), cursorial; apicotibial projection and its armature undifferentiated; tarsal formula 1-2-2; female forelegs with spinous process on trochanter, and spiniform tubercles on tibia and sometimes femur; forewing with complete venation and indication of costal fracture; antennae flagelliform; female with a minute knoblike structure with paired clawlike appendages in ventral abdominal intersegmental membrane 8–9; Oriental .... Megenicocephalinae
   - Minute, length under 2 mm, slender; coloration dull; pronotum in macropterous males with collar well delimited, middle and hind lobes differentiated by sinuation of lateral margin only; pronotum in apterous females formed of collar and a single lobe (Fig. 13.1B); eyes of females with 0–5 ommatidia (Fig. 13.1B); forelegs raptorial, with apicotibial projection and its armature differentiated (Fig. 13.1D); tarsal formula 1-1-1; male foretarsus without spines; forewing venation of males reduced to 2–3 longitudinal veins and 0–1 cross veins; antennae incrassate (Fig. 13.1B); female abdomen strongly sclerotized with sharply delimited sets of dorsal external and inner laterotergites (Fig. 13.1B); Neotropical .... Alienatinae
3. Upper and lower surfaces of fore and hind wings with numerous macrotrichia on veins and membranous areas (only macropterous males known); male genitalia formed by large bulbous closed projecting "phallandrium" with long tapering or subtruncate apex; parameres large, partly projecting out of the pygophore; Oriental .... Phallopiratinae
   - Most macrotrichia on upper surface of forewings occurring on veins and wing margin; lower surface of forewings and both surfaces of hind wings without macrotrichia; macropterous to apterous; male genitalia complex to simple, but not as above 4
4. Micropterous to apterous, no sexual pterygodimorphism; middle lobe of pronotum without paired, inversely Y-shaped impressions; male genitalia complex, never represented by a simple racquet-shaped guide, parameral sclerotizations, and a medial sclerite only; subantarctic islands (Crozet; Auckland), New Zealand, New Caledonia, New Guinea .... Phthirocorinae
   - Macropterous to apterous, sexual pterygodimorphism common; if micropterous to apterous, then either not occurring in SW Pacific area and on subantarctic islands, or with paired, inversely Y-shaped impressions or vestiges on middle lobe of pronotum; male genitalia reduced, usually formed by a usually racquet-shaped guide (Fig. 13.1E), paired parameral sclerotizations (rarely somewhat protruding), and a simple, plate-shaped supradistal lobe, the last sometimes prominent, arcuate or spiniform; worldwide .... Enicocephalinae

ALIENATINAE (FIG. 13.1B). Males with posterior pronotal lobe virtually absent; forewing venation strongly reduced; females apterous (Fig. 13.1B), largely desclerotized; middle and hind tarsi 1-segmented.

This small subfamily comprises only the speciose genus *Alienates* Barber, discovered originally in the Bahama Islands, but widely distributed in the Caribbean, Mesoamerica, and northern South America. These tiny bugs superficially resemble small midges. All species appear to live in soil.

ENICOCEPHALINAE (FIG. 13.1A). Male genitalia greatly simplified (Fig. 13.1E); females sometimes micropterous to apterous; micropterous and apterous Old World enicocephalines restricted to *Oncylocotis* Stål; some taxa with caducous wings, for example, *Nesenicocephalus* Usinger.

This is the most speciose group in the family, with two tribes recognized. At present 26 genera and about 250 species have been described. Most species occur in the tropics and subtropics, but the range of the subfamily extends south to southern Chile and Tasmania and north to

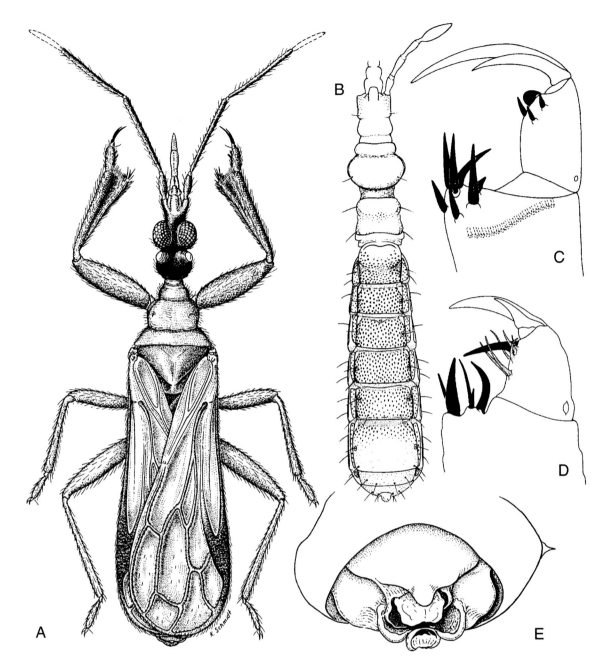

Fig. 13.1. Enicocephalidae (from Wygodzinsky and Schmidt, 1991). **A.** *Hymenocoris brunneocephalis* Wygodzinsky and Schmidt. **B.** *Alienates elongatus* Wygodzinsky and Schmidt. **C.** Foreleg, distal portion, generalized Enicocephalinae. **D.** Foreleg, generalized Alienatinae. **E.** Male genitalia, *Urnacephala californica* Wygodzinsky and Schmidt.

the northern United States and the southern fringes of the Palearctic, Japan being the northernmost part of the range in the Old World.

The most speciose genera within the Enicocephalini are the New World *Neoncylocotis* Wygodzinsky and Schmidt and *Enicocephalus* Westwood and the Old World *Henschiella* Horvath, *Hoplitocoris* Jeannel, *Stenopirates* Walker, and particularly *Oncylocotis* Stål. The last genus includes some micropterous to apterous species or morphs, some of them truly neotenous; an undescribed species from New Caledonia has minute club-shaped forewings similar to dipteran halteres. The genus *Cocles* Bergroth (Madagascar; Seychelles Islands) is holoptic, the eyes occupying most of the head and obscuring the usual distinction between anterior and posterior cephalic lobes.

The Systelloderini comprise only the cosmopolitan genus *Systelloderes* Blanchard, but the group is well characterized only for the New World taxa, for which many of species await description. Some of its species retain plesiomorphic genitalia rather resembling those of the Phthirocorini. It is possible that some of the Old World genera of Enicocephalini actually belong in the Systelloderini.

MEGENICOCEPHALINAE. Large, broad bodied, mostly reddish; costal margin of forewing with a small break resembling an incipient costal fracture; pronotum formed of 2 lobes, resembling superficially the condition in Aenictopecheinae, but collar undifferentiated, and middle and hind lobes normally developed; forelegs strongly developed, essentially cursorial, lacking usual apicotibial armature, but often with variously tuberculate to spinose tibia, femur, and trochanter (sexually dimorphic).

This monotypic subfamily contains only the Oriental genus *Megenicocephalus* (Malay Peninsula; Indonesia).

PHALLOPIRATINAE. Forewings and hindwings generally covered by macrotrichia on both the upper and lower surfaces; foretibia in males compressed in two different planes; male genitalia represented by a bulbous "phallandrium" (Štys, 1985b) produced into a spine, resembling a scorpion telson and including paired elements homologous to genital plates of other Heteroptera.

Represented only by *Phallopirates* Štys with four species, this Oriental group ranges from Malaya and Borneo to the Philippine Islands. Males have been collected at light, but females are unknown and probably nonflying. Štys (1985b) found no gonopore in the males and suggested that "androtraumatic" insemination takes place, the male having to break off the apex of the phallandrium in order to pass the spermatophore to the female.

PHTHIROCORINAE. Apterous or micropterous; male genitalia structurally diverse, sometimes greatly reduced into an enicocephaline-like condition.

Two tribes are recognized from the southwest Pacific and subantarctic islands.

The two genera and three species of Monteithostolini, all from New Caledonia, are apterous. *Monteithostolus genitalis* Štys has the pygophore subdivided into a distinct tergum, laterotergites, and sternum (Štys, 1982b).

The Phthirocorini comprises two genera and four species from New Guinea, New Zealand, Auckland Islands, and Crozet Island. *Phthirocoris subantarcticus* Enderlein occurs on Crozet Island in penguin and albatross nests.

Specialized morphology. The division of the pronotum into 3 apparent lobes, absence of the costal fracture, and the greatly modified and often reduced male and female genitalia are distinctive for the Enicocephalidae.

Natural history. Most species of Enicocephalidae are found in leaf litter or loose soil, in moss, rotting wood, under bark, and in similar microhabitats, usually in the humid tropics or subtropics; species from arid zones and temperate regions probably live mostly in soil crevices. All enicocephalids are undoubtedly generalized predators, but few observations are available. The most thorough review of the biology of the group is that of Wygodzinsky and Schmidt (1991). Hickman and Hickman (1981) studied the life history of *Oncylocotis tasmanicus* (Westwood) and described the nymphal stages.

Adults can be collected by use of Berlese funnels, pitfall traps, or lights. All taxa in which both sexes are capable of flight form nuptial swarms, which can be diurnal or crepuscular. These swarms are mixed or unisexual (Štys, 1981b), but the details of mating habits are unknown. The method of copulation has not been described for any species. The swarms may be formed by an extremely high number of individuals and with a strongly skewed sexual ratio (*Systelloderes*, *Oncylocotis*, *Stenopirates*).

Distribution and faunistics. New World taxa, with the exception of *Systelloderes* spp., can be identified using Wygodzinsky and Schmidt (1991). The Palearctic was treated by Štys (1970b). Villiers (1958, 1969a) monographed the Malagasy and African and Malagasy faunas, respectively. For other areas in the Old World the works of Štys cited above are essential.

EUHETEROPTERA

14
Dipsocoromorpha

by Pavel Štys

General. The Dipsocoromorpha, which have no common name, include the smallest true bugs. They range in length from about 0.5 to 4.0 mm and range in appearance from tiny beetles to small, dull-colored, free-living members of the Cimicoidea.

Diagnosis. Minute, 4 mm or less; head often strongly declivous (Fig. 18.1B); antennae flagelliform, with very short segments 1 and 2, segment 2 at most twice as long as 1, segments 3 and 4 usually very long, much thinner than segments 1 and 2, and with many long, thin, semierect to erect setae (except Stemmocryptidae; Fig. 19.1A); pronotum variable, with 1 or 2 lobes; proepisternum often inflated and produced below ventral margin of eyes, sometimes reduced; forewing usually tegminal (Figs. 15.1A, 17.1, 18.1A, 19.1B); hind wing usually with several deeply incised lobes; pretarsus with 2 equally developed claws, pair of setiform parempodia, and usually on at least one pair of legs with a dorsal and ventral arolium, the arolia sometimes very large; number of abdominal spiracles often reduced to 3–6; male subgenital plate, when present, represented by sternum 7; male genitalia, including pregenitalic abdominal segments, usually (except some Ceratocombidae) asymmetrical and extremely complex (Figs. 15.1D, 18.1D), some laterotergites appendage-like (Fig. 18.1E); bases of parameres directly associated with articulatory apparatus of phallus; nymphs with dorsal abdominal scent-gland openings on as many as 4 segments.

Classification. The most recent classification (Štys, 1970a, 1983a) includes five families: Ceratocombidae, Dipsocoridae, Hypsipterygidae, Schizopteridae, and Stemmocryptidae. Hypsipterygids and stemmocryptids were discovered only relatively recently (Drake, 1961; Štys, 1983a; respectively). Formerly the names Dipsocoridae or Cryptostemmatidae were used either for all the dipsocoromorphans or for Dipsocoridae and Ceratocombidae (e.g., China and Miller, 1959) or for Dipsocoridae, Ceratocombidae, and Hypsipterygidae (Emsley, 1969).

Dipsocoromorphans have been regarded as a group of uncertain affinity or considered jointly with enicocephalomorphans in a group called either Dipsocoromorpha (Miyamoto, 1961a) or Enicocephalomorpha (Popov, 1971). Nonetheless, early on, Reuter (1910, 1912a) and later Štys (1970a) used the series name Trichotelocera for the group, setting these unusual bugs aside from other terrestrial Heteroptera. The usage of Dipsocoromorpha in the present sense was stabilized by Štys and Kerzhner (1975).

The authors who have contributed most to knowledge of Dipsocoromorpha are Reuter, McAtee and Malloch, Wygodzinsky, Emsley, Štys, Linnavuori, Kerzhner, Josifov, Miyamoto, Hill, and Ren.

Reuter (1891) provided the first monograph of the group. Available taxonomic information was summarized, with particular reference to the American fauna, by McAtee and Malloch (1925), who also provided a key to the described genera. In a series of papers on the South American and African faunas Wygodzinsky set a high standard for describing and illustrating structures in the group, particularly the genitalia, and his papers on the fauna of Angola are still essential references (Wygodzinsky, 1950, 1953). Emsley (1969) monographed the schizopterid fauna of Trinidad and provided at the same time an excellent morphological overview of the Schizopteridae and, to a lesser extent, of the whole infraorder. Štys (1970a), independent of Emsley, established a higher classification and surveyed the morphology of the infraorder, in particular the complex morphology of the male abdomen and terminalia. Later, in association with his description of the family Stemmocryptidae, Štys (1983a) provided an account of the comparative morphology of the group.

Key to Families of Dipsocoromorpha

1. Antennae not conspicuously flagelliform, segment 1 rather short, segments 2–4 moderately long, subequal in length, 3 and 4 only slightly thinner than 2, without strikingly long erect setae (Fig. 19.1A); ocelli situated behind eyes; macropterous; forewing tegminal, distal part of claval suture dis-

tant from wing margin, costal fracture long (Fig. 19.1B); male genitalia asymmetrical (Fig. 19.1D); New Guinea .... Stemmocryptidae

– Antennae flagelliform, segments 1 and 2 short and thick, segment 2 at most twice as long as segment 1, segments 3 and 4 strikingly longer and much thinner than 1 and 2, provided with many long, thin, erect to semierect setae; ocelli, if present, interocular; forewing tegminal, hemelytraceous, elytraceous, or reduced; distal part of claval suture usually meeting or closely approaching wing margin, costal fracture usually short or absent (Fig. 15.1A, B) 2

2. Pronotum with 3 longitudinal carinae meeting basally in a well-defined marginal ambient ridge; general facies tingidlike (Fig. 17.1); forewing without costal fracture, with remigial veins forming a system of about 10 large rectangular to polygonal cells (Figs. 16.1E, 17.1); labium straight, thin, segments 1, 2, and 4 minute, 3 strikingly long; head porrect, with distinctly delimited cephalic regions, including frons; macropterous to submacropterous; male genitalia asymmetrical; African and Oriental .... Hypsipterygidae

– Pronotum without longitudinal ridges and without distinct marginal ridge, general facies seldom tingidlike (Figs. 15.1A, 18.1A); forewing not as above; labium not as above, number of segments sometimes reduced; head declivous to porrect, frons never delimited; male genitalia symmetrical or asymmetrical 3

3. Proepisternal lobe narrow in lateral view, not inflated and not extending cephalad; articulation of forecoxa laterally exposed, supracoxal cleft extremely short to hardly indicated, proepimeron large; costal fracture present, except in extreme brachypterous forms 4

– Proepisternal lobe broad in lateral view, mostly inflated and extending below eye; articulation of forecoxa and basal part of latter covered, supracoxal cleft long 5

4. Costal fracture short, just interrupting forewing margin; metapleuron without evaporatorium; male genitalia and abdomen symmetrical or asymmetrical; forewing tegminal to hemelytraceous; macropterous to brachypterous, or rarely coleopteriform .... Ceratocombidae

– Costal fracture reaching to about middle of width of forewing (Fig. 16.1D); metapleuron with evaporatorium; male genitalia and abdomen asymmetrical; forewing tegminal; macropterous to brachypterous .... Dipsocoridae

5. Coleopteriform, forewings elytraceous; proepisternal lobe large, but not inflated; head porrect; male genitalia symmetrical (Fig. 15.1D); foretarsus with ventral spines (male *Feshina* Štys), or head declivous and pronotum strikingly transverse (female *Kvamula* Štys); hind coxae without adhesive pads .... Ceratocombidae (part)

– Macropterous to brachypterous, rarely micropterous, often coleopteriform, forewing tegminal to mostly hemelytraceous, often elytraceous, rarely micropterous; proepisternal lobe usually strongly inflated, and extending below eyes, and head strongly declivous; male genitalia, often including abdomen, always asymmetrical (Fig. 18.1C–E); hind coxae with adhesive pads mesoventrally; costal fracture usually absent, or interrupting costal margin only, long and transversing the remigium in *Guapinannus* Wygodzinsky .... Schizopteridae

15
Ceratocombidae

by Pavel Štys

General. These little-known bugs, which range in length from 1.5 to 3.0 mm, are mostly dull-colored, ranging from yellowish to dark brown. Ceratocombini and Trichotonanninae often resemble small long-legged drymine lygaeids with more or less tegminal forewings and flagelliform antennae; Issidomimini are more similar in appearance to small anthocorids.

Diagnosis. Antennae flagelliform, second segment 1.5–2.5 times as long as segment 1; labium ranging from thin and elongate to short and thick in *Feshina* Štys; pronotum ecarinate; propleuron with reduced proepisternum and exposed procoxal articulation except in rare coleopteroid morphs; forewing in macropterous forms always with distinct, but very short, costal fracture just interrupting marginal vein (Fig. 15.1B), and 2–4 large cells distally; tarsal segmentation often sexually dimorphic, with nearly all possible combinations of 2 or 3 segments

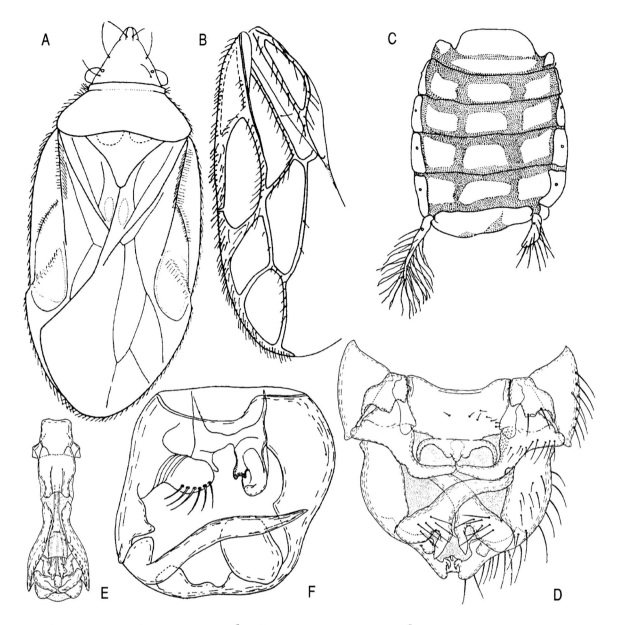

Fig. 15.1. Ceratocombidae. **A.** *Ceratocombus mareki* Štys. **B.** Forewing, *C. mareki* (A, B from Štys, 1977). **C.** Male abdomen, *Trichotonannus dundo* Wygodzinsky. **D.** Male genitalia, *Muatianvuaia barrosmachadoi* Wygodzinsky. **E.** Aedeagus, *M. barrosmachadoi*. **F.** Male genitalia, *T. dundo* (C–F from Wygodzinsky, 1953).

for all pairs of legs; spiracles present on abdominal segments 2–8 or 3–8, usually dorsal; male genitalia sometimes symmetrical (Fig. 15.1D), but laterotergites 9 secondarily associated with tergum 8 and appendage-like, unlike all other Heteroptera; ovipositor well developed; spermatheca always present (Fig. 16.1B, C).

Classification. Štys (1970a) first recognized the uniqueness of the Ceratocombidae, raising the group to family level and establishing a suprageneric classification (modified by Štys [1982a, 1983a]). Štys (1983a) provided updated keys to genera and higher taxa. Previously, members of the group were usually regarded as belonging to the Dipsocoridae (= Cryptostemmatidae) or Dipsocorinae (= Cryptostemmatinae) in a broadly defined Dipsocoridae. Two subfamilies are currently recognized, containing approximately 8 genera and 50 species.

Key to Subfamilies of Ceratocombidae, Based on Males

1. Abdominal segments 7–9 and parameres asymmetrical; laterotergites 7–9 asymmetrical, 7 and 8 appendage-like, 9 transformed into processes of pygophore; head, pronotum, and legs with long, erect, black bristles; eyes with a central bristle; distal part of forewing in macropterous forms with 4 large cells; macropterous to brachypterous; Paleotropical .... Trichotonanninae
– Abdominal segments 7–9 and their laterotergites symmetrical; parameres symmetrical or asymmetrical; laterotergite 7 not appendage-like, 8 variably developed (from appendage-like to tubercle-shaped to undifferentiated), 9 appendage-like and secondarily articulating with tergum 8; if bristles different from other setae, then concolorous and not strikingly strong and long; eyes without a strong central bristle; distal part of macropterous forewing with 2–3 large cells (rarely with additional small ones); macropterous, brachypterous, or coleopteriform .... Ceratocombinae

CERATOCOMBINAE. Macropterous, forewing with only 2–3 large cells distally, in contrast to Trichotonanninae; body devoid of strong bristles; no central eye bristle; male abdomen and laterotergites 7–9 always symmetrical; laterotergite 7 not appendage-like, 8 appendage-like or not, 9 always appendage-like; parameres symmetrical or asymmetrical.

Two tribes are recognized. The Ceratocombini includes the cosmopolitan *Ceratocombus* Signoret, with 3 subgenera and 25 valid species and literally hundreds of undescribed species, particularly from the Indo-Pacific. Many species are pterygopolymorphic (macropterous to strongly brachypterous) and are usually dull-colored (yellowish to dark brown), though some Madagascan, Oriental, and Australian species have distinct color patterns. The genus *Leptonannus* Reuter is known from three macropterous Nearctic, Neotropical, and Afrotropical species. *Feshina schmitzi* Štys from Zaire is coleopteroid, the male resembling a hydraenid beetle; its foretarsus bears an armature of spines similar to that found in females of *Trichotonannus* Reuter.

The Issidomimini includes three genera, all from the Eastern Hemisphere: *Issidomimus* Poppius, one Oriental and one Papuan species; *Kvamula* Štys, four species from Vietnam; and *Muatianvuaia* Wygodzinsky, five Afrotropical species. Numerous Indo-Pacific species remain undescribed. Most Issidomimini are macropterous, but the female of *Kvamula coccinelloides* Štys is coleopteroid and resembles a scymnine coccinellid beetle.

TRICHOTONANNINAE. Superficially similar to robust and strongly setose *Ceratocombus*; conspicuous central eye bristle; macropterous to moderately brachypterous (pterygopolymorphic); forewing with 4 large cells distally; setae on legs; foretarsi sexually dimorphic, 3-segmented in male, 2-segmented in female, with armature of spines on lower surface; metathoracic scent glands absent; male abdomen asymmetrical, laterotergites 7 and 8 appendage-like; pygophore, proctiger, and parameres strongly asymmetrical and remnants of laterotergites 9 developed as pygophoral processes, but never appendage-like.

This subfamily comprises only the genus *Trichotonannus* Reuter, with six species from the Afrotropical, Madagascan, and insular Oriental regions (Philippines, Nicobar Island). Although the general facies resembles that of Ceratocombini, the structure of the male abdomen is different. *Trichotonannus oidipos* Štys from Madagascar has the eyes composed of only a few isolated ommatidia.

Specialized morphology. Ceratocombids are unique among the Dipsocoromorpha in that all Ceratocombini and some Issidomimini retain a symmetrical male abdomen and genitalia. Although some ceratocombids are macropterous, most species are pterygopolymorphic, some possibly always brachypterous. Males and females of Issidomimini are usually macropterous.

Natural history. Most species live in moderately humid leaf litter, decaying wood, mosses, sphagnum, and similar habitats with interstitial spaces. *Ceratocombus* species in Europe are typically found in bracken leaf litter mixed with needles in coniferous woods, drying sphagnum mixed with leaf litter and needles in wet woods, and reed litter. Ceratocombids can best be collected in pitfall traps, at light, and by berlesating litter.

All ceratocombids are probably generalized predators on small arthropods. The labium in *Ceratocombus* suggests a searching predator, while that of some other genera suggest active prey hunting or totally different habits, such as feeding on molds. *Ceratocombus*, *Leptonannus*, and *Trichotonannus* are fast runners but do not jump.

Distribution and faunistics. The family is cosmopolitan, but its diversity is poor in cold and temperate zones, though some species (e.g., *Ceratocombus corticalis* Reuter in the Palearctic) are restricted in cold regions. Maximum diversity is reached in tropical areas, and the number of undescribed species is enormous.

Palearctic *Ceratocombus* species were revised by

Kerzhner (1974). Useful information about the Afrotropical and Madagascan faunas was provided by Wygodzinsky (1953), Linnavuori (1974), and Štys (1977, 1983b). American species were treated and keyed by McAtee and Malloch (1925).

Outdated global treatments of species are available by Reuter (1891) and McAtee and Malloch (1925). Keys to world genera were provided by Štys (1982a, 1983a). Data on morphology and biology are contained in papers by Emsley (1969), Hill (1980), Štys (1958, 1959, 1970a, 1977, 1982a, 1983b), and Wygodzinsky (1953).

16
Dipsocoridae

by Pavel Štys

General. These are usually somewhat flattened, elongate bugs that never exceed 3 mm in length. They have the general habitus of many free-living Cimicoidea.

Diagnosis. Head porrect; labium thick, not exceeding procoxae, 4-segmented, first segment well-developed, other segments subequal in length; pronotum without well defined collar; proepisternum reduced, procoxal articulation exposed; scutellum large; macropterous to moderately brachypterous; forewings tegminal, membrane defined only by absence of veins, claval angle and claval commissure not formed; deep costal fracture almost reaching M (Fig. 16.1D); legs rather thick, tarsal formula usually 3-3-3 or in females of some species 2-2-3, possibly also 2-2-2 (Emsley, 1969); metapleural scent-gland evaporatorium present, in common with Stemmocryptidae; abdominal venter with dense pilosity; mediotergites largely desclerotized, a single (or double) set of laterotergites bearing spiracles ventrally, spiracles developed on segments 3–8, 4–8, or 4–7, the anterior ones probably absent. Nymphs with scent-gland orifices between terga 3/4, 4/5, 5/6, and 6/7; male abdomen always strongly asymmetrical, sinistral in contrast to other Dipsocoromorpha, asymmetry affecting in extreme cases segments 2–9, sometimes some of the pregenital segments symmetrical; one or more of laterotergites 3 left, 6 left, 7 right, and 8 right sometimes appendage-like; parameres asymmetrical; phallus long, sometimes with coiled processus gonopori; female subgenital plate 7 terminating in medial complex laminate structure; ovipositor developed, valvulae with microtrichial armature resembling filtering device.

Classification. The Dipsocoridae as here restricted comprise the genera *Cryptostemma* Herrich-Schaeffer and *Pachycoleus* Fieber (Štys, 1970a). The family name Dipsocoridae Dohrn, 1859, is based on a junior synonym, *Dipsocoris* Haliday, and was later replaced by Cryptostemmatidae McAtee and Malloch, 1925. Nonetheless, the former name has won wide acceptance and is now in general use (Štys, 1970a).

The family was covered in now outdated monographs by Reuter (1891) and McAtee and Malloch (1925). More recent summaries are those of Emsley (1969) and Štys (1970a).

Two genera are recognized. *Cryptostemma* Herrich-Schaeffer, with 3 subgenera and 26 described and many undescribed species, is known from all zoogeographical regions except New Zealand. *Pachycoleus* Fieber (often regarded as a subgenus of *Cryptostemma*) comprises four Palearctic and many undescribed Neotropical species.

Specialized morphology. The presence of a metathoracic scent-gland evaporatory area and the sinistral asymmetry of the male genitalia are both distinctive for the family, although the former also occurs in the Stemmocryptidae.

Natural history. *Cryptostemma* species inhabit a moist zone within gravel-sand banks of clear streams and rarely lakes, and may be found under stones on such a substrate. When disturbed, the bugs immediately exhibit an escape reaction, either creeping into the ground or simultaneously jumping and flying. Habitats of *Cryptostemma* are subject to periodic flooding. The bugs survive either in trapped air bubbles or by using plastron respiration. Dipsocorids are undoubtedly generalized predators of small arthropods and have been observed to feed on dead mayflies.

Pachycoleus spp. in the Palearctic inhabit wet moss and *Sphagnum* on the banks of ponds and lakes and in fens and bogs. The typical microhabitat is a small mound of *Sphagnum* and other mosses formed around tussocks of hygrophilous grasses, sedges, and other plants.

All macropterous dipsocorids may be attracted to lights.

Distribution and faunistics. Regional revisions have been provided by Josifov (1967) for Palearctic and Hill (1987) for Australia, but particularly the latter is not exhaustive since numerous species remain to be described. Systematic collecting in suitable habitats will increase the number of known species many times.

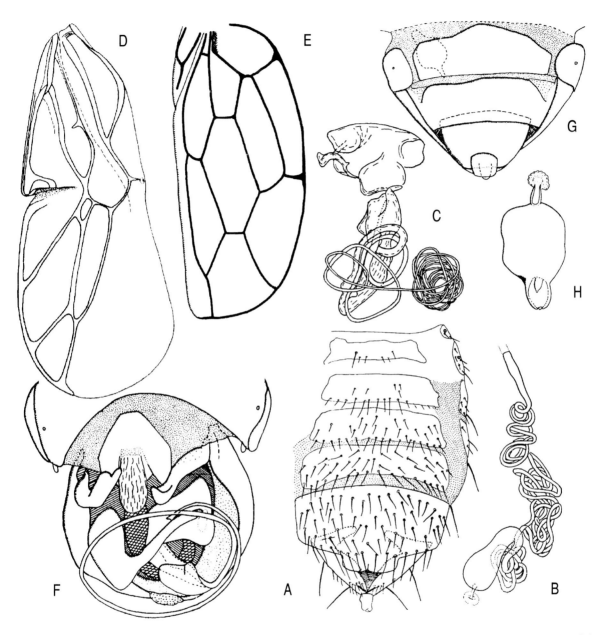

Fig. 16.1. Ceratocombidae. **A.** Female abdomen, *Trichotonannus dundo*. **B.** Spermatheca, *T. dundo*. **C.** Spermatheca, *T. oidipos* (A–C from Wygodzinsky, 1953). Dipsocoridae. **D.** Forewing, *Cryptostemma incurvatum* Štys (from Štys, 1977). Hypsipterygidae. **E.** Forewing, *Hypsipteryx machadoi* Drake. **F.** Male genitalia, *H. machadoi*. **G.** Female genitalia, *H. machadoi*. **H.** Spermatheca, *H. ungadensis* Štys (E–H from Štys, 1970a).

17
Hypsipterygidae

by Pavel Štys

General. Members of this group bear a striking resemblance to many Tingidae. They are rather flattened dorsally and are 2–3 mm long.

Diagnosis. Head porrect; all regions of head capsule sharply delimited; antennal segment 4 relatively short and thick, segment 3 very long (Fig. 17.1); labium straight, reaching or almost reaching base of metasternum, segments 1, 2, and 4 distinct, very short, segment 3 very long; pronotum with expanded lateral margins and 3 longitudinal carinae (Fig. 17.1); forewing broadly explanate, tegminal, clavus sharply delimited, no costal fracture, no medial fracture, all veins raised, forming 9 large cells, entire wing finely areolate; legs strikingly slender; tarsi 2-segmented in both sexes; most abdominal sterna divided into mediotergites and laterotergites; spiracles situated on laterotergites on segments 2–8 in males, on 3–7 in female; male pregenital abdomen symmetrical, pygophore, parameres, and laterotergites 9 asymmetrical, the last appendage-like; phallus complex, with long, coiled processus gonopori (Fig. 16.1F); female with large subgenital plate 7; all components of ovipositor present, modified; spermatheca present (Fig. 16.1H).

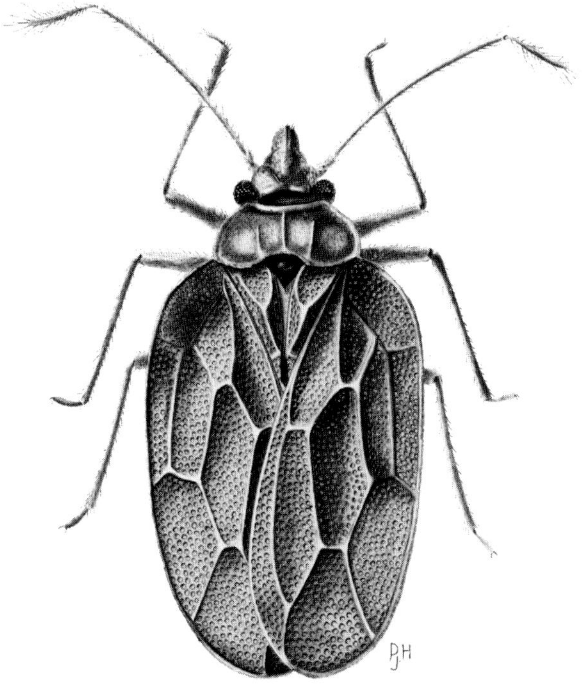

Classification. The family was described by Drake (1961) as a subfamily Hypsipteryxinae [*sic*] of Dipsocoridae, and later it was raised to family rank by Štys (1970a). A single genus *Hypsipteryx* Drake includes *H. ecpaglus* Drake from Thailand, *H. machadoi* Drake from Angola, and *H. ugandensis* Štys from Uganda.

Specialized morphology. The tingidlike appearance, thinness of the legs, and straightness of the labium are unique to the group.

Natural history. *Hypsipteryx ecpaglus*, with fully developed fore- and hind wings, was collected at light. All other *Hypsipteryx* specimens from Africa have the forewings slightly abbreviated and the hind wings minute, laceolate, and nonfunctional, and were collected from leaf litter or decaying wood. The slender legs suggest that *Hypsipteryx* species do not jump. Unlike other dipsocoromorphs from the Paleotropics, hypsipterygids are virtually absent from Berlese samples, suggesting they may have somewhat different life habits.

Distribution and faunistics. This little-known Paleotropical group is treated by Drake (1961) and Štys (1970a).

Fig. 17.1. Hypsipterygidae. *Hypsipteryx ectpaglus* Drake (from Drake, 1961).

18
Schizopteridae

by Pavel Štys

General. These cryptic bugs are sometimes remarkably similar to certain beetles and members of the Omaniidae, owing to their often near-black coloration and uniformly sclerotized, coleopteroid forewings. They range in length from 0.8 to 2.00 mm.

Diagnosis. Small, compact, rotund; integument with pits, tubercles, a dense layer of microtrichia, or other ornamentation; body usually lacking conspicuous elongate vestiture common in most other Dipsocoromorpha; head usually strongly declivent (Fig. 18.1B); compound eyes varying from relatively small to very large, sometimes extending posteriorly along nearly entire lateral margin of pronotum; ocelli present or absent in adults, either proximal to or distant from compound eyes; antennal segments 1 and 2 very short, subequal in length; labium with 3 or 4 segments, varying from short

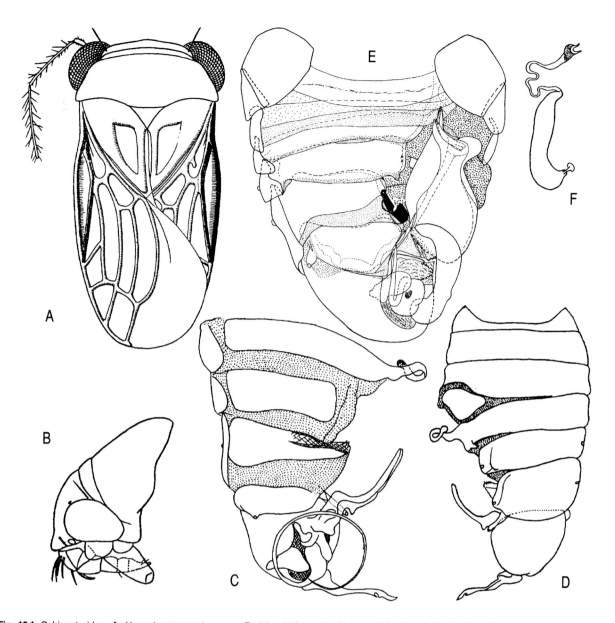

Fig. 18.1. Schizopteridae. **A.** *Hypselosoma matsumurae* Esaki and Miyamoto. **B.** Lateral view head, *H. hirashimai* Esaki and Miyamoto (A, B from Esaki and Miyamoto, 1959c). **C.** Male abdomen, dorsal view, *Machadonannus ocellatus* Wygodzinsky. **D.** Male abdomen, lateral view, *M. ocellatus* (C, D from Wygodzinsky, 1950). **E.** Male genitalia, *Semangananus mirus* Štys (from Štys, 1974). **F.** Spermatheca, *Seabranannus immitator* Wygodzinsky (from Wygodzinsky, 1950).

and not extending beyond forecoxae to very long and reaching well onto abdomen; proepisternal lobe inflated, extending anteriorly to cover posteroventral surface of head; metasternum often with a median spine or pair of V-shaped processes, sometimes extending onto abdomen; hind coxa with adhesive pad on mesal margin, apparently facilitating jumping; macropterous to strongly coleopteroid, forewings tegminal and nearly uniformly sclerotized, medial and costal fractures usually lacking; hind wings developed or not in nonmacropterous forms; nymphal dorsal abdominal scent gland between terga 6/7; male pregenital abdominal segments sometimes all symmetrical, more commonly segments 6–8 showing some modifications on the right-hand side (Fig. 18.1C, D); male genitalia, including parameres, usually strongly asymmetrical (Fig. 18.1C, E); ovipositor developed or not; spermatheca present (Fig. 18.1F).

Classification. Reuter (1891) first accorded the taxon higher-group status, Schizopterina, as a subdivision of the Ceratocombidae. Reuter (1910) treated the group as a family, a designation that has been followed by all subsequent authors. The Schizopteridae, as treated herein, are divided into two subfamilies. Although Emsley (1969), in his monographic work on the family, provided clear diagnoses for his new subfamily Ogeriinae and the Schizopterinae, many genera in his scheme were left incertae sedis and have not yet been placed. At least 35 genera and about 120 species have been described.

Key to Subfamilies of Schizopteridae

1. Eyes exceedingly large, broadly overlapping anterolateral margins of pronotum (Fig. 18.1A); ovipositor well developed, with first and second valvulae retained .... Hypselosomatinae
– Eyes of moderate size, overlapping at most anterolateral angles of pronotum; ovipositor reduced to a varying extent, sometimes completely absent .... Schizopterinae

HYPSELOSOMATINAE. Abdominal spiracles present on segments 2–8; ovipositor well developed.

Three genera were originally included, *Glyptocombus* Heidemann and *Ommatides* Uhler from the New World and *Hypselosoma* Reuter, primarily from the Old World tropics. Hill (1980, 1984) described seven additional genera from the Australian mainland and Tasmania.

SCHIZOPTERINAE. Abdominal spiracles usually present on sterna 6 and 7 and tergum 8, but other patterns known; ovipositor absent or poorly developed.

The remaining genera of Schizopteridae are placed in this subfamily. *Schizoptera* Fieber, with several recognized subgenera, is by far the largest genus. The group is distributed worldwide, excluding the Palearctic.

Specialized morphology. The typically globular bodies, relatively short first and second antennal segments, metasternal spine, and adhesive pads of hind coxae are all characteristics distinctive of the Schizopteridae. As with most other dipsocoromorphs, the male genitalia show strong dextral asymmetry. The male pregenital abdominal segments are often strongly modified, apparently forming secondary genitalia in *Semangananus mirus* Štys.

Natural history. Emsley (1969), in the most complete account of schizopterid biology, showed that many members of the group are ground- and litter-dwelling. Yet, many taxa have never been found by collecting in these habitats but only by collecting at lights. Their attraction to lights suggests broad-scale occupation of arboreal habitats. Schizopterids habitually jump and fly (winged morphs) when disturbed rather than run, as do other dipsocoromorphs. It appears that they develop only one egg at a time, which occupies nearly the entire abdomen.

Hill (1980, 1984) aptly summarized information for the Tasmanian and mainland Australian hypselosomatines, indicating that the microhabitats include leaf litter, moss, grass tussocks, sedge and rush sods, and grass and fern foliage. Known major habitat types include rain forests, other wet forests, tropical palm swamps, elevated bogs, and sod tussock grasslands.

Distribution and faunistics. As do all other dipsocoromorphan groups, the Schizopteridae show by far their greatest diversity in the tropics, although the group is by no means restricted to the strictly tropical latitudes, having been taken in Tasmania and as far north as southern Michigan in the United States. Many taxa remain to be described from the Indo-Pacific.

The major comprehensive works on the group are those of McAtee and Malloch (1925), with keys to genera and emphasis on the North American species, and Emsley (1969), which includes a catalog to the world fauna and details on the fauna of Trinidad, with an excellent treatment of morphology for the group as a whole. Much of the remaining descriptive work has been done by Štys and Wygodzinsky, some of whose works are cited elsewhere in the text. Hill (e.g., 1980, 1984, 1990) has made substantial contributions to the literature on the Australian fauna.

19
Stemmocryptidae

by Pavel Štys

General. Elongate, soft-bodied, light brown, 2.0–2.4 mm long, these obscure insects vaguely resemble some flattened, lightly sclerotized, free-living Cimicoidea.

Diagnosis. Macropterous; no conspicuous sexual dimorphism; head porrect; ocelli postocular, adjoining hind margins of eyes; gula long; antennal segment 2 subequal in length to segments 3 and 4, the latter 2 only slightly thinner and devoid of long, thin, erect setae; labium about as long as head, directed forward (Fig. 19.1A), segment 1 very short, segment 2 short, segments 3 and 4 longer; pronotum simple, subtrapezoidal, without dorsal collar; forewings membranous, venation reduced, medial fracture running in front of R (as in Enicocephalomorpha and Dipsocoridae), costal fracture long (Fig. 19.1D), clavus not delimited; forewings freely overlapping, not forming

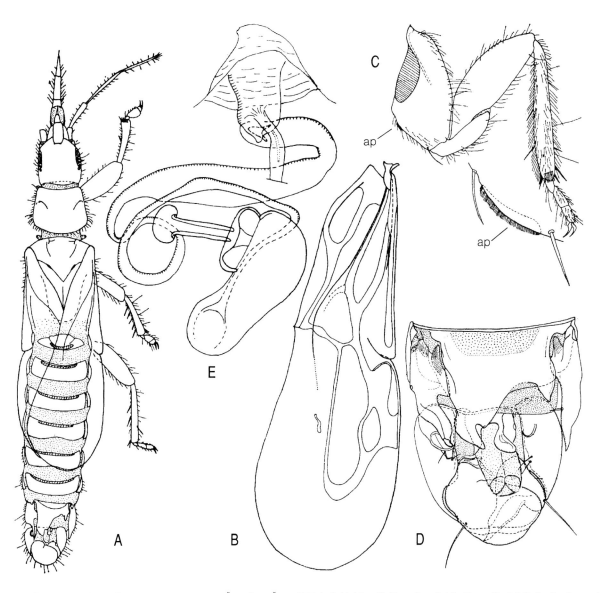

Fig. 19.1. Stemmocryptidae. *Stemmocrypta antennata* Štys (from Štys, 1983a). **A.** Habitus. **B.** Forewing. **C.** Hind leg, with detail of adhesive pad on coxa. **D.** Male genitalia. **E.** Spermatheca. Abbreviation: ap, adhesive pad.

claval commissure; legs short; tarsi swollen, tarsal formula male 3-3-3, female 2-3-3, pretarsus with 2 long setiform parempodia and a dorsal and ventral arolium; metathoracic scent gland unpaired, with single opening, evaporatorium present; abdominal spiracle 1 present in membrane posterior to metapostnotum, spiracles 2–7 on dorsal laterotergites; no remnants in adults of nymphal dorsal abdominal scent glands; male abdominal segment 8 asymmetrical, with appendage-like laterotergites (Fig. 19.1D); pygophore, laterotergites 9, and parameres asymmetrical; phallus complex; female sternum 7 forming subgenital plate; ovipositor plate-shaped, modified; spermatheca present (Fig. 19.1E).

Classification. This is the most recently discovered family of Heteroptera, being based on and including only *Stemmocrypta antennata* Štys, 1983a, from Laing Island, Papua New Guinea.

Specialized morphology. This peculiar group of insects combines characteristics of the Dipsocoromorpha, including the asymmetrical male genitalia, spermatheca, dorsal and ventral arolium, and tegminal forewings, with those of some Cimicoidea, including habitus and antennal structure.

Natural history. *Stemmocrypta antennata* was collected in UV light traps and by berlesating leaf litter. Stemmocryptids are probably searching predators, and their short thick legs suggest they do not jump.

Distribution and faunistics. Known only from New Guinea (Štys, 1983a).

NEOHETEROPTERA

20
Gerromorpha

General. The Gerromorpha, or semiaquatic bugs, have been recognized as a monophyletic group since the time of Dufour (1833), but until recently they generally were called the Amphibicorisae (or Amphibiocorisae). Most members have the ability to walk on the water surface film, and some, such as most Gerridae and many Veliidae, spend nearly the entire active period of their lives on the water surface, a few even living on the open ocean. All taxa are predaceous, feeding upon other insects or arthropods. Most of the following material, including the key, is derived from the monographic work of Andersen (1982a; see also Spence and Andersen, 1994).

The Gerromorpha, and particularly the Gerridae, have been the subject of study for their methods of locomotion (Miyamoto, 1955; Andersen, 1976), methods of ripple communication and prey location on the water surface film (Wilcox, 1972, 1979, 1980; Wilcox and Spence, 1986; J. T. Polhemus, 1990b), wing polymorphism and its determination (Brinkhurst, 1959, 1963; Andersen, 1973; Vepsäläinen, 1974a, b), and spatial competition and coexistence (e.g., Spence and Scudder, 1980; Spence, 1983).

Diagnosis. Head in most families elongate and often more or less cylindrical and protruding distinctly anterior to the eyes, head much shorter in Veliidae and Gerridae, appendages elongate and slender (Fig. 20.2B), and coloration somber. Characters that Andersen (1982a) listed as synapomorphic are, among others: 3 pairs (rarely 4) of cephalic trichobothria inserted in deep cuticular pits in the adult (Fig. 10.8A); epipharynx with a long, narrow, external projection; mandibular levers quadrangular; maxillary levers absent; rostral groove distinct and accepting labium at rest (Fig. 22.1B); labium elongate, segments 1 and 2 very short, segment 3 very long, much longer than segment 4 (Figs. 21.1B, 22.1B), except in Veliidae, Gerridae, and Hermatobatidae, in which labium generally much shorter, segment 3 usually not conspicuously longer than remaining segments, and rostral groove not developed; forewings usually not differentiated into anterior coriaceous and posterior membranous portions (Figs. 20.2B, 21.1C); pretarsus with dorsal and ventral arolium (Fig. 10.5C, D), ventral arolium sometimes highly modified; peg plates (sieve pores) generally distributed on body surface (Fig. 10.7C), except in Gerridae and Hermatobatidae; 2-layered hair pile usually covering head, thorax, and basal segments of the abdomen, consisting of short, densely placed microtrichia, and longer, less densely set macrotrichia; female gynatrial complex with long, tubular, entirely glandular spermatheca and secondary fecundation canal (Figs. 21.2K, 26.2F).

Pterygopolymorphism is common, it sometimes shows continuous variation, or there may be only macropterous and/or micropterous or apterous morphs. Numerous species and many suprageneric taxa are known only from the apterous morph. The details of the structure and function of the specialized body hair layers found in all Gerromorpha have been studied by Andersen (1977b).

Discussion. General treatments include *The Semiaquatic Bugs* (Andersen, 1982a), which is the most comprehensive reference available, including sections on morphology, classification, phylogeny, biogeography, functional adaptations, list of genera and suprageneric taxa, and keys for their identification. Andersen's cladogram of family relationships is shown in Figure 20.1. *The Semiaquatic and Aquatic Hemiptera of California* (Menke, 1979a) is a useful reference for North American

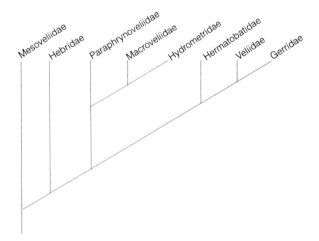

Fig. 20.1. Phylogenetic relationships of families of Gerromorpha (after Andersen, 1982a).

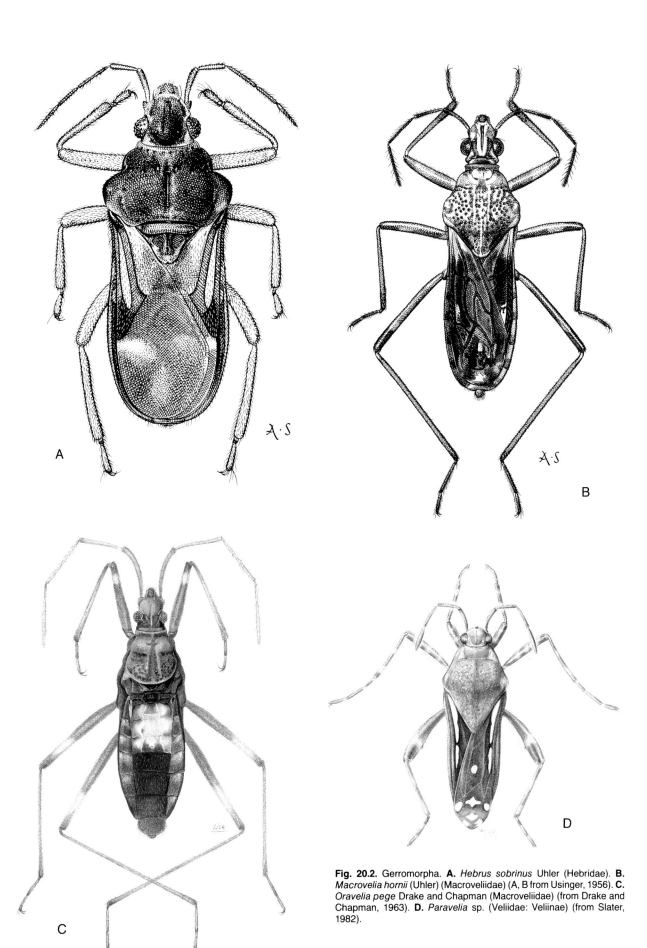

Fig. 20.2. Gerromorpha. **A.** *Hebrus sobrinus* Uhler (Hebridae). **B.** *Macrovelia hornii* (Uhler) (Macroveliidae) (A, B from Usinger, 1956). **C.** *Oravelia pege* Drake and Chapman (Macroveliidae) (from Drake and Chapman, 1963). **D.** *Paravelia* sp. (Veliidae: Veliinae) (from Slater, 1982).

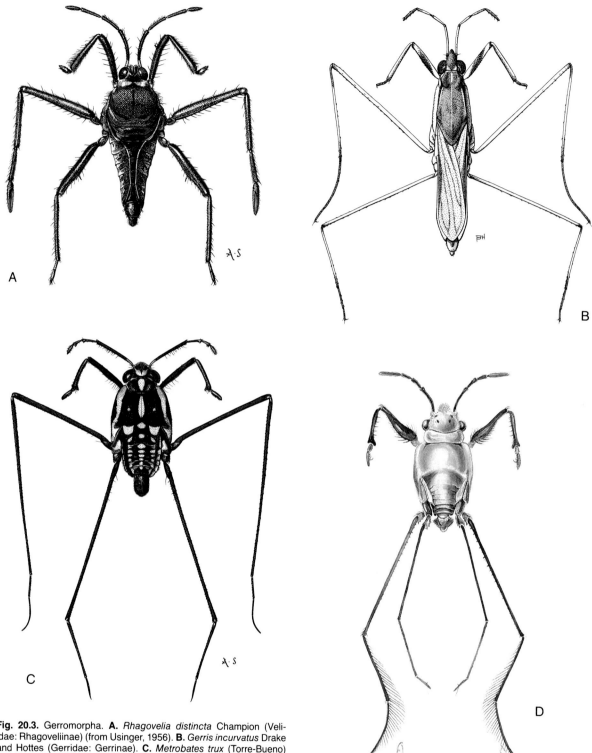

Fig. 20.3. Gerromorpha. **A.** *Rhagovelia distincta* Champion (Veliidae: Rhagoveliinae) (from Usinger, 1956). **B.** *Gerris incurvatus* Drake and Hottes (Gerridae: Gerrinae). **C.** *Metrobates trux* (Torre-Bueno) (Gerridae: Trepobatinae) (from Usinger, 1956). **D.** *Halobates sericeus* Eschscholtz (Gerridae: Halobatinae) (from Zimmerman, 1948).

genera and for the species known to occur in California. Other regional treatments for North America are those of Brooks and Kelton (1967), Bobb (1974), and Bennett and Cook (1981). The West Indian fauna is covered in part by Cobben (1960b, c). The fauna of the western Palearctic is treated in the works of Stichel (1955), Poisson (1957b), Kerzhner and Jaczewski (1964), and Savage (1989) and that of the eastern Palearctic by Kanyukova (1988). Poisson (1957a) prepared a preliminary account of the fauna of South Africa. The work of Linnavuori (1971, 1981, 1986) covers much of northern Africa and the Arabian Peninsula. The last comprehensive work for India was that of Distant (1904, 1910), for Indonesia Lundblad (1933), and for Australia that of Hale (1926). The habitats, habits, and distributions of the marine Gerromorpha were treated by Andersen and Polhemus (1976).

Key to Families of Gerromorpha

1. Macropterous 2
– Apterous or short-winged 7
2. Scutellum exposed, forming subtriangular, rounded, or transverse plate behind pronotal lobe (Fig. 20.2A) 3
– Scutellum not exposed, covered by posteriorly extended pronotal lobe (Fig. 20.2B) 4
3. Ventral surface of head with a pair of well-developed bucculae covering base of labium (Fig. 22.1B); tarsi 2-segmented, first segment very short .... Hebridae (part)
– Bucculae not so well developed and covering base of labium as above; tarsi 3-segmented .... Mesoveliidae (part)
4. Bucculae covering first and sometimes also second labial segments laterally; claws inserted apically on last tarsal segment (Fig. 25.1C) (slightly preapically in *Limnobatodes* Hussey, Hydrometridae) 5
– Bucculae not so well developed and not fully covering basal labial segments; claws inserted distinctly before apex of last tarsal segment (Figs. 27.1G, H) (with a few exceptions) 6
5. Head not distinctly prolonged behind eyes; length of postocular area less than the diameter of an eye; forewing with 6 closed cells (Figs. 20.2B, 24.1B) .... Macroveliidae
– Head distinctly prolonged behind eyes; length of postocular area longer than diameter of an eye; forewing with 3 closed cells (Fig. 25.1A) .... Hydrometridae (part)
6. Head with a distinct longitudinal, median impressed line on dorsal surface (Figs. 20.3A, 27.1A); foretibia of male usually with a grasping comb of short spines along inner margin; hind femur usually stouter than middle femur .... Veliidae
– Head without a median impressed line on dorsal surface; foretibia of male without grasping comb; hind femur usually more slender than middle femur .... Gerridae
7. Abdominal scent-gland orifice absent from metasternum 8
– Abdominal scent-gland orifice present on metasternum 9
8. Head much more than 3 times as long as wide (Fig. 25.1B), eyes far removed from base of head .... Hydrometridae (part)
– Head at most 3 times as long as wide, eyes situated very close to or at base of head 6
9. Pronotum very short, exposing both meso- and metanotum 10
– Pronotum longer, at least covering mesonotum 11
10. Head longer than wide, porrect (Fig. 21.1A); all claws apical (Fig. 21.1F) .... Mesoveliidae (part)
– Head shorter than wide, subvertical (Figs. 26.1A, B); claws of forelegs subapical, claws of middle and hind legs apical (Fig. 26.1C) .... Hermatobatidae
11. Bucculae well developed, covering base of labium; tarsi 2-segmented .... Hebridae (part)
– Bucculae not so well developed and not obscuring basal labial segments; tarsi 3-segmented 12
12. Antennal segment 4 not subdivided in middle by membranous zone 13
– Antennal segment 4 subdivided in middle by a membranous zone .... Paraphrynoveliidae
13. Head not distinctly prolonged behind eyes; length of postocular region longer than diameter of an eye .... Macroveliidae (part)
– Head distinctly prolonged behind eyes; length of postocular region longer than diameter of an eye .... Hydrometridae (part)

Mesoveloidea

21 Mesoveliidae

General. This small group of insects is thought to be the relatively most primitive of all semiaquatic bugs. They vary greatly in habitus (Fig. 21.1A) and degree of wing development and range from 1.2 to 4.2 mm in length. Members of the widely distributed genus *Mesovelia* Mulsant and Rey are sometimes referred to as water treaders.

Diagnosis. Micro- and macrohair layer restricted to head and prosternal region of thorax; ocelli present or absent; head without dorsal indentations and corresponding internal apodemes; base of labium not obscured by bucculae (Fig. 21.1B); pronotum truncate posteriorly; scutellum developed and exposed (Fig. 21.1H); forewing venation reduced (Fig. 21.1C); tarsi 3-segmented (Fig. 21.1E, F); pretarsus inserted apically or subapically; first abdominal mediotergite with a pair of longitudinal ridges in macropterous forms (Fig. 21.1H); aedeagus (Fig. 21.1I) with specialized ejaculatory pump and bulb; parameres symmetrical; ovipositor usually laciniate (Fig. 21.1J); gynatrial complex as in Fig. 21.1K; anterior end of egg obliquely truncate with an egg cap developed in many species; eclosion by means of an embryonic bladder rather than an egg burster.

Classification. The most important single paper on the group is that of Andersen and Polhemus (1980), in which four new genera were described and a phylogeny and a classification for the family were presented, with two subfamilies being recognized. Previously China and Miller (1959) recognized a more broadly conceived group with the subfamilies Mesoveliinae, Mesoveloideinae, and Macroveliinae. Štys (1976) recognized the Macroveliidae as a distinct family, as had earlier authors. The name Madeoveliidae was proposed by Poisson (1959) and is a senior synonym of Mesoveloideinae of China and Miller (1959). Horvath (1929) provided a catalog of the species. Eleven genera and about 39 species are currently recognized.

Key to Subfamilies of Mesoveliidae

1. Macropterous 2
– Wingless .... Mesoveliinae
2. Ocelli lacking; claws inserted subapically on tarsi (Fig. 21.1E); forewing with 2 basal cells and one apical cell (Fig. 21.1C) .... Madeoveliinae
– Ocelli present; claws inserted apically on tarsi (Fig. 21.1F); forewings with 2 or 3 basal cells but no apical cell (Fig. 21.1D) .... Mesoveliinae

MADEOVELIINAE. Head deflected in front of eyes; ocelli absent; scutellum subtriangular; forewing with 2 basal cells and one apical cell (Fig. 21.1C); pretarsus inserted anteapically (Fig. 21.1E).

Included genera are *Mesoveloidea* Hungerford, two species from tropical America, and *Madeovelia* Poisson, one species from Guinea.

MESOVELIINAE. Ocelli present in macropterous forms; scutellum reduced, apically rounded (Fig. 21.1H); pretarsus inserted apically (Fig. 21.1F); forewing with 3 basal cells, no apical cells (Fig. 21.1D); most genera apterous.

Included genera are *Austrovelia* Malipatil and Monteith, two species from Australia and New Caledonia; *Cavaticovelia* Andersen and Polhemus, one species from Hawaii; *Cryptovelia* Andersen and Polhemus, one species from the lower Amazon basin; *Darwinivelia* Andersen and Polhemus, two species from northern South America and the Galapagos Islands; *Mesovelia* Mulsant and Rey, about 25 species, cosmopolitan; *Mniovelia* Andersen and Polhemus, one species from New Zealand; *Nereivelia* J. and D. Polhemus, one species from Singapore and Thailand; *Phrynovelia* Horvath, three species from New Guinea and New Caledonia; and *Speovelia* Esaki, two species from Mexico and Japan.

Specialized morphology. *Cryptovelia terrestris* Andersen and Polhemus, among the smallest of all known Heteroptera, is apterous and has compound eyes consisting of only 3 or 4 facets (Fig. 21.1B). On the other hand *Mniovelia kuscheli* Andersen and Polhemus from New

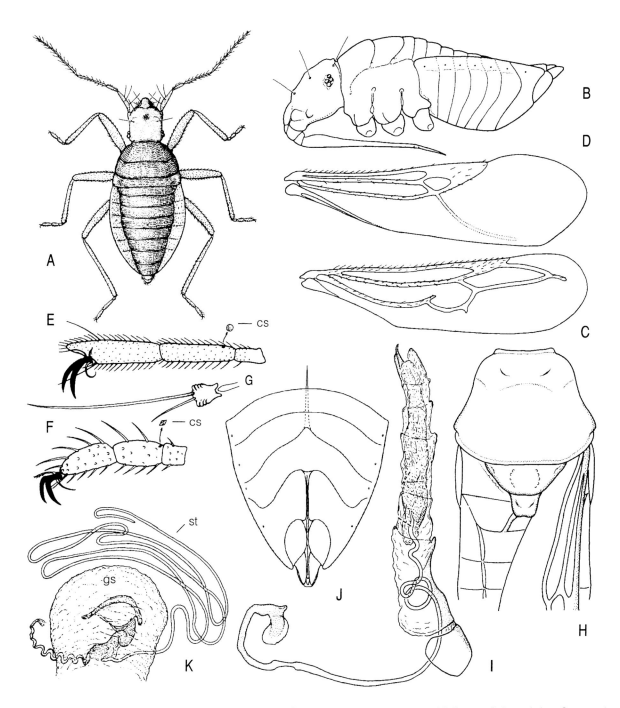

Fig. 21.1. Mesoveliidae (from Andersen, 1982a). **A.** Adult female, *Cryptovelia terrestris* Andersen and Polhemus. **B.** Lateral view, *C. terrestris*. **C.** Forewing, *Mesoveloidea williamsi* Hungerford. **D.** Forewing, *Mesovelia mulsanti* White. **E.** Left middle pretarsus, *Mesoveloidea williamsi*. **F.** Left middle pretarsus, *C. terrestris*. **G.** Unguitractor plate and parempodia, *Mesoveloidea williamsi*. **H.** Dorsal view, thorax and base of abdomen, *Mesovelia mulsanti*. **I.** Aedeagus, *Mesovelia mulsanti*. **J.** Female terminal abdominal segments, ventral view, *Mesovelia mulsanti*. **K.** Gynatrial complex, *Mniovelia kuscheli* Andersen and Polhemus. Abbreviations: cs, campaniform sensillum; gs, gynatrial sac; st, spermathecal tube.

Zealand—also very small—has very large eyes compared with other Gerromorpha.

Three thoracic morphs associated with degree of forewing development were recognized by Galbreath (1975) in *Mesovelia mulsanti* White, but she was not able to demonstrate definitively the mechanism for their determination, although environmental factors appeared to play an important role.

Natural history. The life history of several *Mesovelia* spp. in the Holarctic has been studied in detail (Hungerford, 1917; Ekblom, 1930; Zimmermann, 1984). The females insert the eggs into plant tissue with their elongate ovipositor, laying 100 or more eggs. In very cold regions the bugs apparently overwinter as eggs, whereas in warmer climates adults may be present and active year round. Most *Mesovelia* spp. inhabit the margins of ponds of streams, some leading a rather secretive existence, others spending most of their time in the open. They are extremely agile on the open water surface, although they are usually associated with some form of aquatic vegetation. *Mesoveloidea* has been found living on wet moss near streams; *Cryptovelia, Mniovelia,* and *Phrynovelia* are known to inhabit forest leaf litter, often at a considerable distance from the nearest body of water. *Cavaticovelia* and some *Speovelia* spp. are cave-inhabiting (Esaki, 1929; Gagné and Howarth, 1975). All *Speovelia* spp., some *Mesovelia* spp., *Nereivelia,* and *Darwinivelia* are associated with intertidal marine habitats (Andersen and Polhemus, 1980; Carvalho, 1984a).

Female *Mesovelia furcata* Mulsant and Rey possess symbiotic inclusions in the midgut wall. These are apparently transmitted by the female to the egg shortly after oviposition, as the female taps the opercular end of the egg with the anus (Cobben, 1965).

The macropterous morph in some species of *Mesovelia* (e.g., *M. furcata;* Jordan, 1951) mutilates the wings by tearing the distal portions with the legs, thus losing the ability to fly.

Mesovelia amoena Uhler is parthenogenetic in Hawaii and tropical areas, males having been collected only in North America, Mexico, and Hispaniola (J. T. Polhemus and Chapman, 1979c).

Distribution and faunistics. *Mesovelia* is widely distributed, with a few of the included species also having very wide distributions. All other taxa are of limited distribution and widely scattered. This fact and the basal phylogenetic position within the family suggest an ancient group.

Andersen and Polhemus (1980) included a checklist of species and distributional record and Andersen (1982a) provided much additional information on the family. J. T. Polhemus and Chapman (1979c) supplied a useful review of the group for North America, with emphasis on the fauna of California.

Hebroidea

22
Hebridae

General. Sometimes referred to as velvet water bugs, hebrids are some of the smallest members of the Gerromorpha, ranging in length from 1.3 to 3.7 mm, and having the general appearance of very small veliids (Figs. 20.2A, 22.1A) They are infrequently encountered by the general collector because of their minuteness and often secretive habits.

Diagnosis. Micro- and macrohair layer covering body, except abdomen, and appendages; antennae often rather short, segment 2 subequal to or shorter in length than segment 1 (Fig. 22.1D, E), segment 4 undivided in *Hebrus* (*Subhebrus*), *Lipogomphus, Merragata,* and *Hyrcanus* and with a median constricted membranous zone in all other taxa, giving 5-segmented appearance (Fig. 22.1E); ocelli present, each with a deep pit anterior to it, corresponding to an internal apodeme; bucculae prominent, obscuring 2 basal segments of labium (Fig. 22.1B, C); labium short in *Hyrcanus;* pronotum truncate along posterior margin, usually exposing short transverse scutellum and posteriorly adjoining triangular metanotum (Fig. 20.2A); thorax ventrally with a pair of longitudinal carinae between coxae, forming rostral groove (Fig. 22.1C); adult tarsi 2-segmented, segment 2 representing fusion of segments 2 and 3 of other Gerromorpha (Fig. 22.1G); pretarsus inserted apically (Fig. 22.1G); forewing with 1 or 2 basal cells, no venation on distal portion (Fig. 22.1A, F); genitalia apparently inserted anteapically on abdomen in both male and female; pygophore small; parameres usually symmetrical (Fig. 22.1H); aedeagus as in Fig. 22.1I; ovipositor valves much reduced, weakly sclerotized, more platelike than in Mesoveliidae; gynatrial complex as in Fig. 22.1J.

Classification. Two subfamilies are currently recognized, following the classification proposed by Andersen (1981b, 1982a). Approximately 160 species are placed in seven genera.

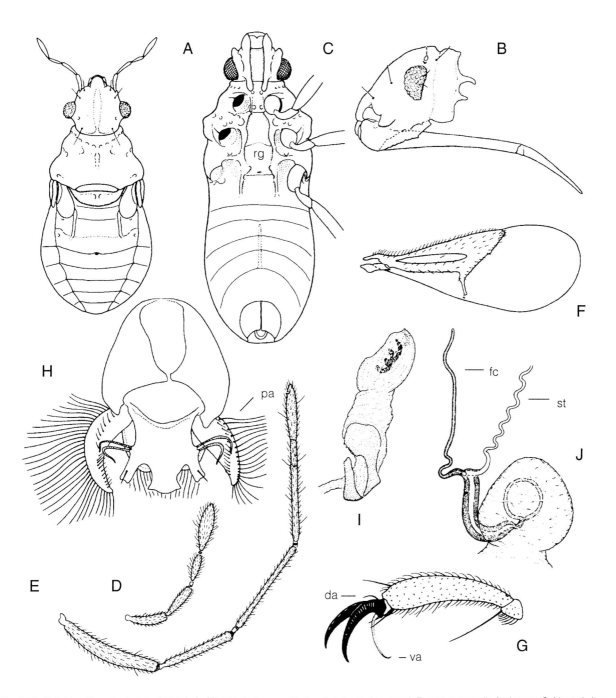

Fig. 22.1. Hebridae (from Andersen, 1982a). **A.** *Merragata brunnea* Drake. **B.** Lateral view, head, *Timasius ventralis* Andersen. **C.** Ventral view, *Hebrus pusillus* (Fallén). **D.** Antenna, *T. ventralis*. **E.** Antenna, *Hyrcanus capitatus* Distant. **F.** Forewing, *H. capitatus*. **G.** Foretarsus, *H. capitatus*. **H.** Dorsal view, pygophore and proctiger, *T. ventralis*. **I.** Aedeagus, *H. capitatus*. **J.** Gynatrial complex, *T. ventralis*. Abbreviations: da, dorsal arolium; fc, fecundation canal; pa, paramere; rg, rostral groove; st, spermathecal tube; va, ventral arolium.

Hebridae 91

Key to Subfamilies of Hebridae

1. Antennal segments slender, segment 4 much longer than segment 1 .... Hebrinae
– Antennal segments stout, segment 4 subequal in length to segment 1 (Fig. 22.1D) 2
2. Large species more than 2.5 mm. in length; head narrow and pointed apically; labium not reaching mesosternum .... Hyrcaninae
– Small species not more than 2.5 mm. in length; head broad not pointed apically; labium reaching metasternum .... Hebrinae

HEBRINAE. Antennal segment 4 with preapical cluster of modified blunt setae; eyes situated near base of head (Fig. 20.2A); antennae distinctly longer than head; labial segment 3 not surpassing hind margin of head, dorsal and ventral arolia subequal in length; parameres symmetrical, except in *Timasius*.

Included genera are *Hebrometra* Cobben, four species from Ethiopia (Cobben, 1982); *Hebrus* Curtis, largest genus in the family, with approximately 110 described and many undescribed species distributed on all major land masses; *Lipogomphus* Berg, four species from South and Central America and the southern United States (Andersen, 1981b); *Merragata* Buchanan-White, several species, cosmopolitan; *Neotimasius* Andersen, one species from Southern India; and *Timasius* Distant, 15 species from Sri Lanka, parts of India, Taiwan, southeast Asia and Java (Andersen, 1981b).

HYRCANINAE. Eyes distinctly removed from anterior margin of the pronotum; length of antenna subequal to or shorter than length of head; labial segment 3 not surpassing hind margin of head; dorsal arolium distinctly shorter than ventral arolium (Fig. 22.1G); parameres symmetrical.

Hyrcanus Distant, with four species from the Oriental region, is the only included genus (Andersen, 1981b).

Specialized morphology. At least in *Hebrus ruficeps*, there exist, in addition to the peg plates, some small (6–10μm) cup-shaped or mushroomlike structures, scattered among the hair layer. In addition, in most hebrids, antennal segment 4 possesses subapically a small group of short, blunt setae of unknown sensory function (Cobben, 1978).

The structure and function of the male and female reproductive systems have been studied in detail for *Hebrus pusillus* and *H. ruficeps* (Heming–Van Battum and Heming, 1986, 1989).

Natural history. Hebrids live on vegetated margins of ponds or similar permanently damp habitats, sometimes deep in mats of moss or frequently in interstices, as well as among sparse vegetation on sloping stream banks. Some species are more specialized in their habitat requirements, such as members of the genera *Hebrometra* and *Timasius*, which live on rocks in streams or torrents or on seeping rock faces or those splashed by waterfalls. A few species tolerate brackish, saline, or marine conditions. *Hebrus ruficeps* can overwinter frozen in ice among *Sphagnum*.

As far as is known, hebrids lay their eggs superficially on a substrate, such as on mosses, attaching them in a lengthwise position with a gelatinous substance.

Distribution and faunistics. The Hebridae are nearly worldwide in distribution, with the greatest generic diversity in the Asian tropics. Drake and Chapman (1958) provided a checklist of New World species. Stichel (1955) treated the European fauna, Miyamoto (1965a) the species from Taiwan, Poisson (1944) some species from Africa, and Cobben (1982) those known from Ethiopia.

23
Paraphrynoveliidae

General. This family, with no common name, includes only one genus and two species, which range in size from 1.7 to 2.4 mm. Its members, which have the general appearance of wingless hebrids or *Microvelia* spp. (Fig. 23.1A), have rarely been collected.

Diagnosis. *Microvelia*-like in general appearance; all known specimens apterous; ocelli absent (Fig. 23.1B); antennae flagelliform, segments 1 and 2 elongate, subequal in length, segments 3 and 4 longer, segment 4 with a medial membranous area with ringlike cuticular thickenings; pronotum short (Fig. 23.1A), produced posteriorly to cover mesonotum but not metanotum; tarsi 3-segmented (Fig. 23.1E); pretarsus inserted apically, with well-developed dorsal and ventral arolia (Fig. 23.1E); abdominal dorsum as in Fig. 23.1D; ovipositor rather weakly developed (Fig. 23.1C); pygophore protruding from end of abdomen; parameres (Fig. 23.1F) symmetrical; gynatrial complex as in Fig. 23.1G.

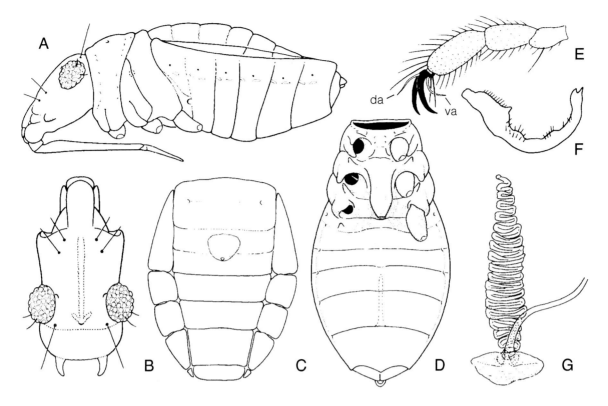

Fig. 23.1. Paraphrynoveliidae. *Paraphrynovelia brincki* Poisson (from Andersen, 1982a). **A.** Lateral view. **B.** Dorsal view, head. **C.** Ventral view, thorax and abdomen. **D.** Internal view, pregenital segments of abdominal dorsum, including scent apparatus. **E.** Middle tarsus. **F.** Left paramere. **G.** Gynatrial complex, dorsal view. Abbreviations: da, dorsal arolium; va, ventral arolium.

Classification. *Paraphrynovelia* Poisson was originally placed in the Mesoveliidae. More detailed study suggested, however, that the two known species should be placed in a distinct family in a clade including the Macroveliidae and Hydrometridae (Andersen, 1978).

Natural history. *Paraphrynovelia* spp. live in the transition zone between land and water in wet debris and water-soaked moss on rocks (Andersen, 1978).

Distributions and faunistics. This group is restricted to southern Africa. Information on the details of distribution were given by Poisson (1957a), and Andersen (1978, 1982a) (see Chapter 9).

24
Macroveliidae

General. Members of this small family have no common name. They are similar to some mesoveliids in general appearance (Fig. 20.2B, C) and range in length from 2.5 to 5.6 mm.

Diagnosis. Micro- and macrohair layer generally distributed, but reaching only onto segment 1 of abdomen; ocelli present in macropterous *Macrovelia* (Fig. 20.2B), reduced in *Oravelia,* and absent in *Chepuvelia;* antennae flagelliform, segments 1 and 2 elongate (Fig. 20.2B, C); labial segments 1 and 2 very short, segment 3 approximately 2.5 times length of segment 4, labium reaching to about mesocoxae (Fig. 24.1A); pronotum in *Macrovelia* extending posteriorly to cover rudimentary scutellum, mesonotum, and metanotum (Fig. 20.2B), but truncate in apterous *Chepuvelia* and *Oravelia* (Fig. 20.2C); forewing venation in *Macrovelia* as in Figs. 20.2B, 24.1B; tarsi 3-segmented; pretarsus inserted apically (Fig. 24.1D); pygophore inserted apically on abdomen; parameres (Fig. 24.1E) symmetrical; aedeagus as in Fig. 24.1F; *Macrovelia* ovipositor plate-shaped, similar to Hebridae, but

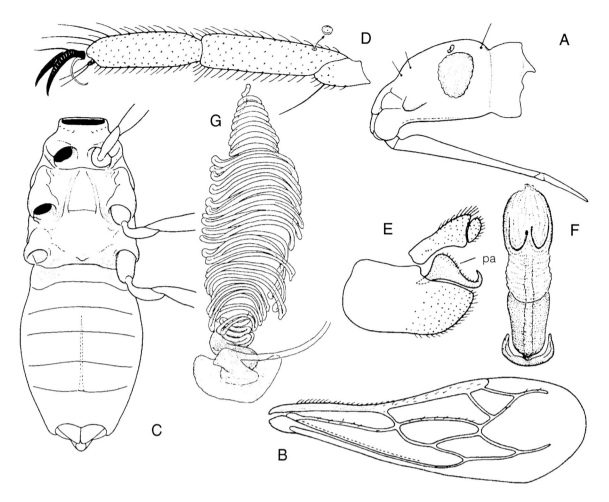

Fig. 24.1. Macroveliidae. *Macrovelia hornii* Uhler (from Andersen, 1982a). **A.** Lateral view, head. **B.** Right forewing. **C.** Ventral view, body. **D.** Left middle tarsus. **E.** Pygophore and proctiger, lateral view. **F.** Aedeagus, ventral view. **G.** Gynatrial complex, dorsal view. Abbreviation: pa, paramere.

located at apex of abdomen (Fig. 24.1C); gynatrial complex as in Fig. 24.1G.

Classification. The composition of the Macroveliidae has varied considerably at the hands of different authors. The current classification was established by Štys (1976) and Andersen (1978, 1982a). It comprises *Chepuvelia* China (1962), one species from southern Chile; *Macrovelia* Uhler, one species from western North America; and *Oravelia* Drake and Chapman, one species from Fresno County, California.

Specialized morphology. *Macrovelia hornii* Uhler may be either macropterous or brachypterous. *Oravelia* and *Chepuvelia* are known only from apterous individuals.

Natural history. *Chepuvelia usingeri* China is found in wet forest litter, all known specimens having been collected by sifting and using Berlese funnels. *Macrovelia hornii* and *Oravelia pege* Drake and Chapman inhabit springs or seeps with abundant vegetation; they are negatively phototrophic and often secrete themselves in shaded areas. *Macrovelia hornii* overwinters as an adult and may be active during warm periods even in midwinter (McKinstry, 1942; Anderson, 1963; Drake and Chapman, 1963; J. T. Polhemus and Chapman, 1979b). The eggs of *Macrovelia* are glued to the substrate in a longitudinal position.

Distribution and faunistics. J. T. Polhemus and Chapman (1979b) summarized available information on the North American taxa.

Hydrometroidea

25
Hydrometridae

General. This is one of the most distinctive heteropteran groups, many members having an extremely elongate body and appendages; all taxa have the eyes far removed from the anterior margin of the pronotum (Fig. 25.1A, B). Commonly called marsh treaders or water measurers, they range in length from 2.7 to 22 mm.

Diagnosis. Extent of micro- and macrohair layer variable; head elongated in front of and behind compound eyes (Fig. 25.1A, B); posterior pair of cephalic trichobothria inserted on tubercles (Fig. 25.1A), except in *Hydrometra;* ocelli present or absent; antennal segment 4 with an apical invagination, generally bordered by modified setae; tarsi 3-segmented (Fig. 25.1C); pretarsus usually inserted apically, ventrally in *Limnobatodes,* dorsal and ventral arolium usually both well developed (Fig. 25.1C); forewing development variable, venation usually somewhat reduced (Fig. 25.1A); pygophore protruding apically from abdomen; phallus as in Fig. 25.1E; parameres symmetrical (Fig. 25.1D); ovipositor reduced in most species; gynatrial complex as in Fig. 25.1F.

Classification. Although it has long been clear that *Hydrometra* was a member of the Gerromorpha (Ekblom, 1926; Andersen, 1982b), the position of all members of the group was not always so obvious. For example, *Heterocleptes* was originally placed in the Reduviidae by Villiers (1948). Three subfamilies—comprising seven genera and more than 110 species—are currently recognized, following the classification of Andersen (1977a, 1982b).

Key to Subfamilies of Hydrometridae

1. Antennal segment 1 subequal to or shorter than segment 2, usually at most barely exceeding apex of head; body length 6 mm or more; metasternum lacking scent-gland orifices ...... Hydrometrinae
– Antennal segment 1 much longer than segment 2, surpassing apex of head by more than half its length; body length 3–5 mm; metasternum with scent gland orifices present 2
2. Ocelli present; posterior pair of cephalic trichobothria inserted on prominent rounded elevations (Fig. 25.1A); head and pronotum clothed only with simple setae; articulation between antennal segments 1 and 2 subapical .... Heterocleptinae
– Ocelli lacking; posterior pair of cephalic trichobothria inserted on small tubercles; head and pronotum clothed with stout black spinules; articulation between antennal segments 1 and 2 apical .... Limnobatodinae

HETEROCLEPTINAE (FIG. 25.1A). Micro- and macrohair layer covering head, thorax, and base of abdomen; posterior cephalic trichobothria very long, inserted on prominent round elevations; ocelli present (even in apterous morphs of *Veliometra*); antennal segment 1 longer than segment 2, antennal segment 2 inserted laterally and somewhat anteapically on segment 1, antennal segment 4 in *Heterocleptes* with a membranous ring distad of midpoint, producing 5-segmented appearance; pronotum relatively short and broad, produced posteriorly over the metanotum, even in apterous morphs; arolia rudimentary, parempodia padlike in *Heterocleptes;* peg plates present.

Included genera are *Heterocleptes* Villiers, four species from Africa and Borneo (Villiers, 1948; China and Usinger, 1949b; China, Usinger, and Villiers, 1950; Andersen, 1982b), and *Veliometra* Andersen, one species from the central Amazon basin of Brazil (Andersen, 1977a).

HYDROMETRINAE. Micro- and macrohair layer covering entire body; anteocular portion of head much longer than length of pronotum; antennal segment 1 subequal to or shorter in length than segment 2; metasternal scent glands absent.

Included genera are *Bacillometra* Esaki, four species from tropical South America (Esaki, 1927); *Chaetometra* Hungerford, one species from the Marquesas Islands; *Dolichocephalometra* Hungerford, one species from the Marquesas Islands; and the cosmopolitan *Hydrometra* Latreille, with at least 80 species (Andersen, 1977a).

LIMNOBATODINAE. Sharing many features with the Hydrometrinae; distinguished by micro- and macrohair layer covering head, thorax, and base of abdomen; anteocular portion of head shorter than pronotal length; antennal segment 1 distinctly longer than segment 2; claws inserted slightly before tarsal apex; metasternal scent glands present.

Limnobatodes paradoxus Hussey, known from Belize,

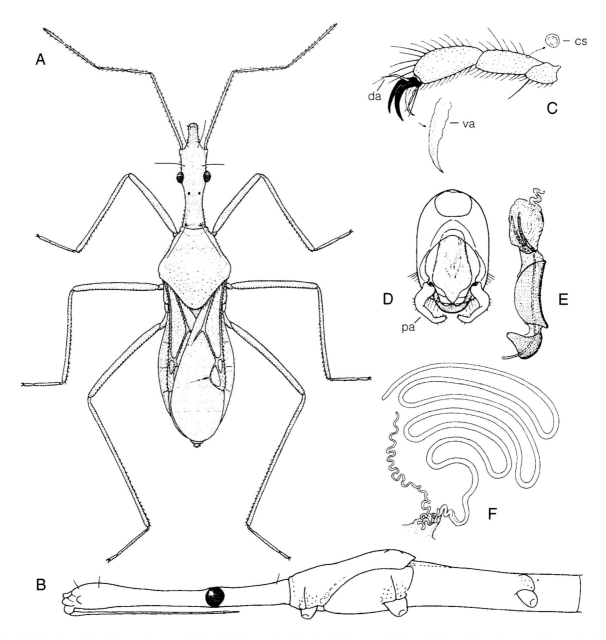

Fig. 25.1. Hydrometridae (from Andersen, 1982a). **A.** *Heterocleptes hoberlandti* China and Usinger. **B.** Lateral view, head and thorax, *Hydrometra* sp., macropterous female. **C.** Left middle tarsus, *Veliometra schuhi* Andersen. **D.** Pygophore and parameres, dorsal view, *V. schuhi*. **E.** Phallus, *Hydrometra stagnorum* (Linnaeus), lateral view. **F.** Gynatrial complex, *Heterocleptes hoberlandti*. Abbreviations: da, dorsal arolium; cs, campaniform sensillum; pa, paramere; va, ventral arolium.

Brazil, and Peru (Hussey, 1925; Andersen, 1977a) is the only included taxon.

Specialized morphology. The hydrometrids, particularly the Hydrometrinae, possess many modifications associated with elongation of the body and appendages, these particularly affecting the head and thorax (Fig. 25.1B). Other novel structures are mentioned in the diagnosis. Males of some *Hydrometra* spp. have a clump of modified setae ventrally on abdominal segment 7 (Fig. 10.7E). A thorough review of the morphology of *Hydrometra martini* Kirkaldy was provided by Sprague (1956).

Natural history. *Veliometra* has been collected in the marginal vegetation of an impondment of an Amazonian forest stream. The scant data available for *Heterocleptes* and *Limnobatodes* suggest that at least the former is semiterrestrial, most collections of the latter being from lights. *Hydrometra* spp. are usually found on or around quiet bodies of water and generally are associated with marginal vegetation but may also be found on damp rock walls. *Hydrometra* spp. can walk on the water surface with great agility and apparent effortlessness.

The eggs of *Hydrometra* are placed some distance above the water level and are cemented to the substrate in an upright position by their base (Sprague, 1956). Similarities in structure suggest a like habit for *Limnobates*. The eggs of *Veliometra* are more similar in structure to those of other nonhydrometrid Gerromorpha and are probably cemented in a horizontal position in closer association with water (Andersen, 1982a).

Distribution and faunistics. The family is most diverse in the tropics, with only species of *Hydrometra* occurring elsewhere. The group is notable for the occurrence of the endemic genera *Chaetometra* and *Dolichocephalometra* in the Marquesas Islands, an area from which nearly all other gerromorphans are absent. Hungerford and Evans (1934) and Drake and Lauck (1959) offer the most comprehensive distributional surveys, for the world and the Western Hemisphere, respectively.

Gerroidea

26
Hermatobatidae

General. Members of this small, seldom-collected group of ovoid-bodied marine bugs closely resemble short-bodied apterous gerrids. Sometimes referred to as coral treaders, they range in length from 2.7 to 4.0 mm.

Diagnosis. Micro- and macrohair layers covering entire body; peg plates absent; head extremely short and broad, anteclypeus and antennal fossae hidden in dorsal view (Fig. 26.1A, B); ocelli absent; 3 pairs of cephalic trichobothria; antennae relatively long, segment 2 longest, inserted relatively close together and far from anterior margin of eye; labium short and stout, segments 1 and 4 subequal in length to segment 3, segment 2 shorter (Fig. 26.1A, B); always apterous; prothorax short; meso- and metathorax indistinguishably fused, except laterally; prothoracic sternum narrow, meso- and metathoracic sterna broad; all coxae inserted in a horizontal position and directed caudad (Fig. 26.1D); forefemur incrassate, foretibia slightly curved, with an elongate apical grooming comb; femur and tibia of middle and hind legs more elongate and slender than foretibia; all tarsi 3-segmented, pretarsus of foreleg inserted anteapically, claws slender, strongly developed, pretarsus of middle and hind legs inserted apically (Fig. 26.1C); dorsal arolium flattened horizontally and tapering, ventral arolium 3-branched (Fig. 26.1C); parempodia long, setiform, and symmetrically developed (Fig. 26.1C); metathoracic scent-gland apparatus present; abdomen highly modified, greatly shortened (Fig. 26.1D); scent-gland apparatus present on abdominal segment 4 in nymphs; parameres reduced, left fused to margin of pygophore, right in form of a small plate loosely connected with pygophoral margin; aedeagus as in Fig. 26.1E; female genital segments simple, ovipositor absent; gynatrial complex as in Fig. 26.1F; 4 ovarioles.

Classification. Carpenter (1892) described *Hermatobates* in the Gerridae. Coutiere and Martin (1901) recognized it as the subfamily Hermatobatinae. Poisson (1965) accorded the group family status, an action supported by more recent authors (Štys and Kerzhner, 1975; Andersen and Polhemus, 1976; Andersen, 1982a). Andersen (1982a) argued that *Hermatobates* Carpenter is the sister group of Veliidae + Gerridae. Most of the eight known species were described from only one or a few specimens.

Specialized morphology. The head of *Hermatobates* is extremely short and broad, only apterous forms are known, and the abdomen is strongly shortened and shows substantial fusion of terga and sterna, especially in females.

Natural history. *Hermatobates* spp. are associated with coral reefs and rubble. They secrete themselves in an air bubble at high tide, coming out to feed during low tide. Those that do not retreat before the rising tide are able to continue skating or resting on the water surface until the next tidal cycle. They move very rapidly on the water surface and can jump a considerable distance. There are apparently only four nymphal instars (Andersen, 1982a; Foster, 1989; D. A. Polhemus, 1990a).

Distribution and faunistics. Most of the known species occur in the Indian or Pacific ocean. A single species, *Hermatobates breddini* Herring, is known from the West Indies. In addition to the work of Andersen (1982a), useful papers on the group include those of Coutiere and Martin (1901), China (1956), Herring (1965), Cheng (1977), and J. T. Polhemus and Herring (1979).

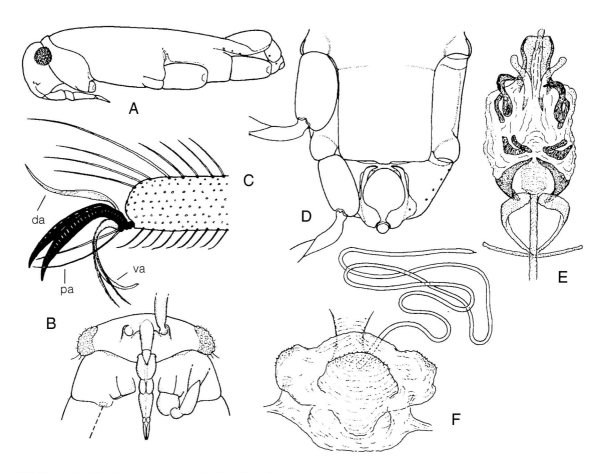

Fig. 26.1. Hermatobatidae. *Hermatobates weddi* China (from Andersen, 1982a). **A.** Lateral view, body. **B.** Ventral view, head and prothorax. **C.** Pretarsus, middle leg. **D.** Ventral view, male pterothorax and abdomen. **E.** Aedeagus, dorsal view. **F.** Gynatrial complex, dorsal view. Abbreviations: da, dorsal arolium; pa, parempodium; va, ventral arolium.

27
Veliidae

General. Sometimes referred to as broad-shouldered water striders, small water striders, or riffle bugs, the Veliidae range from about 1 to 10 mm in length. Most species are rather stout-bodied. They occupy a wide range of habitats, including the surface of the ocean, and, along with the Gerridae, represent the epitome of adaptation for life on the water surface film.

Diagnosis. Entire body surface (except abdomen in *Ocellovelia* China and Usinger), covered with macro- and microhair layer; all members with "grooved setae"; head rather short and broad, with a median impressed dorsal line and with a pair of deep pits near posteromesal angle of eyes, representing external evidence of internal apodeme for attachment of antennal musculature; compound eyes usually large (Figs. 20.2D, 20.3A), sometimes reduced; ocelli absent (except in *Ocellovelia*); cephalic trichobothria as in Fig. 10.8A; antennal sockets usually located beneath eyes and obscured in dorsal view (Fig. 27.1A–C) (in contrast to foregoing families of Gerromorpha); bucculae relatively small, not obscuring basal segments of labium (Fig. 27.1C); labium reaching to mesosternum, segments 1 and 2 short, segment 3 longest, segment 4 short (Fig. 27.1B, C); pronotum enlarged, obscuring meso- and metanotum and part of abdominal mediotergite 1 (Fig. 20.2D) (at least in macropterous morphs, variously reduced in apterous morphs); coxae of meso- and metathorax widely separated; mesosternum with a mesal longitudinal impression, but without a distinct rostral groove (Fig. 27.1I); metasternum with characteristic pair of strongly arched grooves leading from metasternal scent orifice to a subovate swelling with a tuft of long hairs on metasternopleuron; forewing with a variable number of closed cells (Fig. 27.1D–F); alary polymorphism common; all legs usually of similar structure, sometimes metathoracic pair longest; femora variously elongated, dilated, or spinose; male foretibia usually with grasping comb of short, stout spines on posterior, innermost margin; mesotibia usually with row of erect trichobothrium-like setae on posterior surface; tarsi 1-, 2-, or 3-segmented; last tarsal segment always cleft over one-fourth to nearly two-thirds of length, pretarsus situated in cleft (Fig. 27.1G, H); parempodia asymmetrically developed; pretarsus of middle legs often highly modified (e.g., claws flattened and asymmetrical and ventral arolium modified into "swimming fan") (Fig. 27.1H); abdomen generally elongate, sometimes shortened; abdominal mediotergites 1–3 with a pair of medial longitudinal ridges; dorsal abdominal scent-gland apparatus absent in both nymphs and adults; pygophore usually large, protruding from abdomen; parameres usually symmetrically developed (Fig. 27.1J); aedeagus as in Fig. 27.1K; testes as in Fig. 27.1L; female usually with a reduced, plate-shaped ovipositor (Fig. 27.1I), with a median connection between second valvulae as in Gerridae; gynatrial complex as in Fig. 27.1M; 4 ovarioles.

Classification. The Veliidae comprise six subfamilies, following the classifications of Štys (1976) and Andersen (1982a), who treated the controversial Macroveliidae and Ocelloveliinae as a distinct family and subfamily, respectively. *Hebrovelia* Lundblad is treated as belonging to a distinct tribe within the Microveliinae (see Štys, 1976), and the Haloveliinae, which have been variously placed by prior authors, are treated as a subfamily within the Veliidae. The family currently includes 38 genera and nearly 600 species.

Key to Subfamilies of Veliidae

1. Ocelli present; forewings with 6 closed cells (Fig. 27.1D) .... Ocelloveliinae
– Ocelli absent; forewings with fewer than 6 closed cells (Figs. 20.2D, 27.1E, F) 2
2. Middle tarsus deeply cleft with a fan of plumose or hairy swimming fans arising from base of cleft
   .... Rhagoveliinae
– Middle tarsus not so deeply cleft, lacking plumose or hairy swimming fans 3
3. Middle tarsus 3-segmented (first segment short) 4
– Middle tarsus 2-segmented 5
4. Foretarsus 2-segmented; forewing divided into a proximal coriaceous portion with 2 closed cells and membranous distal portion .... Perittopinae
– All tarsi 3-segmented (basal segments short); forewings not divided into coriaceous and membranous portions, but with 4 closed cells (Fig. 20.2D) .... Veliinae
5. Foretarsus 1-segmented; middle tarsus rarely more than twice as long as hind tarsus ....
   .... Microveliinae
– Foretarsus 2-segmented; middle tarsus 3 or more times length of hind tarsus .... Haloveliinae

HALOVELIINAE. Middle legs greatly elongated; posterior pronotal lobe strongly reduced in apterous forms; macropterous *Entomovelia* Esaki and *Stongylovelia* Esaki with forewing venation reduced to some short basal thickenings and pair of obsolete longitudinal veins.

Included genera are *Entomovelia* Esaki, *Halovelia* Bergroth, *Halovelioides* Andersen, *Strongylovelia* Esaki, and *Xenobates* Esaki. All *Halovelia* are intertidal, living on relatively protected oceanic areas such as coral reefs; the remaining taxa live in mangroves and estuaries in the Orient, including New Guinea. Certain authors have placed this highly modified group within the Gerridae (e.g., Esaki, 1924, 1930) or in its own family. *Halovelia* has been studied in detail by Andersen (1989a, c), with 30 species being recognized; Andersen (1991c) offered a revised key to the genera.

MICROVELIINAE. Tarsal formula unique, 1-2-2.
This is the largest subfamily of Veliidae, with 21 genera placed in 3 tribes—Hebroveliini, Veliohebriini, Microveliini (Štys, 1976; Andersen, 1982a). The first two tribes are monotypic, have the pretarsus inserted near the apex of the tarsus (subapical), and are probably largely terrestrial.

The largest genus of Microveliini, *Microvelia* Westwood, contains approximately 170 species. These are the classic small veliids that occur around the shores of ponds and springs and in quiet stream-margin habitats. Drake and Hussey (1955) provided a checklist to the New World species, Poisson (1940) treated some of the species known from Africa, Esaki and Miyamoto (1955) dealt with the Japanese species, and J. T. Polhemus (1974) dealt with the *M. austrina* group from Central America and Mexico. *Xiphovelia* Lundblad from the Orient (Japanese species treated by Esaki and Miyamoto, 1959b), *Xipheveloidea* Hoberlandt from Africa, and *Euvelia* Drake from Bolivia and Brazil inhabit flowing mountain streams or the margins of rivers (J. T. Polhemus and D. A. Polhemus, 1984) and, along with the

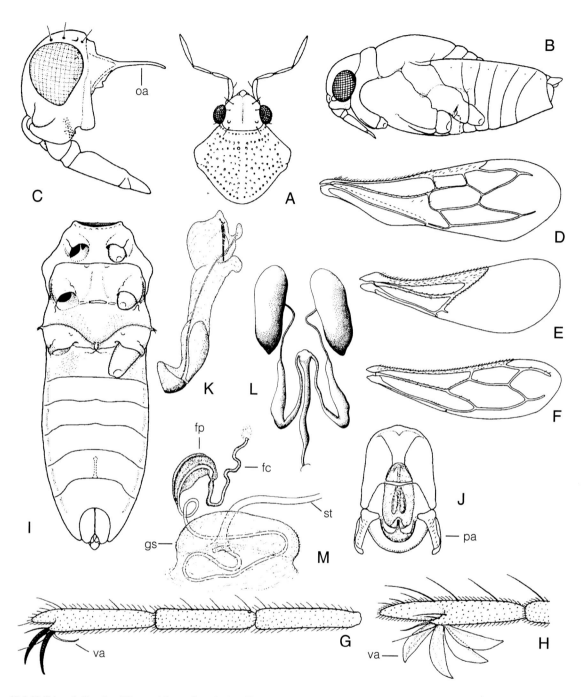

Fig. 27.1. Veliidae. **A.** Head and thorax, *Microvelia reticulata* (Burmeister). **B.** Lateral view, body, *Entomovelia doveri* Esaki. **C.** Lateral view, head, *Ocellovelia germari* (Distant). **D.** Right forewing, *O. germari*. **E.** Right forewing, *Perittopus* sp. **F.** Right forewing, *Microvelia pulchella* Westwood. **G.** Left middle tarsus, *Perittopus* sp. **H.** Left middle pretarsus, *Xiphoveloidea major* (Poisson). **I.** Ventral view, body, *O. germari*. **J.** Pygophore and parameres, dorsal view, *Neoalardus typicus* (Distant). **K.** Aedeagus, lateral view, *Velia caprai* Tamanini (A–K, M from Andersen, 1982a). **L.** Testes, *V. caprai* (from Pendergrast, 1957). **M.** Gynatrial complex, *Tenagovelia sjoestedti* Kirkaldy. Abbreviations: fc, fecundation canal; fp, fecundation pump; gs, gynatrial sac; oa, occipital apodeme; pa, paramere; st, spermathecal tube; va, ventral arolium.

marine Neotropical *Husseyella* Herring, have 4-bladed swimming fans. Andersen (1981a) reviewed the genus *Pseudovelia* Hoberlandt in the Orient, documenting the presence of 11 species, all of which live near the margins of streams or rivers, some on the riffle zone. Four species of *Baptista* Distant and two of *Lathrovelia* Andersen live in secluded aquatic environments and water-filled cavities (Andersen, 1989b). *Veliohebria* Štys from New Guinea is apparently largely terrestrial in habits (Štys, 1976). *Tonkuivelia* Linnavuori (1977) is unusual in the Microveliini, having a subapical insertion of the pretarsus, and is also probably terrestrial.

OCELLOVELIINAE. Ocelli present; forewing with 6 cells (Fig. 27.1D); tarsi 3-segmented; pretarsus inserted preapically.

Ocellovelia China and Usinger (1949a), with two species from South Africa, is a group that at various times has been placed in the Mesoveliidae, Macroveliidae, and an omnibus Veliidae.

PERITTOPINAE. Foretarsus 2-segmented; venation of forewing reduced, with 2 closed cells (Fig. 27.1E).

Contains only *Perittopus* Fieber with about five species from the Orient.

RHAGOVELIINAE. Pretarsus of middle leg with swimming fan formed from dorsal branch of ventral arolium inserted at base of deep cleft in distal segment of tarsus.

Included genera are the widely distributed *Rhagovelia* Mayr, with many species, commonly referred to as riffle bugs; *Tetraripis* Lundblad; and the marine *Trochopus* Carpenter from the Caribbean (Lundblad, 1936; Bacon, 1956; Matsuda, 1956; Hungerford and Matsuda, 1961; J. T. Polhemus and D. A. Polhemus, 1989a; J. T. Polhemus, 1990a).

VELIINAE. Tarsi 3-segmented on all legs; pretarsus simple in most genera; thorax unmodified in apterous morphs; parameres large and symmetrical; female with foretibial grasping comb (Andersen, 1982a).

Included genera are *Angilia* Stål, *Angilovelia* Andersen, *Oiovelia* Drake and Maldonado, *Paravelia* Breddin, *Stridulivelia* Hungerford, *Velia* Latreille (see Tamanini, 1947, 1955, for European taxa), and *Veloidea* Gould. *Velia* spp. and *Paravelia* spp. are often found on slow-moving streams and are some of the largest members of this family of generally rather small bugs.

Specialized morphology. The "pit organs" described in *Halovelia* by Cobben (1978) resemble the punctures common in most Gerromorpha (Fig. 10.7D) (except Gerridae and Hematobatidae). Also the Veliidae possess what Andersen (1977c) referred to as "thornlike outgrowths" (the "grooved setae" of Cobben [1978]), straight or slightly curved structures thickened at the base and tapering apically and projecting above the microhair layer.

The Veliidae, in common with the Gerridae, and as opposed to all other Gerromorpha, do not have an isoradial salivary pump, but rather the pump is inflected from behind on both sides; rather than being entirely membranous, the dorsal surface is sclerotized (Andersen, 1982a).

The posterior margin of the head capsule is provided with a pair of slender occipital apodemes (Fig. 27.1C), which serve as attachments for the maxillary retractor muscles inserting on the prolonged maxillary stylets.

The veliine genera *Angilovelia* Andersen and *Stridulivelia* possess stridulatory structures incorporating the metafemur and the connexival margin (Andersen, 1981a) (see also Chapter 10).

Many veliids possess striking modifications of the pretarsus, and indeed the family contains more variation in these structures than all the other Gerromorpha combined. The claws and dorsal and ventral arolia are often modified, as for example in *Xiphoveloidea* (Fig. 27.1H), where they are in the form of four leaflike structures, whereas in the Rhagoveliinae a branching process arising dorsally from the ventral arolium forms a swimming fan.

In contrast to all other members of the family, males of some *Microvelia* spp. have asymmetrically developed parameres, the left paramere being greatly reduced.

The gynatrial complex has a glandular gynatrial sac, and (except in *Ocellovelia*) the basal thickening of the fecundation canal is modified into a "fecundation groove," or as in the Gerridae, a "fecundation pump" may be present on the fecundation canal (Fig. 27.1M).

Natural history. Most Veliidae live on or near standing water, some are marine, and *Hebrovelia, Tonkuivelia,* and *Veliohebria* are apparently semiterrestrial. The group does not show as strong a relationship with the open water surface as do the Gerridae. At least four species each of *Paravelia* and *Microvelia* live in the tanks of Bromeliaceae or occasionally in tree holes in the New World tropics (J. T. Polhemus and D. A. Polhemus, 1991a). A few species of veliids have been recorded from foam masses in tropical streams. Among these is *Oiovelia spumicola* Spangler (1986) (Veliinae) from southern Venezuela, which lives secreted inside the foam masses; it is a member of a genus whose other two described species are recorded as free-living. *Afrovelia phoretica* J. T. Polhemus and D. A. Polhemus (1988) (Microveliinae) was recorded as occurring on the surface of foam masses caught among debris in rivers in western Madagascar.

The locomotory mechanisms of most genera are like those in the foregoing families of Gerromorpha, whereby the legs are moved alternately and no single pair predominates. Some groups, such as *Velia* live on slow-moving water, have the middle legs slightly elongate, and move them in a rowing motion—in contrast to the alternating triangle gait practiced by most insects—but have retained the ability to walk on land. Many species can move extremely rapidly on the water surface. *Stridulivelia* spp. form fast-swirling swarms. *Rhagovelia* spp., which often form huge schools, represent the extreme of modification within the family, possessing adaptations for life on extremely fast-flowing and disturbed water. In the last group, the middle legs are the primary locomotory appendages, the pretarsus is modified into a swimming fan, and the contralateral legs are moved simultaneously in a rowing motion; the ability to walk on land has been lost (Andersen, 1976).

Many veliids are known to practice "expansion skating." This phenomenon is produced by the ejection of a droplet of fluid from the rostrum; the surface tension of the water is momentarily lowered, causing the bug to move away much more rapidly than is possible simply by the use of the legs (Linsenmair and Jander, 1963; Andersen, 1976).

Veliids and other tiny aquatic bugs are capable of grasping the water surface for ascension of the meniscus, an obstacle often of greater size than the bug itself. This phenomenon was apparently observed independently by Baudoin (1955) and Miyamoto (1955).

Sexual dimorphism is present in some veliids, usually with males being smaller than females. In the most extreme cases the female may be nearly twice the size of the male. Males are sometimes phoretic and may spend long periods riding on the backs of the females. This behavior was described in detail by Miyamoto (1953; see also Andersen, 1982a:310) for *Microvelia diluta* Distant, and has also been noted in *Afrovelia phoretica* (J. T. Polhemus and D. A. Polhemus, 1988), and in *Halovelia* spp. (Kellen, 1959). J. T. Polhemus (1974) described modifications of the thorax in *Microvelia* to accommodate the male, whose ability to grasp his mate is apparently facilitated by combs located on the tibiae of the fore- (and sometimes also middle) legs.

Many species of Veliidae self-mutilate their wings.

The eggs of the Veliidae are glued lengthwise to the substrate, attached by their less convex side, often to marginal vegetation.

Distribution and faunistics. Identification aids for the western Palearctic include those by Stichel (1955; *Velia* prepared by Tamanini; see also Tamanini, 1947) and Kerzhner and Jaczewski (1964). The eastern Palearctic has been treated by Kanyukova (1988). Smith and Polhemus (1978) keyed the North American species, and J. T. Polhemus and Chapman (1979d) keyed the North American genera and California species. The species of the Great Lakes region were keyed by Bennett and Cook (1981). Cobben (1960b) treated the fauna of the Netherlands Antilles. New World *Rhagovelia* were monographed by Bacon (1956).

Useful treatments for Asia include Esaki and Miyamoto (1955) on *Microvelia* and *Pseudovelia* from Japan and (1959b) on *Xiphovelia* from Japan; Lundblad (1936) on Old World *Rhagovelia* and *Tetraripis*; J. T. Polhemus and D. A. Polhemus (1989a) and J. T. Polhemus (1990a) on *Rhagovelia* from the Malay Archipelago and Southeast Asia; and Andersen (1989a, c, 1991c) on *Halovelia* and *Halovelioides*. The African *Microvelia* fauna has been treated by Poisson (1940) and Linnavuori (1977), and Poisson (1957a) dealt with some taxa in the South African fauna.

28
Gerridae

General. The Gerridae, known as water striders, pond skaters, or wherrymen, spend—with few exceptions—nearly their entire lives on the open water surface. They range in length from some diminutive *Rheumatobates* spp. (Fig. 28.1A) of 1.6 mm to the 36-mm long *Gigantometra*. They are always long-legged, and their bodies vary from nearly globular (Fig. 20.3C) to elongate and cylindrical (Fig. 20.3B).

Diagnosis. Entire body and appendages covered with micro- and macrohair layer; peg plates absent; head usually somewhat extended beyond anterior margin of compound eyes, antennal fossae often visible from above (Fig 28.1B); ocelli absent; 4 pairs of cephalic trichobothria (Fig. 28.1C) (rather than 3 as in other Gerromorpha), all inserted in pits in the Rhagadotarsinae and *Eurygerris* (Gerrinae), other groups with fourth pair projecting from body surface as in nymphs; antennal segment 1 usually longer and somewhat stouter than remaining 3 segments (Fig. 20.3C); labium usually surpassing prosternum, segment 1 longer than segment 2, segment 3 longest, segment 4 short (Fig. 28.1B); pronotum without collar, never punctured, extending posteromesally to obscure mesonotum in macropterous forms (Fig. 20.3B); mesothorax elongated relative to the condition found in other Gerromorpha, particularly so in larger taxa, ratio of meso- to metathoracic lengths ranging from 1.2:1 to 10:1 (Fig. 28.1F); lateral channels of metathoracic scent-gland system, when present, similar in structure to those of Veliidae; forelegs relatively short and stout, middle and hind legs slender and greatly elongated, femora and tibiae generally subequal in length (Fig. 20.3B, C); forecoxa small and inserted near midline of thorax; middle and hind coxae rotated and extended caudad (Fig. 28.1F), middle coxa large and usually greatly extended along lateroventral margin of thorax, hind coxa somewhat smaller than midcoxa and not similarly prolonged; all tarsi 2-segmented, segments of varying lengths, with large campaniform sensillum on segment 1, suggesting fusion of primitive segments 1 and 2; pretarsus inserted anteapically, usually laterally; dorsal arolium usually bristlelike and short, ventral arolium usually well developed; parempodia setiform; claws of middle and hind legs usually smaller than those of forelegs; forewing venation variable, never with more than 4 cells, 2 apical cells each terminating in a relatively long free vein (Figs. 20.3C, 28.1D, E); hind wing lacking a distinct first anal vein; alary polymorphism common, macropterous and apterous morphs predominating; abdominal length variable; pygophore usually protruding from apex of abdomen (Fig. 20.3C), sometimes obviously asymmetrical, including segment 8; parameres usually present and symmetrically developed; aedeagus as in Fig. 28.1G; ovipositor generally relatively short with plate-shaped valvulae; 4 ovarioles; gynatrial complex as in Fig. 28.1H.

Classification. Following the classification of Andersen (1982a), we recognize eight subfamilies containing 60 genera and approximately 500 species.

Key to Subfamilies of Gerridae

1. First abdominal sternum present and distinct; ovipositor serrate, elongate; bucculae produced (Fig. 28.1C) .... Rhagadotarsinae
   – First abdominal sternum absent, or fused with metasternum; ovipositor not serrate, short; bucculae not produced (Fig. 28.1B) 2
2. Forewing differentiated into a coriaceous basal portion and membranous distal portion (Fig. 28.1E); middle femur stout, shorter than mesotibia and usually shorter than hind femur; metathoracic scent-gland orifices usually absent .... Trepobatinae
   – Forewing not differentiated into a thickened basal and membranous distal portion (Figs. 20.3B, 28.1D); middle femur slender, usually distinctly longer than middle tibia; metathoracic scent-gland orifices present 3
3. Metasternum greatly reduced, represented by a very short subtriangular plate enclosing the scent-gland orifice; if metasternum reaching metacetabular region, then middle tibia with distinct fringe of setae; claws of hind tarsus straight or S-shaped .... Halobatinae
   – Metasternum well developed, reaching metacetabular region laterally; claws of hind tarsus falcate or absent 4
4. Metacetabular groove distinct, dorsally reaching anterior end of abdominal tergum 1; foretarsus very long, at least one-half length of foretibia .... Ptilomerinae
   – Metacetabular groove often indistinct dorsally, if distinct, not reaching anterior end of abdominal tergum 1; foretarsus less than one-half length of foretibia 5
5. Metacetabular groove distinct, dorsally connected with posterior margin of mesopostnotum; antennal segment 4 curved; labium not extending posteriorly beyond prosternal margin; first segment of foretarsus less than one-half length of second segment .... Cylindrostethinae
   – Metacetabular groove indistinct dorsally; antennal segment 4 straight; labium distinctly surpassing prosternal margin; first segment of foretarsus usually more than one-half length of second segment 6
6. Metasternum with lateral scent channels leading from scent-gland orifices to metacetabular region; secondary straight transverse line present across metanotum; no connexival spines present .... Charmatometrinae
   – Metasternum usually lacking lateral scent channels, if present then insects 30 mm in length; connexival spines present; metanotum without a transverse line 7
7. Pronotal lobe reduced in apterous individuals; arolia present; lateral suture between meso- and metathorax distinct .... Eotrechinae
   – Pronotal lobe usually not reduced in apterous individuals, but if reduced then arolia absent or lateral suture between meso- and metathorax indistinct .... Gerrinae

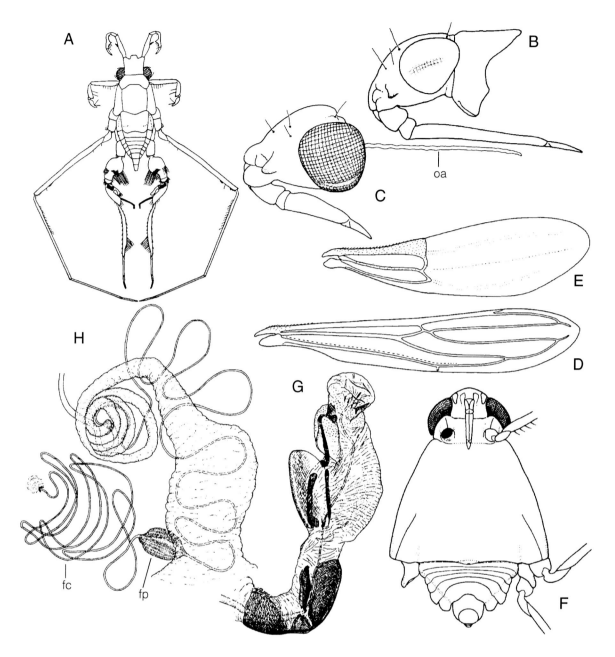

Fig. 28.1. Gerridae. **A.** *Rheumatobates meinerti* Schroeder, male (from Hungerford, 1954). **B.** Lateral view, head and pronotum, *Onychotrechus rhexenor* Kirkaldy. **C.** Lateral view, head, *Rhagadotarsus kraepelini* Breddin. **D.** Right forewing, *Aquarius paludum* (Fabricius). **E.** Right forewing, *Trepobates taylori* (Kirkaldy). **F.** Ventral view, body, apterous male, *Eurymetra natalensis* (Distant). **G.** Aedeagus, inflated, lateral view, *Neogerris parvulus* (Stål). **H.** Gynatrial complex, dorsal view, *Cylindrostethus palmaris* (Drake and Harris) (B–H from Andersen, 1982a). Abbreviations: fc, fecundation canal; fp, fecundation pump.

CHARMATOMETRINAE. Lateral scent-gland channels present (found elsewhere only in Cylindrostethinae and *Gigantometra*); distinct sutures between the meso- and metathorax; secondary transverse line in front of first abdominal tergum.

Included genera are *Brachymetra* Mayr (Shaw, 1933), *Charmatometra* Kirkaldy, and *Eobates* Drake and Harris. The group is restricted to the Neotropics and has been little studied. Available observations suggest its members live on quiet areas of streams.

CYLINDROSTETHINAE. As in Ptilomerinae and Halobatinae, antennal segment 2 with trichobothrium-like setae; labium short, not surpassing posterior margin of prosternum; uniquely with distinctly flattened middle tibia and tarsus; heavily sclerotized second valvulae.

Included genera are *Cylindrostethus* Mayr, *Platygerris*

Buchanan-White, and *Potamobates* Champion. They are found in the Neotropics, with *Cylindrostethus* also occurring in the Old World (Hungerford and Matsuda, 1962). All appear to be stream dwellers, the last two genera occupying broad flat areas of flowing streams.

EOTRECHINAE. Shares with Gerrinae sclerotized and setose inner lobes of first valvulae and 2 distinctly glandular areas in wall of gynatrial sac; unique in possessing spinose setae on tibiae and tarsi of middle and hind legs; pretarsus of *Eotrechus* inserted apically.

Included genera are *Amemboa* Esaki, *Chimarrhometra* Bianchi, *Eotrechus* Kirkaldy, *Onychotrechus* Kirkaldy (Andersen, 1980), and *Tarsotrechus* Andersen (Andersen, 1980); all occur in India and the Orient and are generally hygropetric or semiterrestrial. A generic phylogeny was published by J. T. Polhemus and Andersen (1984).

GERRINAE (FIG. 20.3B). In addition to characters shared with the Eotrechinae mentioned above, also recognized by the following: wing venation relatively complete; dorsal and ventral arolia usually strongly reduced; parempodia both rather short; metathoracic scent-gland channel usually absent; abdomen usually elongate; vesica with sclerotized dorsal plate.

Two tribes are currently recognized. The Tachygerrini comprise the New World genera *Tachygerris* Drake and *Eurygerris* Hungerford and Matsuda. The remaining 10 genera are placed in the cosmopolitan Gerrini and include large commonly encountered members of the genera *Aquarius* Schellenberg (Andersen, 1990), *Gerris* Fabricius, *Limnogonus* Stål (Andersen, 1975), and *Limnoporus* Stål (Andersen and Spence, 1992); all are frequently associated with quiet waters, such as ponds. *Tenagogonus* Stål from the Old World tropics (Hungerford and Matsuda, 1958b), by contrast, is usually found in quiet places on streams. The Western Hemisphere fauna was treated by Drake and Harris (1934).

HALOBATINAE. Head and thorax short and wide; metasternum reduced; abdominal segments shortened; thoracic and abdominal segments strongly fused.

Members of the 10 halobatine genera, placed in two tribes, represent some of the most highly modified Gerridae. The tropicopolitan Halobatini comprise *Asclepios* Distant and *Halobates* Eschscholtz (Herring, 1961; Andersen, 1991b; J. T. Polhemus and D. A. Polhemus, 1991b; Andersen and Foster, 1992), most species of which are marine. The remaining genera—placed in the Paleotropical Metrocorini and comprising, for example, the Oriental *Esakia* Lundblad (Hungerford and Matsuda, 1958a; Cheng, 1966), *Metrocoris* Mayr (Boer, 1965; D. A. Polhemus, 1990b; Chen and Nieser, 1993), and *Ventidius*—are found on slow-flowing waters, usually in heavily shaded areas (Hungerford and Matsuda, 1960a; Cheng, 1965).

An important historical paper on the group is that of Esaki (1926). *Metrocoris* has been revised for the Malay Archipelago and the Philippine Islands (D. A. Polhemus, 1990b), with an additional large and mostly undescribed fauna known from tropical continental Asia.

PTILOMERINAE. Antennal segment 4 with an elongate trough ventrally; first segment of foretarsus longer than second.

This subfamily comprises eight genera distributed from Madagascar to New Guinea. *Ptilomera* Amyot and Serville (Hungerford and Matsuda, 1965), is a group of often large insects, all members of which inhabit swiftly flowing waters, often near waterfalls. *Potamometra* Bianchi was revised by Drake and Hoberlandt (1965).

RHAGADOTARSINAE (FIG. 28.1A). Bucculae well developed (Fig. 28.1C), in contrast to other Gerridae; metathoracic scent apparatus absent (also in some other Gerridae); *Rhagadotarsus* with pretarsus inserted ventrally and with thornlike cuticular outgrowths found in the Veliidae; laterotergites and true abdominal sternum 1 distinct; ovipositor long, laciniate, and serrate.

Rhagadotarsus Breddin, with five species from the Old World tropics (J. T. Polhemus and Karunaratne, 1993), is ordinarily found in large schools on the surface of standing water, as for example in fish ponds and on quiet reaches in streams. *Rheumatobates* Bergroth, comprising more than 30 species, is widely distributed in the New World (Hungerford, 1954; J. T. Polhemus and Spangler, 1989), occupying a variety of habitats, including muddy pools, quiet portions of large rivers, and even estuaries; they often assemble in large swarms. Substantial information on the group was provided by Esaki (1926).

TREPOBATINAE (Fig. 20.3C). Middle femur relatively short and stout, always shorter than middle tibia; forewing with 2, closed, basal cells, and transverse line of weakness (Fig. 28.1E).

This subfamily comprises nineteen genera. A world review was provided by J. T. and D. A. Polhemus (1993). The New World genus *Metrobates* Uhler usually lives on streams or rivers. It is distinctive in the subfamily for its flattened body, modified second and third antennal segments, relatively long middle tibia, and conspicuous claws on the middle and hind legs (Anderson, 1932; Drake and Harris, 1932a). *Trepobates* Uhler (Drake and Harris, 1932b), also from the New World, ordinarily inhabits quiet or slow-moving water. *Telmatometra* Bergroth from the New World tropics was revised by Kenaga (1941).

Specialized morphology. The Gerridae have several novel morphological features in common with the Veliidae, including posterior margin of head capsule with pair of long occipital apodemes for attachment of maxillary retractor muscles (Fig. 28.1C); salivary pump laterally inflected, membranous in forms studied, but sclerotized dorsally in *Eurymetra* (Halobatinae); and gynatrial complex usually with a fecundation pump and glandular gynatrial sac (Fig. 28.1H).

The single most distinctive feature of the Gerridae, one unique in the Heteroptera, is the modification of the thorax in association with life on the water surface film, wherein the mesothorax is always elongated, sometimes greatly so, and the middle and hind coxae are oriented in a horizontal plane (Fig. 28.1F). Details of complex thoracic fusions in apterous morphs and allometric growth of appendages were reviewed in detail by Matsuda (1960).

The antennae of some male *Rheumatobates* spp. are irregularly thickened and bent and covered with tufts of setae (Fig. 28.1A). They are used to grasp the female during copulation, a function performed by the foretibiae in all other members of the family (Hungerford, 1954). The hind legs of *Rheumatobates* are also highly modified (Fig. 28.1A) compared with all other Gerridae (Fig. 20.3B–D).

Males of the genus *Ptilomera* have a brush of long setae on the posterior surface of the middle tibia.

Asymmetry of male terminal abdominal segments occurs in most *Halobates* spp. and in some Cylindrostethinae. Although the parameres are always symmetrically developed, they are very small in the Gerrinae, Eotrechinae, and Cylindrostethinae and are absent in the Rhagadotarsinae and *Halobates*.

Natural history. Gerrids commonly occupy bodies of standing water, ranging from relatively small pools to lakes, and indeed they are most easily observed in such quiet situations. Nonetheless, many taxa live on running waters, some of them capable of gliding easily on the rapids or torrents of mountain streams, in which situations they are much less easily observed. The remaining species inhabit marine environments, the majority found in relatively protected areas such as estuaries, mangroves, and lagoons, with a few species of *Halobates* completing their entire life cycle on the open ocean.

"Ripple communication" is well known in the Gerridae and has been studied in some detail; it is not well known to what degree it plays a role in the lives of other gerromorphans, although it has been recorded in the veliid *Microvelia longipes* (J. T. Polhemus, 1990b). Gerrids use waves on the water surface for prey location and for communicating with each other. Prey on the water surface are located by orienting to the waves they make (Murphey, 1971). Gerrids themselves also produce surface waves of a species-specific frequency by vertical movements of one or more pairs of legs. "Calling signals" are produced by males, and both sexes may produce additional "courtship signals" (Wilcox, 1972, 1980). Such signals allow for sex discrimination within a species (Wilcox, 1979) and play a role in territory delimitation and defense as well as species recognition as a defense against cannibalism (Wilcox, 1980; Wilcox and Spence, 1986). The reception of such surface waves is thought to occur either in the stretch receptors located in the tibiotarsal joint (Murphey, 1971) or in specialized setae located on the tarsi (see Lawry, 1973). Andersen (1982a) suggested that vision is also important in prey location.

Most members of the Eotrechinae, including species of *Chimarrhometra*, *Eotrechus*, and *Onychotrechus*, have moved from life on the water surface film to being virtually terrestrial or hygropetric, living in situations typical of those often occupied by the Hebridae. They are capable of extremely rapid and agile movements in such habitats, even on almost vertical surfaces. Andersen (1982c) described the novel orientation of the middle and hind legs in *Eotrechus*, the tibiae being positioned nearly perpendicular to the substrate, with the weight of the animal resting on the relatively short tarsi. The evidence suggests that the apically inserted claws of the Eotrechinae represent a secondarily derived condition.

Eggs are fixed to floating objects in a lengthwise position, with the convex side uppermost. *Rhagadotarsus* is known to embed its eggs in plant tissue, and the structure of the ovipositor suggests that *Rheumatobates* does the same (Andersen, 1982a).

Description of wing polymorphism and methods of its determination have been intensively studied in the Gerridae. They are discussed in Chapter 6. Whereas most gerrids are either macropterous or apterous, intermediate wing morphs are known in *Gerris*.

Distribution and faunistics. Keys to the higher groups and genera and illustrations of representatives of most genera can be found in Hungerford and Matsuda (1960b) and in Andersen (1982a); Matsuda (1960) provided diagnoses of all genera. Faunistic treatments are available for Europe (Stichel, 1955; Wagner and Zimmermann, 1955), the former Soviet Union (Kanyukova, 1982), California (J. T. Polhemus and Chapman, 1979e), Argentina (Bachmann, 1966), the Malay Peninsula and Singapore (Cheng and Fernando, 1969; see also Miyamoto, 1967), Japan (Miyamoto, 1961b), and Africa and Madagascar (Poisson, 1965), with many more-restricted surveys also present in the literature.

PANHETEROPTERA

29
Nepomorpha

General. The water bugs were first recognized as a group by Latreille (1810) under the name Hydrocorisae. All except some Corixidae are predators. The short antennae, which are concealed—often completely—below the eyes, are common to all families and have traditionally been treated as an adaptation to the aquatic mode of life. The Ochteridae and Gelastocoridae are usually riparian, however, and they do not possess the streamlined bodies and swimming legs found in the remaining nine families. The Nepomorpha range from very small, as in some species of Helotrephidae and Pleidae, to the largest of all bugs, the Neotropical belostomatid *Lethocerus maximus* DeCarlo, with a length of 110 mm. The painful nature of the bite of the aquatic bugs is the subject of folklore—and reality.

Diagnosis. Compound eyes usually very large and occupying nearly entire sides of head in dorsal and lateral views (Fig. 33.1); ocelli absent (except Ochteroidea, Corixidae: Diaprepocorinae); cephalic trichobothria always absent; labium usually short and very stout (Fig. 32.2A) (except Ochteridae: Fig. 33.2A; Aphelocheiridae: Fig. 37.1C), often with 3 apparent segments (except Corixidae), segment 1 greatly reduced; antennae at most as long as head, often thickened (Fig. 29.1A, D), provided with processes (Fig. 31.2B), or with segments fused, situated posteroventrally on head, often hidden in groove or concavity, musculature reduced; middle and hind legs often flattened (Fig. 30.1) and fringed with setae; forelegs often raptorial with tibia apposed to femur; pretarsus, particularly of foreleg (Fig. 10.4A), sometimes with one claw reduced or absent; parempodia setiform,

Keys to Nepomorpha prepared by Pavel Štys.

number variable from 2 to 4 (Fig. 29.2B, E, G); dorsal and ventral arolia usually present in both nymphs and adults (Fig. 29.2A–I); forewings with well-developed wing-to-body coupling mechanism and frena along scutellum and claval commissure, in the form of hemelytra with coriaceous anterior portion formed of distinct corium and clavus and posterior membranous portion (Fig. 33.1), with none to many veins; scolopophorous organs usually present in meso- and metathorax as well as on abdominal tergum 1 (Larsén, 1957; Parsons, 1962; Mahner, 1993); spiracle present dorsally on abdominal tergum 1, spiracles on segments 2–8 usually ventral; abdomen of fully aquatic forms with various modifications for respiration under water; male genitalia symmetrical in Nepoidea, asymmetrical in most other groups except some Naucoridae (Cheirochelini), Potamocoridae, and many Notonectidae, asymmetry often including some pregenital abdominal segments; ovipositor often weakly developed; spermatheca tubular, without a terminal bulb; number of micropyles variable.

Discussion. We recognize 11 families of Nepomorpha, following the classification of Štys and Jansson (1988). China (1955c) proposed a scheme of relationships among them, treating the Ochteridae as the relatively most primitive, on the basis of its possession of ocelli and the respiratory system typical of terrestrial bugs. More recent authors, including Popov (1971), Rieger (1976), and Mahner (1993), have treated the Nepoidea as the sister group of remaining Nepomorpha. Popov (1971) documented that the fossil record for many families dates back to the mid-Mesozoic, and he provided a detailed discussion of the morphology of fossil and recent taxa. The character analyses of Rieger (1976) and Mahner (1993) show a substantial increase in breadth of coverage and methodological sophistication over previous efforts. Their conclusions contradict one another mainly in the placement of the Corixidae and relationships among those families we place in the Naucoroidea, differences that can be appreciated by comparing Figures 29.3 and 29.4 (Figure 29.3 incorporates some modifications of family-group status from the work of Štys and Jansson [1988]).

The taxonomy of parts or all of many nepomorphan families has been the subject of detailed publications, particularly by H. B. Hungerford and his many students. The attention that has been paid to the group probably results in large part because the habitat of most aquatic bugs is obvious, they often have long adult life spans, and in temperate areas they can be collected even in winter. Excellent morphological studies have been published on several groups, and many families are also the subject of detailed life history investigations.

The western Palearctic fauna was treated by Poisson

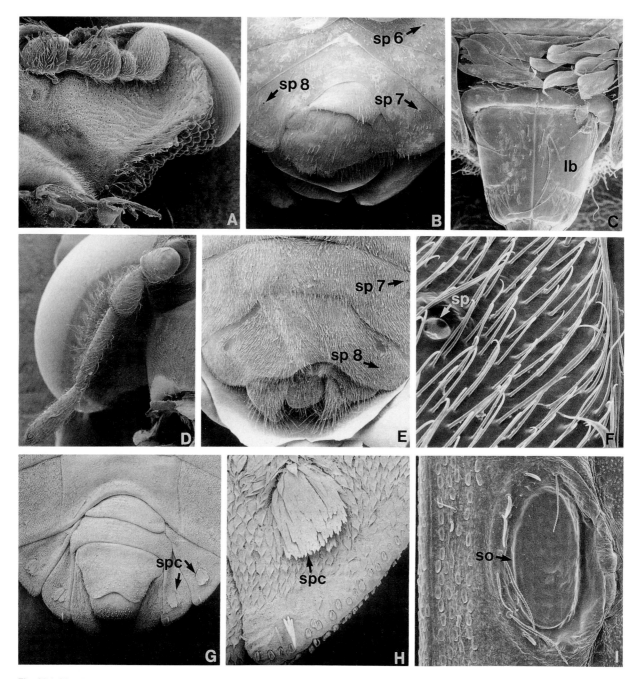

Fig. 29.1. Morphology of Nepomorpha. **A.** Ventral view, head and antenna, *Gelastocoris oculatus* (Gelastocoridae). **B.** Ventral view, male genital segments, *G. oculatus* (Gelastocoridae). **C.** Base of adult labium, *Hydrotrephes* sp. (Helotrephidae). **D.** Ventral view, head and antenna, *Ochterus caffer* (Stål) (Ochteridae). **E.** Ventral view, male genital segments, *O. caffer* (Ochteridae). **F.** Ventral abdominal hair pile, including spiracle, adult *Hydrotrephes* sp. (Helotrephidae). **G.** Abdominal venter, adult male *Cryphocricos latus* (Naucoridae). **H.** Detail, spiracle and modified setae, *C. latus* (Naucoridae). **I.** Abdominal static sense organ, *Ranatra* sp. (Nepidae). Abbreviations: lb, labium; so, static sense organ; sp 6, spiracle 6; sp 7, spiracle 7; sp 8, spiracle 8; spc, spiracle cover.

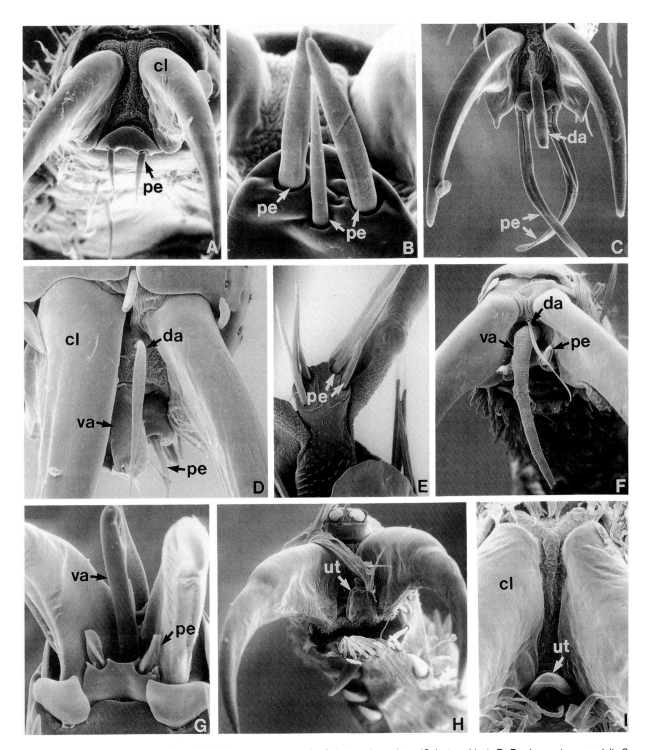

Fig. 29.2. Pretarsus of Nepomorpha. **A.** Foreleg pretarsus, nymph, *Gelastocoris oculatus* (Gelastocoridae). **B.** Foreleg pretarsus, adult, *G. oculatus* (Gelastocoridae). **C.** Foreleg pretarsus, nymph, *Ochterus* sp. (Ochteridae). **D.** Nymphal pretarsus, *Laccocoris hoogstraali* La Rivers (Naucoridae). **E.** Hind leg pretarsus, adult, *Cryphocricos latus* (Naucoridae). **F.** Middle leg pretarsus, adult, *Aphelocheirus lahu* Polhemus and Polhemus (Aphelocheiridae). **G.** Pretarsus, nymph, *Hydrotrephes* sp. (Helotrephidae). **H.** Hind leg pretarsus, nymph, *Diplonychus* sp. (Belostomatidae). **I.** Hind leg pretarsus, nymph, *Laccotrephes* sp. (Nepidae). Abbreviations: cl, claw; da, dorsal arolium; pe, parempodium; ut, unguitractor; va, ventral arolium.

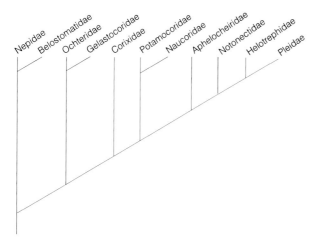

Fig. 29.3. Phylogenetic relationships of families of Nepomorpha (modified from Rieger, 1976).

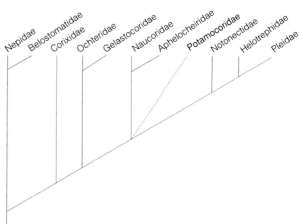

Fig. 29.4. Phylogenetic relationships of families of Nepomorpha (modified from Mahner, 1993).

(1957b), Kerzhner and Jaczewski (1964), and Savage (1989). Lundblad (1933) dealt with the faunas of Java, Sumatra, and Bali. Poisson (many papers) has dealt with much of the fauna of Africa. North America was first treated in detail at the generic level by Usinger (1956) and again by Menke (1979a). Nieser (1975) treated the fauna of Surinam.

Key to Families of Nepomorpha

1. Labium broadly triangular, short, nonsegmented, with transverse sulcation (except Cymatiainae); foretarsus unsegmented (sometimes fused with tibia), spoon-, scoop-, or sickle-shaped (Fig. 34.2A), rarely cylindrical, with ventral fringe of long setae; head overlapping pronotum; pronotum and forewings often with linear or hieroglyphical transverse light and dark patterns (Fig. 34.1) .... Corixidae
   – Labium cylindrical to conical, obviously segmented (e.g., Figs. 31.2A, 32.2A), without transverse sulcation; foretarsus segmented or not, without a fringe of setae; head never overlapping pronotum anterodorsally; coloration never as above 2
2. Apex of abdomen with paired respiratory processes (Figs. 30.1, 30.2, 31.1); body cylindrical or ovoid and flat; medium-sized to gigantic; forelegs raptorial (except *Limnogeton:* Fig. 30.2) 3
   – Apex of abdomen without respiratory processes; body never cylindrical, dorsum flat to extremely convex; very small to medium-sized; forelegs raptorial or cursorial, often modified in males 4
3. Respiratory siphon nonretractile, usually long and filiform (Fig. 31.1), sometimes short and straplike; all tarsi 1-segmented; hind tibiae simple (Fig. 31.1); metacoxae short, free .... Nepidae
   – Respiratory air straps short, retractile, often only apices visible (Fig. 30.1); tarsi 2- or 3-segmented, rarely foretarsus 1-segmented; hind tibiae usually flat (Fig. 30.1), with swimming setae; metacoxae conical, firmly united with metapleuron .... Belostomatidae
4. Ocelli present (Fig. 33.1), if obsolete or absent, then head transverse and eyes pedunculate or subpedunculate (submacropterous Gelastocoridae); legs cursorial 5
   – Ocelli absent; eyes sessile; hind and/or middle legs usually flattened, fringed with setae (except minute Pleidae) 6
5. Antennae relatively long, filiform, partially visible in dorsal view (Fig. 33.1); head moderately transverse; eyes sessile; scutellum flat; legs homomorphous, cursorial (Fig. 33.1) .... Ochteridae
   – Antennae short and incrassate, concealed in cavity between eyes and prothorax (Figs. 29.1A, 32.2A); head strongly transverse; eyes subpedunculate to pedunculate (Fig. 32.1); scutellum tumid; forefemur incrassate, anterior face sulcate to carinate .... Gelastocoridae
6. Dorsum flat to moderately convex (Figs. 35.1A, 36.1, 37.1A); head and prothorax never fused; forelegs strikingly raptorial or antennae relatively long, protruding beyond outline of head 7

- Dorsum usually strongly convex, inversely boat-shaped to tectiform (Figs. 38.1, 39.1, 40.1A), if flattened, then head and pronotum fused and cephalonotal sulcus incomplete; forelegs never raptorial, although sometimes modified in males 9
7. Antennae long, extending beyond lateral margins of head (Fig. 35.1B); labium slender, reaching from prosternum to mesosternum (Fig. 37.1C); head narrow, elongate, strongly produced in front of eyes; foretarsus 3- or 1-segmented, mobile; 2 well-developed claws 8
- Antennae short, not extending beyond head, usually not visible from above (except many Sagocorini and Tanycricini); labium short and thick, not surpassing prosternum (Fig. 36.2A); head usually transverse, anteocular portion only slightly produced in front of eyes (Fig. 36.1); foretarsus 2- or 1-segmented, immobile; 2, 1, or 0 small claws .... Naucoridae
8. Labium just reaching on prosternum, segment 3 shorter than or subequal to segment 4 (Fig. 35.1B); tarsal formula 1-2-2; macropterous to coleopteroid; corium-membrane boundary in macropterous forms sinuate; lateral margins of head, pronotum, and corium proximally with erect setae; Neotropical .... Potamocoridae
- Labium extending at least one-half length of mesosternum, segment 3 at least twice as long as segment 4 (Fig. 37.1C); tarsal formula 3-3-3, tarsomeres sometimes fused in small species; macropterous to micropterous; corium-membrane boundary straight in macropterous forms; outer posterior angle of corium rounded; lateral margins of head, pronotum, and forewings without erect setae; Eastern Hemisphere .... Aphelocheiridae
9. Elongate, wedge-shaped, usually over 4.0 mm long; eyes large (Figs. 38.1, 39.1A, B); vertex narrow; legs often markedly heteronomous; hind legs long, oar-shaped, with reduced and inconspicuous claws (Fig. 38.1); head not fused with prothorax; midline of pterothoracic venter ecarinate; midline of abdominal venter with a flattened to sharp keel, never with an irregular carina .... Notonectidae
- Broadly oval, robustly built species (Figs. 39.1, 40.1A) under 4 mm long; eyes small to medium-sized; vertex broad; legs homonomous; hind legs not oar-shaped, usually with 2 claws; head firmly and immovably associated with prothorax; basal abdominal segments with an irregularly shaped carina 10
10. Dorsum of head not fused with pronotum; head-pronotum boundary straight (Fig. 39.1); antennae 3-segmented; dorsum strongly convex .... Pleidae
- Dorsum of head fused with pronotum; cephalonotal sulcus not straight (Fig. 40.1A); antennae 2-segmented or nonsegmented (Fig. 40.1C); dorsum rather flat to extremely convex ....
.... Helotrephidae

Nepoidea

30 Belostomatidae

General. Often referred to as giant water bugs, this group, whose species range in length from 9 to 110 mm, includes the largest true bugs. All species are flattened dorsoventrally, ovoid to elongate ovoid (Figs. 30.1, 30.2), and brown in color. Only *Limnogeton* Mayr, the African snail predator (Fig. 30.2), lacks the greatly enlarged raptorial forefemora and middle and hind legs adapted for swimming.

Diagnosis. Antennae short, concealed in grooves on underside of head, usually 4-segmented, segments 2 and 3 with lateral projections (Fig. 30.3A); membrane of forewing with reticulate venation; forefemur usually enlarged with tibia apposed, foretarsus 2- or 3-segmented, 1 claw slightly to greatly reduced; middle and hind legs usually flattened (Fig. 30.1) and fringed with setae; abdominal tergum 8 modified into a pair of "airstraps"; nymphs with metepisterna extending posteriorly to cover 2 or 3 proximal abdominal sterna; dorsal abdominal scent glands nonfunctional in nymphs; static sense organs associated with spiracles 2–7; male genitalia symmetrical, aedeagus as in Fig. 30.3B–D; testes as in Fig. 30.3E; spermatheca as in Fig. 30.3F, G.

Classification. This group was first recognized as a higher taxon by Leach (1815) as Belostomida. The classification of Lauck and Menke (1961) recognized three subfamilies comprising nine genera and 146 species.

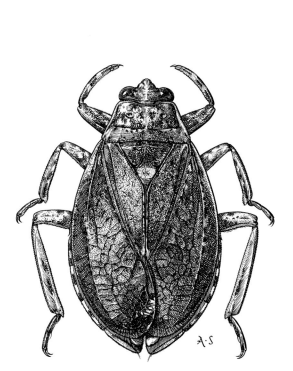

Fig. 30.1. Belostomatidae. *Abedus indentatus* (Haldeman) (from Usinger, 1956).

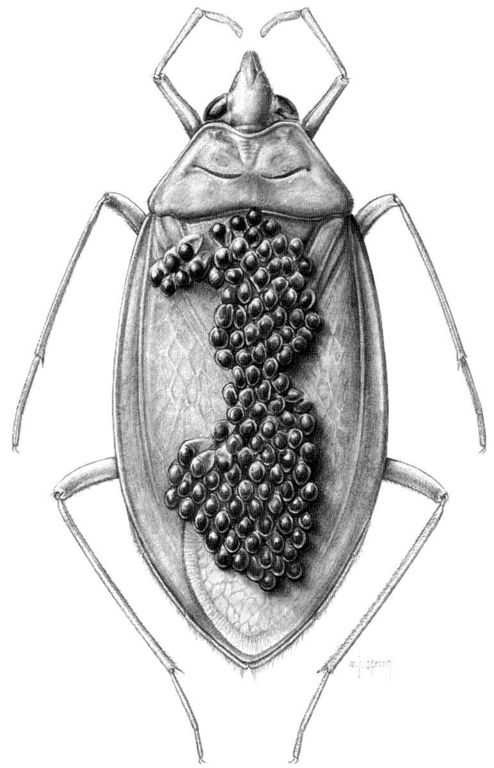

Fig. 30.2. Belostomatidae. *Limnogeton* sp.

Key to Subfamilies of Belostomatidae

1. Abdominal sterna 3–7 with midplate and paired lateral plates, the latter not differentiated by a mesal sulcus; sulci differentiating lateral plates terminating near proximal angles of mesal plate 7; abdominal spiracles located near center of lateral plates, far removed from lateral sulci ...... Belostomatinae
– At least abdominal sternum 6 and usually also some other segments (from 3 to 7) with mesal, sublateral, paired spiracle-bearing lateral plates; lateral sulci terminating near proximal angles of lateral plate 7; abdominal spiracles on mesal margins of lateral plates close to lateral sulci 2
2. Proximal portions of lateral lobes 7 (laterad of subgenital plate) subdivided into sublateral and lateral plates; lateral sulci on lateral lobes 7 either parallel to margins of subgenital plates and terminating discally, or converging toward tip of subgenital plate and disappearing underneath latter; antennal segments 2 and 3 each with one fingerlike projection, 4 with 2; hind tibia and tarsus thinly compressed, much more dilated than middle tibia and tarsus; foretarsus 3-segmented, although often appearing 2-segmented externally; anterior claw of foreleg large, posterior claw vestigial to absent; length 37–115 mm .... Lethocerinae
– Lateral lobes 7 formed entirely or largely by lateral plates, sublateral plates being absent or not visible externally, or developed as minute triangular sclerites; lateral sulci terminating at basilateral angles of subgenital plate (not continuing onto lateral lobes 7) or only slightly distad of them; antennal segment 2 with one fingerlike projection, 3 with a large expanded dorsal lobe, 4 short and transverse; tibiae and tarsi of middle and hind legs similar, flat, narrow; foretarsus 2-segmented (externally appearing 1-segmented); claws of foreleg paired, vestigial; length 25–30 mm .... Horvathiniinae

Key to subfamilies of Belostomatidae adapted from Lauck and Menke, 1961.

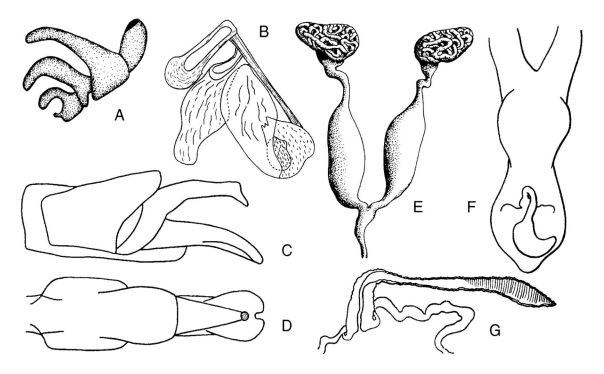

Fig. 30.3. Belostomatidae. **A.** Antenna, *Horvathinia* sp. (from Lauck and Menke, 1961). **B.** Aedeagus, *Diplonychus rusticum* (Fabricius) (from Kumar, 1961). **C.** Aedeagus, lateral view, *Lethocerus* sp. **D.** Aedeagus, sagittal view, *Lethocerus* sp. (C, D from Lauck and Menke, 1961). **E.** Testes, *Diplonychus rusticum*. **F.** Spermatheca, *Lethocerus niloticus* Stål. **G.** Spermatheca, *L. griseus* (Say) (E–G from Pendergrast, 1957).

BELOSTOMATINAE (FIGS. 30.1, 30.2). Relatively small to large, ovoid, elliptical, or elongate; antennal segments 2 and 3 each bearing a long, dorsal, fingerlike process, segment 4 long; middle and hind tibia and tarsus usually flattened, not broadly dilated; pretarsus of hind leg of *Diplonychus* sp. as in Fig. 29.2H; airstraps variable; metathoracic scent glands absent.

Included genera are *Abedus* Stål (10 species, southwestern United States to Central America; Menke, 1960b); *Belostoma* Latreille (approximately 60 species, New World; Lauck, 1962, 1963, 1964); *Diplonychus* Laporte (approximately 20 species, Africa east to Australia; Poisson, 1949); *Hydrocyrius* Spinola (five species, Africa and Madagascar; Brown, 1948; Poisson, 1949); *Limnogeton* Mayr (four species, Africa; Poisson, 1949).

HORVATHINIINAE. Medium-sized, elliptical; antennal segment 2 long, flattened, segment 3 flattened with a large dorsal lobe, segment 4 dorsoventrally elongated (Fig. 30.3A); hind tibia and tarsus flattened, not broadly dilated; airstraps short, mesal margins remote.

Only *Horvathinia* Montandon, with nine nominal species from central and southern South America, is included. Members of the group are rarely encountered in nature, and their habits and habitats are poorly known. The group was revised by DeCarlo (1958).

LETHOCERINAE. Large to very large, elongate ovoid; antennal segments 2–4 with dorsal appendages; hind tibia and tarsus broadly flattened; airstraps long, mesal margins nearly contiguous.

This subfamily includes only the cosmopolitan genus *Lethocerus* Mayr, comprising approximately 25 species. They are the largest of all heteropterans. Cummings (1933) and DeCarlo (1938) dealt with the New World fauna and Menke (1960a) with that of the Old World.

Specialized morphology. The "airstraps" of adults, derived from abdominal tergum 8, are the most distinctive feature of the group. They do not form a tube, as in the Nepidae; instead, air is transmitted to the subhemelytral airstore by a channel, formed by the setae (which converge mesioventrally).

Static sense organs are present in the Belostomatidae. They are associated with the spiracles of abdominal sterna 2–7 (Möller, 1921), and although not as large and complex, apparently serve the same function as those occurring in the Nepidae (Figs. 29.1I, 31.2C).

Natural history. Many species of Belostomatidae live in standing water. Most are excellent swimmers, although they generally lie in wait for their prey. A few taxa, such as *Abedus* Stål in the New World and some *Diplonychus* Laporte in the Old World tropics, live in slow-flowing portions of streams. Most species are probably general predators, eating whatever they are capable of catching, including vertebrates such as fishes and frogs. *Limnogeton*, which is unique in not having any of the legs modified for swimming, is apparently an obligate freshwater snail predator (Voelker, 1966).

Adult belostomatids protrude their airstraps through the surface, while lying motionless in the water. Respiration in adults takes place primarily through the dorsal first abdominal spiracles, the remaining ventral spiracles playing a lesser role (Miller, 1961; Parsons, 1972, 1973). Respiration in nymphs takes place from the airstore on the abdominal venter.

Uniquely among the Insecta, female Belostomatinae deposit their eggs on the dorsal surface of the males (Fig. 30.2). Males that are carrying eggs engage in brooding behavior, including exposing the eggs to atmospheric air (Smith, 1976).

Most belostomatids are capable of ejecting a foul-smelling liquid from the anus, apparently as a defensive reaction (Harvey, 1907). With the exception of the Belostomatinae, male giant water bugs have well-developed metathoracic scent glands and are said to produce a strong smell, although certainly not of the quantity or intensity of many large Pentatomoidea. When *Lethocerus* is eaten by humans, a common practice in Thailand and other parts of Asia, the scent-gland odor and taste are very obvious.

Although many nepomorphans, and in fact most heteropterans, are attracted to lights, the belostomatids have in some areas been dubbed "electric light bugs" because of their conspicuous presence, especially at mercury vapor lamps.

Distribution and faunistics. The Belostomatidae are worldwide in distribution, although their greatest diversity is in the tropics. Aids to identification are cited under the subfamily treatments and Nepomorpha.

31
Nepidae

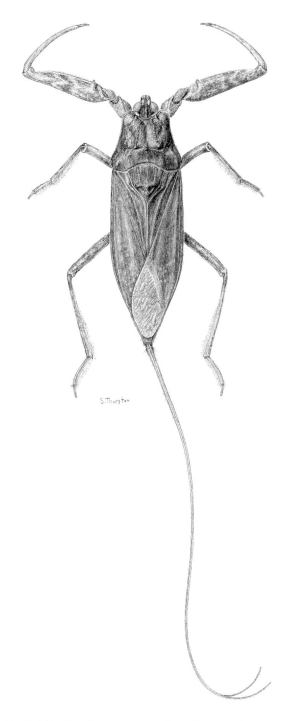

Fig. 31.1. Nepidae. *Laccotrephes* sp. (from Slater, 1982).

General. The water scorpions are brown, elongate-ovoid to tubular-bodied aquatic bugs that range in body length from 15 to 45 mm. Their caudal breathing siphon may be as long or longer than the body (Fig. 31.1).

Diagnosis. Eyes relatively small within Nepomorpha (Fig. 31.2A); antennae usually 3-segmented, segment 2 and sometimes 3 with fingerlike projections (Fig. 31.2B); membrane of forewing with numerous cells (Fig. 31.1); all legs elongate, slender, forelegs raptorial (Figs. 10.4B, 31.1), all tarsi 1-segmented; adult metathoracic and nymphal dorsal abdominal scent glands absent; static sense organs near spiracles on ventral laterotergites 4–6 (Figs. 29.1I, 31.2C); male genitalia symmetrical, parameres and genital capsule as in Fig. 31.2E; aedeagus as

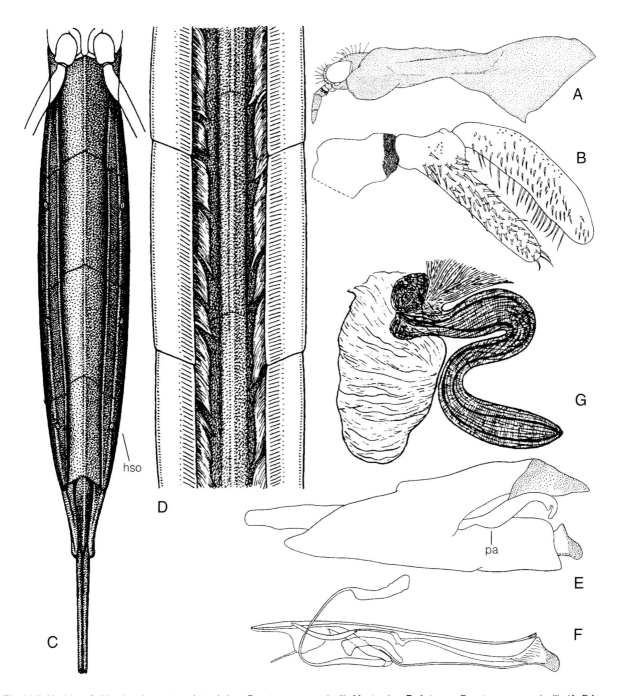

Fig. 31.2. Nepidae. **A.** Head and pronotum, lateral view, *Ranatra compressicollis* Montandon. **B.** Antenna, *Ranatra compressicollis* (A, B from Lansbury, 1974b). **C.** Ventral view, abdomen, *Ranatra* sp. **D.** Ventral view, abdomen, nymphal *Ranatra* sp., showing air channels, (C, D from Menke, 1979a). **E.** Genital capsule, lateral view, *R. drakei* Hungerford. **F.** Aedeagus, lateral view, *R. drakei* (E, F from Lansbury, 1974b). **G.** Spermatheca, *Nepa cinerea* (Linnaeus) (from Larsén, 1938). Abbreviations: hso, static sense organ; pa, paramere.

in Fig. 31.2F; subgenital plate forming an ovipositorlike structure in some taxa; spermatheca as in Fig. 31.2G; eggs uniquely with from 2 to 26 respiratory horns on anterior pole.

Classification. The Linnaean genus *Nepa* was first recognized as a higher taxon by Latreille (1802) as Nepariae. Modern treatments on the higher classification of the group are those of Menke and Stange (1964) and Lansbury (1974a), the latter author having done much recent revisionary work on the group.

Currently two subfamilies comprising 14 genera and 231 species are recognized.

Key to Subfamilies of Nepidae

1. Laterosternites flat, visible, not concealed by ventral laterotergites; ventral laterotergites not longitudinally subdivided; female subgenital plate (= operculum) broad and short, not exceeding apex of abdomen, if triangular, not keeled; body flattened; mesocoxae separated by more than coxal width .... Nepinae
– Parasternites concave, infolded, concealed by ventral laterotergites (Fig. 31.2D); ventral laterotergites longitudinally subdivided; female subgenital plate either laterally compressed and keeled or extending well beyond apex of abdomen (Fig. 31.2C); body nearly cylindrical in cross section; mesocoxae separated by less than coxal width .... Ranatrinae

NEPINAE (FIG. 31.1). Laterosternites not concealed by folding of abdomen; ventral laterotergites divided by a crease tangential to spiracles; female subgenital plate broad and flattened, never extending beyond end of abdomen; eggs with 5 or more respiratory horns.

Two tribes are recognized (Lansbury, 1974a). The Curictini includes only the primarily tropical *Curicta* Stål from the New World. The Nepini includes seven genera, the best known being the paleotropical *Laccotrephes* Stål and the Holarctic *Nepa* Linnaeus.

RANATRINAE. Laterosternites concealed by folding of abdomen (Fig. 31.2D); ventral laterotergites not divided by a crease tangential to spiracles; female subgenital plate laterally compressed, keeled (Fig. 32.1C), and often extending beyond apex of abdomen; eggs with 2 respiratory horns.

Three tribes are recognized (Lansbury, 1974a). The Austronepini and Goondnomdanepini, each including only a single genus from Australia, respectively, *Austronepa* Menke and Stange and *Goondnomdanepa* Lansbury. The Ranatrini comprise the tropical Asian *Cercotmetus* Amyot and Serville and the widespread *Ranatra* Fabricius, the Oriental species of which were reviewed by Lansbury (1972).

Specialized morphology. The water scorpions are noted for their anteriorly projecting head, the elongate prothorax (Fig. 31.2A) with the proximal coxae and the anteriorly directed legs, the respiratory siphon, and the presence of 2–26 respiratory horns on the anterior pole of the eggs. The modified abdominal tergum 8 is formed into a paired structure, forming a tube, which directs air from the atmosphere to abdominal spiracle 8, which is placed dorsally.

The static sense organs in the Nepidae, located near the spiracles on ventral laterotergites of abdominal segments 4–6, are large and conspicuous (Figs. 29.1I, 31.2C) (Baunacke, 1912). These structures have been shown experimentally to function to keep the insect correctly oriented in the water (Thorpe and Crisp, 1947). The morphology of *Nepa cinerea* Linnaeus was treated in detail by Hamilton (1931).

Natural history. Water scorpions are poor swimmers and tend to "crawl" through the water. Most inhabit quiet water in ponds or streams. They often "hang" in vegetation, with the respiratory siphon protruding through the water surface and the forelegs outstretched, waiting for prey. The foretibia has an apical sense organ, which apparently senses prey vibrations, although visual responses seem also to be involved in prey capture (Cloarec, 1976). Life histories of *Ranatra* spp. have been studied by Torre-Bueno (1906) and Radinovsky (1964). The abdominal dorsum of some nepids, including *Laccotrephes* spp., is bright red, in strong contrast with the dull greenish brown of the body surface. When the bugs are in flight the red coloration is obvious, but when they alight they become nearly invisible because the bright color disappears under the hemelytra.

Distribution and faunistics. This worldwide group shows its greatest diversity in the tropics. Sources are noted above.

Ochteroidea

32
Gelastocoridae

General. The toad bugs, which range in length from 7 to 15 mm, like the Ochteridae, are usually found in riparian situations. They very much resemble small Bufonidae, and all are capable of jumping. The coloration and texture often match remarkably the background upon which they are found.

Diagnosis. Body surface often roughened and "warty" (Fig. 32.1); eyes large, with a distinct mesal emargination on dorsal surface, appearing reniform (Fig. 32.1); adults with ocelli; labrum broad and flaplike; antennae 4-segmented, without fingerlike projections (Figs. 29.1A, 32.2A); membrane of forewing sometimes reduced, if developed, usually with numerous veins (Fig. 32.1); forefemora greatly enlarged, interior surface grooved for reception of tibia and tarsus; foretarsus 1-segmented or sometimes fused with tibia, middle tarsus 2-segmented, hind tarsus 3-segmented, claws unequally developed on forelegs, equally developed on middle and hind legs; foreleg pretarsus of nymphs as shown in Fig. 29.2A, adult as in Fig. 29.2B; antennal cleaner on foreleg as in Fig. 10.4I, J; nymphs lacking dorsal abdominal scent glands; spiracles on abdominal sterna 3 and 4 shifted toward midline; male genitalia asymmetrical, abdominal venter as in Figs. 29.1B, 32.2E; aedeagus as in Fig. 32.2B; right paramere as in Fig. 32.2C, D; left paramere reduced or absent; female abdominal venter as in Fig. 32.2F; spermatheca as in Fig. 32.2G.

Classification. This group was first recognized as a higher taxon by Billberg (1820) under the name Galgulides, after *Galgulus* Latreille, a preoccupied name replaced much later by *Gelastocoris* Kirkaldy (1897). Two genera are currently recognized. *Gelastocoris* from the New World, contains approximately 15 species. *Nerthra* Say, which contains approximately 85 species, is cosmopolitan and is a senior synonym of *Mononyx* Laporte. The taxon has frequently been treated as a family group under the name Mononychidae. These genera, which have sometimes been accorded subfamily status, can be separated according to the following key.

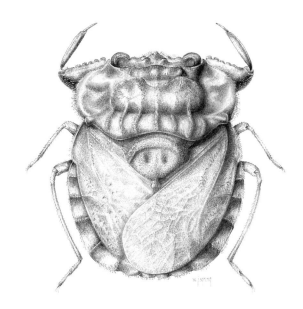

Fig. 32.1. Gelastocoridae. *Nerthra* sp.

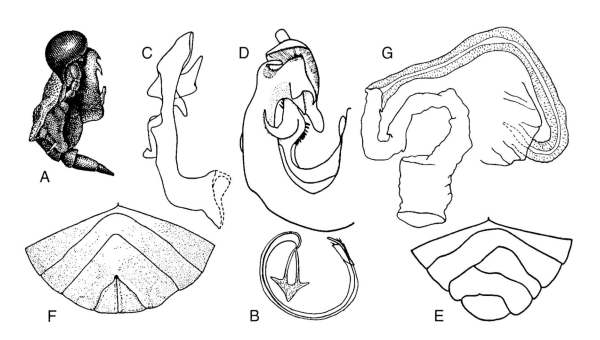

Fig. 32.2. Gelastocoridae. **A.** Lateral view, head, *Gelastocoris oculatus* (from Popov, 1971). **B.** Aedeagus, *G. oculatus* (from Kumar, 1961). **C.** Right paramere, *Nerthra hungerfordi* Todd. **D.** Right paramere, *G. peruensis* Melin. **E.** Male abdomen, *N. amplicollis* (Stål). **F.** Female abdomen, *N. amplicollis* (C–E from Todd, 1955). **G.** Spermatheca, *N. annulipes* (Horvath) (from Pendergrast, 1957).

Key to Genera of Gelastocoridae

1. Foretarsus 1-segmented, freely articulating with tibia, with 2 claws; forefemur moderately incrassate, about twice as long as wide, its apposable face flat and bordered by two rows of short spines; labium arising near apex of head, stout, directed caudad (Fig. 32.2A) .... *Gelastocoris*
– Foretarsus fused with tibia, tibiotarsus terminating in a single claw (Fig. 32.1); forefemur subtriangular, very broad at base, about as wide as long, its apposable face with a flangelike extension projecting over tibiotarsus when apposed; labium appearing to arise from ventral surface of head, slender, inversely L-shaped, apex directed ventrad to anteroventrad .... *Nerthra*

Specialized morphology. The toadlike appearance and the modified forelegs, with unusually large femora (Fig. 32.1), are distinctive for the group. The forewings in some *Nerthra* spp. are immovably fused (Todd, 1955). Parsons reviewed the morphology of the head (1959), thorax (1960b), nervous system (1960a), and abdomen and thoracic scolopophorous organs (1962) of *Gelastocoris oculatus* (Fabricius).

Natural history. Gelastocorids are generally found in riparian situations, although *Nerthra* spp. are not infrequently encountered a long distance from water. Some species can be extremely numerous at a given locality, whereas others seem to lead a solitary existence. Although most *Gelastocoris* spp. are usually found in the open, *Nerthra* spp. are capable of at least some burrowing and are frequently found beneath stones or other objects. The life history of *Gelastocoris oculatus* was reviewed by Hungerford (1922b).

Distribution and faunistics. Although toad bugs occur worldwide, they are most diverse in the tropics. The genus *Nerthra* is especially diverse in Australia and Melanesia.

The major source on the group is by Todd (1955).

33
Ochteridae

General. Sometimes referred to as velvety shore bugs (Menke, 1979b), members of this group—which range in length from 4.5 to 9 mm—appear remarkably like Saldidae with short antennae, generally dark coloration, and pruinose markings on the dorsum.

Diagnosis. Eyes large, mesal margin emarginate dorsally (Fig. 33.1); ocelli present; antennae 4-segmented, visible dorsally, projecting, filiform (Figs. 29.1D, 33.1); clypeus transeversely rugose; labium long, slender, tapering, reaching hind coxae, segment 3 much longer than others (Fig. 33.2A); medial and costal fracture continuous and well developed, membrane of forewing with several closed cells (Fig. 33.1), but without anastomosing veins found in many other Nepomorpha; all legs slender, forefemora not enlarged or otherwise modified, tarsal formula 2-2-3; foreleg pretarsus of nymph as in Fig. 29.2C; nymphs lacking dorsal abdominal scent glands; male abdomen ventrally as in Fig. 29.1E; aedeagus as in Fig. 33.2C; genital capsule and right paramere as in Fig. 33.2B; left paramere reduced; ovipositor reduced; spermatheca as in Fig. 33.2D.

Classification. This group was first recognized as a higher taxon under its current name by Kirkaldy (1906). Latreille (1809) proposed the unnecessary replacement name *Pelogonus* for *Ochterus* Latreille, and the family was often referred to as Pelogonidae in the older literature.

Three genera and 55 species are currently recognized.

Specialized morphology. Ochterids are notable for their saldidlike appearance (Fig. 33.1), but they nonetheless have a preponderance of obvious nepomorphan characters, including the short, ventrally inserted antennae, asymmetrical male genitalia, and tubular spermatheca. Rieger (1976) reviewed the morphology of the head and thorax of *Ochterus marginatus* Latreille in detail.

Natural history. These bugs are usually found along the shores of ponds or streams, and in the tropics they largely replace the Saldidae; they may also inhabit more open sandy habitats. Life histories of ochterids have been published by Takahashi (1923) and Bobb (1951).

Distribution and faunistics. Ochterids are found worldwide, but are most diverse in the tropics. The Western Hemisphere fauna was revised by Schell (1943), the Eastern Hemisphere fauna by Kormilev (1971b), and *Ochterus* in Australia by Baehr (1989).

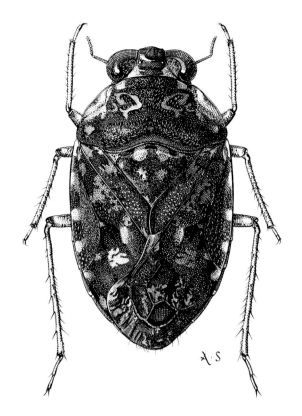

Fig. 33.1. Ochteridae. *Ochterus barberi* Schell (from Usinger, 1956).

Corixoidea

34
Corixidae

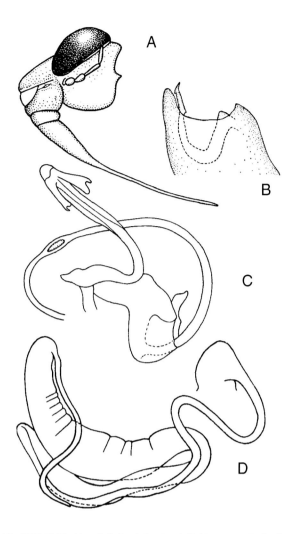

Fig. 33.2. Ochteridae. **A.** Lateral view head, *Ochterus marginatus* (Latreille) (from Popov, 1971). **B.** Genital capsule, *O. barberi* (from Schell, 1943). **C.** Aedeagus, *O. seychellensis* D. A. Polhemus (from D. A. Polhemus, 1992). **D.** Spermatheca, *O. marginatus* (from Pendergrast, 1957).

General. The water boatmen represent the most speciose of the families of aquatic bugs. Because of their oarlike hind legs (Fig. 34.1), they have the general appearance of flattened backswimmers, but swim with the dorsal side up. They range in length from 2.5 to 15 mm.

Diagnosis. Small to medium-sized, dorsum flattened; head broad, strongly hypognathous; ocelli absent except in Diaprepocorinae; antennae 3- or 4-segmented, hidden between eyes and prothorax; labium modified from typical heteropteran condition, broad at base, tapering distally, unsegmented, immovably fused to head, usually with transverse grooves and a longitudinal channel; labrum reduced, covered by labium; scutellum exposed or hidden by pronotum; forewings of uniform texture, membrane area without veins (Fig. 34.1); forelegs relatively short; foretarsus with a single segment, usually modified into a pala in the form of a spoon or a scoop, fringed with long setae (Fig. 34.2A), sometimes fused with tibia; middle legs very long and slender, tarsi 1- or 2-segmented, pair of claws very long (Fig. 34.1); hind legs flattened, oarlike, fringed with setae, tarsi 2-segmented (Fig. 34.1); stridulatory structure formed by field of pegs on basomesal surface of forefemur apposed to edges of maxillary plates (in males only of Micronectinae); metathoracic scent glands present in adults; larval dorsal abdominal scent glands present between terga 3/4, 4/5, and 5/6; male abdomen with a "strigil" posterolaterally on tergum 6 (Fig. 34.2B, C); male distal abdominal terga and sterna and genital segments usually strongly asymmetrical (Fig. 34.2B); aedeagus as in Fig. 34.2D, E; parameres as in Fig. 34.2F; spermatheca as in Fig. 34.2G, H.

Classification. The Corixidae were first treated at the family rank by Leach (1815) under the name Corixida. Their unusual morphology has attracted the attention of many authors, and Börner (1934) went so far as to accord them subordinal rank under the name Sandaliorrhyncha. There seems little doubt, however, that they belong to the Nepomorpha, in spite of their obvious specializations; more recent opinions on their phylogenetic position can be seen in Figs. 29.3 and 29.4. The classification of Hungerford (1948) recognized six subfamilies, which comprise 34 genera and 556 species.

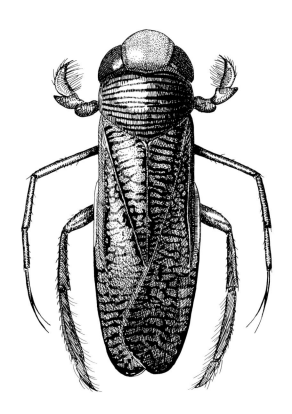

Fig. 34.1. Corixidae. *Trichocorixa reticulata* (Guérin-Méneville) (from Lauck, 1979).

Key to Subfamilies of Corixidae

1. Scutellum broadly exposed, only anterior margin covered by short, transverse pronotum; female foreleg (and sometimes male foreleg) with fused tibiotarsus; color patterns never as below 2
– Mesoscutellum covered by arcuate to subtriangular posterior extension of pronotum (Fig. 34.1), rarely its extreme apex exposed, although often fully exposed in preserved specimens owing to deflection of pronotum; foreleg with tibia and pala separate in both sexes; pronotum and hemelytra mostly with complex mosaic of alternating transverse, linear, undulating, hieroglyphical, or zig-zag light and dark patterns, these rarely obsolete or absent 3
2. Ocelli present; middle tarsus 2-segmented; male foreleg with fused tibiotarsus; antennae 4-segmented; Australia and New Zealand .... Diaprepocorinae
– Ocelli absent; middle tarsus 1-segmented; male foreleg with pala and tibia separate; antennae 3-segmented; tropicopolitan .... Micronectinae
3. Exocorial pruina present on small basal portion of forewing, extremely short, about as wide as and similar in length to endocorial pruina; embolar groove practically undeveloped; proximal sector of Cu on forewing indistinguishable, Cu developed as a distinct ridge only distad to meeting with costal fracture; Afrotropical .... Stenocorixinae
– Exocorial pruina present along costal margin in long embolar groove, long and narrow, much longer than endocorial pruina, usually terminating distad of costal fracture, rarely at costal fracture or just proximad of it (the latter structure absent in some taxa); Cu present from near base of forewing as distinct ridge delimiting embolar groove and exocorial pruina; junction of Cu and costal fracture (or projected intersection) situated at border of exocorial pruina or nearby 4
4. Labium without transverse striation on ventral face; male pala cylindrical or terete, without differentiated palmar face and without palar pegs; male palar claw long, thick, appendage-like, not resembling rake setae; media of forewing usually meeting apex of costal fracture, forming closed

field within pruinose embolar groove, inner angle of field remote from Cu ridge; cosmopolitan
.... Cymatiainae
- Labium with transverse striation on ventral face; male pala variously shaped, palmar face differentiated and palar pegs developed, if pegs, poorly defined or reduced in number, then pala very broad; male palar claw usually resembling a thickened rake hair, if appendage-like then short; exocorial pruina without a closed field within it, as above 5
5. Infraocular (postocular) portion of gena very broad in lateral view, ventrally delimited by a hypocular sulcus arising near subacutely produced inferior angle of eye and running toward posterolateral margin of head; posteroventral margin of eye markedly concave in lateral view; media running very close to and parallel with Cu, often indistinct; Neotropical .... Heterocorixinae
- Infraocular portion of gena very narrow to broad in lateral view, hypocular sulcus present or absent; if infraocular portion of gena broad and hypocular sulcus present, then the latter arising almost midway along posteroventral margin of eye, this margin shallowly concave, inferior angle not subacutely produced, usually broadly rounded; media variously developed, but not as above, usually running midway between costal margin and Cu, only apically turning toward Cu or the inner portion of costal fracture; cosmopolitan .... Corixinae

CORIXINAE (FIG. 34.1). Labium with transverse sulcations; antennae 4-segmented; scutellum completely covered by pronotum or nearly so; embolar groove and nodal furrow (costal fracture) present; female pala spoon-shaped, male pala variable; claw of hind leg inserted subapically.

This is by far the largest of the subfamilies and contains the majority of known species placed in four tribes and 26 genera, with the large and widespread genus *Sigara* Fabricius being divided into many subgenera. The group is worldwide in distribution.

CYMATIAINAE. Labium without transverse sulcations; antennae 4-segmented; scutellum covered by pronotum; embolar groove present, nodal furrow absent; pala elongate, nearly cylindrical in both sexes, not fused with tibia; hind leg claw inserted apically.

This group, whose name has been emended from Cymatiinae because of homonymy, contains only two genera, the Holarctic *Cymatia* Flor and *Cnethocymatia* Jansson from northern Australia and New Guinea.

DIAPREPOCORINAE. Ocelli present; antennae 4-segmented; scutellum exposed; male and female foreleg with fused tibiotarsus; middle tarsus 2-segmented.

These water boatmen have been thought of as the most primitive members of the group because they have ocelli, an unusual trait within the Nepomorpha. The group contains only *Diaprepocoris* Kirkaldy from Australia and New Zealand.

HETEROCORIXINAE. Labium with transverse sulcations; scutellum covered by pronotum; embolar groove present, nodal furrow present; tibia of foreleg nearly as long as palm-shaped pala.

This New World tropical group contains only *Heterocorixa* Buchanan-White.

MICRONECTINAE. Labium with transverse sulcations; antennae 3-segmented, last segment broad and lying in a pocket in head; scutellum exposed; embolar groove short, shallow, its pruinose area not crossed by nodal furrow; pala fused to foretibia in female; claw of hind leg inserted apically.

This group contains the Old World *Micronecta* Kirkaldy —with many subgenera—which is diverse in the Palearctic and the Orient. *Tenagobia* Bergroth is restricted to the Neotropics.

STENOCORIXINAE. Body more elongate and slender than in other Corixidae; labium with transverse sulcations; pala slender; scutellum covered by pronotum; embolar groove absent, nodal furrow present; male abdomen only slightly asymmetrical.

This subfamily is represented by the single African species *Stenocorixa protrusa* Horvath.

Specialized morphology. The morphology of the labium in the Corixidae is unique within the Heteroptera. It was long thought to be associated strictly with feeding habits, but the correlation may not be so strong as supposed. The forelegs also are highly modified, the femur functioning in stridulation, and the tarsus existing in the form of a spoon-shaped pala (Fig. 34.2A), which functions in food gathering. The brown and pale hieroglyphic patterning of the dorsum is characteristic of most members of the family and is important in the recognition of some taxa.

The so-called strigil on abdominal tergum 6 (Fig. 34.2B, C) in the males is misnamed and appears to have nothing to do with stridulation. It has been implicated in holding the female during mating (Larsén, 1938) and in helping to maintain the subhemelytral air stores (Popham et al., 1984).

Natural history. In addition to being the most speciose of all the families of Nepomorpha, the Corixidae are also the most widely distributed, occupy a relatively

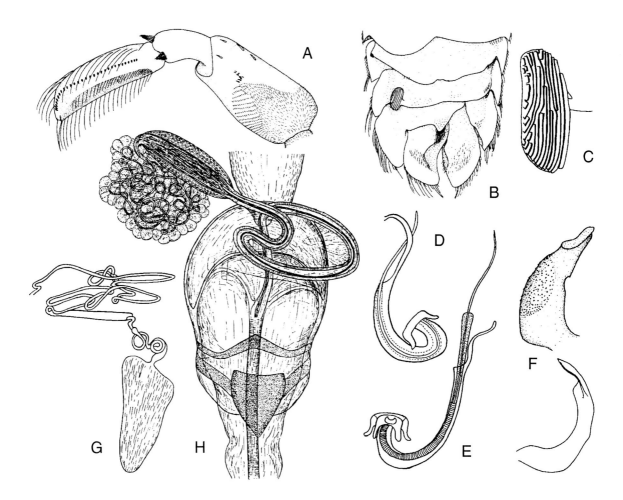

Fig. 34.2. Corixidae. **A.** Male foreleg, *Corixa dentipes* Thunberg. **B.** Male abdominal dorsum, *C. jakowleffi* (Horvath). **C.** Strigil, *C. dentipes* (A–C from Jansson, 1986). **D.** Aedeagus, *Agraptocorixa hyalinipennis* (Fieber). **E.** Aedeagus, *Hesperocorixa interrupta* (D, E from Kumar, 1961). **F.** Parameres (right above, left below), *C. dentipes* (from Jansson, 1986). **G.** Spermatheca, *A. hyalinipennis* (from Kumar, 1961). **H.** Gynatrial complex, including spermatheca, *Sigara sahlbergi* (Fieber) (from Larsén, 1938).

wide range of habitats, and are sometimes extremely numerous. Some are known to inhabit saline waters on an exclusive basis (see Lauck, 1979).

Water boatmen, unlike other nepomorphans, eat primarily plant material, particularly algae, using their palae to forage on bottom ooze. They are also known to feed on animal matter as well, such as nematoceran larvae, including mosquitoes (Lauck, 1979). Several species are eaten by humans in Mexico, and those same species are important in the pet-food trade.

Stridulation in the Corixidae and its role in courtship and mating have been studied in detail by Jansson (1972, 1976).

Distribution and faunistics. The New World fauna was treated in its entirety by Hungerford (1948), and that work is still the single most important reference on the fauna. The European fauna has been reviewed exhaustively by Jansson (1986). The remainder of the world is covered in a somewhat more piecemeal manner, but papers by Hutchinson (1929) for southern Africa and Lundblad (1933) for the Orient should be consulted.

Naucoroidea

35
Potamocoridae

General. These small bugs, which are between 2.5 and 3.0 mm long, have the appearance of small naucorids (Fig. 35.1A).

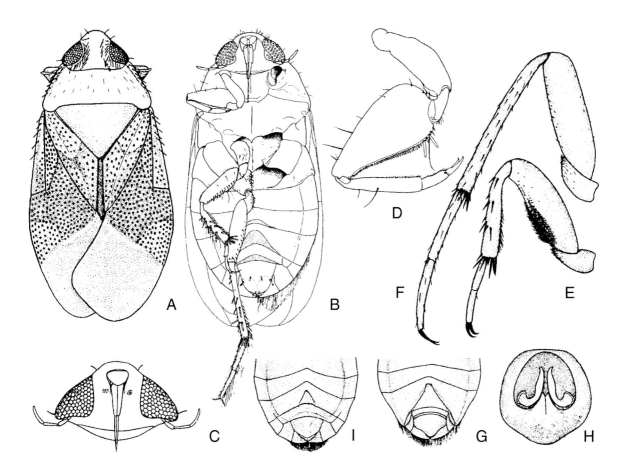

Fig. 35.1. Potamocoridae. **A.** *Potamocoris parvus* Hungerford (from Hungerford, 1941). **B.** Ventral view, body, *P. nieseri* van Doesburg (from van Doesburg, 1984). **C.** Ventral view, head, *P. parvus* (from Hungerford, 1941). **D.** Foreleg, *P. nieseri* (from van Doesburg, 1984). **E.** Middle leg, *P. parvus*. **F.** Hind leg, *P. parvus*. **G.** Male abdomen, ventral view, *Potamocoris parvus*. **H.** Genital capsule, *P. parvus*. **I.** Female abdomen, ventral view, *Potamocoris parvus* (E–I from Hungerford, 1941).

Diagnosis. Small; eyes not overlapping anterolateral angles of pronotum; antennae relatively long, visible dorsally, filiform (Fig. 35.1A, B); labium short, just reaching onto prosternum (Fig. 35.1B, C); membrane of forewing without veins; forefemora only slightly enlarged, with tibia and tarsus opposable, middle and hind femora without obvious swimming modifications; foretarsus 1-segmented, middle and hind tarsi 2-segmented; 2 claws on all legs; nymphs with dorsal abdominal scent glands between terga 3/4 (Hungerford, 1942); male genital segments as in Fig. 35.1G, H; female as in Fig. 35.1I.

Classification. In his original description of *Potamocoris*, Hungerford (1941) noted the unusual combination of characters in this group of bugs. In the same journal issue, Usinger (1941) accorded the group subfamily status in the Naucoridae. Cobben (1978), in a footnote, indicated that the group deserved family status, a position followed by Štys and Jansson (1988).

Two genera, *Coleopterocoris* Hungerford (1942) and *Potamocoris*, each with four species, are recognized.

Specialized morphology. The attributes of the Potamocoridae are seemingly a mixture of some of those possessed by the Aphelocheiridae and Naucoridae, including the elongate antennae of the former and the short labium of the latter.

Natural history. Nothing appears to be known of the biology. Most specimens have apparently been collected at lights, although Hungerford (1942) had access to a nymph.

Distribution and faunistics. This group is restricted to the Neotropics. The main sources are those of Hungerford and Usinger mentioned above, DeCarlo (1968), La Rivers (1950, 1969), J. T. Polhemus and D. A. Polhemus (1982), and van Doesburg (1984).

36 Naucoridae

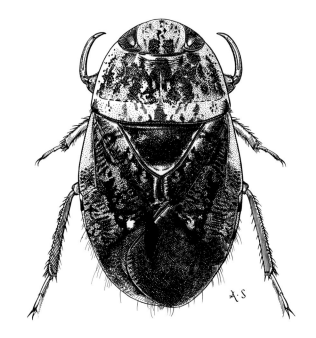

Fig. 36.1. Naucoridae. *Pelocoris shoshone* La Rivers (from Usinger, 1956).

General. Naucorids, or creeping water bugs, are usually ovoid (Fig. 36.1) to slightly elongate, flattened bugs ranging in length from 5 to 20 mm. Most have the general appearance of tiny belostomatids, in both coloration and shape.

Diagnosis. Eyes often overlapping anterolateral angles of pronotum (Fig. 36.1); antennae 4-segmented, short, hidden, not projecting laterally; labium short and stout (Fig. 36.2A); forewing membrane without venation (Fig. 36.1); forefemora conspicuously enlarged (Fig. 10.4A); apex of foretibia as in Fig. 10.7A; hind legs often modified for swimming; foretarsus usually fused with tibiae, 1- or 2-segmented, with 1 or 2 claws or without claws, middle and hind tarsi with 2 distinct segments (basal segment very small or absent), with 2 equally developed claws; pretarsal structure variable (see Fig. 29.2E); nymphal pretarsus as in Fig. 29.2D; metathoracic scent glands present; posterolateral angles of abdominal connexiva sometimes produced; nymphal dorsal abdominal scent gland between terga 3/4; some taxa with paired sublateral sense organs on abdominal sternum 2; male and female venter in copulatory position as in Fig. 36.2B; aedeagus as in Fig. 36.2C; spermatheca vermiform as in Fig. 36.2D.

Classification. This group was first treated at family rank by Leach (1815) as the Naucorida and has been so recognized by most subsequent authors, although the limits of the family have been subject to dispute, as explained in the discussions under Aphelocheiridae and Potamocoridae.

The most recent classification (Štys and Jansson, 1988) included five subfamilies—separated as in the following key—comprising 40 genera and approximately 395 species.

Key to Subfamilies of Naucoridae

1. Labium arising from a deep excavation on underside of head, its insertion markedly distant from apex of head owing to more or less horizontal, laminate extension of latter; Oriental and Papuan . . . . Cheirochelinae
- Labium arising freely, not from an excavation, usually near ventral apex of head; sometimes true apex of head markedly folded over and insertion of labium distinctly ventra 12
2. Lateral margins of pronotum crenulate, dorsal surface roughly granulate; head rather narrow, anterior portion longer than wide, distinctly projecting in front of subglobular eyes; head deeply inserted into anteriorly strongly concave pronotum; maxillary plates well delimited and anteriorly reaching or exceeding level of apex of head; New World . . . . Cryphocricinae (part)
- Lateral margins of pronotum smooth, dorsal surface smooth to punctate, not granulate; head broad, outline streamlined jointly with pronotum; interocular region usually much broader than long (narrow in Ambrysini), only slightly exceeding eyes, the latter never subglobular and usually not projecting beyond outline of head, eyes rarely somewhat rounded and interocular portion slightly projecting; maxillary plates variously developed, often incompletely or not at all delimited, not reaching level of apex of head 3
3. Proepimeral lobes strongly expanded, meeting to cover posterior portion of prosternum; head deeply inserted into concave and markedly trisinuate anterior pronotal margin (Ambrysini) . . . . Cryphocricinae (part)

- Proepimeral lobes not in mutual contact, posterior portion of prosternum exposed; head usually not deeply inserted into pronotum, anterior pronotal margin usually shallowly concave to almost straight 4
4. Foreleg pretarsus in females always with 2 claws, often minute and closely appressed, sometimes resembling a single structure; males of most genera with a tomentose adhesive patch on ventral surface of middle tibiae, weakly developed in females; morphological apex of head folded over, oriented posteriorly, insertion of labium markedly posterior to topographical apex of head; middle and hind femora with 2 longitudinal rows of conspicuous bristles or spinelike setae on ventral face, in addition to 2 usual rows on posterior face .... Laccocorinae
- Foreleg pretarsus with or without 1 minute claw; middle tibia without distoventral adhesive tomentose patch; labium usually inserted apically or subapically on ventral apex of head, rarely on anteroventral surface of head; middle and hind femora without additional rows of bristles or spiniform setae on ventral surfaces, or with just an indication of upper row, or with some scattered additional setae 5
5. Meso- and metasterna with prominent, broad, laterally expanded median longitudinal carinae bearing foveae or otherwise excavate; inner margins of eyes (dorsal view) distinctly diverging anteriorly; gula short; apex of head (anterior view) simply arcuate, not notched; lateral margins of abdomen with fine serration; body broadly oval to subcircular, flattened; New World .... Limnocorinae
- Meso- and metasternal longitudinal carinae inconspicuous, thin, platelike, or absent; inner margins of eyes in dorsal view usually distinctly converging anteriorly, infrequently parallel-sided to slightly diverging; gula usually long; apex of head straight, delimited laterally by distinct notches or indentations laterad of basal angles of labrum; lateral margins of abdomen without fine serration; body more elongate, robust, dorsum moderately convex; worldwide .... Naucorinae

CHEIROCHELINAE. Labium inserted in a deep excavation well removed from anterior margin of head; labrum often greatly reduced; anterior margin of pronotum excavated in region adjoining interocular space of head.

Three tribes are recognized in this Oriental (including New Guinea) group: Cheirochelini (three genera), Sagocorini (eight genera), and Tanycricini (three genera).

CRYPHOCRICINAE (FIG. 36.1). Labium inserted near anterior margin of head, not in a deep excavation; labrum well developed; anterior margin of pronotum excavated in region adjoining interocular space of head.

Three tribes are recognized in this New World group: Ambrysini (two genera), Cataractocorini (one genus), Cryphocricini (two genera).

LACCOCORINAE. Labium inserted near anterior margin of head, not in a deep excavation; labrum well developed; anterior pronotal margin nearly straight; foretarsus always with 2 claws.

Seven genera are currently recognized in this primarily Paleotropical group, which is marginally represented in the Palearctic (Štys and Jansson, 1988).

LIMNOCORINAE. Labium inserted near anterior margin of head, not in a deep excavation; labrum well developed; anterior pronotal margin nearly straight; foreleg with at most 1 claw; meso- and metasterna with prominent, broad, laterally expanded median longitudinal carinae bearing foveae or otherwise excavate; lateral abdominal margins with fine serrations; abdominal terga 2–5 fused; stridulatory structure in all species involving hind femur and connexival margin.

This New World group currently contains two genera.

NAUCORINAE. Labium inserted near anterior margin of head, not in a deep excavation; labrum well developed; anterior pronotal margin nearly straight; foreleg with at most 1 claw; meso- and metasterna not prominent, without expanded longitudinal carina; no stridulatory structures.

This worldwide group appears for the most part to be diagnosed by the absence of characters found in the Limnocorinae, a point of view supported by the character analysis of Mahner (1993). The seven included genera are restricted primarily to the tropics of either the Old or New World.

Specialized morphology. Researchers studying naucorids have tended to treat the Potamocoridae and Aphelocheiridae as having more primitive attributes than the Naucoridae as here construed. Thus, they have considered the short, concealed antennae and short labium to be apomorphic for the Naucoridae, although the antennae project beyond the eyes in the Tanycricini.

The Cryphocricini have plastron respiration (Fig. 29.1G, H) and pressure receptors (Parsons and Hewson, 1975), as possibly do some other Naucoridae.

Natural history. Naucorids live in a variety of aquatic environments. Some, such as *Pelocoris femoratus* (Palisot de Beauvois) in North America live in quiet pond waters. *Limnocoris* spp., and indeed many naucorids and aphelocheirids, are found in marginal riffles in streams on gravel substrates. A few taxa, such as *Cryphocricos* Signoret in the New World tropics and *Idiocarus* Montan-

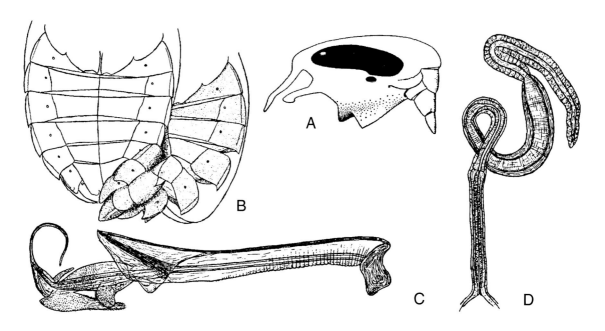

Fig. 36.2. Naucoridae. **A.** Lateral view head, *Ambrysus magniceps* La Rivers (from Parsons, 1969). **B.** Copulation, *Naucoris cimicoides* (Linnaeus). **C.** Phallus, *N. cimicoides*. **D.** Spermatheca, *N. cimicoides* (B–D from Larsén, 1938).

don and *Cheirochela* Hope in the Old World tropics live in torrential streams, situations where it would be difficult to come to the surface at regular intervals, suggesting a lifestyle befitting their plastron respiration.

Distribution and faunistics. Although the group occurs worldwide, the greatest diversity is conspicuously in the tropics. La Rivers (1971) treated some members of the New Guinea fauna and compiled a world catalog, with corrections published by La Rivers (1974, 1976).

37
Aphelocheiridae

General. Aphelocheirids have the general appearance of many naucorids (Fig. 37.1A, B), with their ovoid flattened bodies, but they differ in some characters, particularly the relatively long labium and antennae. They range from 3.5 to 11.5 mm in length.

Diagnosis. Head produced anteriorly, inserted into and surrounded posterolaterally by pronotum (Fig. 37.1A, B); antennae elongate, slender, 4-segmented, filiform; labium relatively long, reaching well onto metasternum, segment 2 very short, segment 3 very long, segment 4 less than one-half length of segment 3 (Fig. 37.1C); alary polymorphism common (Fig. 37.1A, B); all tarsi 3-segmented; all legs with a pair of equally developed claws; middle leg pretarsus as in Fig. 29.2F; metathoracic scent glands lacking, dorsal abdominal scent glands persistent in adults; posterolateral angles of connexiva often produced and spinelike (Fig. 37.1A, B); abdomen with simple dorsal and ventral plates; venter of thorax and abdomen with a plastron; abdominal spiracles 2–7 uniquely in Nepomorpha surrounded by rosettes (see Larsén, 1938:126; Thorpe and Crisp, 1947); abdominal sternum 2 with a pair of sublateral sense organs of unknown function; aedeagus as in Fig. 37.1D; female subgenital plate as in Fig. 37.1E; spermatheca as in Fig. 37.1F, G.

Classification. This group was first treated at family rank by Fieber (1851) as Aphelochirae, within the Naucoroidea. It has been treated as a subfamily of Naucoridae (D. A. Polhemus and J. T. Polhemus, 1988), or a distinct family (Štys and Jansson, 1988), by most subsequent authors. Comparison of the arguments of Hoberlandt and Štys (1979) and Mahner (1993) with the conclusions of Štys and Jansson (1988) and D. A. Polhemus and J. T. Polhemus (1988) indicates that there is still little agreement about the ranking and relationships of the Aphelocheiridae, Naucoridae, and Potamocoridae. The cladogram of Rieger (Fig. 29.3) indicates that the Naucoroidea of Štys and Jansson (1988) is not a monophyletic group, whereas that of Mahner (1993; Fig. 29.4) suggests otherwise but leaves the Potamocoridae unplaced.

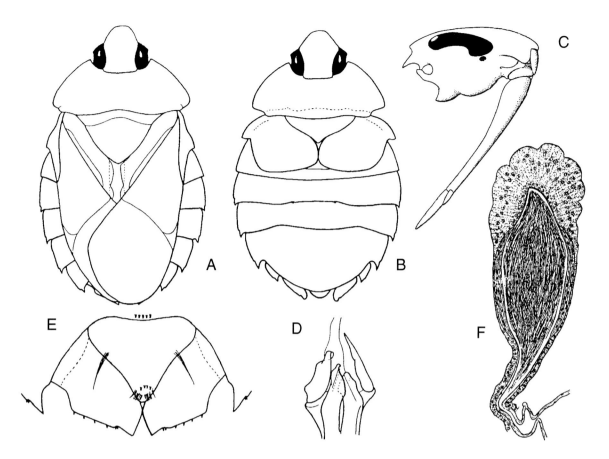

Fig. 37.1. Aphelocheiridae. **A.** Male habitus, *Aphelocheirus malayensis* Polhemus and Polhemus. **B.** Female habitus, *A. malayensis* (A, B from D. A. Polhemus and J. T. Polhemus, 1988). **C.** Lateral view, head, *A. aestivalis* (Fabricius) (from Parsons, 1969). **D.** Phallus, *A. malayensis*. **E.** Female subgenital plate, *A. malayensis* (D, E from D. A. Polhemus and J. T. Polhemus, 1988). **F.** Spermatheca, *A. aestivalis* (from Larsén, 1938).

Only the genus *Aphelocheirus* Westwood, with two subgenera and approximately 55 species, is currently recognized.

Specialized morphology. The Aphelocheiridae have relatively long slender antennae and a long labium, attributes not encountered in most other Nepomorpha. All species have a plastron, and they uniquely possess "rossettes" surrounding the ventral abdominal spiracles (Larsén, 1938) as well as pressure receptors (Thorpe and Crisp, 1947; Parsons and Hewson, 1975).

Natural history. Aphelocheirids inhabit the benthos of streams and lakes. Some species tolerate rather cold temperatures, and some are capable of living several meters below the water surface. Respiration via a plastron enables them to extract oxygen directly from the water and remain submerged for their entire lives (D. A. Polhemus and J. T. Polhemus, 1988).

Distribution and faunistics. This is essentially a Paleotropical group, extending from Africa in the west to Northern Queensland in the east, with several species in Madagascar. As with most such distributions, which are remarkably common in the Heteroptera, a few species occur in the Palearctic. Particularly useful references are by Kanyukova (1974), Hoberlandt and Štys (1979), and D. A. Polhemus and J. T. Polhemus (1988).

Notonectoidea

38
Notonectidae

General. The backswimmers are all elongate and fusiform (Fig. 38.1). Superficially they resemble members of the Corixidae because of the oarlike hind legs. Backswimmers are relatively large among the Notonectoidea, ranging in length from 5 to 15 mm.

Diagnosis. Dorsum strongly convex, venter concave, abdomen with a median keel; compound eyes large; antennae 3- or 4-segmented, partially concealed between head and prothorax (Fig. 38.1); labium 4-segmented, short; membrane of hemelytra without veins, subdivided in 2 parts disposed in a tentlike fashion; fore- and middle legs adapted for grasping, hind legs oarlike, hind tibia and tarsus fringed with long setae (Fig. 38.1); tarsi of fore- and middle legs usually apparently 2-segmented, first segment always greatly reduced, sometimes absent; hind tarsi always 2-segmented; all pretarsi with 2 claws, those of hind legs reduced; abdomen with a median longitudinal keel and heavy fringes of setae medially and laterally forming air chambers; abdominal spiracles 2–8 located lateroventrally on abdomen, within the ventral air chamber; thoracic and first abdominal spiracles elongate and slitlike; abdominal sternum 5 strongly produced anteriorly at midline, sternum 4 very narrow at midline; male genital segments symmetrical or nearly so; aedeagus as in Fig. 38.2D (see also Truxal, 1952); spermatheca vermiform, as in Fig. 38.2F, G; ovipositor present, varying in degree of development (Fig. 38.2E).

Classification. The Notonectidae were first recognized as a family group under the name Notonectariae by Latreille (1802), and until 1928 often were treated as including also the Pleidae (and Helotrephidae).

Two subfamilies are recognized (Hungerford, 1933), comprising 11 genera and 343 species.

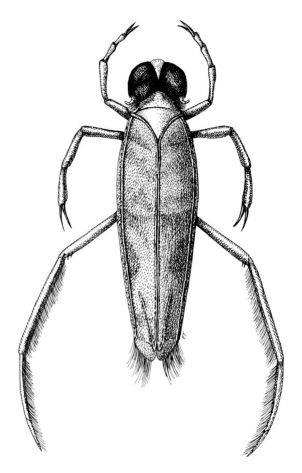

Fig. 38.1. Notonectidae. *Martarega mexicana* Truxal (from Menke, 1979a).

Key to Subfamilies of Notonectidae

1. Claval commissure simple, without a sensory pit; antennae 4- or 3-segmented; male parameres symmetrical or asymmetrical .... Notonectinae
– Claval commissure with a large, proximal, sharply delimited, seta-lined, sensory pit; antennae 2- or 3-segmented; male parameres usually asymmetrical .... Anisopinae

ANISOPINAE. Claval commissure basally with a pit; male foretibia usually (*Anisops*, *Buenoa*) with a stridulatory comb (plectrum) located proximally on interior surface (Fig. 38.2C), labium with an apposing "prong" (Fig. 38.2B); metathoracic scent glands absent.

This group, which was first recognized as a tribe by Hutchinson (1929), contains four genera, most members of which are rather small: *Anisops* Spinola from the Old World, *Buenoa* Kirkaldy from the New World, *Paranisops* Hale from Australia, and *Walambianisops* Lansbury from Australia.

NOTONECTINAE (FIG. 38.1). Claval commissure without a pit basally; males lacking foreleg stridulatory structures; metathoracic scent glands present. Obviously none of the above characters are unique to the Notonectinae, and as noted by Mahner (1993), the monophyly of the group cannot be supported.

Members of the seven included genera are usually somewhat larger than those of the Anisopinae. *Notonecta* Linnaeus, although worldwide in distribution, is most diverse in the New World. *Enithares* Spinola occurs in the Old World. The remaining genera contain many fewer species.

Specialized morphology. The Anisopinae have a conspicuous stridulatory plectrum proximally on the tibia (Fig. 38.2C). In *Anisops* and *Buenoa* the structure is apposed to the rostral prong (Fig. 38.2C). On the forefemur in *Buenoa* the plectrum is apposed to coxal pegs. All Ani-

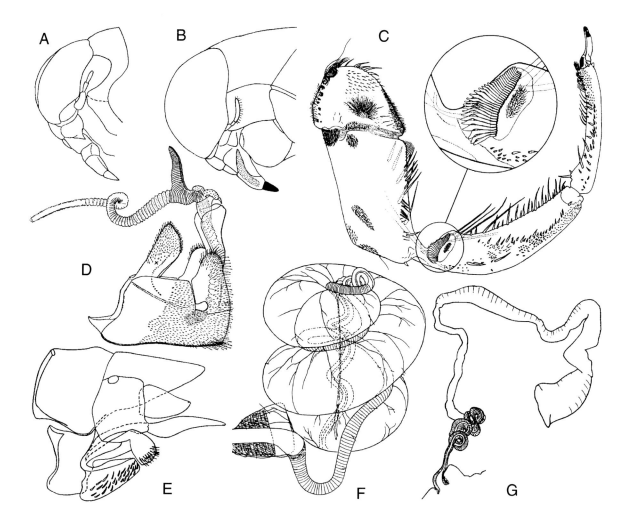

Fig. 38.2. Notonectidae. **A.** Lateral view, head, *Enithares maai* Lansbury (from Lansbury, 1968). **B.** Lateral view head, *Anisops megalops* Lansbury. **C.** Foreleg with detail of stridulatory structure, *Anisops megalops* (B, C from Lansbury, 1962). **D.** Male genitalia, *A. megalops*. **E.** Female genitalia, *Enithares* sp. (D, E from Lansbury, 1968). **F.** Spermatheca, *Notonecta obliqua* (from Pendergrast, 1957). **G.** Spermatheca, *N. glauca* Linnaeus (from Larsén, 1938).

sopinae have a seta-lined pit located at the anterior end of the claval commissure.

Hemoglobin was discovered in *Buenoa* by Hungerford (1922a) and also occurs in *Anisops*. Hemoglobin-containing cells are concentrated in abdominal segments 3–7. The hemoglobin functions in providing neutral buoyancy, allowing the insects to remain at a constant level in the water column (Bare, 1928; Miller, 1966).

Natural history. One of the most obvious features of the Notonectidae (and other Notonectoidea) is their habit of swimming on their backs. The resonant quality of stridulation in *Buenoa* was studied by Wilcox (1975). References to habitat partitioning in temperate-latitude species are given in Chapter 5.

Distribution and faunistics. The back-swimmers are worldwide in distribution and well represented in temperate as well as tropical regions. Keys to the world genera are given by Lansbury (1968). *Notonecta* was revised by Hungerford (1933), Oriental *Enithares* by Lansbury (1968), *Anisops* by Brooks (1951), *Buenoa* by Truxal (1953), and *Paranisops* by Lansbury (1964).

39
Pleidae

General. Sometimes referred to as pygmy backswimmers, members of this group range in length from 1.5

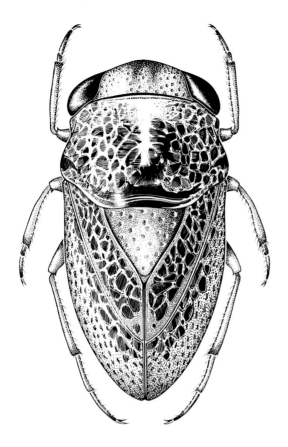

Fig. 39.1. Pleidae. *Plea* sp. (drawn by T. Nolan; from CSIRO, 1991).

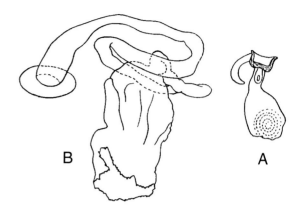

Fig. 39.2. Pleidae. **A.** Aedeagus, *Neoplea striola* (Fieber) (from Kumar, 1961). **B.** Spermatheca, *Plea atomaria* (Pallus) (from Pendergrast, 1957).

to 3.0 mm. They have the general appearance of tiny Notonectidae, but lack the oarlike hind legs (Fig. 39.1).

Diagnosis. Small; body globular (Fig. 39.1), heavily punctured; head very broad and short, immobile relative to thorax; frontoclypeus medially with a distinctive sensory organ (Cobben, 1978; Mahner, 1993); labium 4-segmented, short, with a small apical labellum as in Helotrephidae; antennae 3-segmented; scutellum relatively large; forewings elytraceous, meeting along midline; hind wings developed or not; all legs apparently cursorial; fore- and middle tarsi 2- or 3-segmented, hind tarsi 3-segmented; all legs with 2 claws; nymphal dorsal abdominal scent gland located between terga 3/4; thoracic venter and abdominal venter with a laminate keel on segments 2–5 or 2–6; male genitalia not strongly rotated as in Helotrephidae, and only weakly asymmetrical; aedeagus as in Fig. 39.2A; spermatheca as in Fig. 39.2B; ovipositor well developed.

Classification. Fieber (1851) first recognized the group as a family Pleae, but the group was ranked as a subfamily of Notonectidae by most authors until Esaki and China (1928) argued for recognition of the group as a family. The world fauna includes three genera (*Plea* Leach, Old World; *Neoplea* Esaki and China, New World; and *Paraplea* Esaki and China, widely distributed) and 37 species.

Specialized morphology. Although pleids are in general structure similar to the Notonectidae, they do not have the oarlike hind legs, but rather have all legs somewhat homomorphous in form (Fig. 39.1), the hind tibia and tarsus usually not being fringed with long setae. Wefelscheid (1912) studied the anatomy and biology of *Plea minutissima*.

Natural history. Like the Notonectidae, the Pleidae swim in the inverted position, using the legs in a rowing fashion. They frequent vegetated areas in quiet-water habitats and are reported to feed on mosquito larvae, ostracods, and other small arthropods. The air store functions in a manner similar to that in the Notonectidae (Gittelman, 1974, 1975).

Distribution and faunistics. Pleidae are broadly distributed but show their greatest diversity in the tropics. The main works on the group are those of Esaki and China (1928) for the world and Drake and Chapman (1953) on the New World fauna.

40
Helotrephidae

General. These small, globular to somewhat flattened bugs range in length from 1.0 to 4.0 mm. Their habitus (Fig. 40.1A) is similar to that of members of the family

Pleidae, but they have many unique characteristics, most notable being the more complete fusion of the head and prothorax. They have no common name.

Diagnosis. Body generally globular (Fig. 40.1A); head and prothorax fused into cephalonotum, demarcation indicated by a finely impressed line; compound eyes relatively small (Fig. 40.1B); antennae 1- or 2-segmented, the terminal segment furnished with numerous long setae (Fig. 40.1C); labrum reduced, nearly membranous; labium short, 4-segmented (Fig. 40.1D) with a small apical labellum; scutellum very large (Fig. 40.1A); forewings elytraceous, without distinct veins, slightly overlapping; hind wings often absent; tarsal formula variable from 3-3-3 to 1-1-2; all pretarsi with 2 equally developed claws, a pair of equally developed, short parempodia, and at least a ventral arolium in both nymphs and adults (Fig. 40.1E); venter of thorax and abdominal segments 2–5 or 2–6 with a laminate keel; nymphs with a single dorsal abdominal scent gland; male genitalia strongly asymmetrical (Fig. 40.1F), genital capsule rotated 90° to the left, left paramere lying ventrally, right paramere dorsally; aedeagus as in Fig. 40.1G, H; parameres as in Fig. 40.1K; spermatheca vermiform, as in Fig. 40.1J; ovipositor present (Fig. 40.1I).

Classification. The genus *Helotrephes* Stål was placed in the Notonectidae by its author and subsequent workers, until Esaki and China (1927) recognized members of the group as belonging to a distinct family, primarily on the basis of the nearly complete fusion of the head and prothorax. Esaki and China (1927) recognized two subfamilies to contain the three genera known at that time. Currently this small family is divided into four subfamilies, which comprise 16 genera and 44 species (J. T. Polhemus, 1990c). Papáček, Štys, and Tonner (1988) provided characters for a cladistic analysis of relationships among the subfamilies.

Keys to Subfamilies of Helotrephidae

1. Tarsal formula 3-3-3 or 2-2-3; scutellum wider than long 2
– Tarsal formula 1-1-2; scutellum at least as long as wide 3
2. Tarsal formula 3-3-3; cephalonotum small, posterior margin straight to moderately convex, lateral margins posterior to eyes insinuate; propleuron easily visible in lateral view; Neotropical .... Neotrephinae
– Tarsal formula 2-2-3; cephalonotum extremely large, posterior margin strongly convex and broadly extending over proximal portion of scutellum and hemelytra, lateral margins posterior to eyes convex; propleuron concave, not visible in lateral view; Oriental .... Trephotomasinae
3. Dorsum strongly convex, "pleoid"; lateral cephalic carina distinct; antennae 2-segmented; cephalonotal sulcus complete, W-shaped, moderately postocular in position; scutellum usually moderately longer than wide, rarely about as long or strikingly longer, its sides convex; lateral postocular margins of cephalonotum usually convex .... Helotrephinae
– Dorsum rather flat, "naucoroid"; lateral cephalic carina indistinct; antennae 1-segmented; lateral sectors of cephalonotal sulcus absent, only medial, inversely U-shaped portion situated far behind eyes retained; scutellum strikingly long and narrow, sides straight; lateral postocular margins of cephalonotum insinuate, indented, or sharply constricted .... Idiocorinae

HELOTREPHINAE. Body globose; antennae 2-segmented; tarsal formula 1-1-2.

This group was revised by Esaki and China (1928) and divided into two tribes by J. T. Polhemus (1990c). Currently 11 genera, broadly distributed in the Paleotropics, are included.

IDIOCORINAE. Body at least somewhat flattened; antennae 1-segmented; tarsal formula 1-1-2.

This group was first recognized by Esaki and China (1927) to include the genera *Idiocoris* Esaki and China and *Paskia* Esaki and China from tropical Africa. For both genera, the authors provided exquisite illustrations that have seldom been matched for quality or detail in studies of the Nepomorpha.

NEOTREPHINAE. Cephalonotum small; antennae 2-segmented in macropterous forms, 1-segmented in brachypterous specimens; tarsal formula 3-3-3.

This group, which was first recognized by China (1940), contains the Neotropical genera *Neotrephes* China and *Paratrephes* China.

TREPHOTOMASINAE. Cephalonotum very large; antennae 2-segmented; tarsal formula 2-2-3.

This is the most recently described subfamily of Helotrephidae and contains only the Oriental genus *Trephotomas* Papáček, Štys, and Tonner.

Specialized morphology. The helotrephids are distinguished primarily by the complete or nearly complete fusion of the head and the prothorax into a cephalonotum

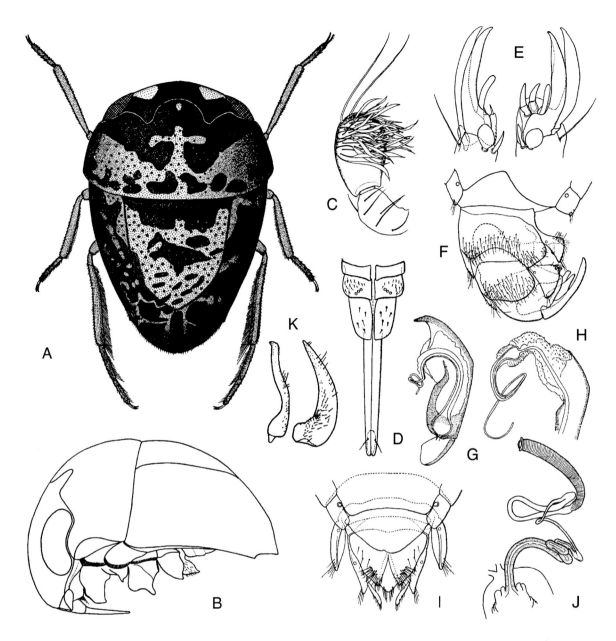

Fig. 40.1. Helotrephidae. **A.** *Heterotrephes admorsus* Esaki and Miyamoto. **B.** Lateral view, *Helotrephes admorsus* (A, B from Esaki and Miyamoto, 1959a). **C.** Antenna, *Helotrephes bouvieri* Kirkaldy. **D.** Labium, *Helotrephes bouvieri* (C, D from Esaki and China, 1928). **E.** Pretarsi, *Idiocoris lithophilus* Esaki and China. **F.** Male genitalia, *I. lithophilus* (E, F from Esaki and China, 1927). **G.** Aedeagus, *Mixotrephes hoberlandti* Papáček, Štys, and Tonner. **H.** Aedeagus, *M. hoberlandti* (G, H from Papáček et al., 1989). **I.** Female genitalia, *I. lithophilus* (from Esaki and China, 1927). **J.** Spermatheca, *Neotrephes usingeri* China (from Pendergrast, 1957). **K.** Parameres, *Helotrephes admorsus* Esaki and Miyamoto (on left dorsal paramere, ventral view; on right, ventral paramere, ventral view) (from Esaki and Miyamoto, 1959a).

(Fig. 40.1A, B). The 1- or 2-segmented antennae, with the covering of elongate setae on the terminal segment are also distinctive (Fig. 40.1C). J. T. Polhemus (1990c) described the occurrence of a stridulatory mechanism in the Helotrephinae, consisting of a serrated lateral margin of the hemelytron plus a ridge dorsally on the hind femur. He also noted that other stridulatory structures might exist in the group (see Miyamoto, 1952).

Natural history. Helotrephids live in a variety of habitats, ranging from the quiet, nearly stagnant waters of ponds to much better aerated backwaters in streams; some have been taken from hot springs. They apparently swim with the venter up, as do other Notonectoidea, or they may swim with the venter down and use only the hind legs for propulsion, in a rowing motion (Miyamoto, 1952). They feed on small invertebrates.

Distribution and faunistics. This group is restricted to the tropics and shows its greatest diversity in the Old World. Although there is no comprehensive reference for identification of species, important works include those of Esaki and China (1927), China (1935, 1940), Esaki and Miyamoto (1959a), Papáček, Štys, and Tonner (1988), and J. T. Polhemus (1990c).

41
Leptopodomorpha

General. Bugs in this group are of very small to modest size. They vary in shape from nearly globose (Fig. 44.1A), to weakly flattened and ovoid (Fig. 43.1B), to elongate and parallel-sided (Fig. 45.1B). All species are predatory. The majority inhabit damp areas adjacent to water, and a relatively large number are intertidal. Many members of the Leptopodidae have no apparent association with water and as a consequence have habitats that are less easily characterized. Those species are more rarely collected.

Diagnosis. Head usually relatively short and broad; compound eyes usually very large, occupying nearly entire side of head (Figs. 43.1A, 45.1A, B); antennae in most groups with all segments of nearly equal diameter, segment 1 short, segments 2–4 longer and of more or less equal diameter, segment length and diameter variable in the Leptopodinae; labium inserted ventrally, short (Fig. 45.2B) or long (Fig. 43.2A), with 3 or 4 obvious segments; ocelli usually present; head with at least 3 pairs of trichobothria dorsally; forewings in the form of hemelytra with a conspicuously coriaceous anterior portion and membranous posterior region, often with a distinct medial fracture combining with costal fracture (Fig. 43.2E) and attaining corial margin in macropterous forms (except Leptopodinae), membrane usually with 3, 4, or 5 closed cells in macropterous forms (Figs. 43.1A, B, 45.1A, B); legs usually relatively short, slender, and mutic, or in Leptopodinae longer, more slender, and femora sometimes armed with spines (Fig. 45.2A); adult tarsal formula usually 3-3-3, 1-2-2 in *Leotichius*, nymphal formula usually 2-2-2; parempodia usually reduced (Fig. 41.2D), sometimes setiform (Fig. 41.2B), accessory parempodia present; small ventral arolium present in some nymphs, absent in most adults; metathoracic scent-gland system with 1–4 reservoirs and 1 or 2 ostioles; abdominal sterna composed of a single plate, dorsal laterotergites present or absent, 8 abdominal spiracles, usually located dorsally on laterotergite (Fig. 45.2D); copulation always in the side-by-side position (Fig. 43.2A), male holding forewing margin of female with grasping apparatus present between laterotergites of abdominal segments 2 and 3, pegs present on this structure in Saldidae (Fig. 41.2J), absent in all other groups; nymphal dorsal abdominal scent-gland opening present between abdominal terga 3/4; male genitalia symmetrical, parameres either hook-shaped with a distinct processus sensualis or club-shaped (Figs. 44.1E, 45.2E); ovipositor valvulae laciniate, or reduced and platelike; spermatheca with bulb and flange (Figs. 43.2I, 45.2G, H).

Discussion. The overall composition of the Leptopodomorpha has been generally agreed upon by most modern authors, although categorical rank and phylogenetic relationships of several taxa have been debated. Following the classification of Schuh and Polhemus (1980b), we recognize four families—Aepophilidae, Saldidae, Omaniidae, and Leptopodidae (including Leptosaldinae and *Leotichius* Distant); their cladogram is shown in Figure 41.1. In contrast Cobben (1971) grouped the Leptopodomorpha as follows: Saldidae (including Aepophilidae, Leptosaldinae), Omaniidae, Leptopodidae, and Leotichiidae. Popov (1985) transferred the Leptosaldinae to the Omaniidae as a subfamily. The catalog of Schuh et al. (1987) is a comprehensive reference to the classification and literature on the group.

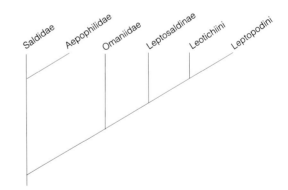

Fig. 41.1. Phylogenetic relationships of family-level taxa of Leptopodomorpha (from Schuh and Polhemus, 1980b).

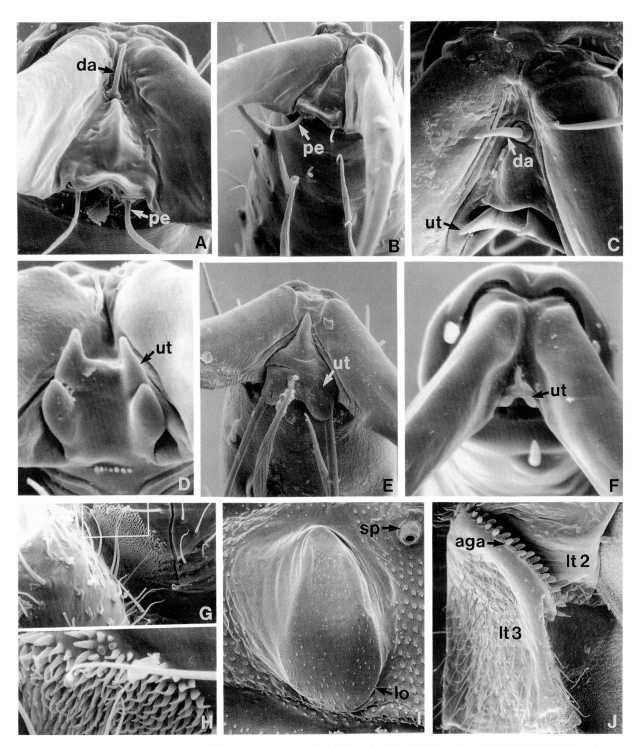

Fig. 41.2. Leptopodomorphan morphology. **A.** Foreleg pretarsus, nymph, with dorsal arolium, *Saldula pallipes* (Saldidae). **B.** Hind leg pretarsus, nymph, *S. pallipes* (Saldidae). **C.** Foreleg pretarsus, adult, with dorsal arolium and reduced parempodia, *S. pallipes* (Saldidae). **D.** Adult pretarsus, *Omania coleoptrata* (Omaniidae). **E.** Foreleg pretarsus, nymph, *Valleriola* sp. (Leptopodidae). **F.** Hind leg pretarsus, adult, *Leotichius shiva* (Leptopodidae). **G.** Metacoxal adhesive pads, *O. coleoptrata* (Omaniidae). **H.** Same as G, detail. **I.** Larval organ on abdominal sternum 3, *S. pallipes* (Saldidae). **J.** Male abdominal grasping apparatus, pegs on laterotergite of abdominal segment 3, *S. pallipes* (Saldidae). Abbreviations: aga, abdominal grasping apparatus; da, dorsal arolium; lo, larval organ; lt 2, laterotergite 2; lt 3, laterotergite 3; pe, parempodium; sp, spiracle; ut, unguitractor plate.

Key to Families of Leptopodomorpha

1. Labium tapering, long, reaching to base of hind coxae or beyond (Fig. 43.2A) 2
– Labium much shorter, reaching at most to apex of forecoxae, often reaching only base of forecoxae (Fig. 45.2B) .... Leptopodidae
2. Compound eyes large, covering most of head in lateral view; forewings covering abdomen, macropterous, submacropterous, or coleopteroid 3
– Compound eyes small (Fig. 42.2), covering a small portion of head in lateral view; wings greatly reduced, in form of pads without obvious venation, covering only anterior portion of abdomen (Fig. 42.1) .... Aepophilidae
3. Compound eyes reaching posteriorly only to level of pronotal collar or very slightly beyond (Fig. 43.1A, B); macropterous or submacropterous, seldom coleopteroid; body length always greater than 2.2 mm .... Saldidae
– Compound eyes reaching posteriorly about one-third length of pronotum, distinctly surpassing pronotal collar (Fig. 44.1A); always coleopteroid; body length always less than 2 mm .... Omaniidae

Saldoidea

42
Aepophilidae

General. The Aepophilidae comprises *Aepophilus bonnairei* Signoret, sometimes referred to as the marine bug. This intertidal creature, slightly more than 2 mm long, has the general appearance of a tiny bed bug (Fig. 42.1).

Diagnosis. Respiratory plastron covering pronotum, scutellum, and hemelytra; compound eyes very small, with approximately 50 ommatidia (Figs. 42.1, 42.2A); ocelli possibly represented by minute remnants; head projecting anteriorly, clypeus visible from above; labium long, segment 1 greatly abbreviated, segment 2 short, segments 3 and 4 subequal in length, each much longer than segment 2; forewings padlike (Fig. 42.1), hind wings absent; as in Saldidae, adult and nymphal pretarsus with "accessory parempodia" (Cobben, 1978) and rudimentary parempodia, ventral arolium present in adult and nymphal stages (Schuh and Polhemus, 1980b); metathoracic scent gland with 2 reservoirs and 2 ostioles (Cobben, 1970); abdomen with both dorsal and ventral

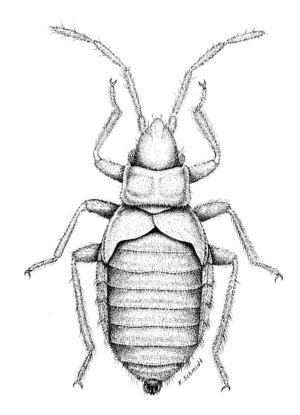

Fig. 42.1. Aepophilidae. *Aepophilus bonnairei* Signoret.

Specialized morphology. Early investigators were not certain by what mechanism *Aepophilus* was able to acquire oxygen when submerged. Most assumed that an air bubble was trapped in the rock crevices in which the bugs secreted themselves. SEM examination reveals, however, that the surface of the pronotum, scutellum, and hemelytra is densely covered with fine, branching microtrichia, which have been observed to maintain a constant air covering and which apparently serve as a physical gill in the form of a plastron (King and Ratcliffe, 1970; King and Fordy, 1984), allowing the bugs to remain submerged for prolonged periods.

Natural history. *Aepophilus bonnairei* lives in rock crevices in the lowest reaches of the intertidal zone—the so-called *Fucus* zone—where it feeds on small invertebrates (e.g., Baudoin, 1946).

Distribution and faunistics. *Aepophilus* is known from the coast of Ireland, the Low Countries, northern France, southwestern England, the Channel Islands, Spain, Portugal, and possibly Morocco. This area is not known for harboring phylogenetic relics, placing *Aepophilus* in something of a unique position.

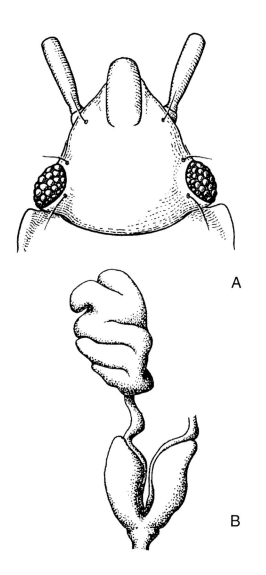

Fig. 42.2. Aepophilidae. *Aepophilus bonnairei.* **A.** Head (from Schuh and Polhemus, 1980b). **B.** Testes (from Pendergrast, 1957).

laterotergites; spiracles located on simple sterna; nymphal abdominal scent glands with 2 reservoirs and 2 ostioles; abdominal grasping apparatus present, but lacking pegs found in Saldidae; eversible glands and nymphal organ found in Saldidae absent; phallus similar to Saldidae but lacking reel system; parameres hook-shaped with a distinct processus sensualis; testes as in Fig. 42.2B; ovipositor valvulae laciniate, serrate.

Classification. *Aepophilus* has often been treated as a member of the Saldidae (e.g., Leston, 1957a; Cobben, 1959), but most work suggests that it is the sister group of all remaining Saldoidea. We therefore follow Schuh and Polhemus (1980b) in giving it family status.

43
Saldidae

General. Shore bugs, which range in length from 2.3 to 7.4 mm, are usually ovoid in outline (Fig. 43.1A, B). As the family name suggests, they can jump, and in fact most are extremely agile through a combination of jumping and flight. Many species vary greatly in the extent of hemelytral pigmentation, and therefore identification is often difficult.

Diagnosis. Compound eyes large, reniform, posterior margin usually reaching, but never surpassing, anterior margin of pronotum (Fig. 43.1A, B); 3 pairs of cephalic trichobothria; labium long, tapering, nearly reaching onto abdomen, segment 1 greatly reduced, segment 3 much longer than either segment 2 or 4 (Fig. 43.2A); medial fracture in forewing well developed and in combination with costal fracture attaining costal margin (Fig. 43.2B); 4 or 5 closed cells in the membrane (Fig. 43.1A, B); subcostal region of the forewing in female modified to accommodate abdominal grasping apparatus of male; parempodia developed and setiform in nymphs,

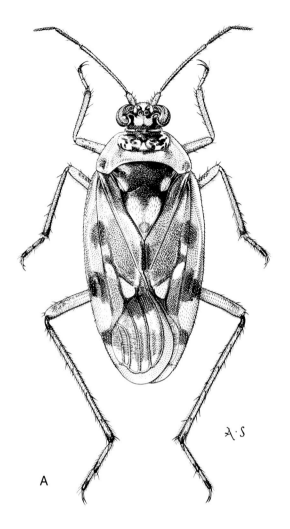
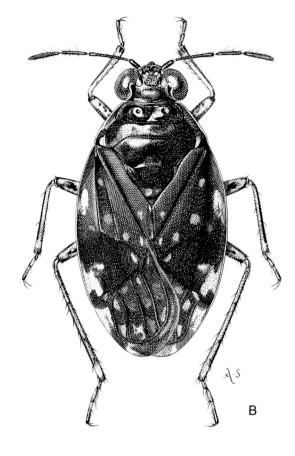

Fig. 43.1. Saldidae. **A.** *Pentacora signoreti* (Guérin-Ménéville) (Chiloxanthinae) (from Usinger, 1956). **B.** *Pseudosaldula chilenis* (Blanchard) (Saldinae) (from Drake, 1962).

greatly reduced and rudimentary in adults, ventral arolium present in nymphs (Schuh and Polhemus, 1980b, contra Cobben, 1978), present on some legs in adults of at least some taxa (Fig. 41.2A–C); metathoracic scent gland with 1 reservoir and 1 ostiole; abdomen with dorsal and ventral laterotergites, spiracles located on sterna; nymphs of most Saldidae (except *Enalosalda* Polhemus and Evans, *Orthophrys* Horvath, and all Saldini) with "larval organ" on either side of midline of sternum 3 near spiracle (Figs. 41.2I, 43.2J–M) (Cobben, 1957); adults with eversible glands located laterally on abdomen between segments 7 and 8; parandria in males well developed and sclerotized (Fig. 43.2F); parameres hooked and with a conspicuous processus sensualis (Fig. 43.2E); phallus with a reel system (Fig. 43.2C, D); spermatheca as in Fig. 43.2I; ovipositor valvulae well developed and serrate.

Classification. The major classical worker on the Saldidae was O. M. Reuter, who described many species and published a catalog of the Palearctic fauna (Reuter, 1895). C. J. Drake described many species from the New World (e.g., 1949a, b, 1955) and published some of the first illustrations of the male parameres. More recent work has been dominated by the contributions of the late R. H. Cobben on the morphology (Cobben, 1957, 1968a) and higher classification of the group (Cobben, 1959).

Nomenclature in the Saldidae was confused well into the twentieth century because of differences of opinion regarding the type genus of the family. Reuter consistently considered *Acanthia* Fabricius to be the type, while most other authors (e.g., Lethierry and Severin, 1896) believed *Salda* Fabricius was the type.

The current classification is primarily the work of Cobben (1959) and is based heavily on the structure of the male genitalia. Two subfamilies comprising 26 genera and 265 species are recognized.

Keys to Subfamilies of Saldidae

1. Hemelytra with long medial fracture reaching at least to level of posterior end of claval suture (fracture lost in *Enalosalda*); female sternum 7 truncate, usually square, or if produced caudad along midline (Fig. 43.2G); membrane with five cells (Fig. 43.1A) .... Chiloxanthinae
– Hemelytra with short medial fracture, not reaching anteriorly more than half the distance from costal fracture to posterior end of claval suture (fracture lost in *Orthophrys*); female sternum 7 produced caudad along midline (Fig. 43.2H); membrane usually with 4 cells (5 cells in *Pseudosaldula*) .... Saldinae

CHILOXANTHINAE (FIG. 43.1A). Membrane with 5 closed cells (Fig. 43.2B); base of penis filum forming at most 1 coil; ductus ejaculatorius with 2 ventral accessory glands; median sclerotized structure of penis paired; apicolateral sclerotized structures of penis absent; sternum 7 in female square.

This group of 22 Recent and four fossil species placed in four Recent and two fossil genera is largely restricted to the Northern Hemisphere and is best known for its association with very high latitudes (*Chiloxanthus* Reuter), saline habitats (*Pentacora* Reuter), or intertidal habitats (*Enalosalda* Polhemus and *Paralosalda* Polhemus and Evans). *Pentacora signoreti* Guérin-Ménéville is probably the largest species of saldid, measuring up to 7.4 mm in length.

SALDINAE (FIG. 43.1B). Membrane with 4 closed cells (except *Pseudosaldula* Cobben with 5; Fig. 43.1B); base of penis filum coiled like a watch spring (Fig. 43.2D); median sclerotized structure of penis unpaired; apicolateral sclerotized structure of penis present; mesal portion of sternum 7 in female produced caudad (Fig. 43.2H).

Three tribes are currently recognized. Most species of Saldini are large, often exceeding 5 mm in length; they are black, with few if any markings, and often are brachypterous. Species of *Teloleuca* Reuter have variegated markings and are usually macropterous. Saldini occur at moderately high latitudes in the Northern Hemisphere in wet freshwater environments, although some species—for example, *Salda littoralis* (Linnaeus)—occasionally may be found in salt marsh or other saline situations.

The Saldoidini are a group of worldwide distribution, with the greatest numbers of species in temperate regions and a lesser number in lowland tropical habitats. Most Saldoidini are associated with fresh water, although *Halosalda* Reuter contains only halophilous species; a few, such as *Saldula laticollis* (Reuter) and *Orthophrys pygmaeum* (Reuter) are intertidal. The genus *Saldula* Van Duzee contains a relatively high proportion of all described species of shore bugs, as well as some of the most complex taxonomic problems in the family; Lindskog and Polhemus (1992) noted that the group is apparently paraphyletic. *Saldoida* Osborn is remarkable among the Saldidae for the pronotal calli produced into conical tubercles and for antlike behavior.

The Saldunculini contain only the Indo-Pacific intertidal genus *Salduncula* Brown, whose seven described species were keyed by J. T. Polhemus (1991b).

Specialized morphology. The well-developed peg plate of the abdominal grasping apparatus in the male is one of the unique features of the Saldidae (Fig. 41.2J). This structure was first described by Drake and Hottes (1951) as stridulatory in function. Cobben (1957) and Leston (1957b) correctly observed that is was actually used during copulation, the male abdomen grasping the anterolateral margin of the female hemelytron. Actual stridulatory structures occur in the genera *Chartoscirta* Stål, *Ioscytus* Reuter, *Lampracanthia* Reuter, *Macrosaldula* Leston and Southwood, *Rupisalda* Polhemus, and *Saldoida* Osborn, all members of the Saldinae. In all cases the stridulitrum is located on the anterior margin of the forewing, and the plectrum is located on the hind femur (J. T. Polhemus, 1985).

The larval organ, of unknown function, is located on abdominal sternum 3 (Figs. 41.2I, 43.2J–M). The eversible glands, also of unknown function, are found between the laterotergites of abdominal segments 7 and 8. Both structures are unique to the saldids, although the former is absent in some taxa.

Sperm ultrastructure has been studied in two species by Afzelius et al. (1985).

Natural history. Many saldids are extremely agile, moving through a combination of jumping and flight. All are predators of small invertebrates, seeking out live food or scavenging on dead individuals. They can be maintained in captivity on immobilized flies or other small arthropods and have been shown to locate subsurface prey through antenna-based chemoreception. Surface prey may be located visually as well as by chemoreception (J. T. Polhemus, 1985).

Most members of the Saldini appear to overwinter as

Keys to subfamilies of Saldidae adapted from J. T. Polhemus, 1985.

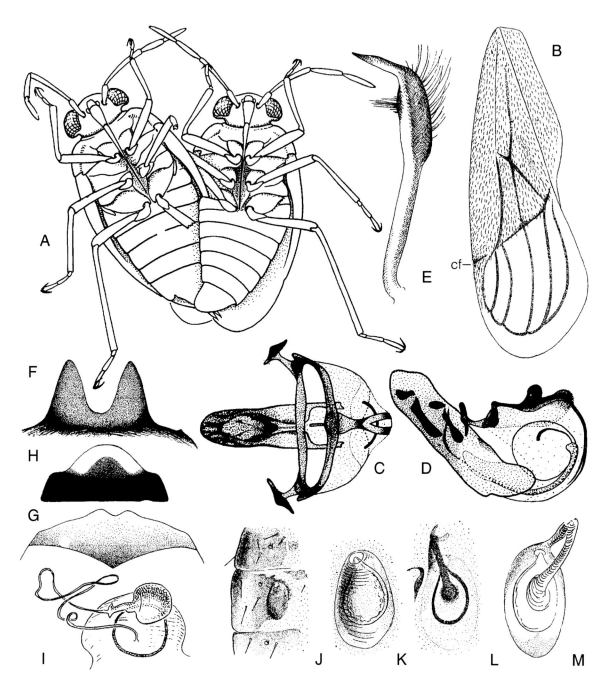

Fig. 43.2. Saldidae. **A.** Mating position, Saldidae (from Schuh and Polhemus, 1980b). **B.** Forewing, *Pentacora grossi* Cobben (from Cobben, 1980b). **C.** Phallus, dorsal view, *Saldula laticollis* (Reuter). **D.** Phallus, lateral view, *S. laticollis* (C, D from J. T. Polhemus, 1985). **E.** Paramere, *S. sibiricola* Cobben (from Cobben, 1985). **F.** Parandria, *S. dentulata* (Hodgden). **G.** Subgenital plate, *P. signoreti*. **H.** Subgenital plate, *S. dentulata* (F–H from Cobben, 1960d). **I.** Gynatrial complex, *S. sibiricola* (from Cobben, 1985). **J.** Larval organ, *Chartoscirta cocksii* (Curtis). **K.** Larval organ, *C. cocksii*. **L.** Larval organ, *C. cocksii*. **M.** Larval organ, *S. fucicola* (Sahlberg) (J–M from Cobben, 1957). Abbreviation: cf, costal fracture.

eggs and produce only a single generation each year, whereas, in the Northern Hemisphere at least, most members of the Saldoidini probably overwinter as adults, some producing multiple generations in a given season. The life cycles of tropical species are poorly known.

Most temperate-region shore bugs are associated with damp substrates along the margins of ponds and streams and occasionally along the margins of larger bodies of water. Some species practice a truly terrestrial existence, as for example members of the *Saldula orthochila* species group and some members of the Saldini. The majority of tropical saldids are saxicolous, usually living in association with large stones, a habitat peculiar to only a few temperate-latitude species. There is a significant high-altitude fauna in the Andes of South America; most of the known species are associated with seeps or pond margins and a few occur on rocks in streams. The eggs of most species appear to be laid in vegetation with the laciniate ovipositor.

Intertidality, one of the most conspicuous adaptations in the shore bugs, occurs in nearly every phyletic line (Schuh and Polhemus, 1980b; J. T. Polhemus, 1985) as well as in the related families Aepophilidae and Omaniidae. Life history studies of the intertidal *Saldula laticollis* in western North America indicated that it can remain submerged for substantial periods (Stock and Lattin, 1976) without the use of a plastron.

Distribution and faunistics. Saldids are known from all major land areas except Antarctica, including remote islands such as St. Helena (one species; Cobben, 1976) and the Hawaiian Islands (eight species; Cobben, 1980a). A preponderance of the described species occurs in the Northern Hemisphere, but only with additional knowledge of phylogenetic relationships within the group at the species level will a more mature understanding of the distributional history of the group come into focus.

Regional treatments exist for the faunas of the Palearctic (Cobben, 1960a, 1985; Kerzhner and Jaczewski, 1964; Péricart, 1990); the Nearctic (e.g., Usinger, 1956; Chapman, 1962; Brooks and Kelton, 1967; Schuh, 1967; J. T. Polhemus and Chapman, 1979a), and Middle America (J. T. Polhemus, 1985). Cobben (1987b) provided an annotated checklist of the African species (see also Cobben, 1987a; J. T. Polhemus, 1981). Only scattered papers are available for South America, Australia, and New Zealand.

Leptopodoidea

44
Omaniidae

General. The Omaniidae, or intertidal dwarf bugs, are the smallest members of the Leptopodomorpha, ranging in length from 1.15 to 1.59 mm. They are similar in size and appearance (Fig. 44.1A) to many Schizopteridae.

Diagnosis. Eyes very large, projecting posteriorly along anterolateral margins of pronotum (Fig. 44.1A); vertex and frons with 4 pairs of trichobothria; labium long, with 4 obvious segments, 1 and 2 short, 4 longer, 3 distinctly longer than 4 (Fig. 44.1B); coleopteroid, hemelytra meeting along midline in the form of elytra, no medial fracture, membrane absent; as in most Saldidae, parempodia well developed and setiform in nymphs, greatly reduced and budlike in adults (Fig. 41.2D); existence of arolium in nymphs unconfirmed; hind coxae with adhesive pads (Figs. 41.2G, 44.1C); metathoracic scent glands with 4 reservoirs in 2 pairs and a single ostiole; dorsal (inner) laterotergites absent; abdominal spiracles dorsal, situated in membrane adjacent to outer laterotergite; abdominal grasping apparatus present between laterotergites of abdominal segments 2 and 3, but without pegs; aedeagus as in Fig. 44.1D; parameres club-shaped (Fig. 44.1E); female gynatrial complex and spermatheca as in Fig. 44.1F; ovipositor reduced and platelike.

Classification. *Omania* Horvath, with a single species from the Red Sea and Gulf of Oman was originally placed in the Saldidae. Cobben (1970) elevated the group to family status, placing four species in a new genus, *Corallocoris* Cobben, which is distributed from Aldabra Atoll off the coast of Africa east to Samoa and Kwajelein Atoll.

Specialized morphology. Although schizopterid in general appearance and size, the leptopodomorphan affinities of this group have been recognized by all workers. The Omaniidae are notable among the Leptopodomorpha for the coleopteroid hemelytra and the presence of adhesive pads on the hind coxae.

Natural history. Although the Omaniidae are all intertidal, they—like many Saldidae—have no obvious morphological specializations for this mode of life (see Cobben, 1970). They apparently feed on small organisms in the intertidal zone and can be found crawling around on rocks at low tide. They secrete themselves in crevices during high tide, apparently using a trapped air bubble

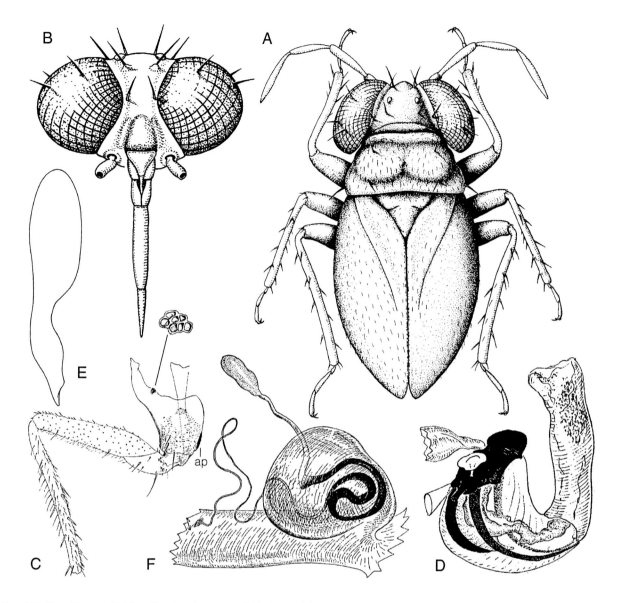

Fig. 44.1. Omaniidae. **A.** Habitus, *Corallocoris nauruensis* Herring and Chapman. **B.** Frontal view, head, *C. nauruensis* (A, B from Schuh and Polhemus, 1980b). **C.** Hind leg, *Omania coleoptrata* Horvath. **D.** Phallus, *O. coleoptrata* (C, D from Cobben, 1970). **E.** Paramere, *C. nauruensis* (from Schuh and Polhemus, 1980b). **F.** Gynatrial complex, including spermatheca, *O. coleoptrata* (from Cobben, 1970). Abbreviation: ap, adhesive pad.

as a source of oxygen. The eggs, which are very large relative to the size of the animal, apparently develop one at a time, and a single egg is laid every few days (Kellen, 1960).

Distribution and faunistics. The basic reference for this Indo-Pacific group is Cobben (1970).

45
Leptopodidae

General. This primarily tropical group has no common name. Its members are very active and agile. Leptopodids range from 1.8 to 7.0 mm in length and are of variable habitus (Figs. 45.1A, B, 45.2A), but they are distinctive for the relatively short, 4-segmented labium and deeply punctured dorsum.

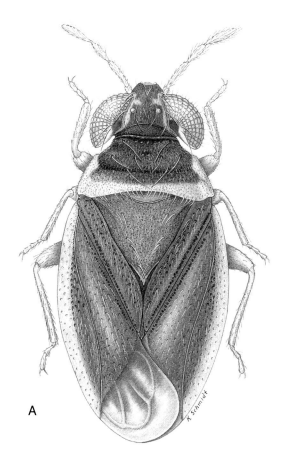
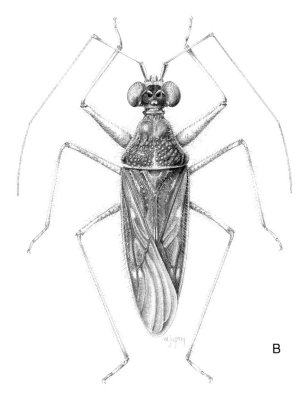

Fig. 45.1. Leptopodidae. **A.** *Saldolepta kistnerorum* Schuh and Polhemus (Leptosaldinae) (from Schuh and Polhemus, 1980a). **B.** *Valleriola* sp. (Leptopodinae) (from Slater, 1982).

Diagnosis. Compound eyes of varied shape and ocelli of varied position; labium 4-segmented, short, at most reaching to apex of forecoxae (Fig. 45.2B); at least clavus, and sometimes most of dorsum, deeply and densely punctured (Figs. 45.1A, 45.2A); parempodia present and setiform in nymphs (Fig. 41.2E), greatly reduced or absent in adults (Fig. 41.2F); arolia absent in those taxa examined; abdominal grasping apparatus present but without pegs; parameres club shaped (Fig. 45.2E); ovipositor platelike (Fig. 45.2D, F); spermatheca with bulb and flange (Fig. 45.2G, H).

Classification. The earliest comprehensive work on the Leptopodidae was that of Horvath (1911). The classification of Schuh and Polhemus (1980b) presented here recognized two subfamilies, some members of which were placed in other families by prior authors. Ten genera and 37 species are included.

Key to Subfamilies of Leptopodidae

1. Ocelli proximate and situated on a low prominence (Figs. 45.1B, 45.2A); compound eyes hemispherical, with ommatidia on dorsal surface of eyes or not .... Leptopodinae
– Ocelli widely separated, not situated on a prominence (Fig. 45.1A); compound eyes distinctly reniform, ommatidia always present on dorsal surface (Fig. 45.1A) .... Leptosaldinae

LEPTOSALDINAE (FIG. 45.1A). Habitus saldidlike; eyes large, reniform, extending well back onto lateral pronotal margins (similar to Omaniidae); 3 cells in membrane; body flattened dorsoventrally; appendages relatively short; spiracles located dorsally on laterotergites.

Leptosalda chiapensis Cobben is known only as an amber fossil from southern Mexico (Cobben, 1971). The other member of the subfamily, *Saldolepta kistnerorum* Schuh and Polhemus, is known from two Recent specimens, one taken from a termite nest in the tropical low-

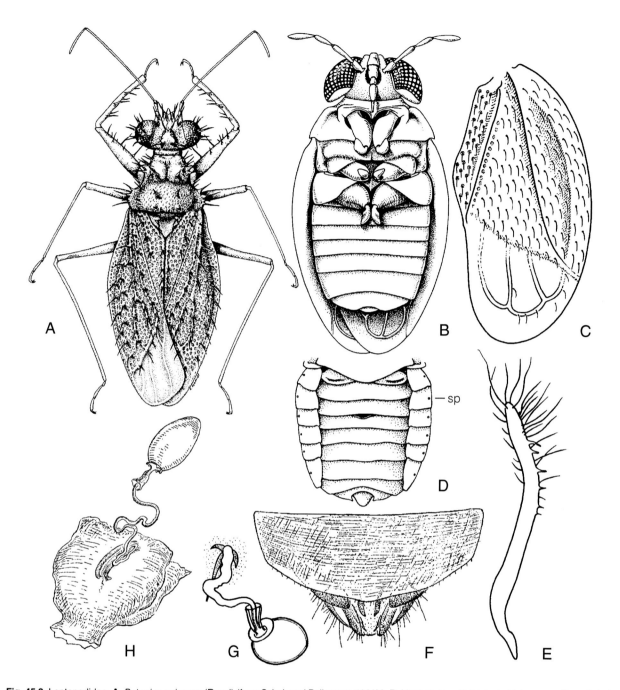

Fig. 45.2. Leptopodidae. **A.** *Patapius spinosus* (Rossi) (from Schuh and Polhemus, 1980b). **B.** Ventral view, *Saldolepta kistnerorum* (from Schuh and Polhemus, 1980a). **C.** Forewing, *Leptosalda chiapensis* Cobben (from Schuh and Polhemus, 1980b). **D.** Dorsal view, abdomen, female *S. kistnerorum* (from Schuh and Polhemus, 1980a). **E.** Paramere, *Erianotus lanosus* (Dufour) (from Schuh and Polhemus, 1980b). **F.** Female genitalia, *P. thaiensis* Cobben (from Cobben, 1968b). **G.** Spermatheca, *Leptopus marmoratus* (Goeze) (from Pendergrast, 1957). **H.** Gynatrial complex, including spermatheca, *P. thaiensis* (from Cobben, 1968b). Abbreviation: sp, spiracle.

lands of western Ecuador (Schuh and Polhemus, 1980a), the other recovered by port inspectors on plant material from Colombia.

LEPTOPODINAE (FIGS. 45.1B, 45.2A). Eyes protruding, nearly hemispherical; ocelli proximate to one another, situated on low prominence; costal fracture absent; forefemora and often dorsum with very long setae or heavy cuticular spines; tarsi elongate, slender; spiracles located dorsally on laterotergites in Leptopodini, ventrally on abdominal sternum in Leotichiini.

Two tribes are recognized, with all members occurring in the Old World, primarily in the tropics. The Leotichiini, with three species of *Leotichius* Distant, range from Burma to Bali.

The Leptopodini comprises seven genera. *Valleriola* Distant includes 11 species, most living on vertical rock surfaces around the shores of ponds and streams, sometimes being very common. Genera such as *Erianotus* Fieber, *Leptopus* Latreille, *Martiniola* Horvath, and *Patapius* Horvath, each with only a few species, are often found under stones in screes or in other dry situations some distance from water. The group is distributed in the southern Palearctic, throughout Africa and Madagascar, and east to New Caledonia.

Specialized morphology. The leptopodids, and particularly the Leptopodinae, are distinctive for their modification of certain body parts. All have at least the clavus, and often much of the dorsum, heavily punctured, in contrast to other Leptopodomorpha. The Leptopodinae have the ocelli situated close together on a low tubercle or prominence (Figs. 45.1B, 45.2A). In *Patapius* Horvath much of the body surface—including the compound eyes—is covered with spines (Fig. 45.2A); in *Valleriola* Distant the legs are very long and covered with a fringe of long setae (Fig. 45.1B); and in *Leotichius* the ommatidia on the upper surface of the compound eyes are not developed.

Péricart and Polhemus (1990) described a stridulatory structure involving abdominal tergum 1 and the vannus of the hind wing for several species of Leptopodinae. They asserted that this structure represented a synapomorphy for the Leptopodinae and that it was absent in *Leotichius* either because that genus represents a distinct subfamily or family or because the structure had been lost. A simpler explanation interprets this structure as a synapomorphy of the Leptopodini sensu Schuh et al. (1987).

Natural history. No simple description serves to categorize the habitats of the Leptopodidae. All members are presumably predatory, although their feeding approach may differ from that of other leptopodomorphans in that they have a very short rostrum.

Leotichius spp. were long thought to inhabit caves, but more recent collections suggest that they are simply denizens of extremely dry tropical habitats. The most extensive collections have been made from an area of ant lion pits under the overhanging roofs of Hindu temples on the island of Bali (J. T. Polhemus and Schuh, 1995).

Distribution and faunistics. The Leptopodidae were long known only from the Old World, until the two New World species, *Leptosalda* and *Saldolepta*—now included in the Leptosaldinae—were described. *Patapius spinosus* Rossi has been introduced near San Francisco, California, in North America. Most species live within the tropical latitudes. Horvath (1911) provided the only comprehensive treatment for the Leptopodini; J. T. Polhemus and D. A. Polhemus (1991c) provided a key to most genera.

46
Cimicomorpha

General. Among the 16 families currently placed in the Cimicomorpha are the two largest families of true bugs, the Miridae and Reduviidae, as well as two of the most obscure, the monotypic Joppeicidae and Medocostidae. Although it can be argued that the primitive cimicomorphan stock was predaceous, the group also contains large numbers of strictly phytophagous species in the families Miridae, Tingidae, and Thaumastocoridae.

The most important classical authors working in the group were Fieber on the fauna of the western Palearctic, Stål on the Nabidae and Reduviidae, and Reuter on the Anthocoridae, Microphysidae, and particularly Miridae. Modern authors have generally become family specialists, because of the sheer numbers of taxa involved.

Diagnosis. Head often prognathous, with labium inserted anteriorly, and clypeus lying more or less dorsally and in a nearly horizontal position, or head not so strongly oriented anteriorly, labium inserted anteroventrally, clypeus positioned anteriorly and more or less vertical in orientation; head often with cephalic trichobothria; antennae often flagelliform; labium with 3 or 4 segments; forewings in macropterous forms always in the form of a hemelytron, with a coriaceous anterior region and a membranous posterior region, often with a long medial fracture and a costal fracture; membrane often with 1, 2, or 3 closed cells, or with a few free veins, or no veins, or veins emanating from closed cells; hind wings usually with a simple, nonbranching distal sector of R+M; tarsi usually with 3 segments, sometimes 2; pretarsus symmetrically developed in most families, lacking arolia in all life stages; claws simple and without pulvilli or other ornamentation (except many Miridae and some Thaumastocoridae and Anthocoridae); metathoracic scent-gland evaporatory area usually with distinctive mushroom bodies (Fig. 10.10A); spiracle 1 sometimes present, remaining spiracles always ventral, located either on distinct laterosternite or on mediosternal plate with laterotergite present or not; spermatheca nonfunctional as a sperm storage organ, being rudimentary, modified into a vermiform gland, or absent; eggs usually with micropyles distinct from aeropyles and both arranged in a ring outside the operculum.

Discussion. First recognized by Leston et al. (1954), the Cimicomorpha have been conceived differently by various authors. Cobben (1968a, 1978), who had serious doubts about the placement of the Reduviidae and Thaumastocoridae within the group, consistently associated the Pachynomidae with the Nabidae rather than with the Reduviidae, as argued by Carayon (1950b), Davis (1969), and Carayon and Villiers (1968). We follow the classification presented by Schuh and Štys (1991). Their cladogram is shown in Figure 46.1.

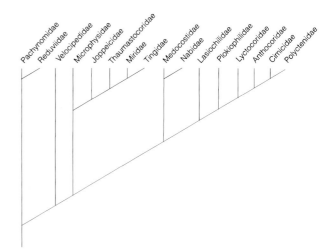

Fig. 46.1. Phylogenetic relationships of families of Cimicomorpha (after Schuh and Štys, 1991).

Key to Families of Cimicomorpha

1. Mandibular plates greatly enlarged and conspicuous, usually exceeding and surrounding apex of clypeus (Figs. 52.1A, 52.2A, 52.3A, B) .... Thaumastocoridae
– Mandibular plates not conspicuously enlarged, never obvious in dorsal view, never exceeding apex of clypeus 2

2. Labium conspicuously 4-segmented, inserted ventrally on head, segment 1 reaching posterior margin of head or nearly so (Figs. 53.3D, 54.3A); fossula spongiosa never present 3
- Labium usually with 3 obvious segments, inserted anteriorly on head (Figs. 47.2A, 49.1B, C), if 4-segmented, segment 1 never approaching posterior margin of head; fossula spongiosa often present on 1 or more pairs of legs 5
3. Pronotum and hemelytra areolate (Fig. 54.2B); hemelytra of nearly uniform texture throughout, without obvious corium-clavus and membrane; antennal segment 2 short (Figs. 54.2B, 54.3A); ocelli always absent; tarsi 2-segmented (Cantacaderinae, Tinginae) .... Tingidae (part)
- Pronotum and hemelytra never areolate, although sometimes heavily punctured; hemelytra usually with obvious corium, clavus, and membrane, although rarely coleopteroid; antennal segment 2 more elongate, usually much longer than segment 1; ocelli present or absent; tarsi 2- or 3-segmented 4
4. Macropterous or brachypterous, rarely coleopteroid; R+M of forewing never raised and keel-like; compound eyes always normally developed; trichobothria present on meso- and metafemora (Fig. 53.3E, F); male genitalia always asymmetrical; tarsi 2- or 3-segmented; ocelli present or absent .... Miridae
- Usually coleopteroid, heavily punctured, compound eyes greatly reduced and ocelli absent (Fig. 54.1), or, if macropterous, R+M in forewing elevated and keel-like, compound eyes developed and ocelli present; trichobothria never present on meso- and metafemora; male genitalia symmetrical; tarsi 2-segmented (Vianaidinae) .... Tingidae (part)
5. Prosternal sulcus present, usually in the form of a stridulitrum (Fig. 10.7F, G), receiving apex of labium; labium usually short, stout, and strongly curving (Fig. 48.3A, C), sometimes more slender and nearly straight (Fig. 48.3B); head necklike behind eyes (Fig. 48.2A–D), frequently with a transverse impression anterior to ocelli; membrane usually with 2 large cells (Fig. 48.2C) (sometimes more) or rarely with a few longitudinal veins (Fig. 48.2A) .... Reduviidae
- No prosternal sulcus receiving apex of labium; labium straight or curving; head not conspicuously necklike behind eyes, never with a transverse impression anterior to ocelli; membrane venation variable 6
6. Antennae with 5 apparent segments (Figs. 47.1, 56.1) 7
- Antennae with 4 segments 8
7. Scutellum with 1–7 pairs of trichobothria laterally; membrane with a "stub," usually most readily visible on ventral surface (Prostemmatinae) .... Nabidae (part)
- Scutellum without lateral trichobothria; membrane without a "stub" .... Pachynomidae
8. Hemelytra usually well developed[1]; never ectoparasitic 9
- Hemelytra always staphylinoid (Fig. 50.1B), in the form of small pads, or absent (Figs. 61.1A, B, 62.1A)[1]; frequently ectoparasitic 18
9. Costal fracture present in macropterous forms, usually demarcating a distinct cuneus (Figs. 50.1A, 53.2A, 57.1, 58.1) 10
- Costal fracture absent in macropterous forms, no cuneus (Figs. 51.2C, 55.1A, 56.1) 16
10. Large species, 10–15 mm long; exocorium expanded into broad embolium (Fig. 49.1A); veins emanating from posterior margin of cells in membrane; penultimate labial segment much longer than other 2 combined .... Velocipedidae
- Much smaller species, usually under 4 mm long; exocorium not greatly expanded; membrane without many emanating veins; proportions of labial segments variable, penultimate segment rarely longer than combined lengths of other segments 11
11. Membrane with a single cell with very heavy veins and a well-developed "stub" on posterior angle (Fig. 50.1D); all tarsi 2-segmented .... Microphysidae (part)
- Membrane with weakly developed venation, stub always close to posterior margin of corium; tarsal formula variable 12
12. Tarsi 2-segmented; male genitalia symmetrical or nearly so, parameres elongate and nearly equally developed .... Plokiophilidae (part)
- Tarsi 3-segmented; male genitalia symmetrical or asymmetrical 13
13. Male genitalia symmetrical or nearly so, parameres elongate, nearly equally developed, genital capsule tubular (Fig. 58.2C); corial glands present (Fig. 10.10D); spider web commensals (*Lipokophila* Štys) .... Plokiophilidae (part)

- Male genitalia strongly asymmetrical, left paramere variable in shape (Figs. 59.1C, 61.1C), right paramere often greatly reduced, genital capsule frequently short and broad; corial glands absent; free-living 14
14. Abdominal terga 1 and 2 with laterotergites, terga 3–8 forming simple plates; no traumatic insemination .... Lasiochilidae
- All abdominal terga with laterotergites; insemination traumatic 15
15. Female with internal apophysis on anterior margin of abdominal sternum 7[2] ...... Lyctocoridae
- Female lacking apophysis on abdominal sternum 7[2] .... Anthocoridae
16. R+M in forewing elevated and keel-like (Fig. 51.2C); small species, about 2.0 mm long .... Joppeicidae
- R+M in forewing not elevated and keel-like; larger species, always over 4 mm long 17
17. Ultimate labial segment longest, labium rather straight (Fig. 55.1B) .... Medocostidae
- Ultimate labial segment not the longest, labium at least weakly curving (Fig. 56.2A) (Nabinae) .... Nabidae (part)
18. All tarsi 2-segmented; hemelytra staphylinoid (Fig. 50.1B) .... Microphysidae (part)
- All tarsi 3-segmented or middle and hind tarsi 4-segmented; hemelytra in form of small pads or absent 19
19. All tarsi 3-segmented; hemelytra in the form of small pads; no ctenidia (Fig. 61.1A, B); compound eyes small; temporary ectoparasites .... Cimicidae
- Middle and hind tarsi 4-segmented; hemelytra absent; ctenidia present (Fig. 62.1A, F, G); compound eyes absent; permanent ectoparasites of bats .... Polyctenidae

[1] Members of many families may be brachypterous or apterous, such as Nabidae and Anthocoridae, although the more common condition is to have well-developed forewings. Thus, specimens with short wings should be run through both halves of couplet 8.

[2] Members of the Lasiochilidae, Lyctocoridae, and Anthocoridae are very similar in general appearance and share many attributes. Correct family placement may require reference to faunistic treatments containing detailed keys for identification of genera and species.

Reduvioidea

47
Pachynomidae

General. This small group of seldom-collected tropical ground-dwelling predators has no common name. They range in length from 3.5 to 15 mm and often have a general facies of prostemmatine Nabidae (Fig. 47.1).

Diagnosis. Dorsum varying from highly polished (*Pachynomus* Klug) to dull, glabrous to highly setose; compound eyes large, head without constricted postocular region (Fig. 47.1); ocelli present or absent; antennae apparently 5-segmented, pedicel subdivided, distal portion usually with a single trichobothrium (Fig. 10.8H), intrapediceloid uniquely present (Fig. 47.1); labium thick, strongly curving, segment 1 obsolete, segment 2 shortest (Fig. 47.2A); bucculae directed anteriorly, obscuring base of labium; gula long; forefemur greatly enlarged; foretibia bearing a fossula spongiosa (Fig. 47.1); forewing with short costal fracture in *Pachynomus* and *Punctius* Stål, membrane with 2 elongate cells with a few to many radiating veins, stub (processus corial) absent (Figs. 47.1, 47.2B, C); metathoracic scent glands with strongly reduced grooves; Brindley's gland present; abdomen with dorsal and ventral laterotergites; abdominal spiracle 1 present, spiracles 2–8 located on mediosternite; pair of trichobothrium-like setae on either side of midline of abdominal segments 3–8 or 4–8 (Fig. 47.2D); parastigmal pits located on ventral laterotergites of abdominal segment 2 (except in *Punctius*); apophysis present on anterior margin of abdominal sternum 7 in female; dorsal abdominal scent-gland scars present on anterior margin of mediotergites 4–6 of adults; abdominal segment 8 in males reduced and telescoped within segment 7; male genitalia symmetrical, pygophore well developed and inserted apically on abdomen (Fig. 47.2E, F); ovipositor platelike; spermatheca absent, paired pseudospermathecae situated on medial ectodermal portion of female gonoducts (Fig. 47.2H).

Classification. *Pachynomus* was treated by its author Klug (1830) as a subgenus of *Reduvius;* Stål later treated *Pachynomus* and *Punctius* Stål as members of the Nabidae, a position followed by many authors (but see

Reuter, 1908), until Carayon (1950b) raised the group to family rank, pointing out that on the basis of reproductive anatomy, the taxon has its closest affinities with the Reduviidae. Two subfamilies comprising 4 genera and about 15 species are recognized, following the classification of Carayon and Villiers (1968).

Key to Subfamilies of Pachynomidae

1. Ocelli absent (Fig. 47.1); posterior margin of corium obliquely angled posterolaterally (Figs. 47.1, 47.2B); corium largely glabrous .... Pachynominae
- Ocelli present; posterior margin of corium truncate (Fig. 47.2C); corium distinctly pubescent .... Aphelonotinae

APHELONOTINAE. Ocelli present; endocorium truncate posteriorly and of same length as clavus; corium pubescent and without a trichobothriiform seta; abdominal sterna 4–8 each bearing a trichobothrium on either side of midline.

Aphelonotus Uhler, the only included genus, is found in tropical America and central Africa. Most species are less than 5–6 mm long and are light brown.

PACHYNOMINAE (FIG. 47.1). Ocelli absent; posterior margin of corium obliquely angled posterolaterally; corium glabrous except for a trichobothriiform seta at anterolateral angle of clavus; abdominal sterna 3–8 each bearing a trichobothrium located on either side of midline (Fig. 47.2D).

The group comprises the genera *Camarochilus* Harris, *Pachynomus* Klug, and *Punctius* Stål, the first being restricted to the New World tropics, the last two to the Old World, ranging from Africa to India. Members of all three genera are relatively large, in the case of *Pachynomus picipes* Klug attaining a length of 10.0–14.0 mm.

Specialized morphology. The antennal pedicel (segment 2) is subdivided, resulting in apparently 5-segmented antennae (Fig. 47.1). Along with the Reduviidae, the Pachynomidae are novel among the Heteroptera for the occurrence of antennal trichobothria (Fig. 10.8H) in most species. They also possess trichobothria on either side of the ventral abdominal midline (Fig. 47.2D), which occur elsewhere only in the Velocipedidae, but the phylogenetic analysis of Schuh and Štys (1991) suggests that they are not homologous in the two groups.

All Pachynomidae except *Punctius* possess parastigmal pits (fossettes parastigmatiques) on the ventral laterotergites of abdominal segment 2; these apparently are not homologous with similar-appearing structures located on the ventral laterotergites of one or more of abdominal segments 3–7 in the Nabidae. Paired pseudospermathecae of the type found in the Reduviidae—and Tingidae—are also present in the family (Fig. 47.2H).

Natural history. Nothing is known of the life habits of the group. Nymphs are unknown, and most specimens have been collected at lights.

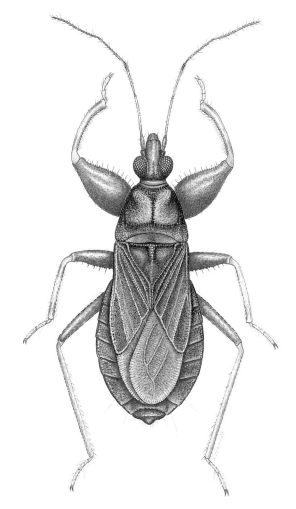

Fig. 47.1. Pachynomidae. *Punctius alutaceus* (Stål).

Distribution and faunistics. Carayon and Villiers (1968) provided a generic revision, synonymic, and distributional information for the known species. Harris (1931b) revised the New World *Aphelonotus* spp., but substantial numbers of species remain to be described.

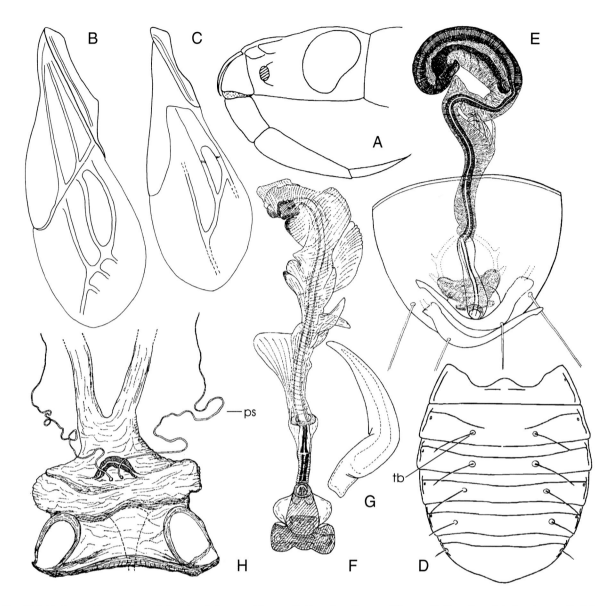

Fig. 47.2. Pachynomidae. **A.** Lateral view, head, *Camarochilus* sp. **B.** Forewing, *Camarochilus* sp. **C.** Forewing, *Aphelonotus* sp. (A–C from Schuh and Štys, 1991). **D.** Abdominal venter, *Camarochilus americanus* (Harris). **E.** Male genitalia, *Aphelonotus africanus* Carayon and Villiers. **F.** Aedeagus, *A. fuscus* Carayon and Villiers. **G.** Paramere, *A. fuscus*. **H.** Gynatrial complex, *A. fuscus* (D–H from Carayon and Villiers, 1968). Abbreviations: ps, pseudospermatheca; tb, trichobothrium.

48
Reduviidae

General. The predatory assassin bugs are one of the largest and morphologically most diverse families of true bugs. Reduviids range in size from relatively small and extremely delicate insects of only a few millimeters in length, such as members of the genus *Empicoris* Wolff (Emesinae), to very large and formidable animals such as *Arilus* Hahn (Harpactorinae), which may attain a length of nearly 40 mm.

Diagnosis. Compound eyes large, protuberant, head behind them usually elongate and constricted behind, usually bearing 2 ocelli (Fig. 48.2A–D); antennae sometimes with flagellar segments fusiform (e.g., Phymatinae; Fig. 48.1), prepedicellite present, not noticeably elongate, pedicel in nymphs and adults with at least 1 trichobothrium, adults sometimes with 20 or more (Fig. 48.3D, E); labium usually with 3 (rarely 4) apparent segments, short, stout, curving, often inflexible (Fig.

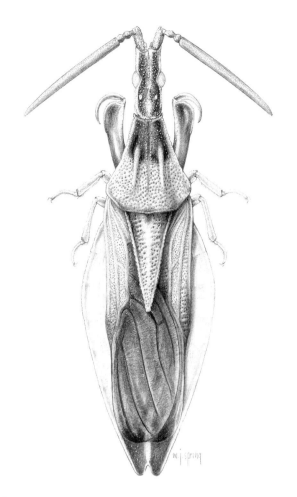

Fig. 48.1. Reduviidae: Phymatinae. *Oxythyreus cylindricornis* (Westwood) (from Slater, 1982).

48.3A, C), sometimes longer, straight, and flexible (Fig. 48.3B); forewings lacking costal fracture, membrane usually with 2 elongate cells and a few veins emanating posteriorly (Fig. 48.2C, D), sometimes cells subdivided or open (Fig. 48.2A); stub (processus corial) absent; fossula spongiosa usually present on one or more pairs of legs; tarsal formula variable, foretarsi sometimes absent; prosternum usually with a transversely striate stridulatory groove (Fig. 10.7F, G); metathoracic scent-gland channels and evaporatory areas strongly reduced or absent; paired Brindley's gland present in first abdominal segment, opening dorsolaterally at thoracicoabdominal junction; paired ventral glands present in some taxa, opening ventrally at thoracicoabdominal junction; dorsal and ventral laterotergites present; abdominal spiracle 1 present (except phymatine complex), spiracles 2–8 located on mediosternite; nymphal dorsal abdominal scent glands—if present—on anterior margin of abdominal terga 4, 5, and 6; abdominal segment 8 in male largely telescoped within segment 7; male genitalia (Fig. 48.3F–J) usually symmetrical; ovipositor usually platelike; female internal genitalia with paired pseudospermathecae and accessory gland (Fig. 48.3L–M); eggs with 3 or more micropyles.

Classification. The classification of the Reduviidae derives originally from the work of Amyot and Serville (1843) and Stål (1859a, 1866a, b). More recent contributions by Usinger (1943), Davis (1957, 1961, 1966, 1969), and Carayon et al. (1958) inquired into the basis of those early systems. N. C. E. Miller and A. Villiers established several subfamilies on the basis of unique characters.

Two up-to-date catalogs by Putchkov and Putchkov (1985, 1986–1989) and Maldonado (1990; see also Kerzhner, 1992) are available, although based on substantially different classifications. Maldonado recognized 25 subfamilies and treated the Elasmodeminae and Phymatinae as separate families; Putchkov and Putchkov recognized 21 subfamilies, including the Elasmodeminae and Phymatinae. We treat the Reduviidae in a manner consistent with the classification of Putchkov and Putchkov, except we place *Carayonia* Villiers in the Saicinae, following Maldonado. The approximately 930 genera and 6500 species are placed in 22 subfamilies, which—except for the phymatine complex—are arranged alphabetically in the following text.

Key to Subfamilies of Reduviidae

1. Antennal segments 3 and 4 incrassate, of greater diameter than segments 1 and 2; forefemora tremendously dilated, the tibiae chelate (Fig. 48.1) . . . . Phymatinae (part)
- Antennal segments 3 and 4 more slender than remaining segments; forefemora not tremendously dilated (Fig. 48.2) 2
2. Membrane with 3 simple veins attaining posterior margin (Fig. 48.2A); body extremely flattened . . . . Elasmodeminae
- Membrane with 1 or more closed cells (Fig. 48.2C, D); body generally not extremely flattened 3

Key to subfamilies of Reduviidae modified from Usinger, 1943.

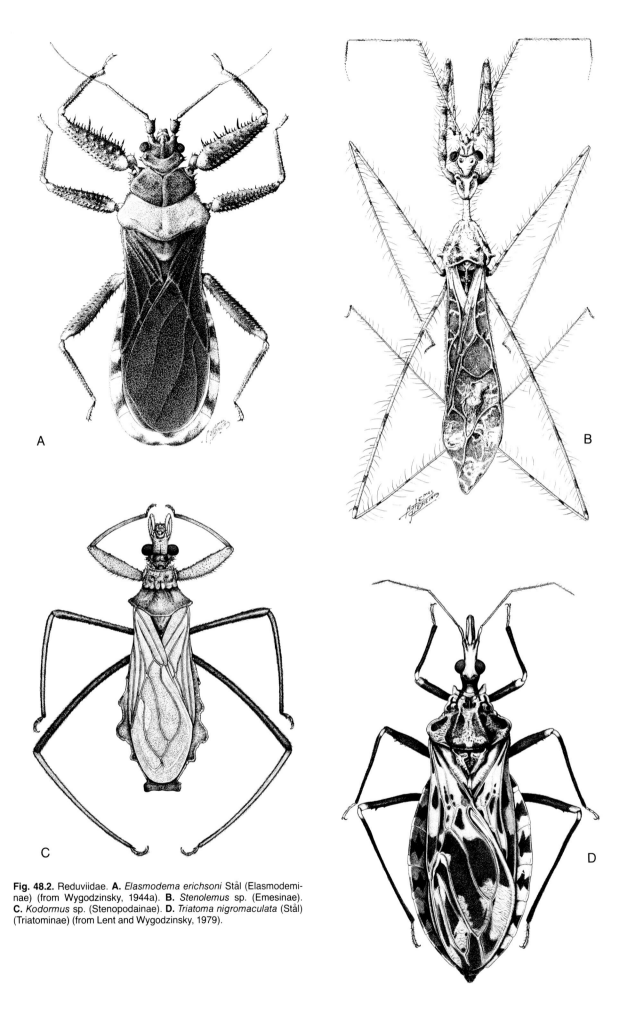

Fig. 48.2. Reduviidae. **A.** *Elasmodema erichsoni* Stål (Elasmodeminae) (from Wygodzinsky, 1944a). **B.** *Stenolemus* sp. (Emesinae). **C.** *Kodormus* sp. (Stenopodainae). **D.** *Triatoma nigromaculata* (Stål) (Triatominae) (from Lent and Wygodzinsky, 1979).

3. Abdominal sternum 3 usually produced anteriorly to form a trichome; antennal segments 3 and 4 inserted proximal to apex of segments 2 and 3, respectively .... Holoptilinae
– Abdominal sternum 3 without trichome; antennal segments 3 and 4 inserted on apex of segments 2 and 3, respectively 4
4. Ventral surface of head produced to form a more or less distinct rostral (buccal) groove 5
– Ventral surface of head not produced to form a rostral groove 6
5. Ocelli absent; antennal segments 2, 3, and 4 usually slender; body usually clothed with a dense vestiture of apically curved setae .... Tribelocephalinae
– Ocelli present; antennal segments 2, 3, and 4 not long and slender; body without a dense vestiture .... Phimophorinae
6. Forecoxae elongate, usually at least four times as long as wide, usually extending well beyond clypeus anteriorly (Fig. 48.2B); body elongate, slender; hemelytra almost entirely membranous except for veins; forelegs raptorial 7
– Anterior coxae usually less than twice as long as broad and not extending beyond clypeus; body, if elongate, not extremely slender; clavus and corium coriaceous, membrane well differentiated; forelegs not conspicuously raptorial 8
7. Ocelli and nymphal dorsal abdominal scent-gland openings absent .... Emesinae
– Ocelli and 3 nymphal dorsal abdominal scent-gland openings present .... Bactrodinae
8. Pronotum with a distinct constriction behind middle 9
– Pronotum without a distinct constriction at or behind middle 10
9. Ocelli present; fore- and middle tibiae not curved distally, provided with fossula spongiosa distally .... Peiratinae
– Ocelli absent, foretibia curved apically and produced beyond insertion of tarsus as a stout spine; fossula spongiosa absent .... Vesciinae
10. Ocelli absent 11
– Ocelli present (except in rare brachypterous forms) 12
11. Second labial segment swollen proximally; membrane with at least 2 closed cells ...... Saicinae
– Second labial segment not swollen proximally; membrane with a single large cell .... Chryxinae
12. Cubitus branching to form an additional four- to six-angled cubital cell between corium and membrane (Fig. 48.2C) 13
– Cubitus simple, not forming such a cubital cell 14
13. Cubital cell usually hexagonal (Fig. 48.2C); antennal segment 1 stout, porrect; nymphs with 2 dorsal abdominal scent glands .... Stenopodainae
– Cubital cell usually quadrangular, antennal segment 1 usually relatively slender; nymphs with 3 abdominal scent glands .... Harpactorinae
14. Antennae with 4 segments 16
– Antennae usually with more than 4 apparent segments 15
15. Antennal segment 2 subdivided into 4–36 pseudosegments; eyes located posteriorly, ocelli placed between them; scutellum without 2 posteriorly projecting prongs .... Hammacerinae
– Antennal segment 3 or segments 3 and 4 usually subdivided, forming a total of 7 or 8 apparent antennal segments; eyes not located posteriorly; ocelli placed posterior to eyes; scutellum with 2 posteriorly projecting prongs .... Ectrichodiinae
16. Head rarely transversely constricted behind eyes, ocelli usually located on oblique elevations at posterolateral angles of long, cylindrical head (Fig. 48.2D); dorsal abdominal scent-gland openings absent in nymphs .... Triatominae
– Head transversely constricted behind eyes, ocelli variously positioned; dorsal abdominal scent gland(s) present in nymphs 17
17. Eyes strongly pedunculate .... Cetherinae
– Eyes not stalked or pedunculate 18
18. Body surface tuberculate or spiny 19
– Body surface not tuberculate or spiny 21
19. Penultimate labial segment very long and straight, much longer than other 2 segments .... Physoderinae
– Penultimate labial segment not elongate, about as long as ultimate segment, antepenultimate segment longest 20

20. Tarsi 3-segmented; fossula spongiosa present on fore- and middle tibiae; head and pronotum bearing large spines (Fig. 48.3A); pronotum without longitudinal carinae .... Centrocneminae
– Tarsi 2-segmented; fossula spongiosa absent from all tibiae; body without large spines, only setigerous tubercles; pronotum with a longitudinal carina on either side of midline .... Phymatinae (Themonocorini)
21. Scutellum with a long posteriorly projecting spine .... Manangocorinae
– Scutellum without a long spine 22
22. Antennae inserted anteriorly, or more commonly laterally, but not on long anteriorly projecting tubercles .... Reduviinae
– Antennae inserted on prominent, anteriorly projecting tubercles at front of head 23
23. Foretarsi 2-segmented; middle and hind tarsi 3-segmented .... Salyavatinae
– All tarsi 3-segmented .... Sphaeridopinae

Phymatine Complex of Subfamilies

Although the phylogenetic relationships of the reduviid subfamilies as a whole are not well understood, four groups—the Phymatinae, Holoptilinae, Elasmodeminae, and Centrocneminae—are drawn together by the following unique characters: presence of ventral glands near posterior border of metathoracic sternum; basal plate struts of male phallus prolonged apically into pair of slender filaments partially or completely enclosed by endosoma; duck-head-shaped parameres; and reduced external female genitalia (Carayon et al., 1958).

CENTROCNEMINAE. Adults with a single trichobothrium on antennal pedicel; labium obviously 4-segmented (Fig. 48.3A); membrane with 2 elongate cells; body highly tuberculate (Fig. 48.3A), as in many other members of phymatine complex, and somewhat flattened.

Currently known from four genera and 33 species distributed from India to the Philippines, the Centrocneminae were revised by Miller (1956b). Most specimens have been found on the trunks of trees, suggesting that the group may be subcorticolous.

ELASMODEMINAE (FIG. 48.2A). Membrane with a few longitudinal veins reaching posterior margin, no closed cells; body strongly flattened, appendages spinose.

Elasmodema Stål, with two Neotropical species, has been treated as a distinct family by some authors (Wygodzinsky, 1944a; China and Miller, 1959; Maldonado, 1990), but its closest affinities are with the Holoptilinae (Carayon et al., 1958). Elasmodemines live under loose bark, apparently preying on other insects that occur there (Wygodzinsky, 1944a).

HOLOPTILINAE. Body and appendages with long to very long setae; membrane venation often with a single cell or 2 longitudinal veins; antennal segments 3 and 4 sometimes fused; abdominal sternum 3 often with a trichome and associated gland.

This group of 15 genera and 78 species is divided among three tribes (Holoptilini, Dasycnemini, and Aradellini) according to Wygodzinsky and Usinger (1963), who keyed the known genera. Most taxa occur in the southern Palearctic, Old World tropics, and Australia (Malipatil, 1985), with *Neolocoptiris* Wygodzinsky and Usinger (1963) occurring in Guyana.

PHYMATINAE (FIG. 48.1). Forefemur often greatly enlarged; foretibia and tarsus fused (except in Themonocorini) and often lying in scrobe on femur (Fig. 10.4G, H), or occasionally absent; forewing membrane often without distinct cells, or if with cells, with several supernumerary veins radiating from them posteriorly.

Commonly known as ambush bugs, this group of 26 genera and 281 species of morphologically distinctive bugs has been monographed by Handlirsch (1897), Kormilev (1962), and Maa and Lin (1956). Four tribal groupings are recognized: Carcinocorini, Macrocephalini, Phymatini, and Themonocorini (Carayon et al., 1958). The most recent source on the group, including keys to tribes, genera, and species of all genera except *Lophoscutus* Kormilev and *Phymata* Latreille, is the conspectus of Froeschner and Kormilev (1989), who, like many other authors, have accorded the group family rank.

Other Reduviid Subfamilies

BACTRODINAE. Elongate, slender; ocelli present; membrane with a single closed cell; forelegs with elongate coxae and spinose trochanters, femora, and tibiae; tarsal segment 3 slightly swollen on middle leg with a short stout spine near middle of ventral surface; claws on all legs unequally developed, one claw large and with large basal tooth, other claw much smaller, about as large as basal tooth of larger claw.

The Neotropical genus *Bactrodes* Stål, with four in-

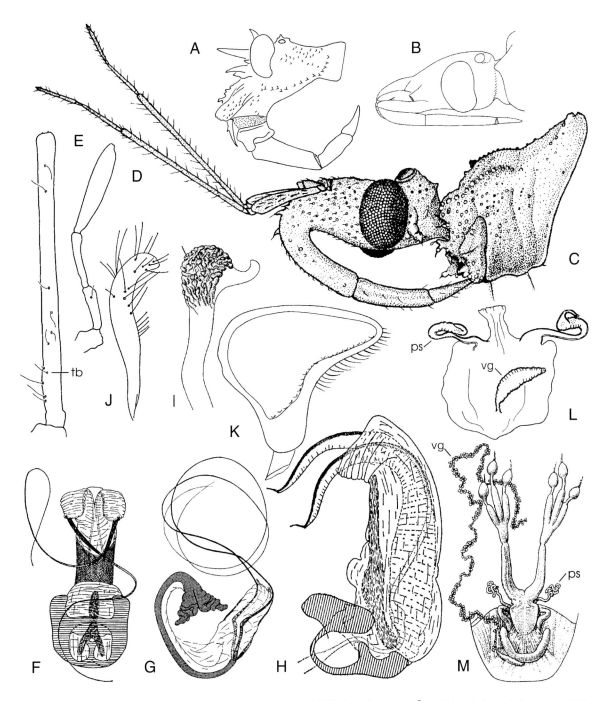

Fig. 48.3. Reduviidae. **A.** Lateral view, head, *Neocentrocnemis signoreti* (Stål) (from Schuh and Štys, 1991). **B.** Lateral view, head, *Triatoma rubrofasciata* (from Lent and Wygodzinsky, 1979). **C.** Lateral view, head, *Kodormus* sp. **D.** Antennal trichobothrium, *Phymata pennsylvanica* Handlirsch. **E.** Antennal trichobothria, *Rasahus sulcicollis* (Serville) (D, E from Wygodzinsky and Lodhi, 1989). **F.** Aedeagus, sagittal view, *Tridemula pilosa* Horvath (from Wygodzinsky, 1966). **G.** Aedeagus, lateral view, *Elasmodema erichsoni* (from Carayon et al., 1958). **H.** Aedeagus, lateral view, *Mesosepis papua* Wygodzinsky (from Wygodzinsky, 1966). **I.** Paramere, *E. erichsoni* (from Carayon et al., 1958). **J.** Paramere, *M. papua* (from Wygodzinsky, 1966). **K.** Right paramere, *Pirates hybridus* Scopoli (from Ghauri, 1964). **L.** Gynatrial complex, *Phymata erosa* (Linnaeus) (from Pendergrast, 1957). **M.** Female reproductive system, *Themnocoris kinkalanus* Carayon, Usinger, and Wygodzinsky (from Carayon et al., 1958). Abbreviations: ps, pseudospermatheca; tb, trichobothrium; vg, vermiform gland.

cluded species, was placed in a monotypic subfamily by Stål (1862). Davis's (1969) analysis indicated that, even though it differed in several characteristics from other reduviids he placed in the Harpactorinae, *Bactrodes* was probably a member of the clade containing the Harpactorini, Rhaphidosomini, and Tegeini. We treat *Bactrodes* as a subfamily, awaiting improved knowledge of character polarity in the Reduviidae. The group was diagnosed and keyed by McAtee and Malloch (1923). Nothing is known of the life habits.

CETHERINAE. Head short, transverse; transverse sulcus running between eyes on vertex; eyes strongly pedunculate.

The group, comprising five genera and 23 species, was first treated as a tribe within the Reduviinae by Jeannel (1919). Putchkov and Putchkov (1985) accorded subfamily status with three tribes—Cetherini, Euphenini, Pseudocetherini. Four genera, including *Cethera* Amyot and Serville, are known from tropical Africa and Madagascar, with only *Eupheno* Gistel occurring in the New World tropics. Villiers (1948) provided a comprehensive treatment of the tropical African fauna. These are active insects, some of which are known to feed on termites (Miller, 1956a).

CHRYXINAE. Head short, transverse, barely projecting in front of compound eyes; corium distinctly delimited from membrane, no cubital cell; membrane with a single cell.

The monotypic Neotropical *Chryxus* Champion was originally given subfamily status (Champion, 1897–1901); subsequently, two additional monotypic genera have been placed in the group, *Lentia* Wygodzinsky (1946) and *Wygodzinskyella* Usinger (1952). Chryxine morphology was well illustrated by Lent and Wygodzinsky (1944) and Wygodzinsky (1946). Wygodzinsky (1946) noted that the movements and coloration of these small insects are reminiscent of some Anthocoridae and particularly *Fulvius quadristillatus* Stål (Miridae), in whose company *Lentia corcovadensis* Wygodzinsky was taken in the field.

ECTRICHODIINAE. Often heavy-bodied with red and black coloration; scutellum with two short prongs projecting posteriorly; sexual dimorphism very common, females often with greatly reduced wings; antennae usually with 7 or 8 apparent segments, resulting from fragmentation of flagellum; 4–20 antennal trichobothria in nymphs and adults (Wygodzinsky and Lohdi, 1989); forewing membrane with 2 or 3 cells; fossula spongiosa present on fore- and middle tibiae.

This circumtropical group comprises approximately 75 genera and 300 species. Cook (1977) dealt with the Asian fauna, and Dougherty (1980) the Neotropical fauna.

EMESINAE (FIG. 48.2B). Foreacetabulum opening anteriorly; pterostigma carried beyond apex of discal cell (Wygodzinsky, 1966); body elongate, slender; appendages often threadlike; wing polymorphism common, often sexual, occasionally both sexes having wings greatly reduced and nonfunctional; ocelli absent except in Australian *Armstrongocoris* Wygodzinsky; fossula spongiosa absent from all legs; claws of foretarsus often unequally developed and parempodia frequently reduced; claws and parempodia on middle and hind legs fully developed; claws sometimes toothed, undersurface sometimes with membranous lamella.

The group was monographed by Wygodzinsky (1966), who recognized six tribes—Collartidini, Leistarchini, Emesini, Ploiariini, Deliastini, and Metapterini—comprising approximately 86 genera, some 20 of which are monotypic and known from a single locality. The Emesinae, the apparent sister group of the Saicinae (Wygodzinsky, 1966), are most diverse in the tropics and show substantial diversity in the islands of the Pacific.

HAMMACERINAE. Head strongly produced anteriorly, not constricted or elongated behind eyes; ocelli present; labium stout, curving, inflexible; membrane 2-celled; fossula spongiosa on fore- and middle tibiae; antennal segment 2 annulated, containing 4–36 pseudosegments.

First proposed by Stål (1859b) as a subfamily level taxon, this group has often been referred to as Microtominae. Hammacerines comprise the New World genera *Homalocoris* Perty and *Microtomus* Illiger with a total of 19 described species. According to Readio (1927) most records indicate they live under bark. *Microtomus* was revised by Stichel (1926).

HARPACTORINAE. Head usually with cylindrical postocular region; antennal segment 1 strikingly elongate; foretibia usually with a well-developed apical spur; metacoxal comb absent; metathoracic scent glands reduced or absent; hemelytron with quadrate or pentagonal corial cell at base of cubital cell, formed uniquely by anterior and posterior cross vein between Cu and Pcu; hind wing with submarginal Sc, postcubital sector relatively broad, Pcu slightly curved and well separated from first anal vein; ventral connexival suture usually poorly formed or absent; vermiform gland absent in females; subrectal gland often present, opening into membrane between styloids and anus; pseudospermathecae usually greatly reduced or absent (Davis, 1969).

First recognized in the classification of Amyot and Serville (1843), this is the largest reduviid subfamily. We follow Davis (1969) and Putchkov and Putchkov (1985) in treating as tribes several groups that have been accorded subfamily status by Maldonado, Miller, and Villiers.

Key to Tribes of Harpactorinae

1. Ocelli lateral, more widely separated from one another than from eyes 2
 - Ocelli less widely separated from one another than from eyes 4
2. Pronotum completely covering scutellum .... Diaspidiini
 - Scutellum at least partly exposed 3
3. Foretarsi 1-segmented; membrane with 3 cells .... Ectinoderini
 - Foretarsi 2-segmented; membrane with 2 cells .... Apiomerini
4. Labial segments 3 and 4 (penultimate and ultimate) slender, straight or slightly procurved, and of uniform thickness or very slightly tapered; labial segment 2 always shorter than 3; antennal segment 1 not strikingly elongate 5
 - Labium usually distinctly recurved and tapering from base to apex; labial segment 2 often subequal in length to segment 3; antennal segment 1 strikingly elongate .... Harpactorini
5. Body and legs long, very slender; postocular region of head long, cylindrical; forecoxal cavities usually closed .... Rhaphidosomini
 - Body more robust, with dense areas of glandular setae; postocular region of head usually short, subhemispherical; forecoxal cavities open .... Tegeini

APIOMERINI. The genera were keyed by Costa Lima et al. (1948). The largest and probably best known genus is *Apiomerus* Hahn (see Costa Lima et al., 1951), with 110 described species, often of conspicuous black and red coloration. Twenty-eight *Heniartes* spp. were treated in the well-illustrated work of Wygodzinsky (1947a). These diurnal predators live on plants, and at least some use sticky material on the anterior tibiae to hold prey.

DIASPIDIINI. A grouping of three Ethiopian genera—*Cleontes* Stål, *Rhodainiella* Schouteden, and *Diaspidius* Westwood—all species of which apparently live in leaf debris or under loose bark.

ECTINODERINI. Containing only the genera *Amulius* Stål and *Ectinoderus* Westwood (Maldonado, 1990), this group of 20 species is distributed from Sri Lanka to New Guinea. Nothing appears to be known of their habits.

HARPACTORINI, INCLUDING RHYNOCORINI VILLIERS (1982). This is a very large group—containing hundreds of genera—that, according to Davis (1969), possesses few distinctive characters. Included are such conspicuous taxa as the wheel bugs of the New World genus *Arilus* Hahn and some of the largest genera in the Reduviidae, such as *Rhynocoris* Hahn and *Sphedanolestes* Stål from the Old World and *Zelus* Fabricius from the New World (see Hart 1986, 1987). The feeding and reproductive behavior of *Rhynocoris* and *Pisilus* spp. have been studied by Edwards (1962) and Parker (1965, 1969), and Louis (1974) described the habits of additional West African species.

RHAPHIDOSOMINI. Comprises the genera *Rhaphidosoma* Amyot and Serville, *Lopodytes* Stål, *Leptodema* Carlini, *Hoffmanocoris* China, and *Harrisocoris* Miller—inhabi-

Key to tribes of Harpactorinae adapted from Davis, 1969.

tants of the deserts and savannas of the Old World—where in the hottest periods of the day they may be found basking at the tips of grasses, fully exposed to the sun.

TEGEINI. Comprising seven genera from the Old World tropics, these diurnal predators inhabit the trunks and foliage of trees and are known to feed on termites (Miller, 1952; China and Miller, 1959).

MANANGOCORINAE. Head short, globose, barely projecting anteriorly beyond relatively small eyes; scutellum with long spine; membrane with 2 cells, inner trapezoidal; tarsi 2-segmented; tibiae without fossula spongiosa.

Containing only the Malaysian species *Manangocoris horridus* Miller (1954a), this taxon may not merit subfamily status. Nothing is known of its habits.

PEIRATINAE. Antennal segment 1 shorter than segment 2; apparent first labial segment (2) shortest; anterior pronotal lobe longer than posterior lobe; forecoxae more strongly enlarged than middle and hind coxae and with lateral surface flattened and weakly concave; foretibia dilated distally, with a fossula spongiosa; middle tibia with or without fossula spongiosa; males of some *Peirates* spp. dimorphic, one morph possessing a large asymmetrical sinistral process on sternum 7 and small symmetrical parameres, the other morph lacking process and possessing large (Fig. 48.3K) asymmetrical parameres (Ghauri, 1964).

This worldwide group (often referred to as Piratinae) comprises approximately 31 genera, including *Ectomocoris* Mayr and *Peirates* Serville—which are diverse and widely distributed in the Old World—*Rasahus* Amyot and Serville from the New World, and the circumtropical *Sirthenea* Spinola.

Peiratines appear to be primarily ground-dwelling, feeding on a variety of arthropods. In general facies, habitat, and behavior they resemble Prostemmatinae (Nabidae). They are often attracted to lights. These are fast-running bugs whose bite can be extremely painful.

PHIMOPHORINAE. Habitus aradid-like; body and appendages covered with patches of waxy secretions; all antennal segments conspicuously thickened with antennal insertions protected by lateral shieldlike structures; bucculae present; labial segment 2 much longer than 3 and 4 combined; forewings largely membranous; legs short and stout, tarsi minute, 2-segmented; prosternum with 2 large shieldlike structures adjacent to stridulatory groove.

Comprises *Mendanocoris* Miller, with two species from the Solomon Islands and Malaya, and *Phimophorus* Bergroth, with a single species from the New World tropics (Usinger and Wygodzinsky, 1964). No information exists on their biology.

PHYSODERINAE. Body surface tuberculate, with spatulate setae; head elongate; eyes relatively small, far removed from the anterior pronotal margin; ocelli present; labial segment 3 (penultimate) very long and straight; apex of pronotum spatulate; membrane with 2 cells; fossula spongiosa absent from all legs; nymphs with 3 dorsal abdominal scent glands.

Comprises 11 Indo-west Pacific genera (including *Physoderes* Westwood with 37 species) and the monotypic *Cryptophysoderes fairchildi* Wygodzinsky and Maldonado (1972) from Panama. The subfamily status of the group was established by Miller (1954b). Known habitats include vegetable debris, the bases of banana and *Pandanus* leaves, hollow trees, and caves.

REDUVIINAE. Recognized primarily by the absence of characters occurring in other reduviids; ocelli usually present; discal cell usually absent; tarsi 3-segmented; fossula spongiosa on fore- and middle tibiae; nymphs with 3 dorsal abdominal scent glands.

This worldwide group comprises at least 140 genera, including *Acanthaspis* Amyot and Serville, *Reduvius* Fabricius, and *Zelurus* Burmeister. Many, if not most, members are thought to be general insect or arthropod predators. Most are nocturnal. Some, such as many *Reduvius* spp., live in animal burrows, whereas *Reduvius personatus* (Linnaeus) is synanthropic.

SAICINAE. Ocelli absent; second apparent labial segment (3) expanded and bulbous basally; opposing surfaces of head and labium armed with stiff setae or spines; fossula spongiosa absent.

This tropicopolitan group was first recognized by Stål (1859b) and currently contains two tribes—Saicini and Visayanocorini (Putchkov, 1987)—and 21 genera, by far the largest of which are *Polytoxus* Spinola from the Old World and *Saica* Amyot and Serville from the New World. It is composed primarily of elongate slender species and was treated as the sister group of the Emesinae by Wygodzinsky (1966). Maldonado (1981) keyed the American genera, and Villiers (1969b) treated the African fauna, including a key to genera. Little appears to be known of their habits, and they have been most commonly collected at lights.

SALYAVATINAE. Head relatively small, somewhat globular, constricted just behind eyes; neck short; eyes relatively small; antennae sometimes apparently 3-segmented (e.g., in *Salyavata*), inserted on prominent, anteriorly projecting tubercles; labium moderately long, weakly curving; pronotum, scutellum, and abdomen often with slender, strongly acuminate spines; fore- and middle tibiae with fossula spongiosa; foretarsi 2-segmented, middle and hind tarsi 3-segmented; membrane with 2 cells; nymphs with 3 dorsal abdominal glands.

This circumtropical group comprises 16 genera, the largest being *Lizarda* Stål from Africa and the Orient; only *Salyavata* Amyot and Serville is known from the New World tropics (Wygodzinsky, 1943). Most species are nocturnal, although at least some *Patalochirus* Palisot de Beauvois are diurnal. Some salyavatine nymphs have a vestiture of sticky setae and cover themselves with debris.

SPHAERIDOPINAE. Head projecting only slightly beyond anterior margin of eyes; eyes large, nearly contiguous ventrally; antennae inserted on anteriorly projecting tubercles; rostrum straight; all tarsi 3-segmented.

This Neotropical group comprises the monotypic genera, *Eurylochus* Torre-Bueno, *Sphaeridops* Amyot and Serville, *Veseris* Stål, and *Volesus* Champion. Pinto (1927) speculated, probably on the basis of the labial structure, that the group fed on vertebrate blood, although there seems to be no direct evidence for this theory.

STENOPODAINAE (FIG. 48.2C). Elongate, dull-colored; large cell, often pentagonal or hexagonal, formed on corium by cubital and postcubital veins, with 2 cells of membrane posteriorly adjacent (Fig. 48.2C); antenniferous tubercles sometimes strongly produced anteriorly; antennal segment 1 elongate, rather strongly developed, remaining segments more slender and in repose folded back against first (Figs. 48.2C, 48.3C); ocelli present; legs long, slender; hemelytral development variable, males macropterous, females often with more reduced forewings.

Approximately 113 genera have been described, most of them from the tropics. Before 1969, this group was referred to as the Stenopodinae. Many species appear to be closely associated with the soil—often being covered with soil or sand—and at least some insert their eggs in the soil. Most species are known from collections made at lights, with males being collected much more commonly than females. It seems certain that the group is largely nocturnal, even though there are almost no available direct observations of their behavior. Barber (1930) revised the New World fauna, and Giacchi (e.g., 1984) has subsequently added considerably to our knowledge of diversity of the group in South America.

TRIATOMINAE (FIG. 48.2D). Labium elongate, nearly straight, with a flexible membranous connection between segments 3 and 4 (Fig. 48.3B); antennal segment 2 in adults bearing from 3 to 7 trichobothria; head not constricted behind compound eyes; dorsal abdominal scent glands absent in nymphs; connexivum often broadly membranous, allowing for abdominal expansion during consumption of blood meal.

Lent and Wygodzinsky (1979) recognized five tribes—Rhodniini, Cavernicolini, Bolboderini, Alberproseniini, and Triatomini—containing 14 genera and 111 species. The group is most diverse in the New World, ranging from the southwestern United States to Argentina. *Triatoma rubrofasciata* (De Geer) is tropicopolitan, five species of *Linshcosteus* Distant are restricted to India, and seven species of *Triatoma* Laporte are restricted to an area ranging from southern India and Burma to New Guinea. The obligate blood-feeding triatomines are well known for their ability to transmit the debilitating Chagas' disease, whose causative agent, *Trypanosoma cruzi,* is widely distributed in the New World (see Chapter 8).

The literature on the Triatominae is immense, including numerous references to *Rhodnius prolixus* Stål, which is an important vector of Chagas' disease and a widely used laboratory animal. Summary sources include compilations by R. E. Ryckman and coauthors (see Literature Cited).

TRIBELOCEPHALINAE. Head, body, and veins of corium densely tomentose, body surface obscured; eyes flattened and not strongly projecting laterally, sometimes nearly contiguous dorsally; ocelli absent; antennal segment 1 thickened and much longer than head; corium narrow and elongate, membrane very broad.

First recognized by Stål (1866a), this broadly distributed paleotropical group comprises 13 genera placed in two tribes—Opistoplatyini and Tribelocephalini—the largest genus being *Tribelocephala* Stål. A few specimens have been taken in leaf litter, but most have been collected at lights, suggesting nocturnal habits.

VESCIINAE. Head sometimes with an elongate median spine between antennae; forefemora strongly swollen; pronotum constricted posterior to midpoint.

First recognized by Fracker and Bruner (1924), this small group comprises five genera from Africa and the Neotropics, the New World *Vescia* Stål being the largest. Nothing is known of the life habits.

Specialized morphology. Assassin bugs are a morphologically diverse group of dramatically varied facies. Not only do they possess a number of phenotypes unique to the group, but they "mimic" such diverse groups as the Aradidae, Berytidae, Coreidae, Enicocephalidae, Hydrometridae, Nabinae, Prostemmatinae, and Pyrrhocoridae.

Most are distinctive for the necklike structure of the head behind the eyes (e.g., Fig. 48.2B–D) and the strongly curving, short, inflexible labium (Fig. 48.3A, C), features that occur consistently elsewhere in the Heteroptera only in the Enicocephalidae and that once were used to unite these two distantly related groups. Reduviids are further distinguished by their almost universal possession of a prosternal stridulatory sulcus (Fig. 10.7F, G).

As predators, reduviids often use their forelegs during prey capture (see Putchkov, 1987). Most Phymatinae have the femora greatly enlarged (Fig. 48.1), as they are to a lesser degree in the Peiratinae. Chelae formed by the juxtaposition of the femur and tibia have developed more than once in the group, and the tarsus is sometimes lost. In groups such as the Emesinae, the forefemur, although serving a raptorial function, is elongate and slender (Fig. 48.2B). The use of glandular setae on the forelegs for prey capture is unique to the assassin bugs. The fossula spongiosa, arising from the apex of the tibiae, shows a great diversity of occurrence and development in the Reduviidae.

The Reduviidae share the possession of antennal trichobothria only with the Pachynomidae and Gerridae (Zrzavý, 1990a). The primitive reduvioid condition appears to be a single trichobothrium (Fig. 48.3D), the situation found in the Pachynomidae and a relatively large number of reduviid subgroups and in the first-instar nymphs of some reduviids whose adults have multiple trichobothria; the maximum number may exceed 20 (Fig. 48.3E) (Wygodzinsky and Lodhi, 1989; Zrzavý, 1990a).

The true spermatheca is transformed into a vermiform gland (Fig. 48.3L, M) and is functionally replaced by paired pseudospermathecae (Fig. 48.3L, M). Many, but not all, reduviids possess Brindley's glands (Brindley, 1930) (an analog of which apparently occurs in the Tingidae) and ventral glands (Carayon et al., 1958).

Egg structure was examined by Haridass (1985, 1986a, b, 1988), including use of scanning electron microscopy. Miller (1956a) gave a survey of the many egg shapes found in the family.

Natural history. All reduviids are predators. Most appear to orient to their prey visually. The Triatominae are distinctive for the blood-sucking habits. Valuable sources on reduviid biology include the works of Miller (1953; based on the fauna of Zimbabwe), Miller (1956a, 1971), Readio (1927) on North America, Putchkov (1987) on the fauna of the Ukraine, and Louis (1974) on reduviids from cacao farms in Ghana. The works of Wygodzinsky (1966) and Lent and Wygodzinsky (1979) contain much valuable information of the Emesinae and Triatominae, respectively.

Most Phymatinae live on vegetation where, well camouflaged, they lie in wait of their prey, toward which they appear to orient visually. Feeding and oviposition habits and nymphal morphology in *Macrocephalus notatus* Westwood have been described by Wygodzinsky (1944b). The life history of *Phymata pennsylvanica americana* Melin has been studied in detail by Balduf (e.g., 1941).

Ectrichodiines are apparently all obligate predators of millipedes, rejecting insects and other arthropods, even when starved (Louis, 1974). Most live in leaf litter and are nocturnal (Miller, 1953; Louis, 1974). *Ectrichodia gigas* Herrich-Schaeffer in the Ivory Coast preys exclusively on the iulid millipede *Peridontopyge spinosissima*. This diurnal predator apparently does not produce a venomous salivary secretion, as the millipedes were not subdued for some time after attack commenced. Interaction between individual reduviids during rest and feeding periods appears to be mediated by chemical stimuli, the bugs producing a distinctive odor, unlike that found in other heteropterans (Cachan, 1952a).

Free-living Emesinae may be found in litter, under stones, on living vegetation, or in dead vegetation such as hanging ferns or palm fronds. Some species are domestic or peridomestic, a few being widely distributed, suggesting transport by humans. Some species are cavernicolous but do not possess characteristics common to many cave dwellers, such as eyelessness or depigmentation. A substantial number of emesines are obligate inhabitants of spider or psocopteran webs, living there without becoming entangled or incurring the wrath of the web makers. Wygodzinsky (1966) thought that most spider-web-dwelling species are not obligate spider predators, but rather opportunists, whereas, Snoddy et al. (1976) reported *Stenolemus lanipes* Wygodzinsky feeding exclusively on the spider *Achaearanea tepidariorum,* and Cobben (1978) concluded that *Stenolemus* spp. are obligate spider predators.

Triatomines commonly inhabit the nests or dwellings of their hosts, such as wood rats or other rodents in the American Southwest and Mexico. They may also be found under fallen logs, in caves, hollow trees, among palm fronds, and in bromeliads. Some species have become domestic or semidomestic, and may be found in chicken coops or the habitations of other domestic animals. In association with humans they often inhabit the thatch of roofs of rustic dwellings or, in more modern structures, live in nearly any dark place where they can secrete themselves. They are strictly nocturnal, taking a blood meal for 20–30 minutes and inflicting no pain on the host. Trypanosomiasis is caused by the spreading of fecal material on the feeding lesions rather than by direct transmission.

The nymphs of some Reduviinae, and some other reduviids (Cetherinae, Salyavatinae, Triatominae), have the habit of covering the body with debris—including soil particles, pieces of termite nest carton, and corpses of ants and other insects—which is affixed by sticking objects to the viscid secretions of specialized setae located on the dorsum (e.g., Odhiambo, 1958a). Camouflaged nymphs of *Acanthaspis petax* Stål stalk their prey; they commonly feed on ants, as well as other insects (Odhiambo, 1958b). *Salyavata variegata* Amyot and Serville was recorded eating the termite *Nasutitermes exitiosus*. Nymphs covered themselves with crumbs of nest carton, sat by perforations in the nest, and with their forelegs captured individual workers as they arrived to repair the nest. For bait, the bugs used the carcasses of termites on which they had fed, dangling them in the opening to attract workers, which were then dragged from the nest and eaten (McMahan, 1982).

Nymphal and adult reduviids stridulate by rubbing the apex of the labium against the transversely striate prosternal sulcus. All other Heteroptera capable of stridulation produce sound only in the adult stage. Reduviid stridulation has usually been regarded as defensive in function; however, reduviids emit other, nondefensive, noises (Gogala and Cokl, 1983; Gogala, 1984), some of which are produced by the labial-prosternal mechanism and some by unknown means. They also probably emit low-frequency vibrations from an abdominal tymbal formed by fused abdominal terga 1 and 2. *Phymata crassipes* (Fabricius) is known for its alternating calls (Gogala and Cokl, 1983; Gogala et al. 1984) imitating in their duration and frequency various acoustic stimuli and ranging in character from other insect calls to the human voice.

Reduviids usually have platelike ovipositors, and many species—particularly in vegetation-dwelling groups such as the Harpactorini—glue their eggs to the plant, often in a group, sometimes covering the eggs with a gelatinous material. A few species are known to guard their eggs. Ground-dwelling species may insert their eggs into the soil, or the eggs may be laid loose. Some reduviines have more well-developed ovipositors, suitable for the insertion of eggs into cracks or crevices.

The glandular secretions from the abdominal trichome of the Holoptilinae are attractive—and apparently toxic—to ants, the prey of these bugs, as reported by Jacobson (1911) for *Ptilocerus ochraceus* Montandon.

Distribution and faunistics. The work of Readio (1927) on the biology of the North American fauna serves not only as a summary of biological information, but also as a review of the literature on the group for the region. The fauna of the western Palearctic is probably best documented in the works of Dispons and Stichel (1959) for Europe and Putchkov (1987) for the Ukraine. Several faunistic works provide valuable aids in understanding the African fauna, notable among them Miller (1953), which contains many field observations as well as nine colored plates of habitus views. Other works include those of Villiers (1948) on West Africa, Schouteden (1931, 1932) and Villiers (1964) on the fauna of the Congo basin, and Villiers (1968, 1979) on the fauna of Madagascar. Many new tropical Asian taxa were described and illustrated by Miller (1940, 1941). The fauna of China was detailed by Hsiao et al. (1981), a volume that includes many photographic illustrations. The catalogs of Putchkov and Putchkov (1985, 1986–1989) and Maldonado (1990) should also be consulted.

Velocipedoidea

49
Velocipedidae

General. The velocipedids appear to have no common name, although they might be referred to as the fast-footed bugs. These tropical Asian bugs range in length from 10 to 15 mm and are distinctive for their elongate head and broadly expanded exocorium (Fig. 49.1A).

Diagnosis. Pronotum, scutellum, corium, and clavus rather heavily punctured; vestiture inconspicuous; head elongate with a neck behind eyes about length of diameter of an eye; eyes hemispherical, protuberant; ocelli present; antennal prepedicellite present; bucculae conspicuously developed, more or less obscuring labial segments 1 and 2; labium straight except for rather sharp angle between segments 2 and 3, reaching to about apex of mesocoxae, segment 1 very short, segment 2 short, segment 3 extremely long, segment 4 short (Fig. 49.1B, C); pronotum collarlike anteriorly, lateral margins of anterior lobe usually produced; exocorium explanate, medial fracture very long, costal fracture present (Fig. 49.1A); membrane with 3 short, closed, basal cells with many simple or bifurcating veins emanating from them, and a stub (processus corial) located near apex of anterior cell (Fig. 49.1D); legs cursorial, femora and tibiae relatively long and nearly cylindrical; no fossula spongiosa; abdomen with distinct dorsal laterotergites in both sexes, ventral laterotergites developed only in males; abdominal spiracle 1 absent, spiracles 2–8 located on discrete ventral laterotergite in males and on sternum in females; abdomen with a pair of trichobothrium-like setae on either side of ventral midline of segments 3–7; pygophore in male well developed and located at apex of abdomen (Fig. 49.1E); parameres symmetrical (Fig. 49.1G); vesica symmetrical, endosoma extremely simple (Fig. 49.1F); ovipositor laciniate, first valvula associated by ramus with first valvifer; spermatheca in form of a vermiform gland (omitted from Fig. 49.1H).

Classification. Stål (1873) described the genus *Scotomedes* and placed it in the Nabidae. Bergroth (1891) subsequently described *Velocipeda*, placing it as a distinct subfamily, Velocipedinae, within the Saldidae. Later it was recognized that the two genera (as well as *Godefridus* Distant of the Reduviidae, Apiomerinae and *Bloeteomedes* van Doesburg) were synonyms. At least six species have been described.

Specialized morphology. The structure of the labium in the Velocipedidae (Fig. 49.1B, C) is similar to that found in the Mesoveliidae, Ochteridae, Aphelocheiridae, Saldidae, and Lasiochilidae (Rieger, 1976; Kerzhner, 1981). The broad expansion of the exocorium (Fig. 49.1A) appears to be virtually unique within the Heteroptera.

Natural history. No direct observations have been published regarding the life habits of this group, although nymphs have been collected. Most known specimens have probably been collected at lights.

Distribution and faunistics. Velocipedids are distributed from Assam in Northeast India to New Guinea. No comprehensive papers exist on the group, although in addition to those cited above, one might see those by Blöte (1945), van Doesburg (1970, 1980), and Kerzhner (1981).

Microphysoidea

50
Microphysidae

General. This group of tiny insects ranges in size from 1.5 to 3.0 mm, with many species resembling small anthocorids (Fig. 50.1A, B). They have no common name and are known in the field to only a few heteropterists. The small size, restricted distribution, and obscure habits of the Microphysidae have left the group little studied.

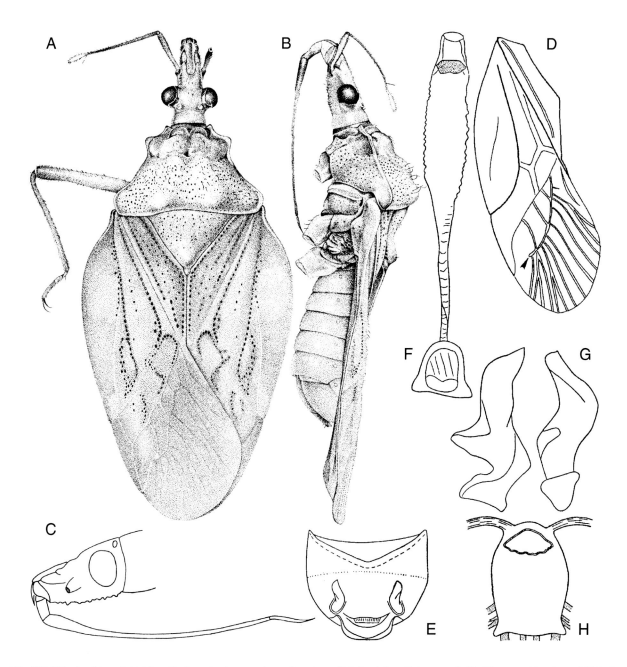

Fig. 49.1. Velocipedidae. **A.** Habitus, *Scotomedes borneensis* van Doesburg. **B.** Lateral view, *S. borneensis* (A, B from van Doesburg, 1970). **C.** Lateral view, head, *S. alienus* (Distant). **D.** Forewing, *S. alienus* (C, D from Schuh and Štys, 1991). **E.** Pygophore, *S. alienus*. **F.** Aedeagus, *S. alienus*. **G.** Parameres, *S. alienus*. **H.** Gynatrial complex, *S. alienus* (E–H from Kerzhner, 1981).

Diagnosis. Head weakly prognathous (Fig. 50.1C); macropterous forms with large ocelli; relative lengths of labial segments usually (shortest first) 1-4-3-2, with segment 1 well developed in Palearctic species, reduced in Nearctic species; metapleuron without ostiolar groove or evaporatory area; single, heavy-veined cell in membrane of forewing, with a distinct stub (processus corial) in posterolateral angle (Fig. 50.1D); costal fracture well developed, forming a distinct cuneus; distal sector of R+M in the hind wing branching to a fork (shared with the Nabinae); females of Palearctic taxa flightless, hemelytra often staphylinoid; tarsi 2-segmented; male genitalia (Fig. 50.1E, F) symmetrical; ovipositor laciniate.

Classification. The Microphysidae were first treated as a separate family by Fieber (1861) but were placed in an omnibus Anthocoridae by Reuter (1884). *Microphysa* Westwood, upon which the family name is based, is a junior synonym of *Loricula* Curtis. Although micro-

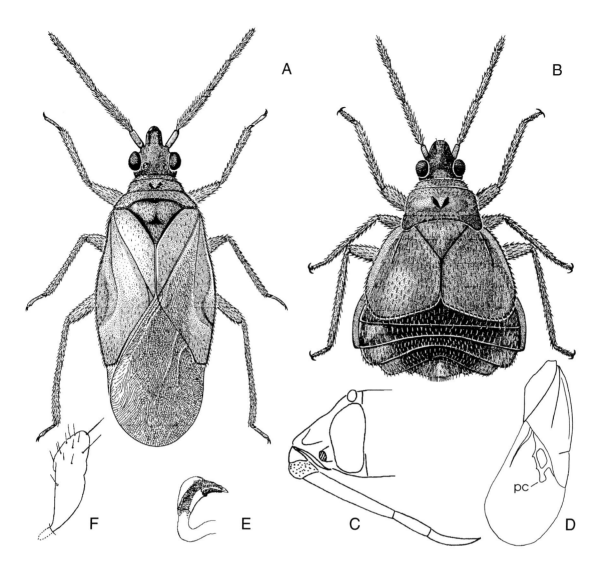

Fig. 50.1. Microphysidae. **A.** Habitus, male, *Loricula pselaphiformis* Curtis. **B.** Habitus, female, *L. pselaphiformis* (A, B from Kelton, 1980a). **C.** Lateral view, head, *Chinaola quercicola* Blatchley. **D.** Forewing, *Myrmedobia exilis* (Fallén) (C, D from Schuh and Štys, 1991). **E.** Aedeagus, *L. pselaphiformis* (from Péricart, 1972). **F.** Paramere, *L. pilosella* (from Miyamoto, 1965b). Abbreviation: pc, processus corial.

physids superficially resemble members of the Cimicoidea, a relatively large number of characters suggest they are actually members of the Miriformes (Schuh and Štys, 1991).

Five genera are currently recognized: *Ciorulla* Péricart (1974) (one species; Uzbekistan), *Loricula* Curtis (12 species; Palearctic), *Myrmedobia* Baerensprung (eight species; Palearctic), *Chinaola* Blatchley (one species; eastern North America), and *Mallochiola* Bergroth (one species; eastern North America and eastern Mexico).

Specialized morphology. The Palearctic genera of Microphysidae are notable for their distinctive sexual dimorphism, manifested particularly in the extreme reduction of the forewings in the females (Fig. 50.1B) of many species with complete loss of the membrane and exposure of 3–7 abdominal terga, giving a staphylinoid or myrmecomorphic habitus. The female abdomen is often greatly broadened and rounded laterally, caudad of the posterior margin of the hemelytra. Nearctic taxa are macropterous in both sexes. The structure of the labium of the taxa from the two areas is also different, with segment 1 greatly reduced in the New World taxa (Fig. 50.1C) and conspicuously present in Palearctic taxa.

Ciorulla is unique among the microphysids in having the labial formula 2-3-1-4.

Natural history. Most female microphysids are encountered on the bark of trees, often those covered with lichens or mosses (Wheeler, 1992). The males and sometimes females can often be collected on herbaceous vegetation under trees harboring the females. Females of the

Palearctic species may also be found in decaying wood, and a few are found in wet litter, particularly *Sphagnum* containing fallen leaves and needles. Males and female are often found in separate habitats except during the period of reproductive activity. All species for which life histories are known are predatory and overwinter as eggs, with nymphal development taking place in the spring (Carayon, 1949; Kelton, 1980a, 1981).

Distribution and faunistics. Published information shows the Microphysidae as being restricted to the Holarctic. Schuh and Štys (1991), however, noted examining specimens of an undescribed species from Namaqualand, South Africa, considerably extending the previously known range.

The work of Péricart (1972) is the most comprehensive treatment available for the family; other useful papers include another study by Péricart (1969). Two Palearctic species, *Loricula pselaphiformis* Curtis and *Myrmedobia exilis* (Fallén), have been introduced into the eastern Nearctic (Kelton, 1980a, 1981).

out stub (processus corial) and with very weakly developed veins (Fig. 51.2C); M-Cu cross vein absent in hind wing (Fig. 51.2D); tarsi 2-segmented, fossula spongiosa absent; metathoracic scent gland with paired reservoirs and widely separated ostioles, grooves present on metapleuron; abdomen broadly membranous at junction with thorax, capable of 90° rotation, terga 1–3 membranous, sternum 2 a narrow strip (Fig. 51.2E), abdominal spiracle 1 absent, spiracles 2–8 located on abdominal sterna; nymphal dorsal abdominal scent glands located on anterior margins of terga 4, 5, and 6 (Cobben, 1978); functional spermatheca absent (Fig. 51.2H); genital capsule distinct, parameres symmetrical; phallus in the form of sclerotized tube (Fig. 51.2F); ovipositor greatly reduced (Fig. 51.2G).

Classification. *Joppeicus* was originally placed in the Aradidae by its author Puton (1881) and later was transferred to the Lygaeidae by Bergroth (1898). Reuter (1910) established a separate family within the Aradoideae and later (1912a) placed it in the Reduvioideae. Schuh and Štys (1991) treated it as a member of the Miriformes.

Joppeicoidea

51
Joppeicidae

General. This family includes only *Joppeicus paradoxus* Puton, a tiny species about 3.0 mm long and of anthocorid-like habitus (Fig. 51.1). The group has no common name and has long perplexed heteropterists with its peculiar combination of morphological attributes.

Diagnosis. Dorsal surface weakly tuberculate; compound eyes relatively small, placed near posterior margin of head (Figs. 51.1, 51.2A, B); ocelli present; antennal segments 1 and 2 relatively short (Fig. 51.2A, B), segment 4 with a stout apical seta; antenniferous tubercles set slightly anterior to eye at about level of ventral margin; labium in repose bent at right angle between segments 2 and 3, segment 1 very small, segment 3 about 3 times length of segment 2 and 1.5 times length of segment 4 (Fig. 51.2A, B); bucculae obscuring base of labium; pronotum carinate laterally, with a median keel on anterior two-thirds (Figs. 51.1, 51.2A); forewing trough-like at Sc, R+M raised and keellike; costal and medial fractures absent (Figs. 51.1, 51.2C); membrane with-

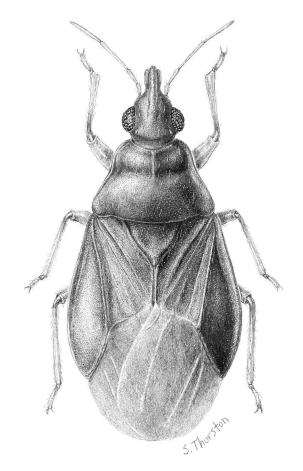

Fig. 51.1. Joppeicidae. *Joppeicus paradoxus* Puton (from Slater, 1982).

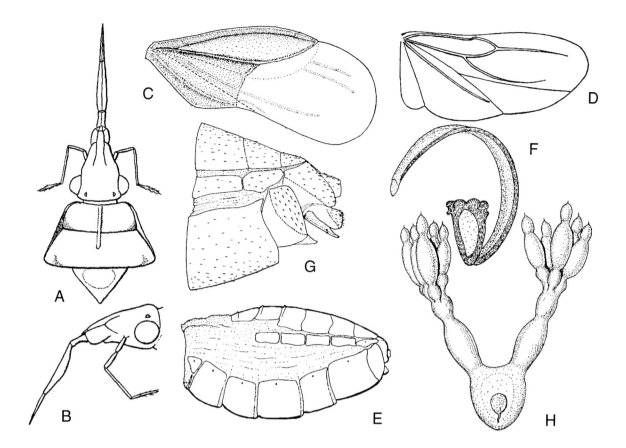

Fig. 51.2. Joppeicidae. *Joppeicus paradoxus*. **A.** Dorsal view, head, pronotum, scutellum. **B.** Lateral view, head (A, B from China, 1955b). **C.** Forewing (from Davis and Usinger, 1970). **D.** Hind wing (from China, 1955b). **E.** Lateral view, abdomen. **F.** Phallus. **G.** Female genital segments, lateral view. **H.** Ovaries and nonfunctional spermatheca (E–H from Davis and Usinger, 1970).

Specialized morphology. Two intensive studies of the morphology of *Joppeicus* (China, 1955b; Davis and Usinger, 1970) revealed that many of the attributes of *Joppeicus*, such as the structure of the wings, phallus, and the large intersegmental membrane between abdominal segments 3 and 4, are unique to the taxon.

Natural history. *Joppeicus* is a general predator, in culture thriving on a wide range of prey items (Davis and Usinger, 1970). The rostrum is held out straight in front of the body when prey is being attacked. In nature, *Joppeicus* prefers marginal, often dusty situations, little frequented by other insects. It has been collected on the surface of the ground, under stones, in shallow caves, under bark, and also in colonies of bed bugs associated with bats (Štys, 1971).

When mating, the male first mounts the dorsum of the female, rotates to the right or left—as seen in profile—then rotates the abdomen 90° at the large intersegmental membrane between segments 3 and 4 and moves the dorsal surface under the abdominal venter of the female, at which time copulation actually takes place (Davis and Usinger, 1970).

Distribution and faunistics. *Joppeicus* is restricted to southern Israel and the Nile and Blue Nile drainages in Egypt and the Sudan. It has been collected most commonly in the Nile delta region (Štys, 1971).

Miroidea

52
Thaumastocoridae

General. This group of morphologically unusual and seldom collected bugs has no common name, although the New World members might be collectively referred to as palm bugs. All species are small, ranging from 2.0

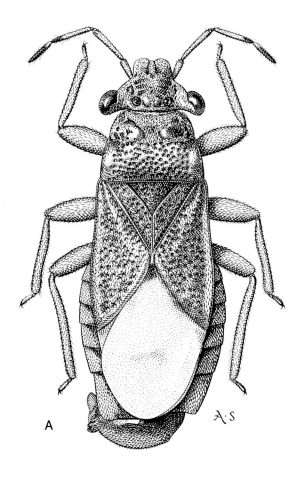

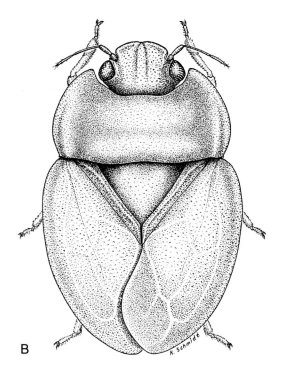

to 4.6 mm, and are distinguished by the strongly anteriorly produced mandibular plates (Fig. 52.1A, B) and the grossly asymmetrical male genitalia.

Diagnosis. Body moderately to strongly flattened; dorsum often heavily punctured, pubescence short and inconspicuous; head with mandibular plates and clypeus projecting in front of eyes, mandibular plates often exceeding clypeus (Figs. 52.1A, 52.3A); ocelli present; antennae relatively short, segments 3 and 4 at most weakly fusiform; labium inserted on ventral surface of head, reaching from anterior margin of prosternum to near base of abdomen (Fig. 52.3A, B); prosternal stridulatory sulcus incorrectly indicated by Drake and Slater (1957; see Fig. 52.3B); legs homomorphous, relatively short, femora mutic; tarsi 2-segmented; macropterous or brachypterous; costal fracture absent; stub (processus corial) absent; metathoracic scent-gland channels present; dorsal and ventral laterotergites absent, abdominal spiracle 1 present; abdominal spiracles 2–8 situated on sternum; dorsal abdominal scent glands present in nymphs on anterior margin of terga 4 and 5; genital capsule grossly asymmetrical (Fig. 52.3F) (usually dextral or sinistral in a given species, rarely direction variable within a species); one or both parameres lost (Fig. 52.2G); phallus as in Fig. 52.2F; ovipositor completely absent; spermatheca absent.

Classification. The first described thaumastocorids were treated as a subfamily of the Lygaeidae (Kirkaldy, 1908). The group was accorded family status by Reuter (1912a). The classification presented here follows Drake and Slater (1957) in placing the Thaumastocoridae in the Cimicomorpha, whereas Cobben (1978) treated the group as of problematic position. Two subfamilies are recognized, the Xylastodorinae and Thaumastocorinae. Viana and Carpintero (1981) treated these as distinct families, an approach rejected by Slater and Brailovsky (1983), because this action failed to establish new criteria for group recognition and obscured the novel morphological features shared by the two groups. Drake and Slater (1957) related the Thaumastocoridae to the Cimicoidea, whereas Kerzhner (1981) treated them as the sister group of the Tingidae, and Schuh and Štys (1991) as the sister group of the Miridae + Tingidae.

Fig. 52.1. Thaumastocoridae. **A.** *Onymocoris izzardi* Drake and Slater (from Drake and Slater, 1957). **B.** *Discocoris drakei* Slater and Ashlock.

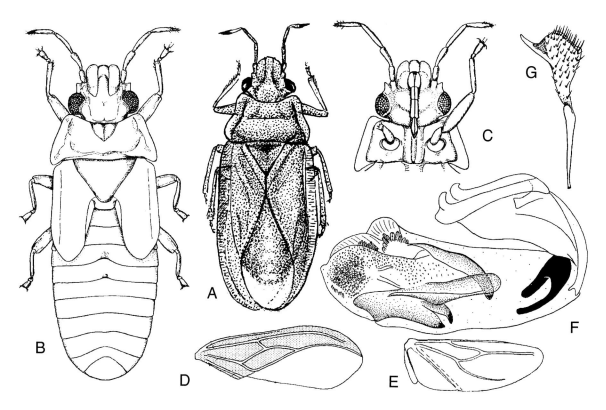

Fig. 52.2. Thaumastocoridae. **A.** Habitus, *Xylastodoris luteolus* Barber (from Slater and Baranowski, 1978). **B.** Dorsal view nymph, *X. luteolus*. **C.** Ventral view head and thorax, *X. luteolus*. **D.** Forewing, *X. luteolus*. **E.** Hind wing, *X. luteolus* (B–E from Schaefer, 1969). **F.** Aedeagus, *Baclozygum depressum* Bergroth. **G.** Paramere, *Thaumastocoris hackeri* Drake and Slater (F, G from Drake and Slater, 1957).

Key to Subfamilies of Thaumastocoridae

1. Claws bearing large, basally attached pulvilli (Fig. 52.3E); tibiae apically without a lobate appendage (tibial appendix); males without parameres; costa extending to near apex of forewing (Figs. 52.1B, 52.2D) .... Xylastodorinae
– Claws without pulvilli; tibiae bearing a lobate appendage apically (tibial appendix) (Fig. 52.3C, D); males with one paramere (Fig. 52.2G); membrane extending greatly beyond apex of corium ......
   .... Thaumastocorinae

THAUMASTOCORINAE. Eyes strongly protuberant (Fig. 52.1A); tibial appendix apically on all tibiae (Fig. 52.3C, D); nymphal dorsal abdominal scent glands with a single opening (Slater, 1973); one paramere lost.

Currently four genera are included: *Baclozygum* Bergroth (two species; Tasmania and mainland Australia), *Onymocoris* Drake and Slater (three species; Australia), *Thaumastocoris* Kirkaldy (three species; Australia), and *Wechina* Drake and Slater (one species; southern India).

XYLASTODORINAE. Antenniferous tubercle conspicuously projecting (Figs. 52.2C, 52.3B); large pulvilli arising from base of claw (Fig. 52.3E); nymphal dorsal abdominal scent glands with paired openings (Fig. 52.2B); both parameres lost.

Two genera are currently recognized: *Discocoris* Kormilev (five species, South America) and *Xylastodoris* Barber (one species, Cuba and Florida). They were originally described in separate subfamilies but treated as belonging to a single subfamily by Drake and Slater (1957).

Specialized morphology. The thaumastocorids are tiny, strongly flattened bugs that hold tenaciously to the substrate (Slater, 1973; Schuh, 1975b), apparently with the aid of the accessory pretarsal structures such as the tibial appendix (Fig. 52.3C, D) in the Thaumastocorinae and the large pulvilli (Fig. 52.3E) in the Xylastodorinae. Most species have greatly enlarged and extended mandibular plates (Fig. 52.1A, B), which surpass the

The key to the subfamilies of Thaumastocoridae is modified from Drake and Slater, 1957.

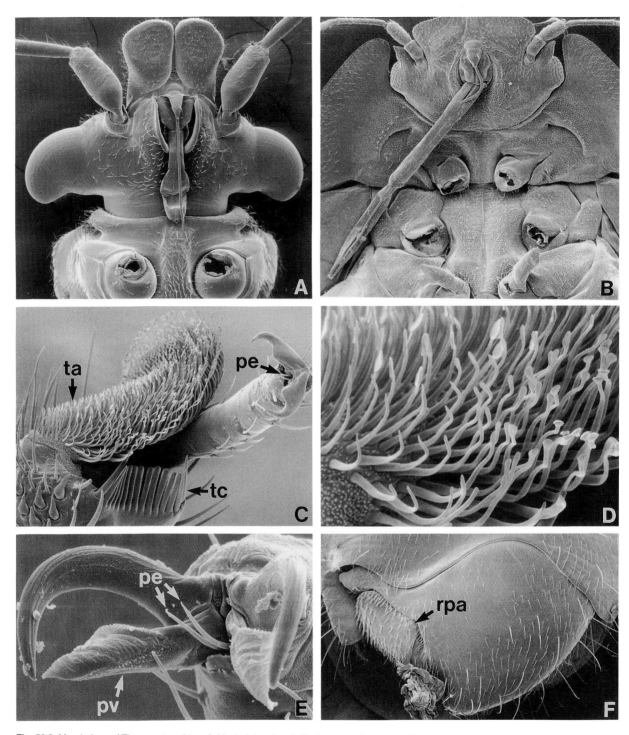

Fig. 52.3. Morphology of Thaumastocoridae. **A.** Ventral view, head, *Baclozygum depressum* (Thaumastocorinae). **B.** Ventral view, head, *Discocoris drakei* Slater and Ashlock (Xylastodorinae). **C.** Apex of tibiae with tibial appendix and comb, *B. depressum* (Thaumastocorinae). **D.** Detail of tibial appendix, *B. depressum*. **E.** Pretarsus, *D. drakei* (Xylastodorinae), showing large accessory parempodia. **F.** Ventral view, male terminal abdominal segments, *B. depressum* (Thaumastocorinae). Abbreviations: pe, parempodium; pv, pulvillus; rpa, right paramere; ta, tibial appendix; tc, tibial comb.

apex of the clypeus, a situation generally found elsewhere in Aradoidea and Pentatomoidea. The asymmetrical and otherwise highly modified male genitalia (Figs. 52.2F, 52.3F), with the parameres being completely lost in the Xylastodorinae, are unique.

Natural history. Thaumastocorids are phytophagous. Thaumastocorinae are known to feed on a variety of dicotyledonous plants. *Baclozygum depressum* Bergroth and *B. brevipilosum* Rose have been collected on *Eucalyptus trachyphloia*, *B. depressum* on *E. globosus* (Myrtaceae) (Hill, 1988), *Onymocoris hackeri* Drake and Slater on *Banksia* sp. (Proteaceae) (Drake and Slater, 1957; Rose, 1965) and *Elaeocarpus obovatus* (Elaeocarpaceae) (Rose, 1965), and *Thaumastocoris australicus* Kirkaldy on *Acacia cunninghami* (Fabaceae) (Kumar, 1964) and *A. maidenii* (Slater, 1973).

Xylastodorinae feed only on palms, *Discocoris* spp. on a variety of taxa, including *Eutorpe edulis* (Kormilev, 1955a), *Phytelephas* sp. (Schuh, 1975b), and *Socratea montana* (Slater and Schuh, 1990), apparently always on the inflorescences (Kormilev, 1955a) or infructescences (Schuh, 1975b; Slater and Schuh, 1990). *Xylastodoris* feeds on the developing fronds of the royal palm, *Roystonea regia* (Barber, 1920; Baranowski, 1958), where it sometimes causes serious damage.

Distribution and faunistics. The work of Drake and Slater (1957) is still the single most important publication on the Thaumastocoridae, but considerable knowledge concerning the diversity and distribution, particularly of the New World fauna, has been gained since the publication of that work, the most important references being those cited above.

53
Miridae

General. The Miridae is the largest family of true bugs. In the United States and Canada members of the group are usually called plant bugs, in Britain capsids, and in Germany *Blindwanzen*. Ranging in size from less than 2 mm to about 15 mm, they are among the most delicate of all bugs, and often their coloration blends well with the foliage, flowers, and bark on which they rest or feed. The habitus varies from simple ovoid to remarkably myrmecomorphic.

Diagnosis. Size and appearance variable; ocelli absent (except Isometopinae); antennal segments 3 and 4 usually

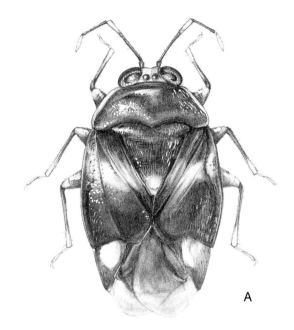

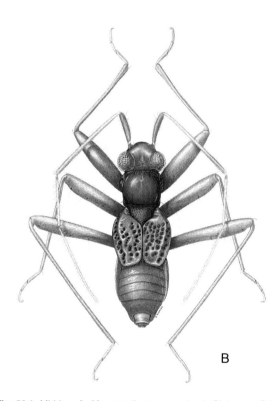

Fig. 53.1. Miridae. **A.** *Magnocellus transvaalensis* Slater and Schuh (Isometopinae) (from Slater and Schuh, 1969). **B.** *Carvalhoma malcolmae* Slater and Gross (Cylapinae) (from Schuh and Schwartz, 1984).

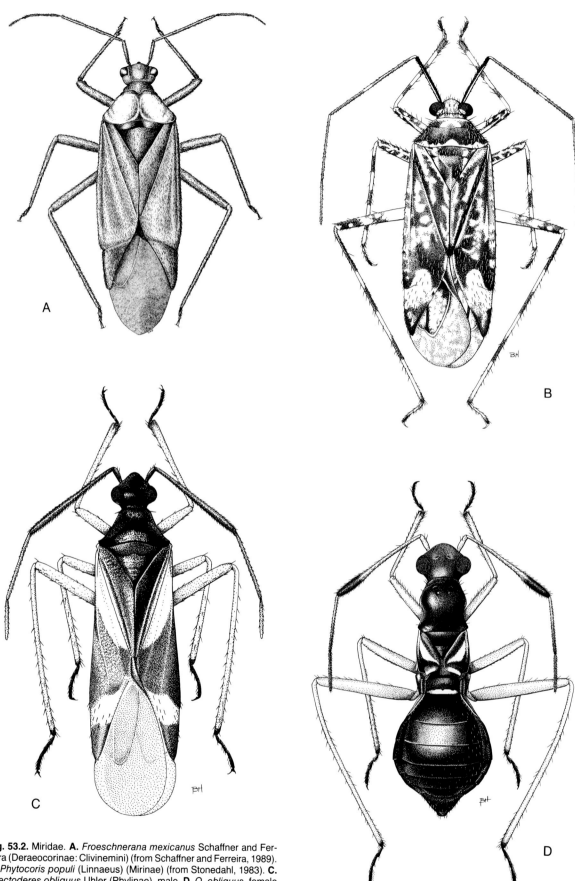

Fig. 53.2. Miridae. **A.** *Froeschnerana mexicanus* Schaffner and Ferreira (Deraeocorinae: Clivinemini) (from Schaffner and Ferreira, 1989). **B.** *Phytocoris populi* (Linnaeus) (Mirinae) (from Stonedahl, 1983). **C.** *Orectoderes obliquus* Uhler (Phylinae), male. **D.** *O. obliquus*, female (C, D from McIver and Stonedahl, 1987b).

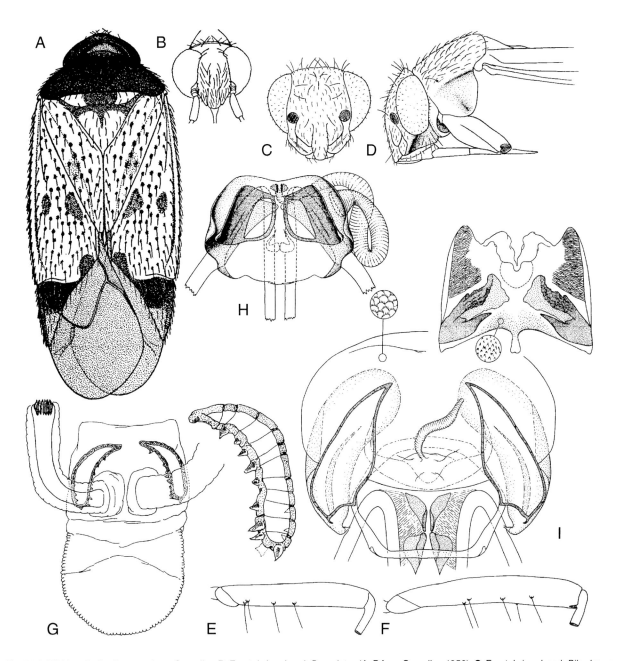

Fig. 53.3. Miridae. **A.** *Psallops oculatus* Carvalho. **B.** Frontal view, head, *P. oculatus* (A, B from Carvalho, 1956). **C.** Frontal view, head, *Pilophorus kockensis* Schuh. **D.** Lateral view, head and prothorax, *P. kockensis* (C, D from Schuh, 1984). **E.** Mesofemoral trichobothria, *Parabryocoropsis* sp. **F.** Metafemoral trichobothria, *Parabryocoropsis* sp. (E, F from Schuh, 1975a). **G.** Gynatrial complex with enlargement of sclerotized rings, *Helopeltis westwoodi* (White) (from Schmitz, 1968). **H.** Gynatrial complex, *Beamerella balius* (Froeschner) (from Henry and Schuh, 1979). **I.** Gynatrial complex, with posterior wall at upper right, *Myrmecophyes oregonensis* Schuh and Lattin (from Schuh and Lattin, 1980).

only slightly smaller in diameter than segment 2, more rarely terete; labium 4-segmented, inserted ventrally on head, usually long and tapering (Fig. 53.3D), sometimes short and stout, diameter of segment 1 greatest; clypeus more or less vertical (Fig. 53.3D); forewing usually with conspicuous costal fracture and cuneus (Figs. 53.1A, B, 53.2A–D), wing often declivent posterior to cuneal fracture; membrane with 1 or 2 closed cells (Fig. 53.6A, B), stub present at most posterior point of cell(s) (see Schuh and Štys, 1991); trochanters 2-segmented, tarsi sometimes 2-segmented, more commonly 3-segmented; claw shape variable, pulvilli often present on either ventral or mesal (inner) surface of claw (Fig. 53.5E); parempodia setiform (Fig. 53.5A, D–F) or modified into fleshy pads (Fig. 53.5C), fleshy pseudopulvilli (accessory parempodia) sometimes arising from base of claw (Fig. 53.5F); meso- and metafemora each with 2–8 (rarely more) trichobothria on lateral and ventral surfaces (Fig. 53.3E,

F); metathoracic scent glands paired, evaporatory area usually present and auricular (Fig. 10.10A), sometimes greatly reduced; nymphal dorsal abdominal scent glands located on anterior margin of terga 4 and 5; pygophore conspicuous; parameres always asymmetrical, left often more strongly developed than right (Fig. 53.4C–D, J–K, O–P); phallus proximally with a sclerotized phallotheca, endosoma often inflatable, sometimes simple (Fig. 53.4A), but frequently with a differentiated distal vesica bearing partially sclerotized lobes often ornamented with spines (Fig. 53.4Q, R), long spicula (Fig. 53.4I), or spinose patches (Fig. 53.4Q), or straplike and relatively rigid (Fig. 53.4E) (Phylinae); spermathecal gland present (Davis, 1955); ovipositor laciniate.

Classification. Hahn (1831–1835) first recognized the Miridae as a family group. Fieber (1861), in his studies of the western European fauna, laid the groundwork for the modern classification of the Miridae based on the structure of the pretarsus. Fieber's scheme was adopted by Reuter (1875, 1878–1896) in his works on the European fauna and later expanded to a worldwide classification (Reuter, 1905, 1910).

Schuh (1974, 1976) published cladograms for mirid subfamilies. These schemes support the sister group relationships of the Phylinae and Orthotylinae and Miridae and Deraeocorinae but do not offer strong evidence for other infrasubfamilial relationships in the Miridae except to suggest that the Isometopinae are the sister group of the remaining subfamilies.

Carvalho's classification, published in his *Catalogue of the Miridae of the World* (Carvalho, 1952, 1957, 1958a, b, 1959, 1960), has been modified by Wagner (1955) and Schuh (1974, 1976), the primary reorganization being of tribes that were based on myrmecomorphic habitus in the subfamilies Orthotylinae and Phylinae and the removal of the Dicyphina from the Phylinae to the Bryocorinae.

At least 1200 genera and 10,000 species of Miridae have been described. The classification presented below places them in eight subfamilies, arranged alphabetically.

Key to Subfamilies of Miridae

1. Tarsi 2-segmented; size small, length usually less than 2.5 mm; ocelli present or absent; claws always with a subapical tooth (Fig. 53.5A); membrane of forewing usually with a single cell (Fig. 53.1A, 53.3A) 2
– Tarsi 3-segmented; size variable, length often much greater than 2.5 mm; ocelli always absent; claws usually without a subapical tooth; membrane of forewing usually with 2 cells (Figs. 53.2B, 53.6B), less commonly with a single cell 4
2. Ocelli present; membrane of forewing always with a single cell (Fig. 53.1A); head frequently compressed anteroposteriorly and elongate dorsoventrally .... Isometopinae
– Ocelli absent; membrane with 1 or 2 cells; head not so strongly compressed anteroposteriorly, or if compressed, not elongate dorsoventrally 3
3. Membrane with a single cell (Fig. 53.3A); head nearly circular in frontal view (Fig. 53.3B), relatively short anteroposteriorly .... Psallopinae
– Membrane with 2 cells; head either elongate anteroposteriorly (*Peritropis* Uhler) or elongate dorsoventrally (*Vannius* Distant) .... Cylapinae (part)
4. Parempodia (or other structures arising between the claws on unguitractor plate) always at least weakly fleshy, flattened and usually either obviously convergent or divergent apically (Fig. 53.5B, C) 5
– Parempodia always setiform (Fig. 53.5A, F), never weakly to strongly flattened and fleshy, although sometimes not perfectly straight; fleshy structures arising from near base of claw sometimes also present (Fig. 53.5F) 9
5. Parempodia always fleshy and divergent apically; rounded or flattened pronotal collar always present; pulvilli usually developed on ventral surface of claw .... Mirinae (part)
– Parempodia (or similarly placed structures) usually convergent apically (Fig. 53.5B, C), rarely nearly straight; pronotal collar present or absent; pulvilli present or absent 6
6. Tarsi dilated distally; "pseudopulvilli" arising on unguitractor plate between claws, recurved and convergent apically, not arising from an alveolus (Fig. 53.5C); rounded pronotal collar always present (Bryocorini) .... Bryocorinae (part)
– Tarsi linear, not dilated distally; true parempodia weakly to strongly fleshy, inserted in an alveolus (Fig. 53.5B); anterior pronotal margin usually fine and reflexed dorsally, or more rarely in form of a rounded or flattened collar 7
7. At least hemelytra (and often thorax and abdomen laterally and ventrally) with some appressed scale-

like silvery setae, often arranged in patches or transverse bands; vesica in form of rigid sclerotized tube; left paramere usually boat-shaped (Pilophorini) .... Phylinae (part)
- Body never with scalelike silvery setae, setae never arranged in patches or bands; vesica never in the form of a rigid tube; left paramere variable in shape 8
8. Parempodia weakly flattened, sometimes weakly curving, and convergent apically; pulvilli present on ventral claw surface; vesica in male sclerotized and rigid; phallotheca (Fig. 53.4F) attached to posterior wall of pygophore; left paramere boat-shaped (Fig. 53.4G); anterior pronotal margin without a collar (*Ellenia* Poppius, *Semium* Uhler, and some other Phylini), or rarely with a distinct collar (*Cyrtopeltocoris* Reuter) .... Phylinae (part)
- Parempodia broadly flattened, always convergent apically as in Fig. 53.5B; pulvilli not present on ventral claw surface; vesica in male not sclerotized and rigid; phallotheca attached to phallobase; left paramere never boat-shaped (e.g., *Falconia* Distant, *Naniella* Reuter); anterior pronotal margin always with a distinct collar .... Orthotylinae (part)
9. Claws with pulvilli or other fleshy structures arising at base (Fig. 53.5F) or from ventral or mesal surface (Fig. 53.5E) 10
- Claws without pulvilli or other fleshy structures arising from basal, ventral, or mesal surface 14
10. Pulvilli arising from ventral surface of claw 11
- Pulvilli covering most of mesal (inner) surface of claw (Fig. 53.5E), or pseudopulvilli arising at very base of claw and free from claw over nearly entire length (Fig. 53.5F); claws sometimes cleft at base (*Dicyphus* Fieber) 12
11. Anterior pronotal margin either not in the form of a collar and reflexed dorsally, or in the form of a flattened collar, never in the form of a rounded collar; vesica in male sclerotized and rigid; phallotheca arising from posterior wall of pygophore; left paramere usually boat-shaped; pulvilli sometimes attached over entire length of claw or occasionally attached only near base ...... Phylinae (part)
- Anterior pronotal margin in the form of a conspicuously rounded collar; vesica in male membranous and inflatable; phallotheca attached to phallobase; left paramere elongate and more or less parallel-sided; pulvilli small; large myrmecomorphic species (*Closterocoris* Uhler; *Cyphopelta* Van Duzee) (Herdoniini, part) .... Mirinae (part)
12. Large flattened pulvillus covering most of mesal (inner) surface of claw (Fig. 53.5E), ventral surface of claw usually with a comb of elongate spicules (Fig. 53.5E); pseudopulvilli absent; tarsi dilated distally (Eccritotarsina) .... Bryocorinae (part)
- Pulvilli absent from mesal claw surface; elongate, fleshy pseudopulvilli arising from base of claw (Fig. 53.5F); tarsi dilated distally or not 13
13. Pretarsus dilated distally; insect relatively large, either elongate (Monaloniina) or heavy-bodied (Odoniellina) .... Bryocorinae (part)
- Pretarsus not dilated distally; small, slender, delicate insects with very small pretarsal structures (Dicyphina) .... Bryocorinae (part)
14. Claws cleft near base, forming a single tooth or projection .... Deraeocorinae
- Claws not cleft near base, although rarely cleft near midpoint 15
15. Claws usually with a subapical tooth (Fig. 53.5A) .... Cylapinae (part)
- Claws cleft near midpoint, forming a distinct tooth on ventral surface (Fig. 53.5D) (Palaucorina) .... Bryocorinae (part)

BRYOCORINAE (FIG. 53.6A). No single feature unequivocally allows for recognition of this group. Tarsus usually swollen distally (except Dicyphina); many—but not all—taxa with a single cell in membrane (Fig. 53.6A), found elsewhere chiefly in Isometopinae.

We treat this group in somewhat greater detail than other subfamilies, because of its morphological diversity and confused classificatory history. Three tribes and five subtribes, comprising about 200 genera are recognized, following the classification of Schuh (1976).

BRYOCORINI. This group of five genera, including *Bryocoris* Fallén and *Monalocoris* Dahlbom from the Northern Hemisphere and *Hekista* Kirkaldy from the Orient, can be recognized by the lamellate pads arising from the unguitractor plate (Fig. 53.5C). The pads do not appear to be parempodial derivatives and were therefore termed pseudopulvilli by Schuh (1976) ("accessory parempodia" of Cobben, 1978). All members of the group appear to feed on ferns (Knight, 1941; Kullenberg, 1944).

DICYPHINI. The subtribe Dicyphina was placed in the

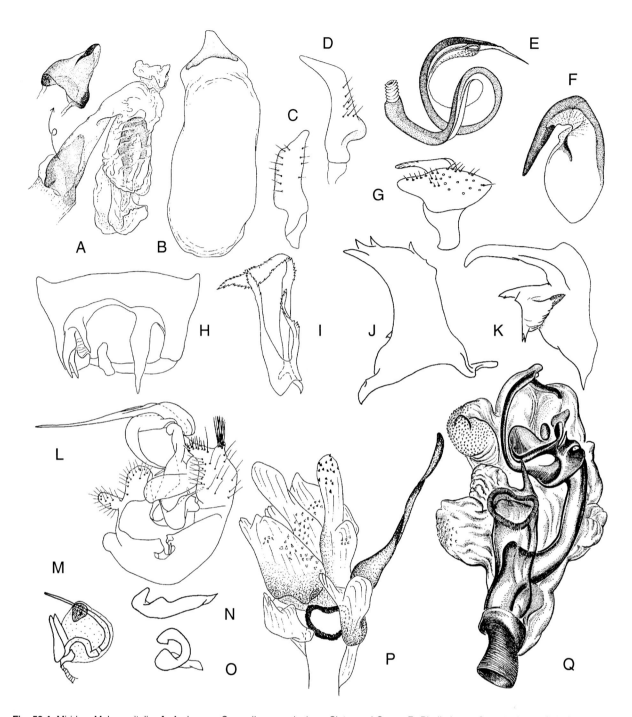

Fig. 53.4. Miridae. Male genitalia. **A.** Aedeagus, *Caravalhoma malcolmae* Slater and Gross. **B.** Phallotheca, *C. malcolmae*. **C.** Left paramere, *C. malcolmae*. **D.** Right paramere, *C. malcolmae* (A–D from Schuh and Schwartz, 1984). **E.** Vesica, *Beamerella personatus* Carvalho. **F.** Phallotheca, *B. personatus*. **G.** Left paramere, *B. personatus*. (E–G from Henry and Schuh, 1979). **H.** Genital capsule, *Oaxacacoris guadalajara* Stonedahl and Schwartz. **I.** Vesical spicula, *O. guadalajara*. **J.** Left paramere, *O. guadalajara*. **K.** Right paramere, *O. guadalajara* (H–K from Stonedahl and Schwartz, 1988). **L.** Genital capsule, *Mertila malayensis* Stonedahl. **M.** Aedeagus, *M. malayensis*. **N.** Left paramere, *M. malayensis*. **O.** Right paramere, *M. malayensis* (L–O from Stonedahl, 1988b). **P.** Vesica, *Phytocoris calli* Knight (from Stonedahl, 1988a). **Q.** Vesica, *Horcias scutellatus* Distant (from Carvalho, 1976).

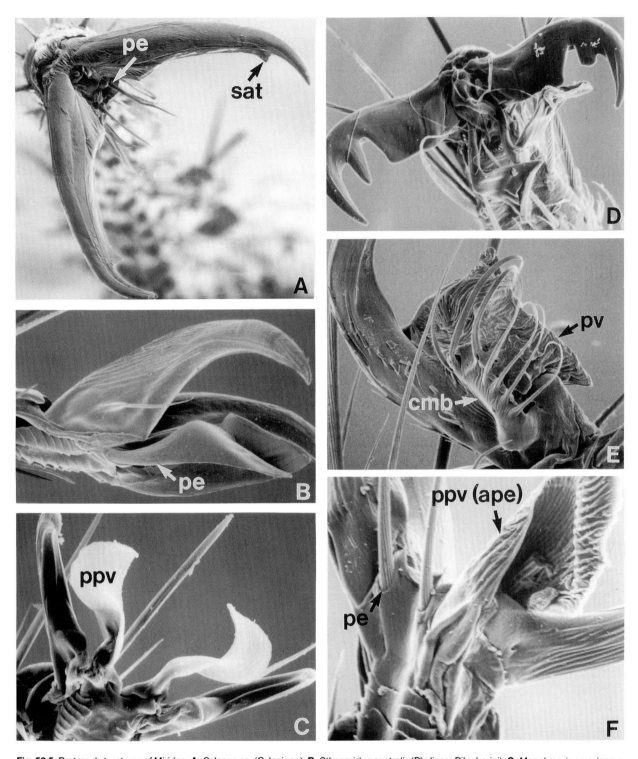

Fig. 53.5. Pretarsal structures of Miridae. **A.** *Cylapus* sp. (Cylapinae). **B.** *Sthenaridea australis* (Phylinae: Pilophorini). **C.** *Monalocoris americanus* Wagner and Slater (Bryocorinae: Bryocorini). **D.** *Palaucoris unguidentatus* Carvalho (Bryocorini: Palaucorina). **E.** *Pycnoderes* sp. (Bryocorinae: Eccritotarsina). **F.** *Cyrtopeltis ebaeus* Odhiambo (Bryocorinae: Dicyphina) (from Schuh, 1975). Abbreviations: ape, accessory parempodium; cmb, comb; pe, parempodium; ppv, pseudopulvillus; pv, pulvillus; sat, subapical tooth.

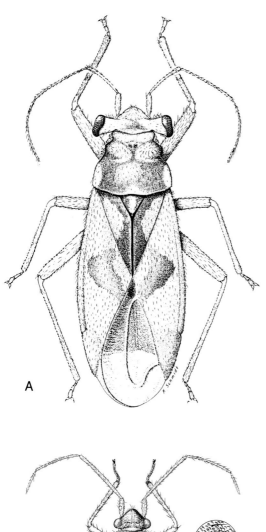

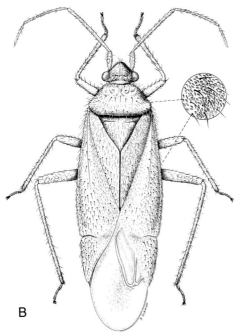

Fig. 53.6. Miridae. **A.** *Stylopomiris malayensis* Stonedahl (Bryocorinae: Eccritotarsini) (from Stonedahl, 1986). **B.** *Pseudopsallus lattini* Stonedahl and Schwartz (Orthotylinae) (from Stonedahl and Schwartz, 1986).

Phylinae by Carvalho (1952, 1958a) and has been treated as a separate subfamily by Knight (1941) and others. The Dicyphina appear to be closely related to the Monaloniina and Odoniellina on the basis of male genitalic structure, distinctive pseudopulvilli (Fig. 53.5F), and greatly reduced scent-gland evaporatory area, although they are much smaller and do not have distally dilated tarsi. An example of the female genitalia in the Monaloniina is shown in Fig. 53.3G. *Pameridea* Reuter, long placed in the Mirinae, was moved to the Dicyphina by Dolling and Palmer (1991). Important works on the Dicyphina include those of Carvalho and China (1952), Odhiambo (1961), and Cassis (1984).

The subtribes Monaloniina and Odoniellina are tropical, with all genera but the Neotropical *Monalonion* Herrich-Schaeffer occurring in the Old World. Important works since the Carvalho catalog include: Carvalho (1981; fauna of New Guinea), Schmitz (1968; African *Helopeltis* Signoret), Stonedahl (1991; Oriental *Helopeltis*), and Odhiambo (1962; various African genera).

ECCRITOTARSINI. The circumtropical subtribe Eccritotarsina is the largest of the bryocorine subgroups and is particularly speciose in the Neotropics. It is recognized by the large disc-shaped pulvillus covering nearly the entire mesal (inner) surface of the claw, asymmetrical parempodia, comb of spicules on the ventral claw surface of most species (Fig. 53.5E), and reduced scent-gland evaporatory area. The male genitalia are sometimes bizarrely modified (e.g., Fig. 53.4M–P). The group was originally recognized by Berg (1884) but was included in an omnibus Bryocorini by Carvalho (1952).

The subtribe Palaucorina (Schuh, 1976), treated as a distinct subfamily by Carvalho (1956, 1984b), is recognized by the globose head, heavily punctured body, short labium, and multiply cleft claws (Fig. 53.5D). All known species are from the western tropical Pacific (Carvalho, 1984b; Ghauri, 1975). Nothing is known about their habits, although the peculiar claws are reminiscent of those of certain emesine reduviids, some of which live in spider webs.

CYLAPINAE (FIG. 53.1B). Head often elongate anteroposteriorly or dorsoventrally; parempodia setiform; pulvilli absent; claws long, slender, usually with a subapical tooth (Fig. 53.5A); antennae long, slender; membrane with 2 cells; legs long; dorsum often heavily punctured (except in *Fulvius* Stål and its relatives) (Schuh and Schwartz, 1984).

Three tribes—Bothriomirini, Cylapini, and Fulviini—were recognized by Carvalho (1957). The last two possess no unique attributes and are almost certainly not monophyletic.

Useful references for this primarily tropical group are Bergroth (1920; a list of genera and species), Carvalho and Fontes (e.g., 1968; descriptive work on the Neotropical fauna), and Carvalho and Lorenzato (1978; treatment of the New Guinea fauna).

DERAEOCORINAE (FIG. 53.2A). Pronotal collar present, rounded; claws cleft basally; pulvilli absent; parempodia setiform.

The general aspect in many species of Deraeocorini is very similar to that of the Mirini, and indeed the similarity of male and female genitalic structure suggests a possible sister group relationship between the Deraeocorinae and Mirinae (Slater, 1950; Kelton, 1959). Many deraeocorines have a distinctly polished and pitted dorsum, setting them apart from most other Miridae, but a few are dull, without punctures (Termatophylini), and others are dull, heavily punctured, and sometimes covered with flocculence (Clivinemini).

Six tribes are recognized: Clivinemini, including 19 genera from the Holarctic; Deraeocorini, a worldwide group of about 50 genera, including the large and diverse *Deraeocoris* Kirschbaum; Hyaliodini, with 23 New World genera distinguished by the hyaline hemelytra; Saturniomirini, comprising three genera from Australia and New Guinea; Surinamellini, a tropicopolitan group comprising 12 myrmecomorphic genera; and Termatophylini, including eight genera of strongly anthocorid-like facies, widely distributed in tropical and subtropical areas.

ISOMETOPINAE (FIG. 53.1A). Ocelli present; membrane with a single cell; subapical claw tooth usually present; tarsi 2-segmented; head strongly flattened anteroposteriorly and often elongate dorsoventrally; antennae sexually dimorphic, segment 2 larger in males than females; most species 2–3 mm in length, with 2 and 3 meso- and metafemoral trichobothria, respectively, but *Gigantemetopus rossi* Schwartz and Schuh (1990) from Sumatra 6.98 mm long, with 5 meso- and 6 metafemoral trichobothria.

The Isometopinae have been treated as a separate family by some authors, because they possess ocelli, and were thus not included in the Carvalho catalog (1957–1960). Carayon (1958) recommended placement in the Miridae on the basis of wing and male and female genitalic structure. McAtee and Malloch (1924, 1932) prepared the only published suprageneric classification of the group (tribes Diphlebini and Isometopini, dividing the latter into Isometoparia and Myiommaria). The 28 genera currently placed in the subfamily are primarily tropical in distribution. The widely distributed *Myiomma* Puton is the most speciose genus, with approximately 40 species.

Useful works include those by Smith (1967; description of African species), Henry (1977, 1979, 1984; analysis of New World fauna), and Eyles (1971, 1972; checklist of world species with a limited bibliography).

MIRINAE (FIG. 53.2B). Parempodia fleshy, lamellate, apically divergent, except in *Closterocoris* and *Cyphopelta*; pronotum with rounded or flattened collar; female genitalia of remarkably uniform structure, males (Fig. 53.4Q, R) showing greatest variation in Herdoniini (Slater, 1950; Kelton, 1959; Schwartz, 1987). Mirines are often large, the resthenine *Callichilella grandis* (Blanchard), with a length of 1.5 cm, being the largest of all Miridae.

The Mirinae—comprising approximately 300 genera—are placed in six tribes: Herdoniini, an almost exclusively New World tropical group comprising about 25 genera of strongly myrmecomorphic habitus (Carvalho, 1973); Hyalopeplini, an Old World tropical group of 15 genera with hyaline hemelytra, showing a vague morphological resemblance to some large wasps, and probably also being good behavioral mimics (Carvalho and Gross, 1979); Mecistoscelini, a group of three grass-feeding genera—with slender bodies and long slender appendages—from the Orient; Mirini, a group of worldwide distribution, comprising at least 250 genera, many of classic ovate mirid habitus, and containing many species of economic importance; Resthenini, a New World group of 17 genera of primarily red and black aposematically colored species, having the scent-gland evaporatory area greatly reduced; and Stenodemini, a widely distributed group of elongate pale-colored grass feeders, comprising about 25 genera.

Important revisionary works on the Mirini to appear since the Carvalho catalog and which also contain valuable comparative morphological information include those by: Kelton (1971, 1975; *Lygocoris* Reuter and *Lygus*, respectively, for North America), Kelton and Knight (1970; *Platylygus* Van Duzee), Schwartz (1984; *Irbisia*), and Stonedahl (1988a; *Phytocoris* for western North America). Carvalho, among his many other papers on the group (1973), provided a comprehensive key for identification of genera of Herdoniini. Carvalho and Gross (1979) revised the world Hyalopeplini. Carvalho and Fontes (1971) provided a key to the genera of Resthenini, among many other papers on the group. Schwartz (1987) provided a generic revision of the Mecistoscelini and Stenodemini.

ORTHOTYLINAE (FRONTISPIECE; FIG. 53.6B). Parempodia fleshy, recurved, apically convergent (as in Fig. 53.5B) (also found in the phyline tribe Pilophorini); male Orthotylini with long slender spicules on vesica (Fig. 53.4), genital capsule sometimes with elongate processes (Fig. 53.4H); female Orthotylini with characteristic flaplike K structures on posterior wall of bursa copulatrix, structure somewhat simpler in Halticini (Fig. 53.3I) and others (Slater, 1950; Schuh, 1974).

Three tribes, comprising approximately 220 genera, are currently recognized: Halticini, a group of about 30 genera of mostly jet black taxa primarily from the Palearctic, but also containing the widespread *Halticus*

Hahn and a few others; Nichomachini, including *Nichomachus* Distant and three other myrmecomorphic genera of similar appearance from the Ethiopian region; and Orthotylini, a diverse but clearly monophyletic group comprising all remaining genera, from which additional tribal groupings—such as the myrmecomorphic Ceratocapsini—have occasionally been segregated, but which have not been consistently recognized by most authors.

Substantial revisionary works published since the Carvalho catalog, which also include valuable comparative information on male genitalic and other structures, are by Asquith (1991; *Lopidea* Uhler north of Mexico), Kelton (1968; *Slaterocoris* Wagner), Carvalho et al. (1983; *Ceratocapsus* Reuter in the Neotropics), Josifov and Kerzhner (1984; *Dryophilocoris*), and Stonedahl and Schwartz (1986, 1988; *Pseudopsallus* Van Duzee and related genera in western North America).

PHYLINAE (FIG. 53.2C, D). Vesica rigid, straplike or tubular (Fig. 53.4E); left paramere usually boat-shaped (Fig. 53.4G); phallotheca attached to posterior surface of pygophore (Fig. 53.4F) rather than to phallobase as in other Miridae; posterior wall in female simple, female genitalia sometimes asymmetrical (Fig. 53.3H). The Phylinae have traditionally been identified by the lack of a pronotal collar (most species) and presence of setiform parempodia, but neither attribute is synapomorphic and certainly neither is restricted to the group.

Five tribes, comprising approximately 300 genera, are currently recognized: Auricillocorini, including five Oriental genera, some strongly myrmecomorphic (Schuh, 1984); Hallodapini, with about 50 genera from the southern Palearctic, Africa, the Orient, and North America, a myrmecomorphic group with many strongly sexually dimorphic taxa; Leucophoropterini, comprising 21 genera of often bizarrely modified myrmecomorphs—as for example *Waterhouseana* Carvalho from New Guinea—with greatest diversity in the Orient, but also occurring in northern and eastern Australia and southern Africa and including the widely distributed *Tytthus* Fieber; Phylini, comprising at least 230 genera, extremely diverse in the Northern Hemisphere, and further subdivided by some authors (e.g., Wagner, 1971–1978); and Pilophorini, a widespread group of 10 genera of variable habitus, but some strongly myrmecomorphic (Schuh, 1991).

Revisionary works published since the Carvalho catalog which also contain important comparative information on male genitalia and other structures include those of Kelton (1964, *Reuteroscopus* Kirkaldy; 1965, *Chlamydatus* Curtis in North America), Schuh (1974, fauna of southern Africa; 1984, fauna of the Indo-Pacific), Carvalho and Gross (1982; *Leucophoroptera* group in Australia), Schuh and Schwartz (1985, *Rhinacloa* Reuter; 1988, New World Pilophorini), Stonedahl (1990; *Atractotomus* Fieber), and Henry (1991; *Pseudatomoscelis* Poppius and *Keltonia* Knight).

PSALLOPINAE (FIG. 53.3A, B). Head nearly spherical in frontal view; tarsi 2-segmented; claws with a subapical tooth; parempodia setiform; membrane with a single cell.

This group comprises only the widely distributed *Psallops* Usinger originally described from the tropical western Pacific (Usinger, 1946; Carvalho, 1956), but with undescribed species from Africa (Schuh, 1974), the Neotropics, and Australia (CSIRO, 1991). Originally placed in the Phylinae (Usinger, 1946; Carvalho, 1956), *Psallops* has also been treated as a part of the Isometopinae (Eyles, 1972) or as a separate subfamily (Schuh, 1976).

Nothing is known of their habits, all known specimens having been collected at lights.

Specialized morphology. Only the Miridae among Heteroptera have trichobothria on the meso- and metafemora (Figs. 10.8I, 53.3E, F). These structures have been examined for all subfamilies (Schuh, 1975a), ranging from a minimum number of 2 mesofemoral and 3 metafemoral trichobothria in most Isometopinae to a modal number of 7 and 8, respectively, with even higher numbers occurring in some Bryocorinae, Cylapinae, and Mirinae. In most cases the "bothrium" has a distinct marginal trichoma (Fig. 10.8I), although it is poorly defined in some taxa and the "trich" is placed on a simple tubercle in the Dicyphini and Stenodomini.

As a group, the Miridae exhibit greater variation in pretarsal structure than any other family of Heteroptera (Fig. 53.5A–F). These structures are fundamental to the classification of the group. The parempodia are typically setiform and symmetrically developed (Fig. 53.5A), although occasionally asymmetrical, as in most Eccritotarsina; in the Mirinae, Orthotylinae, and Pilophorini, they are enlarged, fleshy, and either divergent or convergent apically (Fig. 53.5B). The claws themselves may be simple or variously toothed—subapically in most Isometopinae, Cylapinae, and Psallopinae (Fig. 53.5A), basally in the Deraeocorinae, *Dicyphus* Fieber, Monaloniina, and Odoniellina, and with multiple ventral teeth in the Palaucorina (Fig. 53.5D). Pulvilli occur on the ventral claw surface of many Miridae or, less commonly (Eccritotarsina), on the mesal surface. Fleshy pseudopulvilli (accessory parempodia of Cobben) arise at the junction of the unguitractor plate and claw base (Fig. 53.5F) in the Dicyphini and sublaterally on the unguitractor in the Bryocorini (Fig. 53.5C).

Males of *Harpocera thoracica* (Fallén) from Europe have antennal segment 2 slightly enlarged distally. Flattened setae on the ventral surface apparently serve to grasp the female during copulation (Stork, 1981). In *Hyalochloria* Reuter males, antennal segments 1 and 2 are modified and apparently grasp the antennae of the female during copulation. In many other genera, either antennal segment 1 or 2 is enlarged, flattened, or otherwise of distinctive shape.

Stridulatory structures in the form of a wing-edge stridulitrum–femoral plectrum have been documented in some *Hallodapus* spp. and *Ctypomiris* Schuh (Schuh, 1974, 1984). No direct observations of stridulation have ever been made in these or other taxa that possess stridulatory mechanisms for which there is published and unpublished documentation.

Natural history. Most temperate-region Miridae are univoltine, overwinter in the egg stage, and are host-specific. Life histories of tropical species are less well documented, but as an example, Neotropical *Monalonion* spp. typically show bimodal seasonal population fluctuations, correlated with maximum rainfall and plant growth (Abreu, 1977). Because of the extreme rarity of males, *Campyloneura virgula* (Herrich-Schaeffer) is thought to be parthenogenetic (Lattin and Stonedahl, 1984). Eclosion often occurs in early to late spring in phytophagous species, being tied to growth and often flowering of the host plant. Predaceous taxa often have life cycles synchronized with those of their preferred prey, and they typically appear later in the season than their phytophagous counterparts. Mirids often have a life cycle that lasts no more than six weeks. They normally pass through five nymphal instars, although the number may be as low as three (Liquido and Nishida, 1985). Transformation into the adult form occurs 15–30 days from eclosion. Eggs are inserted into new or senescent plant tissue with only the "respiratory horns" protruding. In temperate latitudes, those species that have more than one generation often overwinter as adults, as for example members of *Lygus* Hahn.

The nymphs of many plant bugs possess an eversible rectum (Knight, 1941; Wheeler, 1980). When nymphs are dislodged from their host plant, they frequently exsert the rectum, which readily sticks to any surface, often attaching them to the plant before they reach the ground (Knight, 1941).

The species of a given genus of Miridae often have a close association with a given genus or family of plants (e.g., Mecistoscelini, Stenodemini, *Irbisia*, *Labops* on grasses, *Platylygus* on *Pinus*). They often feed almost exclusively on new foliage, flowers, or fruits, and Kullenberg (1944) recorded mirids feeding on pollen. Nonetheless, host associations above the species-group or generic level are much less clear-cut, and associations appear to have been determined largely by ecological factors (e.g., *Squamocoris* Knight [Stonedahl and Schuh, 1986]; *Phytocoris* Fallén [Stonedahl, 1988a]; *Pseudopsallus* [Stonedahl and Schwartz, 1986]; Pilophorini [Schuh, 1991]). Some groups may be bivoltine and monophagous (e.g., *Adelphocoris* Reuter) or multivoltine and polyphagous (*Lygus*). Literature on host associations in *Lygus* was compiled by Scott (1977, 1981), Graham et al. (1984), and Young (1986).

Some mirids have been referred to as oligophagous, feeding on both plant and animal material, as for example, many members of the Dicyphina. Different observers have concluded that *Rhinacloa forticornis* Reuter, for example, is either a predator or a plant pest, suggesting that the feeding habits of this species vary greatly depending on local conditions.

Isometopines are usually found inhabiting the bark of trees, where they feed on scale insects (Hesse, 1947; Wheeler and Henry, 1978).

Cylapines are apparently always associated with fungi. *Cylapocoris* Carvalho can be found on soft fungi in the New World tropics. *Cylapus* Say, *Xenocylapus* Bergroth, *Rhinocylapus* Bergroth, and others are usually found on fallen trees infested with hard fungi. Species of *Fulvius* Stål often are found under loose bark, and at least *Corcovadocola* Carvalho and *Carvalhoma* Slater and Gross (1977) live in litter. Herring (1976b) believed that some cylapines feed on other fungus-feeding insects. Wheeler and Wheeler (1994) documented the presence of fungal fragments and spores in the guts of two *Cylapus* spp., the first direct evidence that fungus feeding may exist in the group.

All Deraeocorinae are predaceous, many restricted to a single plant, where they feed on a host-specific phytophage, as for example, the hyaliodine genus *Stethoconus* Flor on the tingid genus *Stephanitis* Stål (Henry et al. 1986). At least some, if not all, members of the Clivinemini feed on scale insects (Miller and Schuh, 1995).

The mirine genus *Phytocoris* Fallén—with about 650 species, the most speciose genus of mirids, and probably of all true bugs—contains predatory members that are usually restricted to a single plant species (Stonedahl, 1988a).

Among bryocorine mirids of the subtribe Eccritotarsina, members of the genus *Neella* Reuter and its relatives feed on inflorescences of the Araceae, and the speciose genera *Eccritotarsus* Stål and *Pycnoderes* Guérin-Méneville, also from the Neotropics, feed on a wide variety of plants. In Mexico and the American Southwest *Halticotoma* Reuter feeds on the leaves of *Yucca* (Liliaceae), and *Caulotops* Bergroth on the leaves of *Agave* (Amaryllidaceae). The stalk-eyed *Hesperolabops* Kirkaldy feeds on the pads of *Opuntia* spp. (Cactaceae). Old World taxa feed principally on the monocot families Musaceae, Orchidaceae, and Pandanaceae (Stonedahl, 1988b). Most, if not all, members of the Dicyphina are associated with plants with sticky glandular hairs.

Many myrmecomorphic taxa may also be at least in

part predatory, as for example the orthotyline *Schaffneria* Knight in eastern North America (Wheeler, 1991). Many members of the genus *Pilophorus* Hahn are known to be partially predatory (Bradley and Hinks, 1968; Schuh and Schwartz, 1988). Unusual among the Miridae is the phyline *Ranzovius* Distant, all species of which live exclusively as scavengers in spider webs (Wheeler and McCafferty, 1984).

Some Miridae, including North American Eccritotarsina and *Hadronema* spp. (Orthotylini) are attracted to cantharadin baits (Pinto, 1978; Young, 1984). *Hadronema uhleri* Van Duzee is known to feed on living and dead meloid beetles, the hemolymph of which contains cantharadin (Pinto, 1978).

Distribution and faunistics. Detailed faunistic treatments for the Palearctic have been available since the time of Fieber's *Die europäischen Hemiptera* (1861) and Reuter's (1878–1896) classic *Hemiptera Gymnocerata Europae,* with its beautiful hand-colored plates. Since the publication of the Carvalho catalog and Carvalho's (1955) now badly outdated keys to the genera for the world, important additions include Wagner (1971–1978; with consistent illustration of the male genitalia in most groups) for western Europe and the Mediterranean and Kerzhner and Jaczewski (1964) for the European portions of the former USSR, and Kerzhner (1988) for the eastern Palearctic.

Slater and Baranowski (1978) keyed the North American genera. Knight (1923, 1941) treated the fauna of eastern North America, and Kelton (1980b) that of the prairie provinces of Canada. Knight (1968) dealt with a portion of the fauna of western North America.

Linnavuori (1975) treated the Sudan, and Wagner (1971–1978) dealt with North Africa. The works of Poppius (1912, 1914), Odhiambo (1961, 1962, and many others), and Schuh (1974) are the most comprehensive available on the Ethiopian fauna. Schuh (1984) provided a detailed review of the phyline fauna of the Indo-Pacific (less Africa). The Australian fauna remains almost unknown, except for the work of Carvalho and Gross (1980) on the Stenodemini and the *Leucophoroptera* group of genera (Phylinae) (Carvalho and Gross, 1982). One must refer to the primary literature, mostly to the literally hundreds of papers by J. C. M. Carvalho, for access to information on the Neotropical fauna.

Carvalho's catalog of the Miridae provided distributional information for all included species. These data suggested that all subfamilies and many tribes are cosmopolitan. Recent work (Schuh, 1974, 1984, 1991) confirmed that conclusion for the subfamilies, but showed that at least in the Phylinae the tribal groupings Hallodapini and Pilophorini as recognized by Carvalho were not monophyletic and when reconstituted have more restricted distributions. Detailed distributional studies exist for some monophyletic groups: *Irbisia* (Schwartz, 1984), Indo-Pacific Phylinae and Eccritotarsini (Schuh and Stonedahl, 1986), *Pseudopsallus* (Stonedahl and Schwartz, 1986), and Pilophorini (Schuh, 1991).

54
Tingidae

General. Members of this family are generally referred to as lace bugs. They range in length from about 2 to 8 mm, and most species are characterized by the distinctive lacelike network of areoles adorning the hemelytra and pronotum (Fig. 54.2B).

Diagnosis. Dorsum and hemelytra with some areas either heavily punctate (Fig. 54.1) or areolate (Fig.

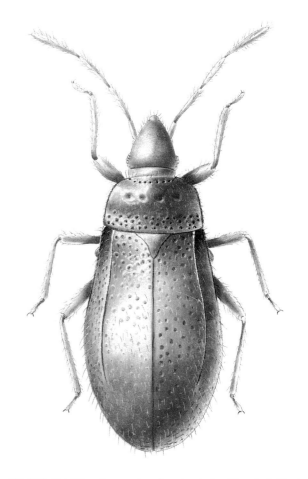

Fig. 54.1. Tingidae. *Anommatocoris coleoptratus* (Kormilev) (Tingidae: Vianaidinae) (from Drake and Ruhoff, 1965).

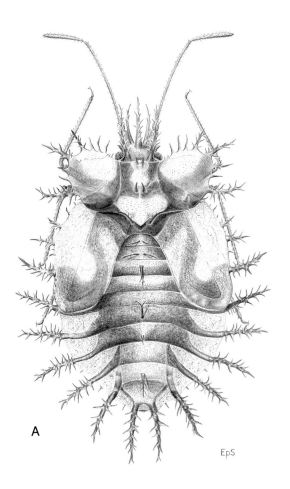 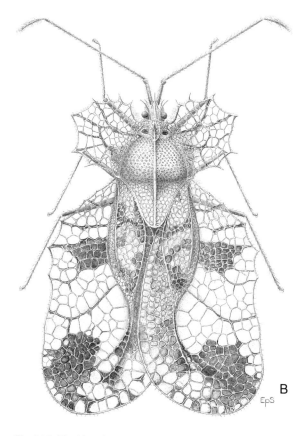

Fig. 54.2. Tingidae. *Ammianus alberti* (Schouteden) (from Drake and Ruhoff, 1965). **A.** Fifth-instar nymph. **B.** Adult.

54.2B); head short to moderately elongate; compound eyes varying from absent to large; ocelli absent in all but macropterous Vianaidinae; antennae usually with all segments of about equal diameter, segments 3 and 4 at least weakly fusiform; labium long, 4-segmented, inserted ventrally on head, clypeus more or less vertical; bucculae long, percurrent, sometimes joining each other anteriorly; hypocostal lamina well developed; thoracic sternum with a conspicuous rostral groove (Fig. 54.3A); tarsi 2-segmented; parempodia small, setiform (Fig. 10.5F) (Schuh, 1976: fig. 3; Cobben, 1978; contra, e.g., Drake and Davis, 1960, Goel and Schaefer, 1970); metathoracic scent glands with paired reservoirs (Carayon, 1962; contra Drake and Davis, 1960); Brindley's glands located dorsolaterally at thoracicoabdominal junction (Carayon, 1962); dorsal laterotergites present, ventral laterotergites absent; abdominal spiracle 1 absent, spiracles 2–8 located on abdominal sterna (Fig. 54.3A); nymphal dorsal abdominal scent gland(s) present on terga 4 and 5 or 5 only; male genitalia symmetrical (Fig. 54.3B, C); phallus with some sclerotized eversible, usually symmetrical appendages (Fig. 54.3D) (Lee, 1969); parameres as in Fig. 54.3B, E; ovipositor laciniate (Fig. 54.3F), connection between first valvula and first valvifer lost, as in Miridae; spermatheca absent, sperm-storage function performed by paired pseudospermathecae (Fig. 54.3G, H).

Classification. The family Tingidae (sometimes spelled Tingitidae) was established by Laporte (1833). The classical descriptive work on the group is that of Fieber (1844). The basic family classification was established by Stål (1873), who, like most earlier authors, included the Piesmatidae in the Tingidae. Tullgren (1918) recognized that the Tingidae lacked—and the Piesmatidae possessed—abdominal trichobothria, but it was not until Leston et al. (1954) summarized the attributes of the Cimicomorpha that the Piesmatidae were consistently divorced from the Tingidae and placed in the Pentatomomorpha.

C. J. Drake (with his many coauthors, particularly F. A. Ruhoff) is the single most prolific author on the group. Drake's later works were lavishly illustrated with habitus views of various tingids, although few of his works contain aids to identification. Drake and Davis (1960) noted that the male genitalia of the Tingidae were not of value in taxonomy; however, Lee (1969) showed that the phallus is more complex and variable than had been generally thought and that some systematically useful variation also existed in the shape of the parameres.

Three subfamilies are currently recognized. Approximately 1900 species are placed in about 250 genera.

Key to Subfamilies of Tingidae

1. Antennal segment 2 longer than segment 1, subequal in length to segment 3 (Fig. 54.1); dorsum finely to coarsely punctate; macropterous and with well-developed compound eyes and ocelli and forewing with a raised keel-like media and distinct membrane, or coleopteroid and with compound eyes greatly reduced and ocelli absent; metathoracic scent-gland evaporatory area extensive, covering entire metapleuron; scutellum moderately large and exposed (Fig. 54.1) . . . . Vianaidinae
– Antennal segment 2 much shorter, subequal to or shorter than segment 1, shorter than segment 3 (Fig. 54.2B); dorsum finely to coarsely areolate (Fig. 54.2B); forewings never with a differentiated membrane, venation obscured and without keel-like media; compound eyes always developed, ocelli always absent; metathoracic scent-gland evaporatory area not nearly so extensive as above; scutellum small, often obscured by posteriorly projecting pronotum (Fig. 54.2B) 2
2. Head projecting anteriorly beyond compound eyes, not strongly declivent (Fig. 54.3A); pronotum sometimes obscuring scutellum, but never triangularly prolonged posteriorly; abdominal sterna 2 and 3 fused . . . . Cantacaderinae
– Head strongly declivent, seldom projecting much beyond anterior margin of compound eyes; pronotum triangularly produced posteriorly and completely obscuring scutellum (Fig. 54.2B); abdominal sterna 2–4 fused . . . . Tinginae

CANTACADERINAE (FIG. 54.3A). Head relatively elongate, extending well anterior of compound eyes; head and hemelytra areolate, the latter entirely coriaceous; antennal segments 1 and 2 both short and not surpassing apex of clypeus; posterior margin of pronotum never prolonged posteriorly; clavus always exposed; nymphal dorsal abdominal scent glands present on anterior margin of terga 4 and 5; abdominal sterna 2–3 fused.

This primarily Southern Hemisphere group, first recognized by Stål (1873), currently contains two tribes, the Cantacaderini with five genera, and Phatnomini, with 15 genera, the tribal classification having been established by Drake and Davis (1960).

TINGINAE (FIG. 54.2A, B). Head usually very short, apex of clypeus distinctly surpassed by elongate antennal segment 1, except in Ypsotingini, where head elongate; antennal segment 2 very short; pronotum commonly hood-shaped, always extending posteriorly to completely cover scutellum and clavi; head and pronotum areolate, the latter entirely coriaceous; nymphal dorsal abdominal scent glands present on anterior margin of terga 4 and 5; abdominal sterna 2–4 fused.

The overwhelming majority of lace bugs, placed in some 230 genera, belong to the Tinginae. As noted by Drake and Ruhoff (1965), all species of lace bugs that are of any economic significance as a result of their attacking cultivated plants belong to this group. Three tribes are currently recognized—Litadeini, Tingini, and Ypsotingini—following the classification of Drake and Ruhoff (1965) with modified diagnoses provided by Froeschner (1969); the Agrammatinae (= Serenthiinae), which had been recognized in previous classifications, was synonymized with the Tinginae by Drake and Davis (1960).

Among the Tinginae, a few stand out for their extreme delicacy and beauty, such as some of the elaborately lacelike species of *Galeatus* Curtis, some for their truly bizarre appearance, such as *Diconocoris capusi* (Horvath), and others for their great species diversity, such as *Stephanitis* Stål, with no fewer than 64 species from the Old World, and *Corythuca* Stål with at least 70 species from the New World.

VIANAIDINAE (FIG. 54.1). Head elongate, extending anteriorly; dorsum highly polished and punctate (Fig. 54.1); ocelli and normally developed compound eyes present in macropterous forms; R+M, in forewing, elevated and keel-like in macropterous forms, membrane without veins; coleopteroid forms most commonly collected (Fig. 54.1); scent-gland evaporatory area greatly enlarged, covering entire metathoracic pleuron; nymphal dorsal abdominal scent gland located on anterior margin of tergum 5, with distinctive lateral channels (Drake and Davis, 1960); abdominal sterna 2–5 fused.

This small Neotropical group was accorded family status by Kormilev (1955b), who recognized its relationship to the Tingidae, whereas most later authors have treated it as a subfamily of the Tingidae (Drake and Davis, 1960; Drake and Ruhoff, 1965; Schuh and Štys, 1991).

Key to subfamilies of Tingidae adapted from Drake and Ruhoff, 1965.

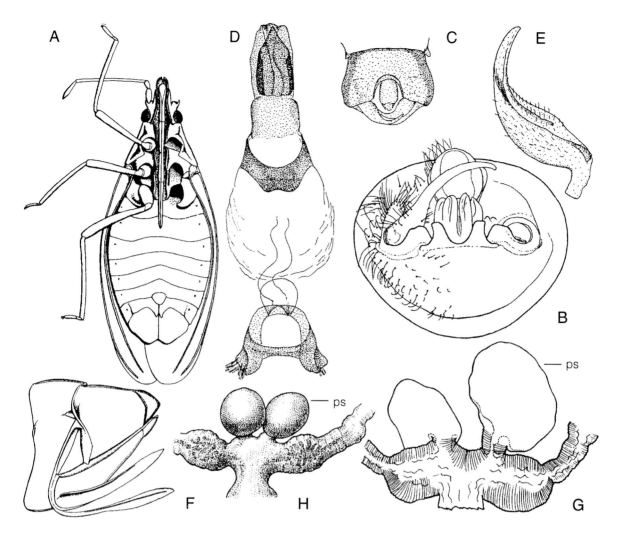

Fig. 54.3. Tingidae. **A.** Ventral view, *Cantacader quadricornis* (Le Peletier and Serville) (from Drake and Davis, 1960). **B.** Genital capsule, *Thaumamannia vanderdrifti* van Doesburg (from van Doesburg, 1977). **C.** Genital capsule, *Leptophya capitata* (Jakovlev). **D.** Phallus, *L. capitata*. **E.** Paramere, *L. capitata* (C–E from Lee, 1969). **F.** Lateral view female genital segments, *Tingis cardui* (Linnaeus) (from Drake and Davis, 1960). **G.** Paired pseudospermathecae, *Tingis ampliata* (Herrich-Schaeffer) (from Pendergrast, 1957). **H.** Paired pseudospermathecae, *Stephanitis pyri* (Fabricius) (from Carayon, 1954). Abbreviation: ps, pseudospermatheca.

Two genera are included, *Anommatocoris* China (a senior synonym of *Vianaida* Kormilev) (three species) and *Thaumamannia* Drake and Davis (two species) (Drake and Ruhoff, 1965; van Doesburg, 1977). Most known specimens are eyeless, coleopteroid, and flightless, but macropterous specimens collected in light traps and flight-intercept traps are now known in collections, revealing the presence of well-developed compound eyes, ocelli, an elevated keel-like R+M, and lack of membrane venation.

Specialized morphology. The Cantacaderinae and Tinginae are notable for the lacelike (areolate) sculpturing on the bucculae, pronotum, hemelytra, and portions of the thoracic pleuron. The pronotum in these groups is variously expanded, always completely obscuring the scutellum in the Tinginae, and often with large variously shaped paranota and occasionally a hood that may partially or completely obscure the head in dorsal view. The forewings are of uniform texture throughout, making recognition of veins and wing regions difficult and subject to varied interpretation (Livingstone, 1967). Sperm storage is assumed by paired pseudospermathecae (Fig. 54.3G, H), the true spermatheca having been lost (Drake and Davis, 1960).

Nymphal tingids also possess distinctive cuticular structures (Fig. 54.2A) that are of use taxonomically because of variation from species to species. Certain of these structures are apparently glandular, producing droplets from subhypodermal secretory cells (Southwood and Scudder, 1956), a phenomenon referred to as "sweating" (Livingstone, 1978), the latter author suggesting they may function in osmoregulation. They may also be sensory (Rodrigues et al., 1982; Pupedis et al., 1985) or defensive (Tallamy and Denno, 1981a). In both nymphs and adults of some species, waxy secretions may form a white bloom or flocculence on the body surface (Southwood and Scudder, 1956), suggesting an alternative function for some of the cuticular outgrowths.

Natural history. All members of the subfamily Vianaidinae appear to be myrmecophilous inquilines (Drake and Davis, 1960; Drake and Froeschner, 1962). Nymphs and coleopteroid adults of *Anommatocoris coleoptrata* (Kormilev) were found in the nest of the ant *Acromyrmex lundi* (Guérin-Ménéville). Some were observed feeding on the fine roots of *Gleditsia triacanthos* (Kormilev, 1955b), indicating that they are phytophagous as are other tingids, in spite of their inquilinous habits.

Available information suggests no pattern of host associations for the Cantacaderinae, the Drake and Ruhoff catalog (1965) listing such diverse plants as bananas, grasses, *Lantana*, *Vernonia*, and mosses.

Most species of Tinginae are free-living and are normally found on a single host species or on a group of closely related species, usually feeding on the undersides of leaves. The majority feed on angiosperms, and most of those on woody dicots, many feeding on densely pilose or spinose plants or on those producing sticky glandular secretions; a very small number feed on monocots (Bailey, 1951). The more than 35 recognized species of *Acalypta* Westwood feed largely on mosses (see Drake and Lattin, 1963; Drake and Ruhoff, 1965). Leaf-feeding species commonly cause serious necrosis. At least the genus *Coleopterodes* Phillipi is subterranean, feeding on the roots of *Baccharis* and *Acacia*. Some species occur on the ground and are associated with the upper parts of the roots or lower parts of stems.

Members of the Old World genera *Copium* Thunberg and *Paracopium* Distant form flower galls, all species apparently feeding on pollen. In the former group, each bug occupies an individual chamber from eclosion to becoming adult, at which time it emerges and becomes free-living. In the latter group, several nymphs live in a single chamber (Monod and Carayon, 1958; Péricart, 1983).

Temperate-zone taxa often overwinter as adults, although a few species overwinter as eggs and some as nymphs. Eggs are often laid after plant development is well progressed, and feeding usually takes place on mature plant tissues, in contrast to the Miridae.

The nymphs of many free-living species are gregarious. Most tingids are slow-moving and, unlike the Miridae, which take flight or run rapidly when disturbed, are incapable of such activity. In addition to gregariousness, at least some tingids exhibit maternal care, which has a significant positive effect on the survival of the young (Tallamy and Denno, 1981b).

Distribution and faunistics. The comprehensive catalog of Drake and Ruhoff (1965) includes distributional as well as host and systematic information. Useful faunistic studies include those of Bailey (1951) and Hurd (1946) for eastern North America and Kerzhner and Jaczewski (1964), Péricart (1983), Putchkov (1974), and Golub (1988) for the Palearctic.

Naboidea

55
Medocostidae

General. The Medocostidae contains only *Medocostes lestoni* Štys (Fig. 55.1A), which ranges in length from 8.3 to 9.5 mm and is known from only a handful of specimens. The group has no common name; the name *Medocostes* Štys is an anagram of *Scotomedes,* type genus of the Velocipedidae.

Diagnosis. Head relatively short, moderately declivent (Fig. 55.1B); ocelli present; antennal prepedicellite present; clypeus declivent; bucculae large, obscuring base of labium; gula short; labium reaching between hind coxae, segment 1 virtually absent, segment 4 longer than segments 2 and 3 combined (Fig. 55.1B); pronotum flattened, with a narrow collar; scutellum large; mesepimeron and metepisternum almost completely occupied by scent-gland evaporatory area (Fig. 55.1C); medial fracture long, no costal fracture; membrane with 3 long basal cells and many simple or bifurcating emanating veins, and a stub (processus corial) apparently located near the base of the anteriormost cell (Fig. 55.1A); fossula spongiosa absent; scent-gland scars located on anterior margin of mediotergites 4, 5, and 6 of adults; abdominal spiracle 1 absent, spiracles present on abdominal sterna 2–8; abdominal sternum 7 with an anterior medial apophysis; parameres symmetrical (Fig. 55.1F), directed dorsomesad, inserted in posterolateral wall of pygophore (Fig. 55.1D); phallus with a phallotheca produced into a long tapering process (Fig. 55.1E); ovipositor laciniate (Fig. 55.1G); gynatrium simple, saclike, with a large, unpaired sclerotized ring gland (Fig. 55.1H); spermatheca in the form of a vermiform gland (Fig. 55.1H).

Classification. *Medocostes* was described—and ac-

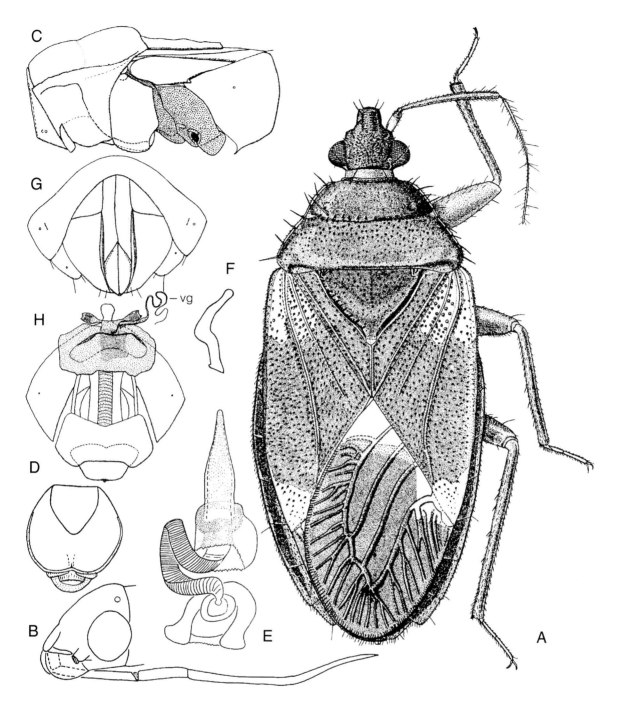

Fig. 55.1. Medocostidae. *Medocostes lestoni* Štys. **A.** Habitus (from Štys, 1967b). **B.** Lateral view, head (from Schuh and Štys, 1991). **C.** Lateral view, thorax (from Štys, 1967b). **D.** Pygophore. **E.** Aedeagus. **F.** Paramere (D–F from Kerzhner, 1981). **G.** Ventral view, female genitalic segments. **H.** Female genitalia including vermiform gland (G, H from Štys, 1967b). Abbreviation: vg, vermiform gland.

corded family status—with two species by Štys (1967b), one of which was later synonymized by Kerzhner (1989). Štys (1967b) and Schuh and Štys (1991) concurred with Štys (1967b) on treatment of *Medocostes* at the family level, although the group has been treated as a subgroup of the Velocipedidae (Kerzhner, 1971) or as a subfamily within the Nabidae (Carayon, 1970; Kerzhner, 1981).

Specialized morphology. *Medocostes* is particularly distinctive for the proportions of the labial segments, segment 4 nearly as long as segments 2 and 3 combined (Fig. 55.1B), a situation unique in the infraorder, and possibly within the Heteroptera.

The structure of the female genitalia was described in detail by Štys (1967b); the structure of the male genitalia is known only from brief statements and schematic illustrations given by Kerzhner (1981).

Natural history. Limited observations suggest that *Medocostes lestoni* is found on the bark of dead trees, from which Kerzhner (1989) concluded that they feed on subcorticolous insects.

Distribution and faunistics. *Medocostes* is known only from tropical west Africa and the Congo basin.

56
Nabidae

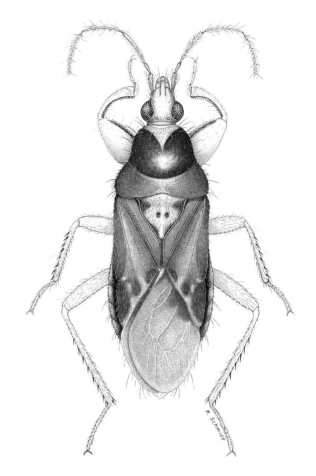

Fig. 56.1. Nabidae. *Pagasa luticeps* (Walker).

General. Members of this family are sometimes referred to as damsel bugs. Most species are of moderate size, only occasionally exceeding 10 mm in length. Many are elongate and of drab coloration, whereas others are more stout-bodied and occasionally possess distinctive red and black color patterns.

Diagnosis. Labium flexible and usually curving, reaching onto pro- or mesothorax, with 4 distinct segments (Fig. 56.2A, B); antennal prepedicellite present, variable in length (Fig. 56.1); fossula spongiosa present on fore- and middle tibiae of most taxa (Fig. 10.4D–F); costal fracture present or absent; membrane with 2 or 3 elongate cells, usually with emanating veins, and a stub (Fig. 56.1); abdominal spiracles 2–8 present, located either on laterosternites (Nabinae) or mediosternites (Prostemmatinae) (Fig. 56.3G–I); trichobothria present on abdomen of nymphs and/or adults of some taxa (Fig. 10.8F); parastigmal pits present in most taxa; internal apophysis present on abdominal sternum 7 of female (except *Arachnocoris* Scott, *Pararachnocoris* Reuter, and *Metatropiphorus* Reuter); nymphal dorsal abdominal scent glands present between terga 4/5, 5/6, 6/7; Ekblom's Organ usually present in males (Fig. 56.3A–F) (see Specialized Morphology); male genitalia usually symmetrical (Fig. 56.3A, D), sometimes parameres (Fig. 56.2D, F) (some Prostemmatinae) or phallus (Fig. 56.2C, E) asymmetrical; ovipositor laciniate (reduced in *Arachnocoris*); spermatheca in form of vermiform gland (Fig. 56.2G, H); eggs with 1 micropyle.

Classification. Early classifications of the Heteroptera included the Nabidae in an omnibus Reduviidae. Fieber (1861) first recognized the family. Stål (1872) and Reuter and Poppius (1910) included the Pachynomidae in the Nabidae; modern authors such as Carayon (1970) and Kerzhner (1981) excluded the Pachynomidae, but have included the Medocostidae and Velocipedidae. We follow the classification of Schuh and Štys (1991) in defining the Nabidae as including only the subfamilies Nabinae and Prostemmatinae, and we follow the tribal classification of Kerzhner (1981). Approximately 20 genera and 500 species are currently recognized.

Key to Subfamilies of Nabidae

1. Scutellum with 1–7 pairs of conspicuous trichobothria along lateral margins (Fig. 56.1); forefemur usually strongly incrassate, provided with a tooth or angular process on ventral surface; opening of pygophore either caudad or ventrad; parastigmal pits situated on sternum 3 or absent (Fig. 56.3G);

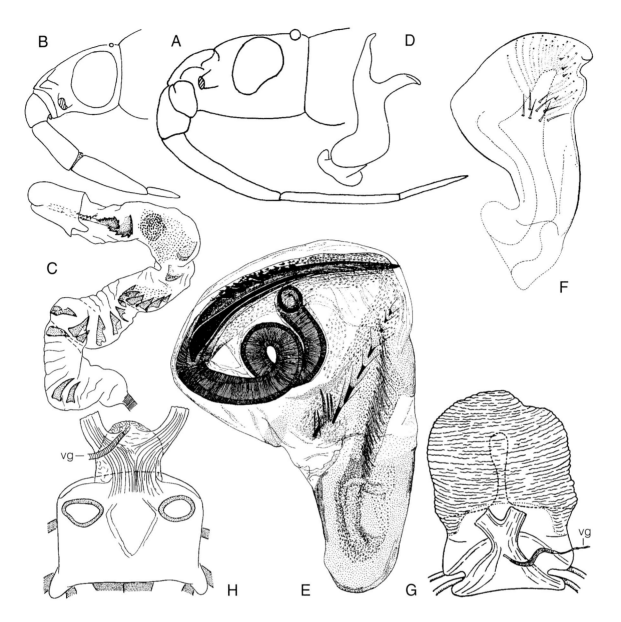

Fig. 56.2. Nabidae. **A.** Lateral view, head, *Nabis ferus* (Linnaeus). **B.** Lateral view, head, *Prostemma guttula* (Fabricius) (A, B from Schuh and Štys, 1991). **C.** Aedeagus, *Arbela carayoni* Kerzhner. **D.** Paramere, *A. carayoni* (C, D from Kerzhner, 1986a). **E.** Aedeagus, *Alloeorhynchus mabokei* Carayon. **F.** Paramere, *A. mabokei* (E, F from Carayon, 1970). **G.** Gynatrial complex, *Nabis americoferus* Carayon (from Carayon, 1961b). **H.** Gynatrial complex, *Arbela carayoni* (from Kerzhner, 1986a). Abbreviation: vg, vermiform gland.

labium relatively short and stout (Fig. 56.2B); Ekblom's Organ always present, male hind tibia distally usually with fewer than 10 stiff setae (Fig. 56.3D–F) .... Prostemmatinae
- Scutellum without trichobothria; forefemur at most only moderately incrassate, without denticles; opening of pygophore oriented dorsad; parastigmal pits located on one (seventh) or several abdominal sterna or absent; labium generally relatively slender and elongate (Fig. 56.2A); Ekblom's Organ present or absent, if present, male hind tibia distally with 30–40 stiff setae (Fig. 56.3A–C) .... Nabinae

NABINAE. Costal fracture absent; parastigmal pits on several abdominal segments (Fig. 56.3I) (absent in Arachnocorini, *Carthasis*, and Gorpini); pygophore opening oriented dorsally (Fig. 56.3A); aedeagus as in Fig. 56.2C; parameres as in Fig. 56.2D; abdominal segment 8 in males normally developed and exposed or telescoped within segment 7; 30–40 stiff setae on apex of hind tibia of most species (Fig. 56.3C) (part of Ekblom's Organ).

Kerzhner (1981) recognized four tribes. The Neotropical Arachnocorini contains *Arachnocoris* Scott, with 11 species (Kerzhner, 1990). The Carthasini comprises two Neotropical genera. The Gorpini contains *Gorpis* Stål from the Old World tropics (Harris, 1930b, 1939a) and *Neogorpis* Barber from Central America and Puerto Rico (Barber, 1924; Kerzhner, 1986b).

The Nabini are worldwide in distribution, but particularly diverse in the Northern Hemisphere. Remane (1953, 1962, 1964) recognized many new species of *Nabis* Latreille in the western Palearctic through the use of the male genitalia as taxonomic characters; internal female genitalia were employed as taxonomic characters by Carayon (1961b). Generic limits were redefined by Kerzhner (1963, 1968, 1981). *Nabis* has undergone a significant radiation in the Hawaiian Islands (Zimmerman, 1948). *Stenonabis* Reuter from the Old World tropics was revised by Kerzhner (1970a). Other Old World nabines from the Southern Hemisphere have been treated by Kerzhner (1970b).

PROSTEMMATINAE (FIG. 56.1). Prepedicellite between antennal segments 1 and 2 greatly elongated, giving appearance of 5-segmented antenna (Fig. 56.1); trichobothria laterally on abdominal terga 4, 6, and 9 of nymphs; 1–7 pairs of trichobothria laterally on scutellum of adults (Fig. 10.8B) (Carayon, 1970); parastigmal pits on abdominal sternum 3 (Fig. 56.3G, H); apex of hind tibiae in males with 10 or fewer setae associated with Ekblom's Organ (Fig. 56.3E, F); abdominal segment 8 telescoped within abdominal segment 7; opening of pygophore either posterior or ventral; aedeagus as in Fig. 56.2E; parameres as in Fig. 56.2F; insemination traumatic via puncture of vaginal wall (Carayon, 1952).

Two tribes are recognized (Kerzhner, 1981). Phorticini contains the pantropical *Phorticus* Stål and the Oriental *Rhamphocoris* Kirkaldy. Prostemmatini includes the circumtropical *Alloeorhynchus* Fieber (Harris, 1928), *Pagasa* Stål from the New World (Harris, 1928), and *Prostemma* Laporte from the Old World.

Specialized morphology. Males of most Nabidae, with the exception of some Nabinae, possess a structure known as Ekblom's Organ, named by Kerzhner (1981) for its discoverer. It consists of two diagonal grooves surrounded by specialized setae, situated behind the posterior foramen of the pygophore (Fig. 56.3A, B, D, E), and a group of specialized setae at the posterodistal margin of the hind tibia (Fig. 56.3C, F) that are rubbed across the pygophoral portion of the organ to distribute attractant pheromones from rectal glands (Carayon, 1970).

The parastigmal pits ("fossettes parastigmatiques") are small depressions containing a concentration of apparently secretory setae (Fig. 56.3G–I), located adjacent to the spiracles on one or more of the ventral laterotergites of abdominal segments 3–7 in many Nabinae and sternum 3 in many Prostemmatinae (Carayon, 1948a, 1950b). Their function is unknown.

Pterygopolymorphism is common in many species, particularly those living at higher latitudes, as for example in *Nabicula propinqua* (Reuter) (Asquith and Lattin, 1990).

Natural history. Lattin (1989) reviewed the bionomics of the Nabidae. Most members of the Nabinae appear to be general predators on small arthropods. They have simple foretibiae and forefemora, which give little indication of their predatory nature. *Arachnocoris* is exceptional; all species live in the webs of spiders, probably scavenging on trapped insects. In contrast, the Prostemmatinae appear to prey exclusively on other Heteroptera (Kerzhner, 1981); some *Alloeorhynchus* species feed on members of the rhyparochromine lygaeid genus *Stilbocoris* Bergroth (Carayon, 1970). The foretibiae and forefemora of all prostemmatines are enlarged and armed with heavy spines, forming a formidable apposable grasping apparatus.

All Nabidae practice vaginal copulation, with fertilization taking place in the mesodermal oviducts near the base of the ovarioles or near the pedicels. Embryonic development takes place after the eggs are laid. Insemination is normal in the Nabinae. In the Prostemmatinae insemination is hemocoelic. Some species inject sperm into the vagina, from where they migrate through the hemocoel toward the ovarioles. Others practice traumatic intravaginal insemination; the vaginal wall is penetrated by the apex of the phallus, and the sperm are injected into the hemocoel or a mesospermalege (see summary in Carayon, 1977).

Distribution and faunistics. Kerzhner's (1981) work (in Russian) on the Palearctic Nabidae is the single most comprehensive volume ever published on the group, bringing together a large amount of previously scattered information. Péricart (1987) documented the fauna of the western Palearctic. The Nearctic and much of the Neotropical nabid fauna was monographed by Harris (1928) with additional information provided in other papers (Harris, 1930a, 1931a, 1939b). Carayon's (1970) work on *Alloeorhynchus* in tropical Africa contains much useful information on the morphology of the Nabidae. New species and distributional information for the fauna of the Philippines and Bismark Archipelago was provided by Kerzhner (1970). The relatively speciose nabine fauna of Hawaii was treated in detail by Zimmerman (1948), although many species remain to be described.

A phylogenetic and biogeographic analysis of *Limno-

Fig. 56.3. Morphology of Nabidae. **A.** Genital capsule and Ekblom's Organ, *Nabis* sp. (Nabinae). **B.** Detail of Ekblom's Organ, *Nabis* sp. **C.** Hind tibia of male *Nabis* sp. showing spines of Ekblom's Organ. **D.** Genital capsule and Ekblom's Organ of *Prostemma* sp. (Prostemmatinae). **E.** Detail of Ekblom's Organ of *Prostemma* sp. **F.** Hind tibiae of male *Prostemma* sp. showing spines of Ekblom's Organ. **G.** Parastigmal pit on anterior margin of abdominal sternum 3, *Alloeorhynchus* sp. (Prostemmatinae). **H.** Parastigmal pit between abdominal sterna 2 and 3, *Prostemma* sp. (Prostemmatinae). **I.** Parastigmal pit on abdominal sternum 7, *Nabis* sp. (Nabinae). Abbreviations: EO, Ekblom's Organ; pm, paramere; psp, parastigmal pit; sp, spiracle.

nabis Kerzhner (Asquith and Lattin, 1990), suggested a late Cretaceous age for the genus, predicting a much older age for the group as a whole.

Cimicoidea

57
Lasiochilidae

General. The lasiochilids have no common name. They are tiny (length 3–4 mm), generally ground-dwelling predators. The habitus (Fig. 57.1A) is very similar to that of Lyctocoridae, Anthocoridae, and some Microphysidae.

Diagnosis. Cephalic trichobothria present; compound eyes large; ocelli present; labium usually long and slender, segment 1 obsolete, segment 3 often much longer than segments 2 and 4; forewing with a distinct, long costal fracture, membrane with a stub (processus corial) located anteriorly near corium-membrane boundary, with 4 or 5 free veins; fossula spongiosa present on at least foretibia; metathoracic scent-gland grooves present and nearly straight, gland with unpaired reservoirs; abdomen with a single pair of dorsal laterotergites on abdominal segments 1 and 2, remaining segments with a simple dorsal plate; sternum 7 of female with punctations on posterior margin (Fig. 57.1C); abdominal spiracle 1 absent, spiracles 2–8 located on simple sternal plate; nymphal dorsal abdominal scent glands located on anterior margin of terga 4–6; male genitalia asymmetrical, left paramere more or less sickle-shaped, not grooved for receiving phallus (Fig. 57.1B); no traumatic insemination; ovipositor laciniate; spermatolytic bodies absent; spermatheca in form of vermiform gland (Fig. 57.1D, E).

Classification. Members of this group have usually been included in a broadly conceived Anthocoridae (e.g., Carayon, 1972a). We accord the group family status, following the phylogenetic work of Schuh and Štys (1991).

The Lasiochilidae includes the widely distributed *Lasiochilus* Reuter, with approximately 50 described species; the small Neotropical genera *Dolichiella* Reuter, *Eusolenophora* Poppius, *Lasiocolpus* Reuter, *Plochiocoris* Champion, and *Whiteiella* Poppius; the monotypic *Iella* Carayon from Mauritius; and the monotypic *Lasiella* Reuter from Southeast Asia, with several other genera being treated as incertae sedis (Carayon, 1972a).

Specialized morphology. The lasiochilids have the fewest novel features of all of the cimicoid families. Although the presence of dorsal laterotergites on only abdominal segments 1 and 2 appears to be unique.

Natural history. Lasiochilids live on the ground, under bark (Kelton, 1978), or on vegetation (Herring, 1967).

In contrast to other cimicoids, insemination in the Lasiochilidae takes place in the vagina of the female; however, typical of all cimicoids, fertilization takes place in the vitellarium, and at least some embryonic development takes place before the eggs are laid.

Distribution and faunistics. Lasiochilids are widely distributed, with the greatest diversity in the tropics and virtual absence from the Palearctic. Herring (1967) keyed the six species of *Lasiochilus* known from the islands of Micronesia.

58
Plokiophilidae

General. The web lovers have an anthocorid-like habitus (Fig. 58.1). They range in length from 1.2 to 3.0 mm.

Diagnosis. Head with a short to rather long necklike extension behind eyes; compound eyes relatively small; ocelli present; cephalic trichobothria long; labium reaching almost to abdomen, segment 1 obsolete, relative lengths of segments (shortest first) 2-3-4; costal fracture well defined (Figs. 58.1, 58.2A); exocorium and/or cuneus covered with corial glands (Figs. 10.10D, 58.2B) (see also Specialized Morphology); membrane with a distinct stub and 1 or 2 free veins or an elongate cell or no veins (Fig. 58.2A); fossula spongiosa on foretibia (Fig. 58.2E); tarsi elongate, 2- or 3-segmented; claws and parempodia possibly always asymmetrically developed (Fig. 10.5E); scent-gland grooves greatly reduced on metapleuron, scent gland with unpaired reservoirs; dorsal laterotergites present; ventral laterotergites fused with sternal plate; abdominal spiracle 1 absent, spiracles 2–8 located on sterna; nymphal dorsal abdominal scent

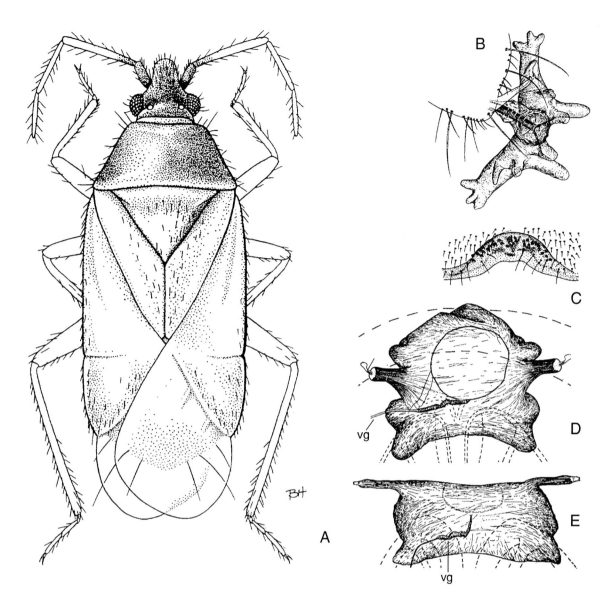

Fig. 57.1. Lasiochilidae. **A.** *Lasiochilus* sp. **B.** Male genitalia, including terminal abdominal segment, left paramere, and inflated aedeagus, *Lasiochilus* sp. **C.** Seventh sternum of female *Lasiochilus* sp., showing punctations. **D.** Bursa copulatrix including vermiform gland, *Whiteiella rostralis* Poppius. **E.** Bursa copulatrix including vermiform gland, *Lasiochilus pallidulus* Reuter (B–E from Carayon, 1972a). Abbreviation: vg, vermiform gland.

glands present on anterior margins of terga 4–6; ovipositor absent; pygophore appendage-like (Fig. 58.2C), capsule often protruding posteriorly and upwardly; parameres simple, elongate, symmetrical or weakly asymmetrical (Fig. 58.2C); distal portion of aedeagus ("processus gonopori" of Štys or "acus" of Carayon [1974]) in form of an elongate sclerotized spine (Fig. 58.2C); spermatheca reduced and nonfunctional; copulatory tubes as in Fig. 58.2D.

Classification. Plokiophilidae were originally treated as aberrant members of the Microphysidae (China and Myers, 1929; China, 1953). Carayon (1961c) raised the group to family rank and recognized two subfamilies on the basis of type of embryonic development. The structure and function of external genitalia and reproductive system make it evident that the group has its closest relationships with the Cimicoidea rather than with the Microphysidae and other Miriformes.

Key to Subfamilies of Plokiophilidae

1. Forefemur without spines .... Plokiophilinae
 – Forefemur with spines (Fig. 58.2E) .... Embiophilinae

EMBIOPHILINAE (FIG. 58.1). Forefemur generally bearing some spines; scutellum broader at base than long (Fig. 58.1); cuneus usually not covered with corial glands; living in embiid webs.

Comprises only the genus *Embiophila* China. China (1953) described *Embiophila myersi* from Trinidad. Carayon later erected the subgenus *E. (Acladina)* for an African species from Katanga. Additional undescribed species are known to occur in India and Malaya.

PLOKIOPHILINAE. Forefemur devoid of spines; scutellum about as long as wide at base; corial glands present on cuneus (Figs. 10.10D, 58.2B); living in spider webs.

This group contains spider-web inquilines in the genera *Plokiophila* China, *Plokiophiloides* Carayon, and *Lipokophila* Štys. The first species to be discovered was *Arachnophila cubana* China and Myers (1929) from Cuba; the preoccupied generic name was later changed to *Plokiophila* by China (1953). Štys (1967a) described *Lipokophila chinai* from southern Brazil, the genus distinctive in the family for its 3-segmented tarsi. The first plokiophilines recorded from the Old World were members of the genus *Plokiophiloides* Carayon from tropical Africa, with one species known from Madagascar (Štys, 1991).

Specialized morphology. The claws of the Plokiophilidae are ordinarily unequally developed, and one or both of the parempodia may be greatly reduced (Fig. 10.5E) (Eberhard et al., 1993). The male and female genitalia are modified for traumatic insemination; the parameres are simple and at most weakly asymmetrical, and the phallus is symmetrical and free (Fig. 58.2C). The females often have modifications of the abdominal dorsum for insertion of the male intromittent organ in the form of paired "copulatory tubes," or, if such are not developed, the phallus punctures the soft dorsal abdominal integument, leaving scars known as "copulatory cicatrices" (Carayon, 1974). The ovipositor, which is well developed in many cimicomorphans, is completely lost, with only a transverse slit present apically on the abdomen.

The exocorium or cuneus or both are rather uniformly covered (either partially or completely) with the conical openings (Fig. 10.10D) of the corial glands, structures unique within the Heteroptera (Carayon, 1974; Eberhard et al., 1993; "tubercular sense organs" of China, 1953). As documented by Carayon (1974), these are connected by a simple duct to large unicellular glands (Fig. 58.2B) of unknown function, which certainly cannot be sensory, as implied by the terminology of China.

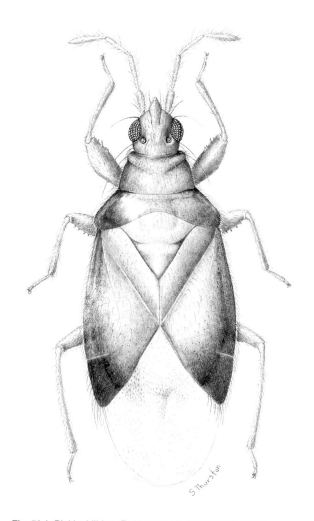

Fig. 58.1. Plokiophilidae. *Embiophila* sp. (from Slater, 1982).

Natural history. All members of the Plokiophilinae are inquilines in the webs of spiders. China and Myers (1929) recorded *Plokiophila cubana* from the webs of *Diplura macrura* (Therophosidae), with adults and several nymphal instars living together on the spider silk. *Lipokophila* spp. from Costa Rica are known from the webs of *Tengella radiata* (Tengellidae) (Eberhard et al., 1993). The bugs eat insects trapped in the spider webs, sometimes feeding on the prey at the same time as the spider. They do not elicit a prey-capture response from their hosts and can walk on the web without becoming entangled. In some cases *Lipokophila* is known to cohabit webs with one or more species of *Arachnocoris* (Nabinae) (Eberhard et al., 1993).

The Embiophilinae are inquilinous scavengers in the

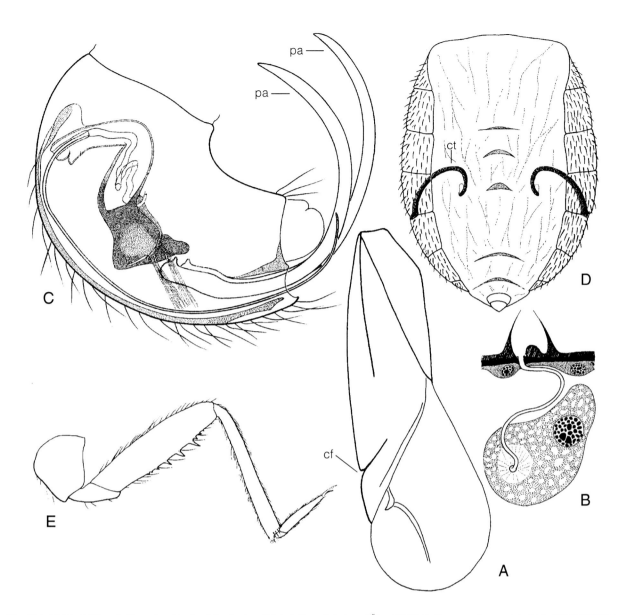

Fig. 58.2. Plokiophilidae. **A.** Forewing, *Lipokophila eberhardi* Schuh (from Schuh and Štys, 1991). **B.** Cross section, corial gland, *Plokiophiloides asolen* Carayon. **C.** Lateral view, genital capsule and aedeagus, *L. chinai* Štys. **D.** Dorsal view, female abdomen, *L. chinai*, showing copulatory tubes. **E.** Middle leg, *Embiophila africana* Carayon, showing femoral spines (B–E from Carayon, 1974). Abbreviations: cf, costal fracture; ct, copulatory tube; pa, paramere.

webs of Embiidina, the web spinners. They feed on the eggs of the embiids as well as dead or dying individuals and possibly mites. Like most spider-web inquilines, embiophilines do not become entangled in the web and do not attract the attention of the host.

Embryonic development in the Embiophilinae begins after the eggs have been laid, whereas in the Plokiophilinae embryonic development occurs in the ovaries, and nearly completely formed first-instar nymphs hatch from the eggs.

Distribution and faunistics. The Plokiophilidae display a distribution pattern commonly found in the tropics, with the New World members occurring in mainland tropical America and the Greater Antilles—in this case Cuba—and the Old World members ranging from tropical Africa to Southeast Asia.

The only comprehensive paper on the plokiophilids is that of Carayon (1974), including many original observations and a key to the species.

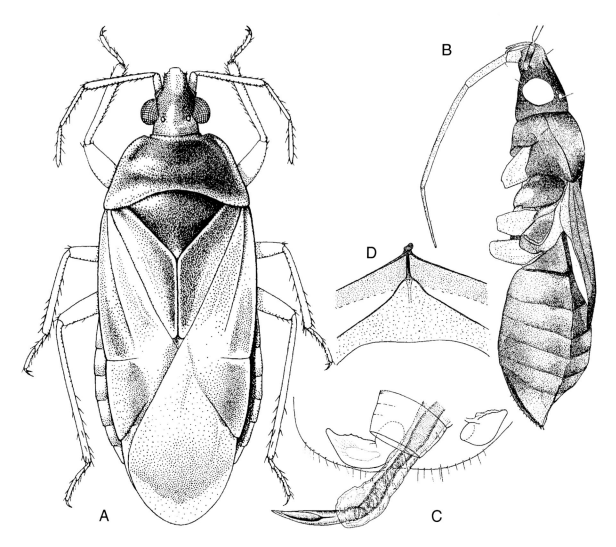

Fig. 59.1. Lyctocoridae. **A.** *Lyctocoris tuberosus* Kelton and Anderson (from Lattin and Stanton, 1992). **B.** *L. menieri* Carayon, lateral view (from Carayon, 1971b). **C.** Male genitalia including terminal abdominal segment, aedeagus, and left paramere, *L. campestris* (from Carayon, 1972a). **D.** Seventh abdominal sternum of female, *L. menieri* (from Carayon, 1971b).

59
Lyctocoridae

General. This is a small group of cimicoids ranging in length from 2.0 to 6.0 mm. Their general appearance (Fig. 59.1A) is like that of most Anthocoridae and Lasiochilidae.

Diagnosis. Head porrect; labium straight, reaching at least to base of abdomen, segment 1 very short, segments 2 and 3 much longer, segment 3 distinctly longer than segments 2 and 4 (Fig. 59.1B); antennal prepedicellite present, short; metathoracic scent-gland grooves present on metapleuron, gland with a single reservoir; fossula spongiosa present on foretibia; forewing with a distinct costal fracture (Fig. 59.1A), membrane with a distinct stub anteriorly near corium-membrane boundary, with 1 or a few free veins; dorsal laterotergites present, ventral laterotergites fused with sternum; abdominal spiracle 1 absent; female with internal apophysis on anterior margin of abdominal sternum 7 (Fig. 59.1D); scent glands in nymphs located on anterior margin of abdominal terga 4–6; male genitalia asymmetrical; parameres subequal in size and weakly asymmetrical; aedeagus not received in groove in left paramere, apex of aedeagus in form of acus (needle of injection) (Fig. 59.1C); ovipositor laciniate; spermatheca absent; spermatolytic syncytial bodies present; traumatic insemination via puncture of intersegmental membrane between terga 6 and 7 on right side of abdomen; fertilization in vitellarium with some embryonic development prior to oviposition; eggs without micropyles.

Classification. Schuh and Štys (1991) restricted the taxon, formerly treated as a subfamily of the Anthocoridae, to contain only the genus *Lyctocoris* Hahn and elevated the group to family status. Twenty-seven *Lyctocoris* species are currently recognized.

Specialized morphology. This group lacks many of the specialized features of the Anthocoridae, Cimicidae, and Polyctenidae. The left paramere does not serve as a copulatory organ, but rather the vesica, modified into an "acus" (needle of injection) (Fig. 59.1C), serves to penetrate the abdominal wall of the female. The female possesses an apophysis internally on the anterior margin of sternum 7 (Fig. 59.1D), and the seminal conceptacles are formed by the epithelium of the genital duct rather than in the peritoneal sheath of the ovariole.

Natural history. Most members of the genus *Lyctocoris* are predatory on other insects and mites in decaying vegetable matter or sometimes under bark (Kelton, 1967). *Lyctocoris nidicola* Wagner in northern and central Europe lives in the nests of birds, and *L. campestris* (Fabricius) inhabits the nests and burrows of small rodents and several times has been reported sucking the blood of humans and domestic animals (Štys and Daniel, 1957). The bionomics of *L. beneficus* (Hiura) as a potential control agent of the rice stem borer in Japan were studied in detail by Chu (1969).

Distribution and faunistics. This group is widely distributed and is particularly speciose in the Palearctic. *Lyctocoris campestris* (Fabricius) is reported as cosmopolitan, probably as a result of transportation in commerce. The western Palearctic fauna was treated by Péricart (1972), and that of the Nearctic by Kelton (1967).

60
Anthocoridae

General. Members of this group are sometimes referred to as flower bugs or minute pirate bugs. They range in length from about 1.4 to 4.5 mm. The habitus varies from typically anthocorid (Figs. 60.1A, 60.2A) to strongly flattened, elongate, and almost enicocephalid-like in some subcortical scolopines.

Diagnosis. Labium straight, of variable length, segment 1 short to rudimentary, segment 3 usually the longest; antennal segments 3 and 4 sometimes distinctly fusiform, prepedicellite present, short; metathoracic scent-gland grooves present on metapleuron, evaporatory area variously shaped, gland with a single reservoir and a single opening; forewing with a distinct costal fracture, membrane with a distinct stub anteriorly near corium-membrane boundary, usually with 4 free veins (Fig. 60.1A); fossula spongiosa usually present on foretibia (Fig. 60.1G), sometimes greatly reduced or absent; dorsal laterotergites present, ventral laterotergites fused with sternum; abdominal spiracle 1 absent; scent glands in nymphs located on anterior margin of abdominal terga 4–6; male genitalia asymmetrical, right paramere greatly reduced or absent, left paramere often sickle-shaped (Fig. 60.1B–E), aedeagus received in groove in left paramere, paramere serving as copulatory organ (Fig. 60.1B–D), except in Anthocorini (Fig. 60.1E); usually with 2 testicular lobes, sometimes 7 (Fig. 60.1F); ovipositor laciniate to greatly reduced; traumatic insemination via puncture of abdominal wall (Fig. 60.2B) or via copulatory tubes (Fig. 60.2C) (see Specialized Morphology); spermatheca absent (Fig. 60.2C), seminal conceptacles formed from hemochrisme or spermatic pocket formed from anterior wall of vagina; fertilization in vitellarium with some embryonic development prior to oviposition; eggs without micropyles.

Classification. The family name Anthocoridae has been applied to several groups of very different composition. Fieber (1861) used the term in a relatively restricted sense, excluding the Microphysidae and Cimicidae, whereas Reuter (1884) included the Microphysidae, but excluded the Cimicidae. Stål (1873) treated the "Anthocoridae" as a subfamily of the Cimicidae, which also included the Schizopteridae. Southwood and Leston (1959) classified all groups of Anthocoridae sensu lato within the Cimicidae. The Anthocoridae of Carayon (1972a) comprised taxa we assign to the Lasiochilidae, Lyctocoridae, and Anthocoridae. We present an infrafamilial classification at the tribal level derived from the work of Carayon (1972a) and Schuh and Štys (1991).

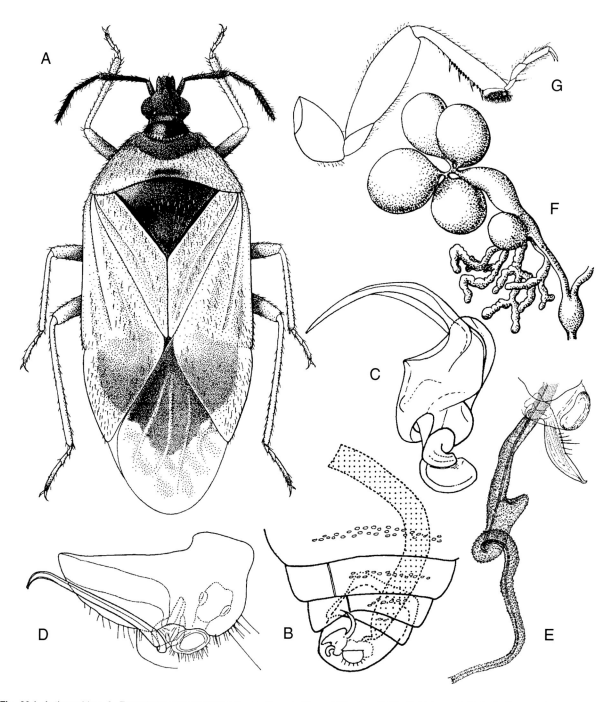

Fig. 60.1. Anthocoridae. **A.** *Tetraphleps latipennis* Van Duzee (from Lattin and Stanton, 1992). **B.** Male genitalia including terminal abdominal segments, left paramere, and ejaculatory bulb (dotted), *Wollastoniella ferruginea* Carayon. **C.** Left paramere, *W. obesula* (Wollaston) (B, C from Carayon, 1958). **D.** Male genitalia, *Xylocoris cacti* Carayon (from Carayon, 1972b). **E.** Aedeagus and left paramere, *Acompocoris pygmaeus* (Fallén) (from Carayon, 1972a). **F.** Testis, *Anthocoris nemorum* (Linnaeus) (from Pendergrast, 1957). **G.** Middle leg, *X. afer* (Reuter), showing tibial spines (from Carayon, 1972b).

Key to Tribes of Anthocoridae

1. Clavus, pronotum posteriorly, and scutellum partially, distinctly, and rather heavily punctate; dorsum covered with long, slender setae .... Almeidiini
– Clavus and pronotum impunctate; dorsum never with long slender setae 2
2. Claws with large pulvilli; fossula spongiosa absent in both sexes; pronotum frequently with distinct macrochaetae laterally .... Oriini
– Claws without pulvilli; fossula spongiosa frequently present in at least males; pronotum with or without macrochaetae laterally 3
3. Hind tibia bearing several heavy spines (Fig. 60.1G); males with large fossula spongiosa on foretibia, smaller on middle tibia .... Xylocorini
– Hind tibia, if with heavy spines, only at apex; fossula spongiosa often present on fore- and middle tibiae, but not of conspicuously unequal size as above 4
4. Males with a unique glandular opening on sternum 4 or 5 (Fig. 60.2D) (occasionally vestigial) .... Scolopini
– Males without glandular opening on sternum 4 or 5 5
5. Labium short, reaching at most to the middle coxae, segment 3 short, no more than one-third the length of segment 4; fossula spongiosa absent or very small on fore- and middle tibiae of males .... Dufouriellini
– Labium usually longer, segment 3 not much shorter than segment 4; fossula spongiosa usually present on the fore- and middle tibiae of males and females 6
6. Macrochaetae present and conspicuous on head and pronotum; antennal segments 3 and 4 filiform, with long slender setae .... Blaptostethini
– Macrochetae absent or indistinct on head and pronotum; antennal segments 3 and 4 generally fusiform, length of setae less than twice diameter of segment .... Anthocorini

ALMEIDINI. Posterior margin of pronotum, part of scutellum, and most of clavus and adjacent corium distinctly punctured; dorsal vestiture long, erect; fossula spongiosa greatly enlarged on foretibia of males, small on middle tibia; ovipositor laciniate.

Included genera are *Almeida* Distant, *Australmeida* Woodward, and *Lippomanus* Distant.

ANTHOCORINI (FIG. 60.1A). Antennal segments 3 and 4 fusiform; left paramere distinctive; endosoma of phallus long, spinule covered, membranous (Fig. 60.1E), sliding into long female copulatory tube; testes as in Fig. 60.1F.

This group, containing 12 genera, is strongly represented in the Holarctic, especially by the large genus *Anthocoris* Fallén with at least 50 described species, but also by other well-known taxa such as *Tetraphleps* Fieber.

BLAPTOSTETHINI. Males with brush of setae on or near posterior margin of left side of abdominal sternum 4; females with pair of copulatory tubes located side by side in membrane between abdominal segments 7 and 8.

This group occurs primarily in the tropics of the Old World, containing the genera *Blaptostethus* Fieber and *Blaptostethoides* Carayon, with a total of about seven described species.

DUFOURIELLINI. Rostrum short, at most reaching middle coxae, segment 3 short, no more than one-third length of segment 4; fossula spongiosa in males greatly reduced or absent on foretibia, absent or vestigial on middle tibia; ovipositor greatly reduced (Carayon, 1972a).

Proposed as Cardiastethini by Carayon (1972a; see Štys, 1975), the Dufouriellini is composed of eight or more genera, with a primarily tropical distribution, the largest being *Cardiastethus* Fieber with approximately 40 species. Some taxa placed in this group by Carayon (1972a)—including all members of the genus *Buchananiella* Reuter and some members of *Brachysteles* Mulsant and Rey and *Cardiastethus* Fieber—possess the "omphalus" (copulatory tube with medioventral opening on sternum 7), but the rest do not, suggesting that the tribe as conceived by Carayon may not be monophyletic.

ORIINI. Claws usually short and straight with large pulvilli; fossula spongiosa absent from all tibiae; foretibia frequently with a row of denticles in males; abdominal sterna 6–8 asymmetrical in males; left paramere of male in form of a spiral, endosoma of aedeagus short (Fig. 60.1B, C); copulatory tube single, short (Fig. 60.2C).

Known to most workers by the cosmopolitan genus *Orius* Wolff, with some 75 described species. Carayon (1972a) placed 15 additional genera in the tribe. Several genera of the Oriini—e.g., *Bilia* Distant, *Bilianella* Carvalho, *Biliola* Carvalho, and *Wollastoniella* Reuter—do

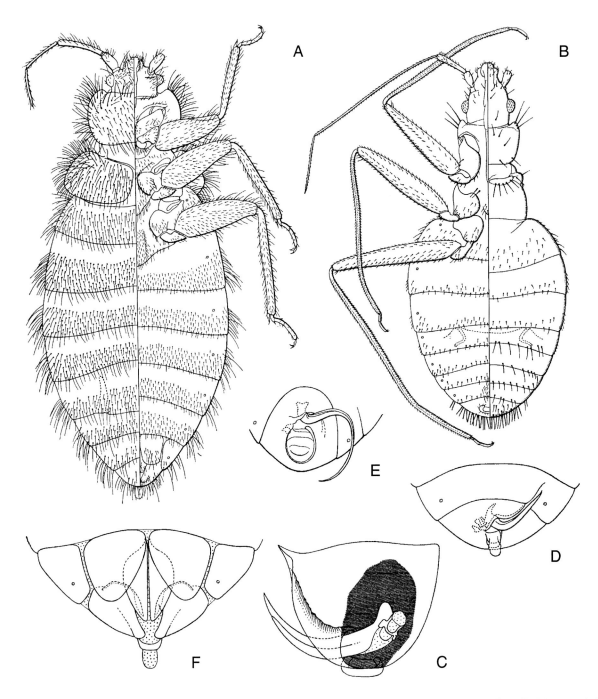

Fig. 61.1. Cimicidae. **A.** *Cacodmus bambusicola* Ueshima (Cacodminae) (on left, dorsal view; on right, ventral view) (from Ueshima, 1968). **B.** *Leptocimex duplicatus* Usinger (Cacodminae) (on left, ventral view; on right, dorsal view). **C.** Male genitalia, *Cimex lectularius* Linnaeus (B, C from Usinger, 1966). **D.** Male genitalia, *Cacodmus bambusicola* (from Ueshima, 1968). **E.** Male genitalia, *L. duplicatus*. **F.** Female abdomen, *Cimex lectularius* (E, F from Usinger, 1966).

Key to Subfamilies of Cimicidae

1. Tibiae mottled; tarsi with several stout spines at inner apex in apposition to claws; labrum over twice as long as wide; body length over 7 mm .... Primicimicinae
- Tibiae not mottled; tarsi without stout spines in apposition to claws; labrum short and broad; body length usually less than 7 mm 2

2. Bristles at sides of pronotum minutely serrate on outer surfaces, or rarely only at obliquely truncate tips; females with paragenital sinus always ventral; metasternum commonly forming a flat plate between coxae .... Cimicinae
– Bristles at sides of pronotum not minutely serrate on outer surfaces or at oblique tips, minutely cleft at tips or acute; females with paragenital sinus usually dorsal, rarely ventral or absent; metasternum a rounded lobe more or less compressed between coxae 3
3. Middle and hind tibiae with short or very long fine bristles, never with additional short, stout, spinelike bristles 4
– Middle and hind tibiae with short, stout, spinelike bristles as well as finer bristles 5
4. Tibiae bent subapically; paragenital sinus dorsal and on females only, rarely lacking .... Cacodminae
– Tibiae straight; paragenital sinus large, left-ventral near base of abdomen in both sexes .... Afrocimicinae
5. Hemelytral pads folded downward at sides; paragenital sinus right-ventral near base of abdomen .... Latrocimicinae
– Hemelytral pads not folded downward at sides; paragenital sinus right-ventral near lateral margin of abdominal segment 6 or 7 .... Haematosiphoninae

AFROCIMICINAE. Left-ventral paragenital sinus large in both males and females.

The only included genus is *Afrocimex* Schouteden (two species; East and Central Africa).

CACODMINAE (FIG. 61.1A, B). Tibiae bent subapically.

Included are six genera comprising 28 species, all Ethiopian and Oriental.

CIMICINAE. Pronotal bristles serrate; metasternum in form of a plate.

Included genera are *Bertilia* Reuter (one species; Chile, Argentina), *Propicimex* Usinger (two species; central and southern South America), *Cimex* Linnaeus (16 species; primarily Holarctic), *Oeciacus* Stål (one species, New World; two species, Old World), and *Paracimex* Kiritshenko (10 species; Orient, New Caledonia).

HAEMATOSIPHONINAE. Paragenital sinus right-ventral on abdominal segment 6 or 7.

The seven included genera contain 10 species widely distributed in the Western Hemisphere from Canada to Argentina.

LATROCIMICINAE. Hemelytral pads downfolded at sides.

Latrocimex Lent is the only included genus (one species; Neotropical region).

PRIMICIMICINAE. Size large; tarsal spines stout.

Included genera are *Primicimex* Barber (one species; Texas to Guatemala) and *Bucimex* Usinger (one species; southern Chile).

Specialized morphology. The complete loss of the hind wings, the rudimentary forewings, and the reduced number of ommatidia in the compound eyes are some of the most obvious characteristics of bed bugs. The abdomen is capable of tremendous expansion. Like other advanced cimicoids, all bed bugs practice traumatic insemination by puncturing the abdomen of the female with the left paramere, which is modified into a copulatory organ (Carayon, 1966), a phenomenon which was first noticed and studied in detail in the easily cultured and observed *Cimex lectularius* Linnaeus and *C. hemipterus*.

Natural history. Bed bugs are temporary blood-feeding ectoparasites. They move onto the host to feed and, when engorged, return to the nest of the host or other secluded location. Most species feed on bats, the remainder on swifts, swallows, and a few other birds. Within the Cimicinae, *Cimex lectularius* feeds on humans and many other hosts, and some other *Cimex* spp. feed on birds. Members of the genus *Oeciacus* feed on birds in the family Hirundinidae, and those of *Paracimex* on birds of the family Apodidae. All members of the Haematosiphonidae feed on birds in several families, with the majority of the species on the Apodidae and Hirundinidae.

The cimicids were at one time some of the most notorious of the true bugs because *C. lectularius* was a regular inhabitant of human dwellings. With the advent of modern insecticides, they have all but disappeared from the view of the general public, or when present in developed countries are considered a disgrace or a mystery. The general natural history and reproductive biology of the Cimicidae are reviewed in detail in Usinger (1966).

Distribution and faunistics. Cimicids are most numerous in the tropics and subtropics, but some species occur in the temperate latitudes, with the presumably most primitive taxa—Primicimicinae—occurring in southern Chile and from Texas to Guatemala. Keys to all described genera and species can be found in Usinger (1966), along with excellent illustrations and bibliography. A more recent bibliography and checklist of the Cimicidae of the Americas and oceanic islands has been published by Ryckman et al. (1981). The fauna of the western Palearctic was treated in detail by Péricart (1972).

Key to subfamilies of Cimicidae modified from Usinger, 1966.

62
Polyctenidae

General. The bat parasites are unique among the Heteroptera in their adaptations—both morphological and reproductive—to a permanent ectoparasitic life. They have the general appearance (Fig. 62.1A) of Diptera Pupipara. They range in length from about 3.0 to 5.0 mm, are rarely encountered, and are unrepresented in most collections.

Diagnosis. Body with some simple setae and some ctenidia, the latter variously placed on antennal segments 1 and 2, gena, posterior margin of head, posterior margin of pronotum, mesonotal lobes, prosternum, and abdominal segment 1 (Fig. 62.1A, C); no compound eyes or ocelli; labrum and clypeus forming large lobelike structures (Fig. 62.1A); antennae 4-segmented, segment 1 and sometimes 2 distinctly enlarged; labium inserted near midpoint of ventral surface of head, 4 segmented, segment 1 sometimes very short (Fig. 62.1A); hypostomal region with modified setae (Fig. 62.1D); pro- and mesothorax lobular as viewed from above; prothoracic sternum grooved (Fig. 62.1E); wings absent; legs short to moderately long; tibiae unevenly sclerotized, annulated, foretibia in males with a brush of slender spines; fossula spongiosa absent; tarsal formula 3-4-4 in adults (Fig. 62.1A), 2-3-3 in nymphs, claws of foreleg often unequally developed and bizarrely modified (Fig. 62.1F), claws of other legs sometimes unequally developed (Fig. 62.1G); condition of metathoracic scent glands unknown; metathoracic pleuron almost totally modified into evaporatorium, scent-gland groove apparently absent; dorsal laterotergites absent, ventral laterotergites present; abdominal spiracle 1 absent, spiracles 2–8 located on ventral laterotergites; abdominal sternum 1 weakly developed in at least some species; dorsal abdominal scent gland present on posterior margin of abdominal terga 4, 5, and 6 in nymphs of some, but not all, taxa; male genitalia grossly asymmetrical, left paramere of male modified into a copulatory organ, right paramere absent; vesica membranous, incorporated into paramere (Fig. 62.1H); ovipositor absent, sternum 8 in form of large plate; fertilization hemocoelic, penetration via right metacoxal membrane.

Classification. The first described species of Polyctenidae were placed in the Nycteribiidae (Diptera) (Giglioli, 1864); the group later was transferred to the Anoplura (Westwood, 1874). Consensus was not reached until Speiser (1904) argued in a well-illustrated paper that the group belonged to the Heteroptera on the basis of morphology and metamorphosis type. A comprehensive revision of the group was published by Ferris and Usinger (1939).

Two subfamilies are currently recognized, following the classification of Maa (1964).

Key to Subfamilies of Polyctenidae

1. Antennal segment 2 flattened, dorsally with a longitudinal, more or less arcuate ctenidial comb on posterior margin; middle and hind tarsi with 3–5 peglike setae on ventral surface of distal segment; claws of middle and hind legs with very small basal projection . . . . Hesperocteninae
– Antennal segment 2 cylindrical, without dorsal comb (Fig. 62.1B); middle and hind tarsi without peglike setae on ventral surface of distal segment (Fig. 62.1B); claws (at least outer ones) of middle and hind tarsi with very distinct basal projection (Fig. 62.1A, G) . . . . Polycteninae

HESPEROCTENINAE. Antennal segment 2 flattened; middle and hind tarsi with 3–5 peglike setae on ventral surface of distal segment; claws of middle and hind legs with a small basal projection.

This subfamily includes the genera *Androctenes* Jordan (three species; Horn of Africa, New Guinea, Cape York Peninsula, Australia), *Hesperoctenes* Kirkaldy (16 species; tropical and subtropical New World), and *Hypoctenes* Jordan (four species; Congo basin, New Guinea).

Key to subfamilies of Polyctenidae modified from Maa, 1964.

POLYCTENINAE (FIG. 62.1A–H). Antennal segment 2 cylindrical; tarsi, unmodified; claws of middle and hind legs with a distinct basal projection.

The subfamily includes the genera *Polyctenes* Giglioli (one species; India, China) and *Eoctenes* (six species; tropical Africa to Solomon Islands).

Specialized morphology. The Polyctenidae possess several novel morphological features, including: absence of compound eyes; presence of ctenidia (Fig. 62.1A, C); tarsal formula 2-3-3 in nymphs, 3-4-4 in adults; and annulate tibiae. They practice traumatic insemination as do

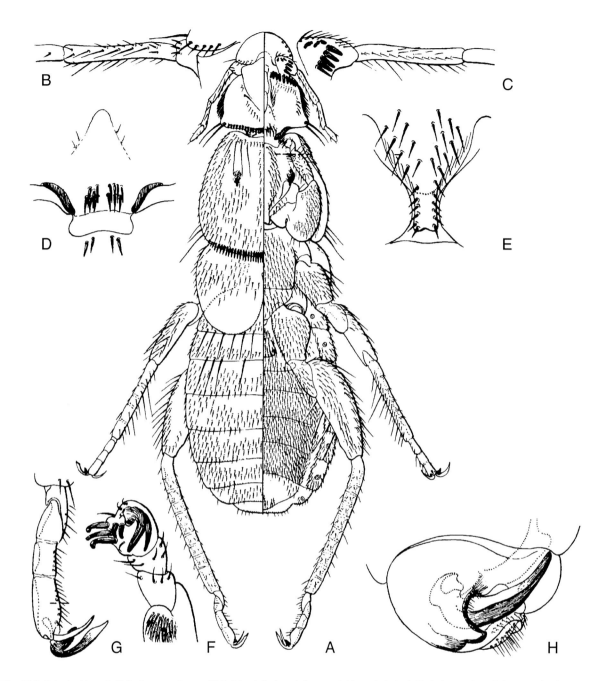

Fig. 62.1. Polyctenidae. **A.** *Polyctenes molossus* Giglioli (on left, dorsal view; on right, ventral view). **B.** Antenna, dorsal view, proximal segments, *P. molossus*. **C.** Antenna, ventral view, proximal segments, *P. molossus*. **D.** Hypostomal region, *P. molossus*. **E.** Prosternal region, *P. molossus*. **F.** Tarsus and pretarsus, foreleg, *P. molossus*. **G.** Tarsus and pretarsus, hind leg, *P. molossus*. **H.** Male genitalia, *Hesperoctenes* sp. (A–H from Ferris and Usinger, 1939).

other higher Cimicoidea, and it is the structures associated with this unusual method of sperm transfer that have helped to establish their phylogenetic affinities.

Natural history. The Polyctenidae are permanent ectoparasites on bats. They are seldom found on bat specimens, suggesting that they are actually rare. But Marshall (1982), in a detailed ecological study of *Eoctenes spasmae* (Waterhouse), found in Malaysia an 85% infestation rate of the host *Megaderma spasmae*, with an average of 13.7 bugs per host. This polyctenid species occurs on all parts of the host body but is particularly common in the hollow formed by the head and shoulder blades, where it is well protected from grooming by the bats.

Detailed study of *Hesperoctenes fumarius* (Westwood)

provides insight into embryonic development and pseudo-placental viviparity in the group (Hagan, 1931). Female polyctenids mate before they become mature. The left paramere—including the aedeagus—of the male is introduced into the right metacoxal membrane, from where the sperm migrate posteriorly. Each female contains a total of 10 embryos in different stages of maturity. The eggs are extruded from the germarium, without a yolk, chorion, or shell, developing as they pass down the oviduct. Nutrition appears to be derived initially from the follicular epithelium, later from a body of nurse cells situated at the posterior end of the developing embryo, and possibly later through the abdominal pleuropodia. The nymphs lie in the vagina with the head pointed anteriorly. After birth they pass through three nymphal instars (Štys and Davidova-Vilimova, 1989).

Polyctenids appear to parasitize only members of the Microchiroptera. Bat families for which credible evidence exists for actual parasitism are Emballonuridae, Nycteridae, Megadermatidae, Rhinolophidae, Hipposideridae, and Molossidae (Maa, 1964).

Distribution and faunistics. The most recent information on the New World fauna comes from the work of Ronderos (1960, 1962) and Ueshima (1972), the latter author including a key to the species of *Hesperoctenes*. Maa (1964) reviewed the Old World fauna. Ryckman and Casidin (1977) prepared a checklist and bibliography, and Ryckman and Sjogren (1980) published a comprehensive catalog.

63
Pentatomomorpha

General. This large grouping consists chiefly of phytophagous species, some of which feed on fungi, many of which feed on seeds, and others on the plant vascular system. It includes some of the largest terrestrial bugs.

Diagnosis. Antennae 3- to 5-segmented, terminal 2 segments usually weakly terete or fusiform (Fig. 68.1A, B), never flagelliform; labium always with 4 obvious segments, segment 1 usually well developed (e.g., Fig. 68.1B); scutellum generally enlarged in Pentatomoidea, sometimes covering entire abdomen (e.g., Fig. 68.1A); forewing in macropterous forms always in form of hemelytron, with anterior coriaceous portion and posterior membranous region; costal fracture never present; venation of membrane sometimes lacking (many Aradidae), when veins present, usually at least 5 in number (e.g., Fig. 83.4), often numerous and in the form of an anastomosing network (e.g., Fig. 70.1); pretarsus always with claws equally developed, usually smoothly curved with well-developed pulvilli (except some Aradidae) subdivided into stalklike basipulvillus and lamellate distipulvillus, parempodia setiform (Fig. 10.5G–I); metathoracic scent-gland channels leading to a peritreme surrounded by an evaporatory area usually elaborated with distinctive mushroom bodies (Fig. 10.10A); except in Aradoidea, abdominal sterna 3–7 laterally with 2 trichobothria (Pentatomoidea, except, e.g., some Podopinae and a few other taxa with 1; Figs. 68.1B, 71.2H, 78.2F, 79.2B) or sterna 3 and 4 laterally or submedially with 2 or more and sterna 5–7 laterally 2 or more (other groups) (Figs. 81.2, 82.2, 83.1A, B), usually short, trichobothria, numbers rarely reduced or trichobothria completely absent; accessory salivary glands tubular; third valvulae absent; spermatheca nearly invariably with an apical storage bulb with flange(s) and associated pumping muscles (Figs. 64.2K, 67.2, 83.1, 87.2G; except Aradidae: Aneurinae), a type consistently occurring elsewhere only in the Leptopodomorpha; midgut usually with 2–4 rows of gastric caeca (Figs. 87.2A, 88.2A; except in Aradoidea); eggs usually with 3 or more micropyles and associated micropylar processes (Fig. 10.14D, E).

Discussion. The current conception of this group was established by Leston et al. (1954), the first authors to consistently ally the Aradoidea with the Trichophora. Almost no work has been done on phylogenetic relationships among the 29 families we recognize in the Pentatomomorpha. Trichobothrial patterns, important in the classification of the group, were analyzed by Schaefer (1975).

Key to Families of Pentatomomorpha

1. Legs not visible from above; body ovoid, scalelike and flattened (Fig. 65.1A); in termite nests .... Termitaphididae
– Legs generally readily visible from above; body of various shapes but not scalelike; usually free-living 2
2. Mandibular plates enlarged, covering or almost covering antennae (Fig. 74.1); compound eyes divided into 2 parts on each side of head; antennae 3-segmented (Fig. 74.2A); large, cryptic, South American bark-dwelling insects .... Phloeidae
– Mandibular plates, if enlarged (e.g., Fig. 70.1), not as above; compound eyes not divided as above; antennae 4- or 5-segmented; occasionally bark-dwelling 3
3. Abdominal venter with 1 or 2 pairs of disc-shaped organs; head, pronotum, and part of costal margin of corium laminately produced ventrally and slightly recurved, producing tortoiselike appearance (Fig. 71.1) .... Lestoniidae
– Abdominal venter lacking disclike organs; costal margin of corium not laminately produced ventrally; if somewhat tortoise-shaped, then forewing elbowed and folded beneath scutellum 4
4. Antennae 5-segmented, pedicel subdivided (e.g., Fig. 66.1) 5
– Antennae 4-segmented 16

5. Length 3.5 mm or less; body elongate; scutellum small, trianguloid, not covering most of abdomen, and remote from posterior end (Fig. 78.1) .... Thaumastellidae (part)
   – Size never less than 4 mm, usually much larger; scutellum frequently enlarged, often covering most of abdomen 6
6. Tarsi 2-segmented 7
   – Tarsi 3-segmented 10
7. Forewings much longer than abdomen, elbowed between membrane and corium (Fig. 75.2B), and folded below scutellum in repose (Fig. 75.1); abdominal venter with a straight black sulcus on either side at level of trichobothria .... Plataspidae (part)
   – Forewings at most only slightly longer than abdomen, not elbowed at juncture of corium and membrane; abdominal sternum lacking a straight, black, transverse sulcus on either side at level of trichobothria 8
8. Scutellum covering entire abdomen; tibiae with heavy black bristles .... Cydnidae (part)
   – Scutellum not covering entire abdomen; if hind tibiae with spines, then these not appearing as heavy black bristles 9
9. Prosternum usually with a large compressed median keel; abdominal segment 8 in males large, exposed .... Acanthosomatidae
   – Prosternum usually lacking a large compressed medial keel; abdominal segment 8 in males much smaller, mostly concealed .... Pentatomidae (part)
10. Tibiae armed with 2 or more rows of heavy black or brown spines (Fig. 69.1A, B, D) ....
    .... Cydnidae (part)
    – Tibiae with setae but without rows of heavy spines 11
11. Scutellum large, convex, covering most of abdomen (Fig. 68.1A) 12
    – Scutellum usually trianguloid, sometimes large, but not covering abdomen (Fig. 79.1) 14
12. Paramere biramous .... Aphylidae
    – Paramere variously shaped but never biramous 13
13. Antennal segment 2 much shorter than segment 1, nearly globose; anterior margin of pronotum meeting lateral margins in a rounded arc; body nearly black, hemispherical (Fig. 68.1A) ....
    .... Canopidae
    – Antennal segment 2 not shorter than segment 1; anterior and lateral margins of pronotum angulate; body not totally black, of varied shape (Fig. 76.1A, B) .... Scutelleridae
14. Spiracles on abdominal segment 2 (first visible segment) fully exposed and well separated from posterior margins of metapleura .... Tessaratomidae (part)
    – Spiracles of abdominal segment 2 completely or almost completely concealed beneath metapleura (except in *Coridius* and *Aspongopus* Laporte) 15
15. Ocelli placed near midline of vertex, often contiguous with one another (Fig. 79.2A); antennae inserted on lateral margin of head .... Urostylidae
    – Ocelli well separated from one another, not placed close to midline of vertex (Fig. 73.1); antennae inserted below lateral margin of head .... Pentatomidae (part)
16. Scutellum large, covering nearly entire corium, reaching or almost reaching end of abdomen (Figs. 72.1A, 75.1) 17
    – Scutellum much smaller, never attaining end of abdomen, never almost covering entire abdomen, sometimes not visible 18
17. Abdominal sterna with a straight transverse sulcus on each side .... Plataspidae (part)
    – Abdominal sterna without a transverse sulcus on each side (Fig. 72.1B) .... Megarididae
18. Clypeus broad, bearing 4–5 distinct toothlike or peglike spines (Fig. 69.1B) (except *Adrisa* Amyot and Serville); tibiae with strong spines .... Cydnidae (part)
    – Clypeus usually without spines, but if spines present never arranged along broadened anterior margin as a row of pegs; tibiae lacking strong spines 19
19. Tarsi 2-segmented 20
    – Tarsi 3-segmented 24
20. Ocelli absent or vestigial 21
    – Ocelli present 23
21. Length 2 mm or less .... Lygaeidae (part)
    – Length at least 4 mm and often much greater 22

22. Forewings elytralike, meeting along midline of body; body and labium not as below .... Lygaeidae (part)
– Forewings variously formed, sometimes almost completely absent, but never meeting along midline of body; body usually rough and often granular and rugose; antennae short and stout; labium usually short, not extending posteriorly to mesocoxae .... Aradidae
23. Length never over 2 mm; forewings never with numerous closed, lacy cells (Fig. 85.1A) .... Lygaeidae (part)
– Sometimes small, but length always greater than 2 mm; forewings frequently composed of numerous closed (often lacy) cells .... Piesmatidae
24. Ocelli absent or vestigial 25
– Ocelli present 27
25. Metathoracic scent-gland auricles and evaporatoria present, well developed .... Lygaeidae (part)
– Metathoracic scent-gland auricles and evaporatoria absent or much reduced 26
26. Pronotum laterally reflexed (Fig. 87.1); abdominal sternum 7 entire in female .... Pyrrhocoridae
– Pronotum not laterally reflexed; abdominal sternum 7 of female split medially .... Largidae
27. Body elongate, slender, at least 8 times as long as maximum pronotal width 28
– Body of various shapes, but if elongate and slender, not more than 5 times as long as maximum pronotal width 30
28. Abdominal spiracles ventral .... Alydidae (part)
– Abdominal spiracles dorsal 29
29. Metathoracic scent-gland auricle produced laterally; distal ends of femora thickened (clubbed) (Fig. 80.1A, B) .... Berytidae (part)
– Metathoracic scent-gland auricle not produced laterally; distal ends of femora not conspicuously thickened (Fig. 181.1) .... Colobathristidae
30. Veins of corium with rows of erect, slightly recurved, spines .... Berytidae (part)
– Veins of corium lacking rows of erect, recurved spines 31
31. Membrane of forewing with at most only 4 or 5 veins (Fig. 84.1B), sometimes corium and membrane not differentiated 32
– Membrane with numerous often anastomosing veins 36
32. Connexiva on abdominal segments 5–7 produced into conspicuous dentate lobes (Fig. 84.1A, D); trichobothria conspicuous on sternum 5 (Fig. 84.1D) .... Malcidae
– Connexiva on abdominal segments 5–7 not prominently produced; trichobothria present or absent 33
33. Female external genitalia platelike (Fig. 78.2D); abdominal sternum 7 of female entire, not divided mesally 34
– Female external genitalia laciniate; abdominal sternum 7 of female divided mesally 35
34. Small, length not over 2–3 mm; macropterous forms with corium divided by a fracture into endo- and exocorium, membrane extending between them along apical corial margin; antennae actually 5-segmented, but articulation between segments 2 and 3 not capable of flexion (Fig. 78.1) .... Thaumastellidae (part)
– Length greater than 3 mm; usually elongate, slender, with base of abdomen constricted; corium often partially hyaline or cellular, but never completely separated into endo- and exocorium; antennae clearly 4-segmented (Fig. 81.1) .... Colobathristidae
35. Abdomen with 7 trichobothria on each side of sterna 3 and 4 (Fig. 82.2B) .... Idiostolidae
– Abdomen with at most 3 or 4 trichobothria on each side of sterna 3 and 4 ...... Lygaeidae (part)
36. Metathoracic scent-gland auricles absent or greatly reduced .... Rhopalidae
– Metathoracic scent-gland auricles large and conspicuous 37
37. Bucculae extending posteriorly beyond bases of antennae (Fig. 90.2B) 38
– Bucculae not extending posteriorly beyond bases of antennae 39
38. Metathoracic scent-gland auricles bristly; ovipositor laciniate; tibiae sulcate; abdominal pore-bearing organs present (Fig. 92.2D) .... Hyocephalidae
– Metathoracic scent-gland auricles never bristly; ovipositor usually flattened and platelike, if laciniate (some Pseudophloeinae) then tibiae not sulcate and abdominal pore-bearing organs lacking .... Coreidae
39. Body shape ovoid-elliptical; 2 trichobothria posterior to spiracles on sterna 3–7 40

- Body more elongate, shape never obviously ovoid-elliptical; 3 trichobothria on sterna 3–7, posterior to spiracles on sterna 5–7 41
40. Membrane of forewing with reticulate venation (Fig. 70.1) .... Dinidoridae
- Membrane of forewing lacking reticulate venation (Fig. 77.1) .... Tessaratomidae (part)
41. Interocular distance greater than width of anterior margin of scutellum (Fig. 88.1B); ovipositor platelike .... Alydidae (part)
- Interocular distance less than width of anterior margin of scutellum (Fig. 92.1A); ovipositor laciniate .... Stenocephalidae

Aradoidea

64
Aradidae

General. This is a large and remarkable family whose members are generally referred to as flat bugs or bark bugs. Most species are extremely flattened dorsoventrally, of elliptical, oval, or rectangular shape, and are black or brown, with stout antennae. They range from 3 to 11 mm in length. Many tropical species are wingless, the dorsal surface frequently appearing granular or rugose (Fig. 64.1B). The family shows variability in many structures that are relatively constant in other pentatomomorphan families.

Diagnosis. Mandibular and maxillary stylets extremely elongate and coiled within head (Figs. 10.2A, 64.2B), a feature shared with Termitaphididae; ocelli absent; labium usually short and stout with 4 distinct segments (Fig. 64.2B); trochanters sometimes fused with femora; tarsi 2-segmented; nymphal abdominal scent glands present between terga 3/4, 4/5, and 5/6 or between 4/5 and 5/6, or less commonly 3/4; all spiracles usually ventral (but see Chinamyersinae); aedeagus as in Fig. 64.2D, F, G; parameres as in Fig. 64.2E, H; testes as in Fig. 64.2I; abdominal trichobothria absent; ovipositor laciniate, abdominal sternum 7 split mesally, except in Aneurinae; spermatheca variable, usually with 2 pump flanges and differentiated bulb (Fig. 64.2K), flanges or entire spermatheca sometimes absent; copulatory method unique, male lying below female; males with 1 or 2 pairs of parandria developed as outgrowths of pygophore and gripping female during copulation (see Leston, 1955c).

Classification. The Aradidae are placed in the Pentatomomorpha on the basis of the following characters: phallus with distinct, sclerotized phallotheca and endosoma consisting of conjunctiva and vesica (Fig. 64.2D); pulvilli, when present, similar in structure to those found in Trichophora (Fig. 10.5G), but absent in Aradinae (Vasarhelyi, 1986); salivary glands tubular; eggs with micropylar processes; spermatheca with bulb and pump flange(s) (Fig. 64.2K). Unlike other pentatomomorphans, the Aradidae lack trichobothria. Most authors have treated the Aradidae as the sister group of the Termitaphididae, on the basis of the long, coiled mandibular and maxillary stylets. Certain groups here recognized as subfamilies of Aradidae (e.g., Aneurinae, Mezirinae) have been accorded family status by some authors.

The genus *Aradus* was established by Fabricius (1803). The group was first recognized as a higher taxon by Brullé as "Aradiens," including members now placed in other families. Spinola (1837) recognized "Aradites" as a family. Amyot and Serville (1843) treated the group as the tribe "Corticolae" with two subgroups the "Brachyrhynquides" and "Aradides." Stål (1873) established the basic higher classification in which he recognized three subfamilies—Aradina, Brachyrhynchina, and Isodermina—and described 24 genera. Subsequent work by Bergroth, Kiritshenko, Parshley, and many others was largely descriptive until Usinger and Matsuda (1959) produced the fundamental monograph of the family. The work of Kormilev and Froeschner (1987) provided a helpful introduction to the systematic literature on the group.

Miller (1938) recognized that the apterous tropical flat bugs that had previously been considered nymphs were in fact wingless adults. This was a revolutionary advance in understanding morphology and taxonomy in the group.

Vasarhelyi (1987) and Grozeva and Kerzhner (1992) discussed recent contributions to understanding of relationships, and each provided a cladogram of subfamily relationships. Jacobs (1986) provided a detailed morphological discussion, based on African *Aneurus* Curtis, with extensive use of scanning electron microscopy.

At least 211 genera containing about 1800 species are currently placed in eight subfamilies.

 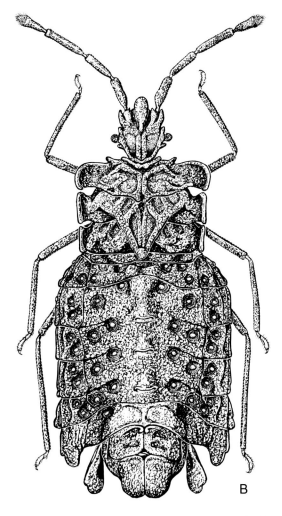

Fig. 64.1. Aradidae. **A.** *Isodermus planus* Erichson (Isoderminae) (drawn by T. Nolan; from CSIRO, 1991). **B.** *Kumaressa scutellata* Monteith (Chinamyersiinae) (drawn by S. Monteith; from CSIRO, 1991).

Key to Subfamilies of Aradidae

1. Mandibular plates produced anteriorly to exceed apex of clypeus, forming a cleft or emarginate anterior margin of head (Fig. 64.2A); first dorsal abdominal scent-gland orifice (or scar thereof) large, usually strongly displaced posteriorly, second orifice rarely well developed, third always obsolete or undifferentiated 2
– Mandibular plates not produced anteriorly to exceed apex of clypeus (Fig. 64.1B), head not appearing cleft or emarginate anteriorly; 3 dorsal abdominal scent-gland orifices (or scars thereof), equal in size, not displaced posteriorly 4
2. Labium arising from open area of bucculae ("open atrium"); anterior dorsal abdominal scent-gland orifice, or scar, not or only slightly displaced posteriorly .... Aneurinae
– Labium arising from nearly closed area of bucculae ("closed atrium"); opening of first dorsal abdominal scent-gland orifice (or scar) displaced posteriorly to middle or posterior margin of segment 4 3
3. Metathoracic scent-gland orifices with a well-developed, usually channel-like, evaporatory area extending to lateral margin of thoracic metapleuron; body usually not incrustate above .... Mezirinae

Key to subfamilies of Aradidae adapted from Usinger and Matsuda, 1959.

- Metathoracic scent-gland orifices lacking a well-developed channel extending to lateral margin of metapleuron; body frequently covered with a heavy incrustation obscuring dorsal surface .... Carventinae
4. Labium arising near apex of clypeus; base of labium exposed; forewings with line of weakness at level of apex of scutellum and often broken off at this level (Fig. 64.1A) .... Isoderminae
– Labium arising well posterior to apex of clypeus; base of labium bordered by well-developed bucculae; wings lacking a line of weakness at level of apex of scutellum 5
5. Metathoracic scent-gland orifices conspicuously present 6
– Metathoracic scent-gland orifices absent 7
6. Metathoracic scent-gland auricle with a long curved seta; mandibular plates well developed, extending on either side of clypeus for nearly half its length .... Prosympiestinae
– Metathoracic scent-gland auricle with a raised ridge, curved at apex; mandibular plates short .... Chinamyersiinae
7. Scutellum greatly enlarged, covering all but narrow lateral margins of abdominal dorsum .... Calisiinae
– Scutellum much smaller, usually triangular, never covering all but lateral areas of abdomen .... Aradinae

ANEURINAE. Body usually elongate oval, often with lateral margins subparallel, either granular or smooth, usually wrinkled and polished; clypeus relatively broad, genae well developed, surpassing apex of clypeus; labium short and broad, arising well behind apex of head from open atrium; scutellum broad, rounded behind; hemelytra usually complete, when present, sclerotized only at extreme base; hind wings reduced even in forms with well-developed forewings; metapleuron without visible scent-gland opening or evaporatory area; trochanters distinct; pulvilli present; abdominal terga 1 and 2 partially to completely fused; abdominal segment 8 in females forming a distinct transverse plate dorsally with short, lateral, spiracle-bearing lobes; venter extremely flattened, platelike; sternum 7 of female unique among Aradidae with no median longitudinal split and nearly straight posterior margin; first valvula in *Aneurus* elongate, directly connected with anteromesal angle of first valvifer, second valvula attached to anteromesal angle of second valvifer.

Seven genera are currently recognized. *Aneurus* contains 95 species and is known from all major zoogeographic regions. *Aneuraptera cimiciformis* Usinger and Matsuda is a monotypic New Zealand isolate, with small flaplike wings and a broadly ovate body.

ARADINAE. Body generally granular, without pubescence; clypeus bulbous, mandibular plates inconspicuous; labium arising well behind apex of head; bucculae well developed; hemelytra usually well developed, occasionally reduced; metapleuron lacking evaporative areas or channel but with a very small hole before hind coxae; trochanters partially fused to femora; claws lacking pulvilli; abdominal terga 1 and 2 fused; ventral surface of abdomen with a median groove.

Four genera are recognized: *Aradus* Fabricius, with nearly 200 described species, the majority occurring in the Holarctic; *Aradiolus* Kormilev, with two Mexican species; *Miraradus* Vasarhelyi, containing four Oriental species; and *Quilnus* Stål, with 10 Palearctic and Nearctic species.

CALISIINAE. Body with coarse granules or blunt spines; clypeus bulbous, with mandibular plates encroaching on either side; labium arising well behind apex of head with bucculae well-developed; rostral atrium open in front, narrowed posteriorly; labium not reaching base of head; scutellum coarsely punctate, greatly enlarged, covering most of abdominal disc and all of hemelytra except thickened costal margins, broadly elevated at base with a distinct longitudinal carina at middle; clavus reduced to narrow sclerotized area along its inner margin; metapleuron lacking a distinct scent-gland opening; legs short and stout; trochanters distinct; pulvilli present; abdominal terga 1 and 2 distinct; abdominal mediosternites separated from laterotergites by sutures and elevated carinae.

Six genera and approximately 100 species are recognized: *Calisiopsis* Champion (three Neotropical species), *Paracalisiopsis* Kormilev (one African species), *Aradacanthia* Costa (three species from the Orient and New Guinea), *Heissia* Kormilev (two African species), *Paracalisius* Kormilev (one African species), and *Calisius* Stål (approximately 90 Palearctic, southern Nearctic, Neotropical, Ethiopian, Australian, Polynesian, and Micronesian species).

CARVENTINAE. Widespread aptery and body fusion resulting in bizarre shape and form; body surface usually incrustate; clypeus usually well developed, mandibular plates short; labium short, constricted subbasally, usually arising well behind apex of head, emerging from closed atrium through longitudinal slit; hemelytra, when present, with very short corium, thickened, laterally dilated at base; metapleuron without a conspicuous scent-gland

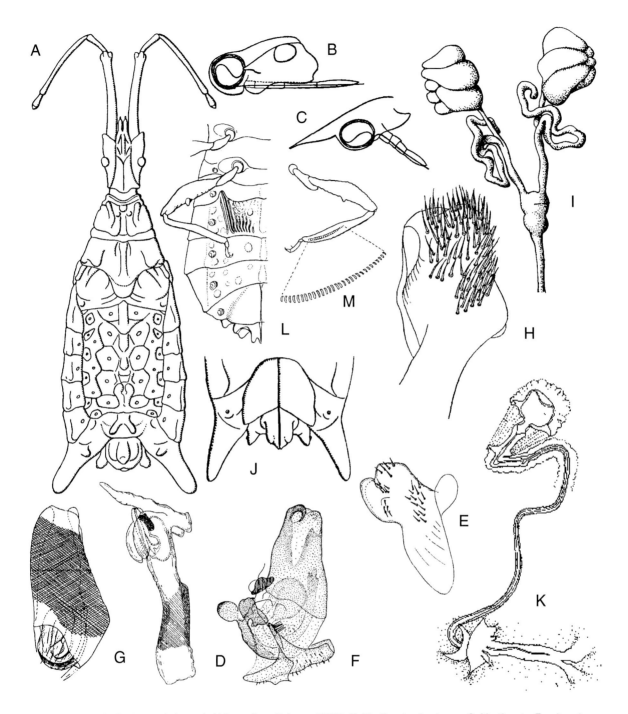

Fig. 64.2. Aradidae. **A.** *Chelonocoris javensis* Usinger (from Usinger, 1954b). **B.** Mouthparts, *Aradus* sp. **C.** Mouthparts, *Termitaradus panamensis* Myers (B, C from China, 1931). **D.** Aedeagus, lateral view, *Glyptocoris fluminensis* Wygodzinsky. **E.** Paramere, *G. fluminensis* (D, E from Wygodzinsky, 1948b). **F.** Aedeagus, lateral view, *Mezira tremulae* (Germar) (from Vasarhelyi, 1982). **G.** Aedeagus, lateral view, *Asterocoris australis* Drake and Harris. **H.** Paramere, *A. australis* Drake and Harris (G, H from Wygodzinsky, 1948b). **I.** Testes, *Aneurus laevis* (Fabricius) (from Pendergrast, 1957). **J.** Female terminal abdominal segments, *Chelonocoris ferrugineus* Usinger (from Usinger, 1954b). **K.** Spermatheca, *Aradus cinnamomeus* (Panzer) (from Pendergrast, 1957). **L.** Stridulatory structures, *Artabanus lativentris* Esaki and Matsuda. **M.** Stridulatory structures, *A. lativentris* (L, M from Usinger, 1954a).

opening or channel; legs usually long and slender; trochanters distinct; pulvilli well developed; abdomen usually with terga 1 and 2 fused.

This primarily tropical group contains approximately 60 genera, only three of which are known in the macropterous condition.

CHINAMYERSIINAE (FIG. 64.1B). Body more or less granular, broad, subflattened; labium arising well behind apex of clypeus, basally with well-developed bucculae on either side; mandibular plates not evident, not encroaching on clypeus; eyes pedunculate or stylate with preocular spines or tubercles; hemelytra in macropterous forms with well-differentiated clavus, corium, and membrane, last with anastomosing veins forming cells, apterous forms known; prosternum platelike between forecoxae; metathoracic scent-gland openings in form of pit, each opening with a prominent pale ridge emerging and curved at tip; trochanters distinct; femora moderately enlarged, lacking spines; tarsi 2-segmented; claws with well developed pulvilli; abdominal terga 1 and 2 separate; at least some abdominal spiracles dorsal, unique condition within Aradidae.

Monteith (1980) believed this to be an ancient lineage comprising two groups that were recognized as tribes by Kormilev and Froeschner (1987). The Chinamyersiini, containing *Gnostocoris* Kormilev (one species from New Hebrides and New Caledonia) and *Chinamyersia* Usinger (two species from New Zealand), are macropterous, have unique "figure-eight" coiling of the stylets in a hump-backed clypeus, a rather conventional spermatheca with an apical bulb and at least a distal pump flange, lack a vermiculate pattern of ridges on the internal evaporative area of the abdominal scent gland (Monteith, 1980), and have abdominal spiracles 3, 4, and 5 dorsal.

The Tretocorini, containing *Tretocoris* Usinger and Matsuda (one species; New Zealand) and *Kumaressa* Monteith (three species; eastern Australia), are apterous, have apparently 3-segmented tarsi (unique in the Aradidae), "clockwise" stylets with about 4 or 5 turns, a unique spermathecal condition in *Kumaressa* with an apical, thick-walled, pyriform chamber that leads to a thick-walled contorted neckpiece without pump flanges, an adult scent-gland internal evaporatory area with a vermiculate pattern of ridges (Monteith, 1980), and dorsal abdominal spiracles.

ISODERMINAE (FIG. 64.1A). Body polished; labium usually entirely free, exposed at base, arising at or just ventrad of apex of head (sometimes arising caudad of apex of head with base enclosed by small bucculae); mandibular plates well developed, extending on either side of clypeus for about half its length; hemelytra fully developed with a line of weakness at level of apex of scutellum, membrane deciduous; hemelytra and hind wing with a single vein; metapleural scent-gland channel small but distinct, reaching to lateral margin of metapleuron; prosternum forming an elevated plate between forecoxae; trochanters not fused to femora, latter inflated and spurred; pulvilli well developed; abdominal terga 1 and 2 distinct; nymphs with dorsal abdominal scent-gland openings prominent, equally developed; spiracles ventral, sublateral.

Only a single genus, *Isodermus* Erichson, is known, with one species in Chile, three in New Zealand, and two in Australia and Tasmania.

MEZIRINAE (FIG. 64.2A). Body granular, punctate; labium arising well behind apex of clypeus from atrium, which is usually closed except for longitudinal slitlike opening; mandibular plates often surpassing clypeus to form a cleft or emarginate apex; aptery common but macropterous forms also widespread; hemelytra, when developed, with 2 prominent longitudinal veins in corium, latter well developed usually reaching to level of apex of scutellum; metapleural scent-gland channel distinct behind middle acetabulum, reaching lateral margin (a definitive method of distinguishing members of the subfamily from species of Carventinae); trochanters usually distinct; abdominal terga 1 and 2 usually distinct.

This is the largest subfamily of Aradidae, containing at least 119 genera and 308 species. It is known from all major zoogeographic regions, but most taxa are tropical or subtropical and the majority of the apterous genera are markedly restricted in distribution. As in the Carventinae, aptery has apparently arisen independently a number of times. Stridulatory mechanisms occur and appear to have arisen independently in several phyletic lines. Kormilev (1971a) monographed the fauna of the Indo-West Pacific.

PROSYMPIESTINAE. Body either polished or dull; robust rather than extremely flattened; sometimes rather lygaeid-like in general appearance; clypeus projecting beyond mandibular plates; labium arising well behind apex of head with first segment partially enclosed by bucculae at base, remaining segments not in a rostral groove; hemelytra in macropterous forms lacking a line of weakness; a single submarginal vein on corium and part of membrane; prosternum not elevated into a broad plate between forecoxae; metapleural scent-gland openings with conspicuous deep pits, each with a stiff curved bristle arising from middle; evaporative area dull and granular; trochanters not fused with femora; pulvilli distinct; abdominal terga 1 and 2 distinct in macropterous forms, more or less fused in brachypterous individuals.

Kormilev and Froeschner (1987) recognized two tribes: the Llaimocorini for the monotypic *Llaimocoris penai* Kormilev from Chile and the Prosympiestini with three included genera—*Adenocoris* Usinger and Matsuda (two species from New Zealand), *Neadenocoris* Usinger and Matsuda (six species from New Zealand), and *Prosympi-*

estus Bergroth (four species from Australia and Tasmania).

Specialized morphology. In addition to the greatly elongated maxillary and mandibular stylets that are coiled within the head (Figs. 10.2A, 64.2B), the most conspicuous morphological modifications in the Aradidae involve wing loss. Wingless forms often have various body parts grotesquely modified (Figs. 64.1B, 64.2A) (Usinger and Matsuda, 1959) and may also be covered with coarse incrustations that almost completely obscure the body texture and shape in a way unparalleled elsewhere in the Heteroptera. Until recently none of these extremely modified tropical apterous forms had been known to have a flying morph (see Usinger, 1950). Monteith (1969) demonstrated that New Guinea forms that had been considered separate genera are actually macropterous and apterous morphs of the same species, noting that this situation may be true of other taxa. Wing reduction occurs in over 50% of all known genera and in seven of the eight subfamilies. In the Aradinae the membrane of the forewing is only rarely lost, and the wings are never shorter than the scutellum. Wing reduction is sexually dimorphic, the female usually being brachypterous and the male macropterous, although occasionally the female exists in both forms and rarely the male is also dimorphic.

In the essentially tropical Mezirinae (Fig. 64.2A) and Carventinae, wing reduction is almost invariably extreme and usually no trace of wings remains. It is in these two subfamilies that most of the other bizarre changes in thoracic and abdominal shape occur. Monteith (1969) noted that 60 of the nearly 120 genera of Mezirinae contain only macropterous species, 12 are all brachypterous or micropterous, and 33 are completely apterous.

In the genus *Isodermus* Erichson (Fig. 64.1A) both males and females have a line of weakness near the base of the wing which enables them to "shed" their wings. This character is apparently found elsewhere in the Heteroptera only in the Gerromorpha.

Stridulatory structures have developed several times (Fig. 64.2L, M). Their occurrence was summarized by Usinger (1954a).

Natural history. Many flat bugs are mycophagous, as suggested by the fact that the majority of species live either under the bark of decaying trees or under bark chips, twigs, or debris on the floor of moist forests. Macropterous species tend to live on a restricted number of host trees, a list of which was provided by Usinger and Matsuda (1959). *Aradus cinnamomeus* Panzer, however, feeds on the phloem, cambium, and xylem of living *Pinus* and *Larix* spp. and causes stunting of the growth of these trees (Strawinski, 1925, and numerous later studies). Aneurinae and Calisiinae—and probably some other groups—feed on the sap of dying or living trees.

Flat bugs sometimes live with termites, as for example the West Australian *Aspisocoris termitophilus* Kormilev, a cylindrical species with reduced eyes. Five other Australian species also occupy this habitat (Kormilev and Froeschner, 1987). *Mezira reducta* Van Duzee lives in the nests of *Zootermopsis nevadensis* Hagen in western North America (Usinger, 1936).

Aradids have also been reported living in the nests of birds and rodents and in the galleries of wood-boring beetles. Presumably in many of these habitats they are feeding on fungi.

Macropterous Aradinae almost invariably are associated with fungal mycelia in and on dead wood. They live subcortically, feeding on the fungus that grows in the moist environment beneath the lifting bark of recently dead logs and trees. The comparatively short life of this food source suggests the need for a dispersal flight to find and colonize a new wood source in a suitable state of decay.

Flightless Mezirinae and Carventinae are inhabitants of the wet rainforest floor, where they have a luxuriant growth of mycelia on dead logs and branches. This habitat provides a relatively constant microclimate at the leaf-litter boundary, offering an almost continuous supply of fungi, and the loss of a need for a dispersal flight to find a new food source. Monteith (1969) believed that camouflage is also associated with wing loss in these tropical aradids. He noted that modifications of structure and general shape, plus a rough surface that holds layers of dirt and debris, are important antipredator survival adaptations in these litter-dwelling species.

McClure (1932) reported prenatal care by *Neuroctenus pseudonymus* Bergroth. He stated that the female leaves the egg mass, and another individual, presumably the male, crawls onto the eggs and remains for two weeks or until the eggs hatch and even remains with the young nymphs for one or two days.

Distribution and faunistics. Aradids are represented in all major faunal regions, with five of the eight subfamilies being essentially cosmopolitan, but with three of the eight being restricted to the Australia–New Zealand–South American arc. Nearly half (785) of all aradid species occur in the Oriental-Pacific area, with over 200 being known from the New Guinea mainland alone (Kormilev and Froeschner, 1987). Degrees of generic and specific endemicity according to Monteith (1982a) vary considerably throughout the Orient-Pacific and reach their peaks in Australia, New Zealand, New Caledonia, and mainland Asia. The distribution suggests a very ancient group of which a careful cladistic study of the relationships would be most rewarding.

Picchi (1977) revised *Aneurus* for most of the Western Hemisphere, and Jacobs (1986) for Africa. Putchkov

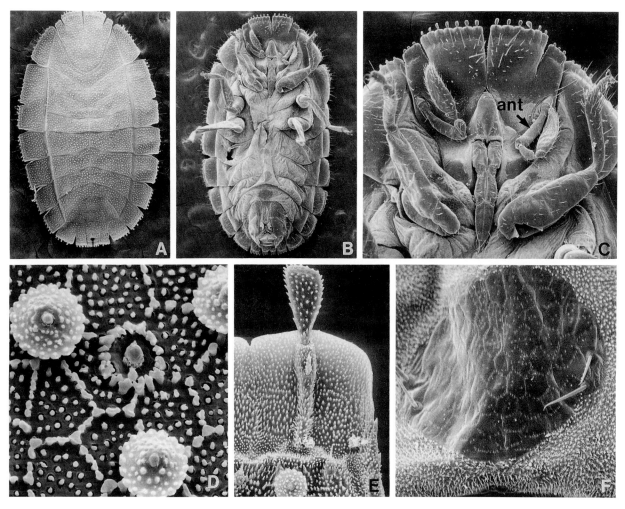

Fig. 65.1. Termitaphididae. *Termitaradus guianae*. **A.** Dorsal view. **B.** Ventral view. **C.** Ventral view, head and prothorax. **D.** Detail of dorsal surface. **E.** Detail of dorsal surface of marginal setae. **F.** Detail of medial area of abdominal sternum 5. Abbreviation: ant, antenna.

(1974) provided keys to the aradid fauna of the Ukraine. Kormilev has produced most of the descriptive work on the world fauna subsequent to the Usinger and Matsuda (1959) monograph. (See bibliography in Kormilev and Froeschner, 1987.)

65
Termitaphididae

General. These curious, scalelike insects range in length from 2 to 3 mm. The body is strongly flattened, ovoid or elliptical, with the lateral lamellae fringed with modified setae (Fig. 65.1A). They have no common name.

Diagnosis. Mandibular and maxillary stylets elongate, coiled within head (Fig. 64.2C); eyeless; no ocelli; wingless; legs small, not visible from above; body covered with small globular nodule-like setae (Fig. 65.1D); antennae geniculate (Fig. 65.1C); males usually with 12 marginal laminae (Fig. 65.1A, B), females with 13 (meso- and metathoracic lobes fused in males); parempodia with pulvillus as in other Pentatomomorpha (Fig. 10.5H; Myers, 1924); no ovipositor; no nymphal dorsal abdominal scent-gland openings; abdominal sternum with some glabrous, polished areas (Fig. 65.1F).

Classification. Termitaphidids are placed in a superfamily with the Aradidae, primarily because of the extremely elongate stylets that are coiled within the head (Fig. 64.2C) (Myers, 1924). Two genera are recognized: *Termitaphis* Wasmann (one Neotropical species) and *Termitaradus* Myers with eight species (five Neotropical; one

Australian; one African; one Oriental) (Usinger, 1942a). A fossil species from Mexican Chiapas amber, dated as being approximately 25 million years old, was described by Poinar and Doyen (1992).

The taxon was first recognized as a higher group by Silvestri (1911) under the name Termitocoridae. Before that time only a single species was known. It had been described by Wasmann (1902) and placed in the Homoptera as an aberrant aphid. Myers (1924) noted that Silvestri's family name was not based upon a genus and proposed Termitaphididae as a replacement. He also discussed the relationships of the group.

Specialized morphology. Distinctive features include the loss of wings, eyes, and ovipositor, the flattened and enlarged marginal laminae, the geniculate antennae (contra Poinar and Doyen, 1992), the short legs, and the peculiar nodule-like setae (Fig. 65.1A–E).

Natural history. These extremely modified insects are known to live only in the nests of termites as do some members of the Aradidae. They lay their eggs among those of the host. The similarity in many features to the Aradidae suggests that all species will prove to be mycophagous.

Myers (1932:371) gave a charming account of *Termitaradus jamaicensis* Myers in a colony of *Heterotermes convexinotatus* under white limestones that lay on leaves of a dry shrub forest floor in the Liguanea Hills of Jamaica:

> When the stones were turned over the *Termitaradus* were as active and nearly as quick as their hosts in running from the light. Their motion over the extremely rough under surface of the stone, covered here and there with irregular carton galleries and concretions of the same packed with termites, was delightful to watch. The exceedingly flexible edges of the marginal lobes underwent a wave-like motion, rippling as it were on the ground, and always completely hiding the legs. The only visible appendages were the outstretched antennae, and these were usually at once withdrawn when, as often happened, a termite ran over the insect. At such a time the *Termitaradus* squatted and sank, so that every part of its periphery fitted the irregularities of the substratum, just like a *Chiton*, and in proportion to its size just as difficult to remove with brush or forceps without such force as would injure it. At first attempts, and always when termites were running over it, it always squatted closer, visibly straining the marginal lobes against the ground, but after being chivvied some time with a brush it usually ended by moving. When turned over on its back it arched the side-plates in the form of a half cylinder to right itself, the legs being quite unable to reach ground.

Distribution and faunistics. Termitaphidids are tropical and subtropical and found in the Neotropical, Ethiopian, Oriental, and Australian regions. The basic references on the group are those of Myers (1924) and Usinger (1942a). The latter gives a key to genera and species.

Pentatomoidea

66
Acanthosomatidae

General. Pentatomoidea generally have a more elongate ovoid to kite-shaped body than many members of the superfamily. The scutellum is not broadened and does not cover the corium, and the tarsi are 2-segmented. They do not have a common name and range in length from 6 to 18 mm.

Diagnosis. Head keeled laterally; antennae 5-segmented; antenniferous tubercles not visible from above; scutellum triangular, usually about half or less than half as long as abdomen, with frena extending along at least basal 7/10 of lateral margins; mesosternum with a strongly projecting keel-like median carina; abdominal sternum 3 usually with long, spinelike, anteriorly projecting process; spiracle on abdominal segment 2 concealed; paired trichobothria on sterna 3–7 transverse, mesad of spiracular line; nymphs with dorsal abdominal scent-gland openings between terga 3/4, 4/5, and 5/6, the anterior opening sometimes small; male abdominal segment 8 large and exposed; pygophore as in Fig. 66.2A; phallus as in Fig. 66.2B; paramere as in Fig. 66.2C; Pendergrast's organs usually present in females, these depressed, round or elongate, located laterally on sterna 5–7 or sternum 7 only; females with posterior margin of sternum 7 deeply emarginate; spermatheca as in Fig. 66.2D, E.

Classification. This group has often been considered a subfamily or a tribe of an inclusive Pentatomidae. It was first recognized as a higher taxon by Signoret (1863) as the Acanthosomites, and he also recognized the Ditomotarsites as having equal status. Stål (1864–1865) and Breddin (1897) considered the group to be a subfamily of Pentatomidae. Mulsant and Rey (1866) treated it as a family (the Acanthosomiens). Kirkaldy (1909), however, gave the taxon only tribal status. Recent authors such as Leston (1953b), Southwood (1956), Miyamoto (1961b),

Cobben (1968a), Gapud (1991), and especially Kumar (1974a)—who revised the world fauna—all treated it as a distinct family. The group comprises 45 genera and 180 species.

Key to Subfamilies of Acanthosomatidae

1. Abdominal spine absent; caudolateral angles of sternum 7 never produced into processes .... Ditomotarsinae
– Abdominal spine usually present, when absent either caudolateral angles of sternum 7 produced into processes or lateral margins of pronotum thin 2
2. Sternal carina usually absent, but if present only as a raised wedge at junction of pro- and mesosterna (may extend slightly forward and backward); when both pro- and mesosternal carinae present then invariably poorly developed and never continuous (in such cases abdominal spine very well developed and sometimes reaching anterior end of prosternal carina) .... Blaudusinae
– Sternal carina usually well developed and receiving at its posterior end the generally distally concave abdominal spine, latter closely applied to sternal carina on left-hand side (or extending over it, or completely fused with it) .... Acanthosomatinae

Acanthosomatinae (Fig. 66.1). Usually recognizable by characters given in preceding key, although some taxa do not have all diagnostic features (see Kumar, 1974a).

Twelve genera are currently recognized, containing many species, in contrast to the monotypy of most of the genera in the other subfamilies. The Acanthosomatinae are distributed in all of the major zoogeographic regions. The Ethiopian fauna is represented by the monotypic *Catadispon* Breddin in West Africa and *Mahea* Distant (two species) in Madagascar and on the Seychelles. *Oncacontias* Breddin and *Rhopalomorpha* Dallas occur only in New Zealand. There are five Australian genera, but two of these are also Oriental and Holarctic. Three genera (containing 39 species) are Palearctic, but all three of these genera are also represented in the Old World tropics and two reach North America.

BLAUDUSINAE. Characterized chiefly by features given in preceding key.

Two tribes are recognized. The Blaudusini comprise 10 or 11 genera, all but three of which are monotypic, the others each containing only two species. Two South American genera are found in Colombia, Bolivia, Ecuador, and Paraguay; *Xosa* Kirkaldy is South African; and the remaining seven genera are Australian.

The Lanopini comprise 12 genera, six of which are known only from Chile or Chile and Argentina, with two additional genera from Brazil. The genus *Abulites* Stål is known from South Africa, and *Noualhieridia* Breddin from Madagascar. Two genera are Australian. All genera are monotypic except *Abulites* (two species).

DITOMOTARSINAE. Abdominal spine absent (most distinctive feature); sternal carina also usually absent, when present in form of a thin, flat, poorly developed ridge.

Two tribes are recognized. The Laccophorellini are a small group containing the monotypic genera *Agamedes* Stål, *Aesepus* Stål, and *Laccophorella* Horvath, confined

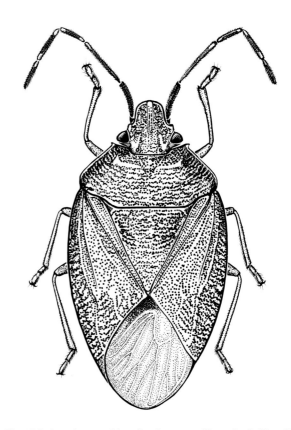

Fig. 66.1. Acanthosomatidae. *Amphaces* sp. (drawn by S. Monteith; from CSIRO, 1991).

Key to subfamilies of Acanthosomatidae adapted from Kumar, 1974a.

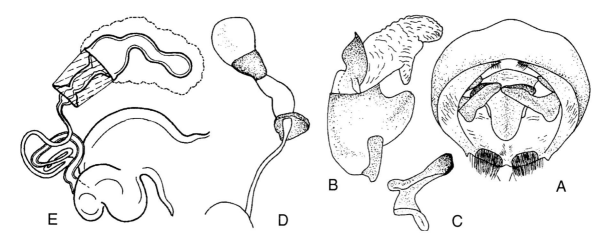

Fig. 66.2. Acanthosomatidae. **A.** Genital capsule, *Elasmostethus cruciatus* (Say). **B.** Aedeagus, lateral view, *E. cruciatus*. **C.** Paramere, *E. cruciatus*. **D.** Spermatheca, *E. cruciatus* (A–D from McDonald, 1966). E. Spermatheca, *Acanthosoma haemorroidale* (Linnaeus) (from Pendergrast, 1957).

to Africa, and the single species of *Sangarius* Stål from Australia. The Ditomotarsini are a somewhat larger complex of 11 genera, five from Chile, two from other areas in South America, and four from Africa—all but two of which are monotypic, these latter two having only two species each.

Specialized morphology. The large mesosternal keel, ventral abdominal spine, 2-segmented tarsi, and Pendergrast's organ are specialized features. In most taxa a maxillary-plate tubercle is also present.

Natural history. Southwood and Leston (1959) summarized the host plants of the relatively well understood Holarctic fauna, most species of which live on trees or shrubs. Comparatively little is known of the biology of most the austral species. Several have been reported to feed on *Ficus,* and others are also associated with different trees. *Sangarius paradoxus* Stål feeds on *Hakea* spp. (Proteaceae) in Australia (Kumar, 1974a).

Several species show marked maternal care of eggs and young nymphs. This phenomenon is known in both Nearctic and Palearctic species (Bequaert, 1935). The most recent studies have been by Kudo (1990) and Kudo et al. (1989), who studied *Elasmucha putoni* Scott and *E. dorsalis* Jakovlev, respectively, in Japan. Females of both species guard the eggs and early-instar nymphs. In some cases they remain with the nymphs into the fifth instar even when the nymphs move from an oviposition site to infloresences some distance away. Maternal care in these Japanese species is primarily a defense against predators, especially the ant *Myrmica ruginodis*. When females were removed from the egg masses, 100% mortality occurred. Female defensive behavior consisted of a series of actions, such as simple jerking, rapid and repeated body jerkings, tilting the body toward the source of disturbance (thus shielding the eggs or nymphs), direct movement toward the source of disturbance, and wing fanning so strong that small predators such as the *Myrmica* and small nabids were actually blown off the plants; the sequence of behaviors was often in the above order. The scent emitted by injured nymphs caused an immediate defensive response by females of both species. They were unable to recognize injury to their own nymphs, however, and responded in general to nymphal scent after injury.

Distribution and faunistics. Two of the subfamilies are dominated by austral elements that include southern Africa as well as Chile and Australia. The primarily austral subfamilies have no species in New Zealand, although *Rhopalomorpha* Dallas (Acanthosomatinae) is restricted to New Zealand. Southwood and Leston (1959) suggested that this is an old austral group that has "revived" in the Oriental and Palearctic regions during the Tertiary. This may well be true but has not yet been demonstrated phylogenetically.

Kumar (1974a) treated the world fauna. Rolston and Kumar (1975) gave keys to the genera of the Western Hemisphere.

67
Aphylidae

General. These small, convex bugs with a greatly enlarged scutellum (Fig. 67.1) range in length from 4 to 5 mm and somewhat resemble the Lestoniidae. They have no common name.

Diagnosis. Head very little extended forward beyond eyes; labium extending posteriorly beyond metacoxae;

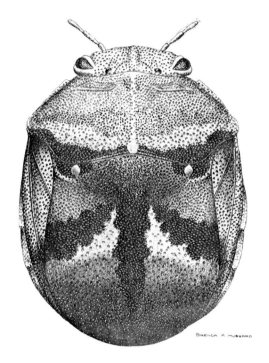

Fig. 67.1. Aphylidae. *Aphylum syntheticum* Bergroth (from Gross, 1975–1976).

pronotum and hemelytra laterally acute, slightly explanate, not broadly laminately expanded; enlarged scutellum not concealing outer half of corium or abdominal laterotergites, latter enlarged, directed ventrally; caudolateral margins of pronotum deeply excised; meso- and metanota visible as small lateral plates behind pronotal excision; thoracic sterna not sulcate; hind wing as in Fig. 67.2A; tarsi 3-segmented; abdomen lacking ventral suckerlike structures as found in the Lestoniidae; abdominal trichobothria transverse, on spiracular line; aedeagus as in Fig. 67.2B; parameres composed of 2 articulated sections; spermatheca invaginated, with median wall sclerotized as in Pentatomidae (Fig. 67.2C).

Classification. First established as a subfamily of Pentatomidae by Bergroth (1906), the taxon has subsequently been considered as a distinct family. Gross (1975–1976) returned it to subfamily status, but in so doing placed all of the other taxa generally considered as subfamilies of the Pentatomidae within a very inclusive Pentatominae.

The family consists of one genus and two species, *Aphylum syntheticum* Bergroth and *A. bergrothi* Schouteden.

Specialized morphology. The parameres are formed of 2 articulated sections. The greatly enlarged convex scutellum, together with the rounded convex pronotum and small head, gives the insects an evenly rounded dorsal surface so that they resemble small chrysomelid beetles.

Natural history. Both known species occur for the most part under *Eucalyptus* bark. *Aphylum syntheticum* Bergroth was taken under the bark of *Eucalyptus camaldulensis*, where the bugs resembled small chrysomelid beetles of the genus *Paropsis*, which are also common in the same habitat (Gross, 1975–1976).

Distribution and faunistics. Members of the Aphylidae are restricted to Australia. A major source on the group is the work of Gross (1975–1976).

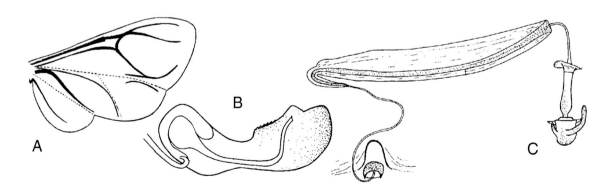

Fig. 67.2. Aphylidae. **A.** Hind wing, *Aphylum bergrothi* Schouteden (from China, 1955a). **B.** Aedeagus, lateral view, *A. syntheticum*. **C.** Spermatheca, *A. syntheticum* (B, C from McDonald, 1970).

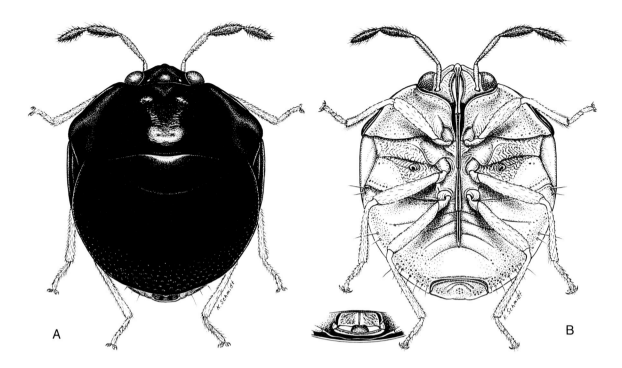

Fig. 68.1. Canopidae. *Canopus caesus* (Germar). **A.** Dorsal view, male. **B.** Ventral view, male. Inset, female terminal abdominal segments.

68
Canopidae

General. These are medium sized (5–7 mm), shining, black bugs, often with purple or greenish reflections, strongly convex dorsum, flattened venter, and semiglobose, obovate body somewhat resembling certain Cydnidae (Fig. 68.1A). They do not have a common name.

Diagnosis. Head short; antennae 5-segmented, length of segment 2 subequal to diameter; prosternal sulcus and strongly laminate propleural carinae present (Fig. 68.1B); scutellum greatly enlarged, convex, covering abdomen and most of forewings (Fig. 68.1A); exocorium exposed at most laterally; forewing elongate, twice as long as abdomen, with line of weakness for folding at end of costa, membrane with well-developed series of parallel veins (Fig. 68.2A); hind wing with lobate posterior margin; tibiae setose, nonspinose; tarsi 3-segmented (Fig. 68.1B); sutures between abdominal sterna obsolete laterad of spiracles (Fig. 68.1B); abdominal trichobothria on sterna 3–7 arranged longitudinally, mesad of spiracular line (Fig. 68.1B); nymphs strongly convex, polished, heavily sclerotized, with three pairs of dorsal abdominal scent-gland openings, between terga 3/4, 4/5, and 5/6, with anterior gland opening twice width of other 2; nymphal sterna 2 and 3 divided mesally; male genitalia with well-defined conjunctival appendages together with conjunctival processes associated with endophallic duct (Fig. 68.2C, D); parameres as in Fig. 68.2E; spermatheca complex, with well-developed pumping mechanism (Fig. 68.2F); valvulae with paired interlocking rami on each side.

Classification. McAtee and Malloch (1928) first established this taxon as a subfamily of Pentatomidae. McDonald (1979) raised it to family status. He believed the group to be related to the Scutelleridae because of the paired interlocking rami on the valvulae and noted that the paired accessory glands resemble those found in some Cydnidae. Gapud (1981) also associated the Canopidae with the Scutelleridae because of the presence of a prosternal sulcus and the strongly laminate propleural carinae. McDonald (1979) concluded that, despite the specialized wing folding and mutual possession of parallel veins in the wing membrane, canopids are not closely related to the Old World Plataspidae, apparently considering these structural resemblances to be convergences. He also believed that the Megarididae are not closely related, although his evidence appeared to be based on the simplified nature of several structural areas in the Megarididae, which he considered primitive.

Only a single genus, *Canopus* Fabricius, is known. It comprises eight species.

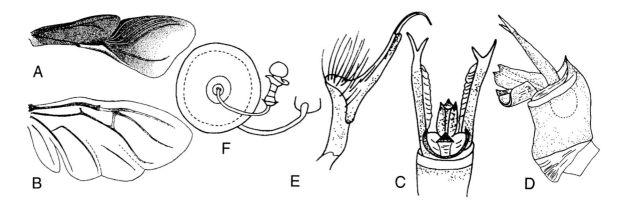

Fig. 68.2. Canopidae. **A.** Forewing, *Canopus caesus*. **B.** Hind wing, *C. orbicularis* Horvath (A, B from McAtee and Malloch, 1928). **C.** Aedeagus, sagittal view, *C. impressus* (Fabricius). **D.** Aedeagus, lateral view, *C. impressus*. **E.** Paramere, *C. orbicularis*. **F.** Spermatheca, *C. burmeisteri* McAtee and Malloch (C–F from McDonald, 1979).

Specialized morphology. The folding of the forewing (Fig. 68.2A), the strongly enlarged and convex scutellum, and the complex spermatheca (Fig. 68.2F) are all specialized features of the group.

Natural history. McHugh (1994) reported nymphs and adults of *Canopus* spp. occurring on the sporophores of polypore fungi at several locations. He also determined that spores of the fungi were present in their guts, strongly suggesting the fungus-feeding habit.

Distribution and faunistics. The basic references on the taxonomy of this strictly Neotropical group are those by McAtee and Malloch (1928) and McDonald (1979).

69
Cydnidae

General. Most species are black or brown, often with a smooth, glossy or shining surface (Fig. 69.1A–D) and a hard integument; a few are brightly colored. Cydnids are ovoid and convex, ranging in length from 2 to 20 mm, and many have a broad, flattened head and legs adapted for digging. They are often called burrower bugs or negro bugs.

Diagnosis. Antennae usually 5-segmented; head quadrate, semicircular, somewhat explanate; ostiolar groove elongate; a strigil of closely spaced sclerotized teeth present on ventral surface of metathoracic wing close to base of Pcu; comblike row of flattened setae or bristles present on distal margin of coxae, adpressed to surface of trochanters; tibiae spinose (Fig. 69.1A, B, D); tarsi 3-segmented; abdominal trichobothria on sterna 3–7 usually oblique or longitudinal, on spiracular line; second abdominal spiracle placed on a differentiated, usually membranous anterolateral portion of sternum; nymphal scent glands present between abdominal terga 3/4, 4/5 and 5/6; genital capsule as in Fig. 69.2A, B; aedeagus as in Fig. 69.2C–F; parameres as in Fig. 69.2G; female ventral laterotergites 8 fused; first valvifers large and broad (Fig. 69.2H, I); spermatheca small with 2 flanges (Fig. 69.2J–M).

Classification. The taxon was first recognized by Billberg (1820) as the Cydnides and has subsequently been treated by most authors either as a distinct family or as a subfamily of Pentatomidae. Since the early twentieth century, family status has been recognized almost universally. Dolling (1981) revised the higher group relationships. His family concept was broad, including the Thyreocorinae and Corimelaeninae. Dolling also included the Thaumastellinae as the sister group of the other Cydnidae. We follow Jacobs (1989) and treat the group at the family level.

Over 110 genera and 600 species (Dolling, 1982) are placed in eight subfamilies, the most recent coming with the elevation of the genus *Parastrachia* Distant to subfamily status by Schaefer et al. (1988).

Fig. 69.1. Cydnidae (from Froeschner, 1960). **A.** *Sehirus cinctus albonotatus* Dallas. **B.** *Amnestus spinifrons* (Say). **C.** *Scaptocoris divergens* Froeschner. **D.** *Cydnus aterrimus* (Forster).

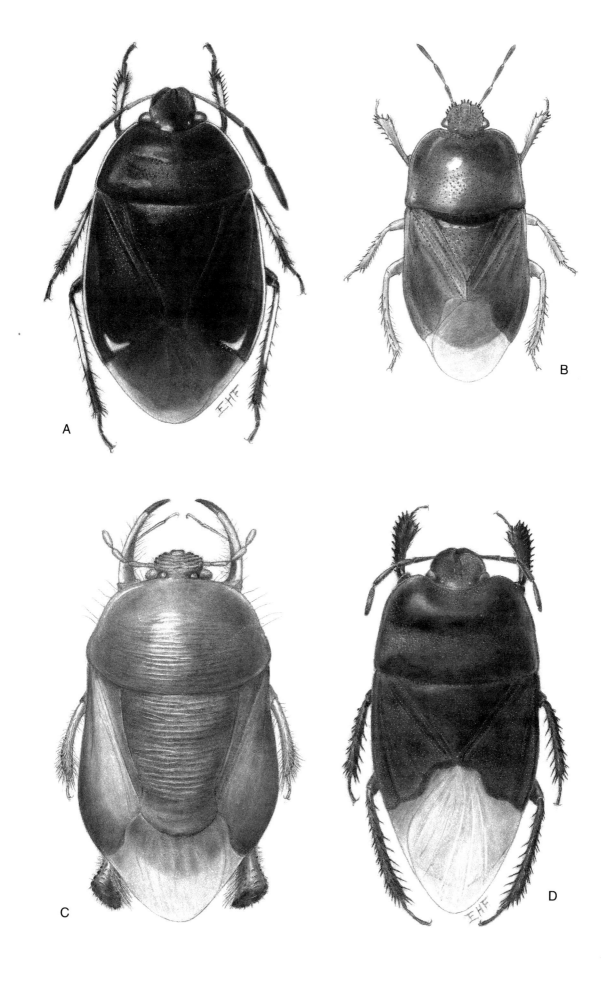

Key to Subfamilies of Cydnidae

1. Coxal comb composed of bristles; bugs relatively large, red and black .... Parastrachiinae
- Coxal comb composed of flattened setae; if relatively large, never red and black 2
2. Claval commissure present (Fig. 69.1B) .... Amnestinae
- Claval commissure absent (Fig. 69.1A, C, D) 3
3. Foretibia strongly cultrate, much produced beyond tarsal insertion, tarsus appearing to arise at middle of tibial length (Fig. 69.1C) .... Scaptocorinae
- Foretibia not cultrate, tarsus arising at or very near apex of tibia 4
4. Scutellum triangular, apex narrowly or sometimes broadly or very broadly rounded 5
- Scutellum semi- to subcircular, apex broadly rounded 7
5. Pronotum with a lateral, submarginal row of setigerous punctures (Fig. 69.1D); tarsal segment 2 subequal in diameter to segments 1 and 3 .... Cydninae
- Pronotum lacking a lateral, submarginal row of setigerous punctures; tarsal segment 2 distinctly narrower than segments 1 and 3 6
6. Pronotum lacking a fine, distinctly impressed subapical groove (Fig. 69.1A); abdominal trichobothria on segments 3–7 arranged in transverse pairs .... Sehirinae
- Pronotum with a fine, distinctly impressed subapical groove paralleling anterior margin; abdominal trichobothria on segments 3–7 arranged in longitudinal pairs .... Garsauriinae
7. Metathoracic wing with jugal lobe entire; Eastern Hemisphere .... Thyreocorinae
- Metathoracic wing with an oval perforation in jugal lobe; Western Hemisphere .... Corimelaeninae

AMNESTINAE (FIG. 69.1B). Very small (1.6–4.5 mm) reddish brown to blackish brown; easily recognizable by distinct claval commissure extending beyond small triangular scutellum.

A single Western Hemisphere genus, *Amnestus* Dallas, containing 24 species, is recognized. The group was revised by Froeschner (1960).

CORIMELAENINAE. Scutellum strongly convex, enlarged, covering most of forewings; many species with pale-colored exocorium, remainder of body surface contrastingly dark; jugal lobe of hind wing perforated; foretibiae not expanded or cultrate; tibiae bearing, in addition to setae, numerous spines along length; reticulation of eyes not attaining ventral surface of head; antennae 5-segmented; tarsi 3-segmented; trichobothria usually arranged transversely.

The status of this Western Hemisphere group has varied. Its members have often been considered a distinct family, sometimes under the name above, although much of the literature is under the name Thyreocoridae. Approximately nine genera and 200 species are recognized. They are known as negro bugs. Keys to the species can be found in McAtee and Malloch (1925).

CYDNINAE (FIG. 69.1D). Usually black or very dark brown; pronotum with a lateral submarginal row of setigerous punctures; vein M of hind wing without a spur or lobe projecting into radial cell; hamus absent; foretarsus arising from distal end of tibia; trichobothria on sterna 3–7 laterally with ventralmost trichobothrium on sterna 3–5 located mesad of or anterior to spiracle, those of sternum 7 (also usually 6) both located posterior to spiracle.

This is the largest and most diverse of the cydnid subfamilies, comprising approximately 90 genera and 300 species (Dolling, 1982). They are found in all major zoogeographic regions, but are most diverse in tropical and subtropical areas. Froeschner (1960) revised the Western Hemisphere fauna.

GARSAURIINAE. Anterior margin of pronotum with distinctly impressed subapical groove; 3 veins arising independently from apex of radial cell (R and M arising independently); abdominal trichobothria on sterna 3–7 in longitudinal rather than transverse pairs as in Sehirinae.

This group, which was first recognized as a higher taxon by Froeschner (1960), contains *Garsauria* Walker, with six species and *Garsauriella* Linnavuori with two species, both genera from the Eastern Hemisphere.

PARASTRACHIINAE. Bucculae meeting posteriorly; antennae 5-segmented; scutellum long, slender, tapering; prosternum with median longitudinal sulcus; coxal combs absent; metathoracic scent-gland opening and peritreme absent, evaporative area greatly reduced; all femora and tibiae mutic and terete; tarsi 3-segmented; female sternum 7 entire; spermatheca a simple rounded bulb with 2 flanges.

This taxon, which contains only the genus *Parastrachia* Distant with two Oriental species, was first recognized by Oshanin (1922) as a tribe of Pentatomidae. Schaefer et al. (1988) elevated it to subfamily rank in the Cydnidae and discussed its checkered taxonomic history and relationships in detail. This group may well merit family status. For example, it lacks the definitive characters of the Cydnidae, as for example the spinose tibiae and coxal

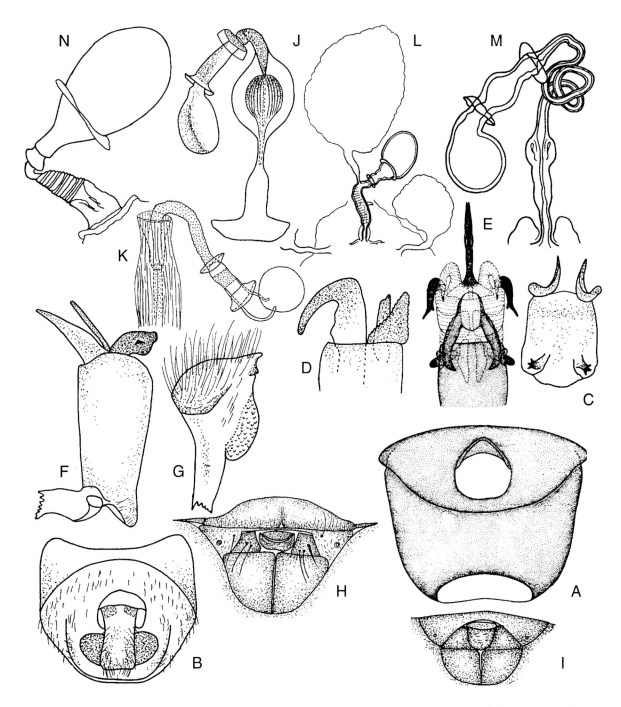

Fig. 69.2. Cydnidae. **A.** Genital capsule, *Corimelaena lateralis* (Fabricius) (from Ahmad and McPherson, 1990). **B.** Genital capsule, *Cyrtomenus crassus* Walker. **C.** Phallotheca, dorsal view, *Corimelaena pulicaria* (Germar). **D.** Conjunctival appendages, *C. pulicaria* (B–D from McDonald, 1966). **E.** Aedeagus, sagittal view, *C. lateralis* (from Ahmad and McPherson, 1990). **F.** Aedeagus, lateral view, *Cyrtomenus crassus*. **G.** Paramere, *C. crassus*. **H.** Female genitalia, ventral view, *C. crassus*. **I.** Female genitalia, *Sehirus cinctus*. **J.** Spermatheca, *Cyrtomenus crassus* (F–I from McDonald, 1966). **K.** Spermatheca, *Thyreocoris scarabaeoides* (Linnaeus) (from Štys and Davidova, 1979). **L.** Spermatheca, *Galgupha ovalis* Huss. **M.** Spermatheca, *Macrocytus brunneus* Fabricius (L, M from Pendergrast, 1957). **N.** Spermatheca, *S. cinctus* (from McDonald, 1966).

combs. It also has the abdominal segment 8 in the male broadly exposed (as in the Acanthosomatidae and Urostylidae). It differs from acanthosomatids, however, by the 3-segmented tarsi and from urostylids in the lack of a claval commissure. Its association with the Cydnidae is based on the resemblance of the thoracic sterna to those of the Sehirinae.

SCAPTOCORINAE (FIG. 69.1C). Robust, usually convex reddish brown; foretibiae cultrate, foretarsus anteapical; hind tibiae broadened, clavate, distally truncate; hind tarsi absent in some species.

This subfamily contains three genera and approximately 15 species that are restricted to the Neotropical, Palearctic, and Oriental regions. The Western Hemisphere species were revised by Froeschner (1960).

SEHIRINAE (FIG. 69.1A). No setigerous setae along lateral margins of pronotum; shape of radial cell in hind wing distinctive, R and M arising from common origin; hamus present; trichobothria on sterna 3–7 placed in transverse row on spiracular line posterior to spiracle.

The subfamily contains about nine genera, comprising at least 60 species. It is most diverse in the Palearctic, with a few species in the Oriental and Ethiopian regions and one widespread species in the Western Hemisphere.

THYREOCORINAE. Scutellum strongly convex, enlarged; jugal lobe entire; genitalic details as noted by Dolling (1981; see also McAtee and Malloch, 1933).

Until recently this subfamily has been considered to include Western Hemisphere species here treated in the Corimelaeninae. Three genera—*Thyreocoris* Schrank, *Strombosoma* Amyot and Serville, and *Carrabas* Distant—and five species are recognized, all from the Old World.

Specialized morphology. Although cydnids are thought to be relatively primitive pentatomoids, most nonetheless possess a rich assortment of derived features, associated primarily with fossorial habits. These include the flattened anteriorly spined heads; the broadened, flattened, and heavily spinose forelegs; the smooth, shining body surface; and the sometimes strongly thickened, stout, heavily armed hind femora. In some species, the tarsi are greatly reduced or absent. Froeschner (1960) noted that many specimens show wear on the legs, particularly the cultrate anterior tibiae, presumably from extended digging.

The comblike setae on the distal coxal margin are thought to prevent particles of dust from entering the coxotrochanteral joint when the insects are burrowing (Dolling, 1981). If they do, then the presumption must be that burrowing is the ancestral condition because these setae occur in all species (except *Parastrachia*), including those that do not burrow.

Many cydnids have a stridulatory mechanism involving the postcubital vein and the abdominal tergum, which Schaefer et al. (1988) believed to be plesiomorphic in the Pentatomoidea, as it occurs in a number of different higher-group taxa.

Natural history. Many species, probably the majority, are fossorial. Some are known to feed on roots of plants and may live as much as 57 inches below the soil surface.

Froeschner and Chapman (1963) noted that *Scaptocoris castaneus* Perty burrows by scraping the soil with the scythelike front tibiae, the other legs being used to push the soil thus loosened into the burrow behind the insect, enclosing the individual in a small "moving chamber." This striking insect has recently been established in the southeastern United States, where it is so abundant that the ground is said to literally "stink" with the odor from the scent glands (Froeschner and Steiner, 1983) (see Chapter 8).

Members of the Corimelaeninae are often mistaken for small beetles. At least three South American species are associated with ants. The life histories of several species have been studied, and in these both adults and nymphs live on the plants well above the ground and eggs are laid on the plants. Sometimes these insects become minor pests of garden vegetables and flowers.

Little is known of the biology of most species of Amnestinae, but they apparently are not strongly fossorial because there are several records of specimens taken by sweeping and under debris and stones. Nymphs of all instars and adults of *Amnestus subferrugineus* (Westwood) have been reported as abundant in guano in bat caves in Panama (Caudell, 1924), suggesting they may be seed feeders. Many species come to lights in large numbers.

The Sehirinae apparently are not fossorial and are often swept from vegetation. Some species of *Sehirus* Amyot and Serville, both in Europe and in North America, prefer species of Lamiaceae, Boraginaceae, and related plants. *Sehirus cinctus* (Palisot de Beauvois) in North America and *S. bicolor* (Linnaeus) in Europe lay their eggs in cavities in the ground in a shallow hole about 2 cm deep and 1 cm in diameter. The clusters of more than 100 eggs are protected by the female, which sits on top of them. Parental care extends to the young nymphs, which aggregate with the female (Southwood and Hine, 1950; Sites and McPherson, 1982). In at least one species, stridulation occurs during the mating ritual.

The Thyreocorinae live on plants above the ground as do the Sehirinae, *Thyreocoris scarabaeoides* (Schrank) feeding on *Viola* spp.

Parastrachia japonensis (Scott) females also excavate an egg chamber and guard the eggs and young nymphs. This species lives above ground on the host plant *Schoepfia jasminodora*, where it feeds on the endosperm of the fruit (Tachikawa and Schaefer, 1985; Schaefer et al. 1991).

Fortunately, burrower bugs frequently come to lights,

so the species are reasonably well understood taxonomically. As with many insects, the flight period may be very short, and in some species it is known to be restricted to the first hour after sunset.

Distribution and faunistics. The Cydnidae are distributed worldwide, and the family is well represented in temperate as well as tropical regions. The group appears to be an ancient one, with isolation of several taxa in one hemisphere or even in one major faunal region as noted in the subfamily treatments.

Froeschner (1960) is the standard work for identification of the Western Hemisphere fauna. There is no comprehensive modern work of an extensive Eastern Hemisphere fauna, but the work of Signoret (1881–1884) is still of value and the work of Linnavuori (1993) on a portion of the African fauna will be indispensable to anyone studying the group.

70
Dinidoridae

General. Species of this family are large to very large (9–27 mm), ovoid-elliptical, heavy-bodied, with laterally keeled heads (Fig. 70.1), and short scutellum. They have no common name.

Diagnosis. Antennae 4- or 5-segmented with at least 2 preapical segments flattened; antenniferous tubercles placed below lateral head margins, not visible from above (Fig. 70.1); bucculae elevated, lobelike, short, closed posteriorly; labium short, not extending posteriorly beyond forecoxae; scutellum medium-sized with a blunt or rounded apex reaching about one-half length of abdomen; membrane of forewing often with reticulate venation (Fig. 70.1); tarsi 2- or 3-segmented; nymphs with dorsal abdominal scent-gland openings only between terga 4/5 and 5/6 (but scar present between terga 3/4); spiracles of abdominal segment 2 fully exposed, removed from posterior margins of metapleura or concealed by posterior margins of metapleura; abdominal trichobothria usually paired on sterna 3–7 (except *Eumenotes* Westwood with 1 trichobothrium per segment, contra Durai, 1987), arranged transversely on large callus mesad of spiracles; ninth paratergites greatly enlarged; gonangulum reduced; aedeagus as in Fig. 70.2A–C; parameres as in Fig. 70.2D,E; spermatheca as in Fig. 70.2F.

Classification. The taxon was first established as a subfamily of Pentatomidae by Stål (1870) as Dinodorida. Lethierry and Severin (1893–1896) called it Dinidoridae, but many of their "family" terminations are actually used in a subfamily sense. Leston (1955a) treated it as a distinct family and was followed by Miyamoto (1961b), Kumar (1962, 1965), and Cobben (1968a, 1978), among others. Durai (1987), in a world revision with keys to genera and species, recognized two subfamilies.

Key to Subfamilies of Dinidoridae

1. Anterolateral angles of scutellum without depressions; posterolateral angles of abdominal connexiva neither tuberculate nor lobed; spiracles of abdominal sternum 2 usually covered by metasternum
 .... Dinidorinae
– Anterolateral angles of scutellum with depressions; posterolateral angles of abdominal connexiva either tuberculate or lobed; spiracles of abdominal sternum 2 not covered by metasternum ....
 .... Megymeninae

DINIDORINAE. Body streamlined, dorsum smooth; pronotum and abdominal connexiva without lateral processes; anterolateral angles of scutellum without depressions; abdominal connexiva not tuberculate; spines on legs often reduced to spicules.

Two tribes are recognized. The Dinidorini contains seven genera and 65 species, including taxa with both 4-segmented (*Cylopelta* Amyot and Serville; *Dinidor* Latreille) and 5-segmented antennae. The Thalmini contains three monotypic genera, all of which have only two tarsal segments.

MEGYMENINAE. Lateral margins of abdomen and often also head and pronotum produced into tubercles, lobes, toothlike processes, or spines; pronotum dorsally often bearing an anteromedian tuberosity, or elevated posteriorly as a transverse ridge; legs heavily and strongly spined, or antennae and legs setulose.

Two tribes are recognized. The Eumenotini contains only the monotypic genus *Eumenotes* Westwood. The Megymenini contains *Megymenum* Guérin-Méneville, with 15 described species, and the monotypic *Doesbergiana* Durai.

Classification. The group was first described as a subfamily of Plataspidae by China (1955a) and was raised to family status by China and Miller (1959).

McDonald (1970) believed that the Lestoniidae are most closely related to the Plataspidae because of the similar aedeagus. He also noted (erroneously) that the lack of spermathecal flanges occurs elsewhere in the Pentatomoidea only in the Scutelleridae and Cydnidae. Schaefer (1993) also supported a relationship with the Plataspidae but stated that the Scutelleridae and Cydnidae have spermathecal flanges.

Two species are known, *Lestonia haustorifera* China and *L. grossi* McDonald (1969).

Specialized morphology. The disc-shaped organs on the female abdomen are unique (Fig. 71.2A). The scale-like body shape, greatly enlarged scutellum, laminately produced hemelytral margins, and lack of spermathecal flanges (Fig. 71.2G), are also specialized features.

Natural history. Both adults and nymphs of *L. haustorifera* China have been collected on Australian cypress (*Callitris preissii*), where they were concentrated on the growing tips and said to resemble scales or small chrysomelid beetles (McDonald, 1970; Gross, 1975–1976).

Distribution and faunistics. A basic source for this strictly Australian group is the work of Gross (1975–1976).

72
Megarididae

General. These are small (5 mm or less), ovoid, strongly convex insects, with scutellum enlarged, globose, and covering the abdomen and wings (Fig. 72.1A, C). They have no common name.

Diagnosis. Anterior margin of head and anterior and lateral margins of pronotum carinate and slightly reflexed; antennae 4-segmented with many setae as long as diameter of segments in females and much longer in males; scutellum completely covering abdomen and forewings (Fig. 72.1A); forewing about twice length of abdomen with thin areas at about middle of costa, adapting wing for folding, membrane lacking parallel veins or with a single major vein (Fig. 72.1A, D); hind wing as in Fig. 72.1E; tarsi 2-segmented; abdominal trichobothria arranged transversely, on spiracular line; male genitalia simple, consisting of conjunctiva surrounding a tubelike endophallic duct (Fig. 72.1F, G); parameres as in Fig. 72.1H; spermatheca reduced and simple, saccular or globular, lacking flanges and pumping mechanism (Fig. 72.1J); ovipositor platelike (Fig. 72.1I); nymphs strongly convex, polished, heavily sclerotized, with most abdominal tergal sutures obliterated.

Classification. McAtee and Malloch (1928) first rec-

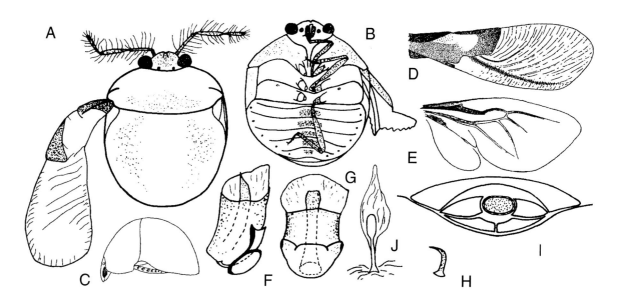

Fig. 72.1. Megarididae. **A.** *Megaris rotunda* McDonald. **B.** Ventral view, body, *M. rotunda* (A, B from McDonald, 1979). **C.** Lateral view, body, *M. hemisphaerica* McAtee and Malloch. **D.** Forewing, *M. laevicollis* Stål. **E.** Hind wing, *M. laevicollis* (C–E from McAtee and Malloch, 1928). **F.** Aedeagus, lateral view, *M. rotunda*. **G.** Aedeagus, sagittal view, *M. rotunda*. **H.** Paramere, *M. rotunda*. **I.** Female terminal abdominal segments, *M. laevicollis*. **J.** Spermatheca, *M. laevicollis*. (F–J from McDonald, 1979).

ognized these insects as a subfamily of Pentatomidae. McDonald (1979) elevated them to family status, believing that despite a superficial resemblance to the Canopidae and Plataspidae the three are not closely related. Rolston and McDonald (1979) believed that megaridids are a very primitive stock, probably representing an early offshoot from the pentatomoid line of evolution. The evidence however appears to be based on the simple nature of several structures that might well be interpreted as derived loss conditions.

The small Southern Hemisphere pentatomoid families such as the Megarididae, Canopidae, Lestoniidae, and Aphylidae, as well as the more widely distributed Plataspidae, could benefit from a cladistic analysis. Such an analysis would not only clarify the relationships among the taxa but also provide evidence for determining rank within the hierarchy.

Only a single genus, *Megaris* Stål, is known. It contains 16 species.

Specialized morphology. The shortened body, greatly enlarged and convex scutellum (Fig. 72.1C), elongate male antennal setae (Fig. 72.1A), and possibly the simplified male and female genital structures (Fig. 72.1F, G, J) are specialized conditions.

Natural history. *Megaris puertoricensis* Barber and *M. semiamicta* McAtee and Malloch have been reported feeding on species of *Eugenia* (Myrtaceae) (respectively, Barber, 1939; Wolcott, 1936). Wolcott (1936) noted that *M. semiamicta* feeds on the flowers.

Distribution and faunistics. The family is restricted to the Neotropics. *Megaris majusculus* McAtee and Malloch is endemic to Cuba. The most comprehensive work on the taxonomy of the group is still that of McAtee and Malloch (1928).

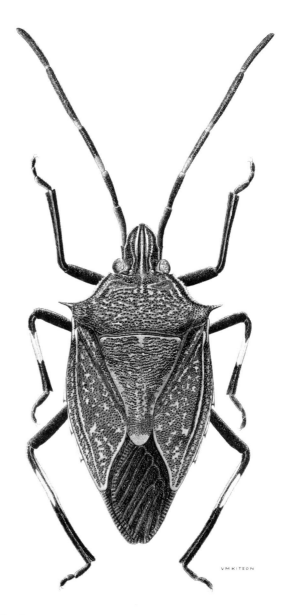

Fig. 73.1. Pentatomidae. *Poecilometis eximus* Stål (from Gross, 1972).

73
Pentatomidae

General. Most stink bugs are of moderate to large size, ranging in length from 4 to 20 mm, and generally ovoid or broadly elliptical in shape (Fig. 73.1). There are, however, some relatively elongate slender species, most of which are associated with grasses.

Diagnosis. Body usually broad and ovoid; antennae usually 5-segmented (Fig. 73.1) (some species with 4 segments); scutellum large, usually triangular or subtriangular, frena present (Fig. 73.1); claval commissure reduced or absent; mesosternum usually lacking a median carina; tarsi 3-segmented (2-segmented in Cyrtocorinae); spiracles of abdominal segment 2 concealed by metapleura except in a few genera of large body size; abdominal trichobothria arranged transversely behind spiracles on spiracular line; nymphal dorsal abdominal scent glands paired between terga 3/4, 4/5, 5/6; genital capsule as in Fig. 73.2A, B; aedeagus as in Fig. 73.2C–E; parameres as in Fig. 73.2G, H; ejaculatory reservoir (Fig. 73.2F) fixed in phallotheca; spermatheca 2-flanged with a well-developed pump, invaginated before pump, with a sclerotized median wall (Fig. 73.2I); ovipositor generally platelike, never truly laciniate; eggs barrel-shaped with a detachable cap (pseudoperculum).

Classification. The family was established by Leach (1815) as "Pentatomides." Stål (1866a) included the Scutelleridae and was followed by many subsequent authors. Even today, the limits of the Pentatomidae are not widely agreed upon. Some authors place taxa that we treat as families as subfamilies; others recognize some of our subfamilies either as distinct families or as tribal units. Clarification of relationships of this great complex is a major need in heteropteran classification.

Miller (1956a) recognized 11 subfamilies, only eight of which are considered here as subfamilies of the Pentatomidae. China and Miller (1959) recognized 13 subfamilies. They included the Tessaratominae, Dinidorinae, Scutellerinae, Acanthosomatinae, Canopinae, and Megaridinae, all treated herein as families. By contrast, Gross (1975–1976) recognized only two subfamilies, the Aphylinae and Pentatominae, the former treated here as a distinct family. Gapud (1991) recognized nine subfamilies; however, in his cladogram an additional subfamily —Halyinae—is recognized. We treat two of Gapud's subfamilies (Megarididae and Aphylidae) as separate families, recognize the Edessinae—considered a tribe by Gapud (1991)—at the subfamily level, and treat the Halyini at the tribal level. Gapud (1991) summarized classification schemes of additional previous authors. Hasan and Kitching (1993) provided a cladistic analysis of the tribes, but did not propose a revised formal classification.

Approximately 760 genera and 4100 species are known. The Pentatomidae are thus one of the four largest families of Heteroptera. The last world catalog was that of Kirkaldy (1909).

Key to Subfamilies of Pentatomidae

1. Labium very short, not extending posteriorly beyond posterior margin of forecoxae .... Phyllocephalinae
- Labium more elongate, considerably exceeding forecoxae 2
2. Basal labial segment thickened, not concealed between bucculae .... Asopinae
- Basal labial segment not noticeably thickened; concealed between bucculae 3
3. Antennae 4-segmented 4
- Antennae 5-segmented 5
4. Bucculae obsolete, shorter than first labial segment; scutellum as wide as long .... Serbaninae
- Bucculae well developed, at least as long as first labial segment; scutellum longer than wide .... Cyrtocorinae
5. Metasternum produced anteriorly onto mesosternum or rarely prosternum; labium not surpassing mesocoxae .... Edessinae
- Mesosternum rarely produced anteriorly onto mesosternum, if so then labium extending onto abdomen; labium usually reaching at least to metacoxae 6
6. Trichobothrium nearest spiracle on sternum 7 laterad of spiracular line by distance at least equal to greatest diameter of spiracular opening 7
- At least one trichobothrium on sternum 7 on or mesad of spiracular line 8
7. Base of abdominal venter with mesal tubercle; metasternum produced, flattened .... Pentatominae (part)
- Abdominal venter rarely tuberculate at base, if so then metasternum thinly carinate mesally .... Discocephalinae (part)
8. Labium arising on or behind imaginary line traversing head at anterior limit of eyes and/or superior surface of tarsal segment 3 of hind legs shallowly excavated in females .... Discocephalinae (part)
- Labium arising before such a line; superior surface of tarsal segments convex or flattened 9
9. Tibiae sulcate on outer surface; labial segment 1 longer than bucculae; trichobothria paired; frena one-third or more length of scutellum; scutellum not reaching apex of abdomen .... Pentatominae (part)
- Tibiae not sulcate on outer surface; labial segment 1 not longer than bucculae; trichobothria single; frena short, less than one-third length of scutellum; scutellum usually U-shaped, reaching apex of abdomen .... Podopinae

ASOPINAE. Labial segment 1 thickened, usually free from bucculae, rarely lying within bucculae for its entire length, in which case foretibiae foliate (*Cecyrina* Walker, *Heteroscelis* Latreille); males often with pair of large

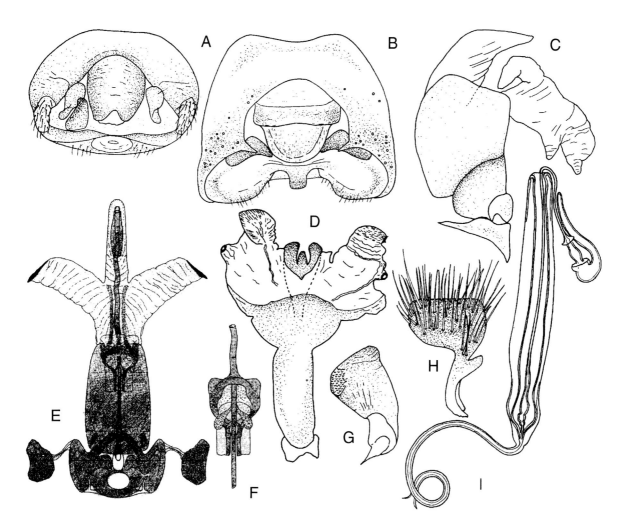

Fig. 73.2. Pentatomidae. **A.** Genital capsule, *Heteroscelis lepida* Stål. **B.** Genital capsule, *Sciocoris microphthalmus* Flor. **C.** Aedeagus, lateral view, *H. lepida*. **D.** Aedeagus, sagittal view, *S. microphthalmus* (A–D from McDonald, 1966). **E.** Aedeagus, sagittal view, *Stagonomus amoenus* (Brullé). **F.** Ejaculatory reservoir, *S. amoenus* (E, F from Seidenstücker, 1965b). **G.** Paramere, *H. lepida*. **H.** Paramere, *Sciocoris microphthalmus* (G, H from McDonald, 1966). **I.** Spermatheca, *Theseus modestus* (Stål) (from Pendergrast, 1957).

pilose sensory patches, these often extending across at least part of last 3 pregenital abdominal sterna; pygophoral plate located interior to (mesad of) each paramere.

This group was first recognized by Amyot and Serville (1843) as the "Spissirostres," although most authors credit it to Spinola (1850) as the Asopoideae.

Asopines are found in all faunal regions. There are approximately 63 genera and 357 species known (Gapud 1981). The work of Thomas (1992) is the most current for the New World fauna. All whose biology is known are predatory.

CYRTOCORINAE. Cryptically colored, resembling tree bark; antennae 4-segmented; pronotum and abdomen strongly expanded laterally, mostly covered by enlarged scutellum, the latter produced mesally into a strong, projecting spine; adults bright black, but appearing dull brown because of dense secretion covering body; tarsi 2-segmented.

First recognized as a higher group by Distant (1880–1893), the systematic position of this group is ambiguous. We treat it as a highly modified subfamily of Pentatomidae, although Brailovsky et al. (1988), who provided a modern description, treated it as a distinct family. The group is Neotropical and consists of four genera and 11 species.

DISCOCEPHALINAE. Body flattened; coloration mottled brown, black, fuscous, or brown and black mottled; antennae either 4- or 5-segmented; labium usually arising on or posterior to an imaginary line traversing head at anterior margin of eyes; metasternum not produced anteriorly onto mesosternum; tarsi 3-segmented; dorsal surface of tarsal segment 3 of hind legs usually excavated in females; trichobothrium nearest spiracle on sternum 7 usually laterad of spiracular line; phallotheca, ejaculatory duct, median penial lobes, and conjunctival appendages (when present) heavily sclerotized, last fused to margin of phallotheca and permanently exserted.

First recognized as a higher group by Fieber (1861), the Discocephalinae, which are speciose in the Neotropics, comprise 71 genera and at least 263 species. Two tribes are recognized: the Ochlerini, erected by Rolston (1981) (23 genera), and the Discocephalini (48 genera).

EDESSINAE. Large to very large; smooth, polished dorsal surface; appearance rather streamlined; antennae either 4- or 5-segmented; metasternum strongly produced, extending anteriorly onto mesosternum and sometimes prosternum and laterad between meso- and metacoxae; anterior projection of metasternal plate bifid (except in *Pantochlora* Stål), labium terminating in this notch; posterior margin of metasternum notched to receive mesal tubercle of abdomen; tarsi 3-segmented.

First recognized as a higher group by Amyot and Serville (1843), edessines until recently were considered a tribe of the Pentatominae, but were elevated to subfamily status by Rolston and McDonald (1979), although Gapud (1991) treated them at a tribal level. These large, robust stink bugs are abundant and diverse in the Neotropics, where four genera and 269 species are known.

PENTATOMINAE (FIG. 73.1). Varied in form and color, usually obovate, often with prominent caudolateral pronotal angles; scutellum never attaining apex of abdomen; frena extending two-fifths or more length of scutellum; metasternum rarely produced anteriorly onto mesosternum, when so labium reaching metacoxae; trichobothria, usually at least one or a pair on each side of sterna 3–7, on or near spiracular line; dilation of spermathecal duct fusiform. (For more details regarding genitalic structures see Rolston and McDonald, 1979.)

This is the largest subfamily of Pentatomidae. Eight tribes are recognized: Aeptini (11 genera, 30 species); Diemeniini (13 genera, 47 species); Halyini Amyot and Serville (82 genera, 361 species); Lestonocorini (5 genera); Mecideini Distant (1 genus, 17 species); Myrocheini (14 genera, 45 species); Pentatomini Leach (404 genera, 2207 species); Sciocorini Amyot and Serville (10 genera, 107 species). Tribal relationships are under active investigation.

PHYLLOCEPHALINAE. Large, flattened; labium not extending beyond posterior margin of forecoxae.

Thirty-one genera and 175 species are known. Most species appear to live on the bark of trees. Miller (1956a) stated that *Tetroda histeroides* (Fabricius), an important rice pest, belongs to this subfamily.

PODOPINAE. Coloration generally dark yellow brown to dark brown or nearly black; antenniferous tubercles visible from above; antennae either 4- or 5-segmented; lateral margins of pronotum usually toothed or tubercle-bearing; tarsi 3-segmented; trichobothria usually paired, single in some Podopini, behind each spiracle on or near spiracular line; scutellum enlarged, always attaining membrane of forewing and often reaching apex of abdomen, covering most of forewings; frena well developed but extending less than one-third length of scutellum; hamus absent; R+M and Cu of hind wing parallel (Leston, 1953a); parameres biramous; pygophore often with appendages on caudolateral margin.

This widely distributed subfamily has had a checkered history varying from subtribe to family status. Many literature references are under the name Graphosomatinae, a polyphyletic group with some of its members belonging to the Podopinae and most of the others to the Pentatominae. Schaefer (1981c) discussed details of the group, recognizing two tribes, the Podopini and the Graphosomatini. Sixty-four genera and 255 species are known.

Podopines are found chiefly in damp marshy or muddy habitats. Some species frequently come to lights. Certain species of *Scotinophara* Stål are serious pests of rice and other cereal crops in Asia and Africa.

SERBANINAE. Body flattened, with broad lateral foliation on head, pronotum, and abdomen; coloration cryptic (resembling Phloeidae in habitus); antennae 4-segmented; tarsi 3-segmented; compound eyes divided into dorsal and ventral portions as in Phloeidae; male conjunctival appendages largely membranous, paired, phallotheca basally membranous, distally sclerotized.

The subfamily was originally described by Distant in the Phloeidae but later was removed and established as a higher taxon within the Pentatomidae by Leston (1953c). A single species is known, *Serbana borneensis* Distant, from Borneo. Nothing appears to be known of the habits of these rare insects, but they probably will be found to live on the bark of trees.

Specialized morphology. Pentatomids (and aphylids) are distinguished from all other Pentatomomorpha by the form of the spermatheca, which is invaginated proximal to the pump with a sclerotized median wall (Figs. 67.2, 73.2I). The scutellum of pentatomids is triangular (Fig. 73.1), although it is sometimes enlarged to cover almost the entire abdomen, as in Podopinae. The 5-segmented antennae found in most taxa are also a specialized feature, as is the flattened platelike female ovipositor. Some stink bugs have stridulatory structures involving the abdominal dorsum and hind wing.

Wing reduction is very rare in the Pentatomidae. In the Western Hemisphere it occurs in only two species—*Brachelytron angelicus* Ruckes from Brazil and *Alathetus haitiensis* Rolston from Haiti. The latter species has short, truncate padlike wings. Rolston (1982a) stated that members of the tribe (Ochlerini) live in the canopy of Neotropical forests and surmised that with the deforestation of Haiti *A. haitiensis* may be extinct. But because it was collected by P. J. Darlington at 5000 feet, the presumed arboreal habitat is in need of verification. Miller (1956a) illustrated *Tahitocoris cheesmanae* Yang (Asopinae), which is cimicoid in outline and apparently apterous.

Natural history. Most pentatomids are plant feeders and have a distinct preference for immature fruits and seeds. Many also feed in the plant vascular system. Only members of the Asopinae are predaceous, apparently secondarily. They feed on a great variety of insects and other small arthropods; the North American *Podisus maculiventris* (Say) has been reported feeding on over 90 species of insects. The majority prefer the larvae of Lepidoptera and Coleoptera. Several species of Asopinae are of considerable importance as biological control agents of destructive insects. *Perillus bioculatus* (Fabricius) is extremely variable in color, the red color being a carotin pigment obtained from the Colorado potato beetle. If the pigment is not metabolized rapidly enough, the body color will change from white or yellow to red. The deposition of the pigment is primarily controlled by temperature, which affects the physiological activity of the bug (Knight, 1922, 1924; Palmer and Knight, 1924).

The phytophagous pentatomine *Thyanta calceata* (Say) has a green summer generation with short pubescence and a brown autumnal-vernal generation with long pubescence. This variation is due to a developmental photoperiod response. The adults are capable of color reversal in both directions (McPherson, 1977).

One of the striking biological features of some pentatomids (and other pentatomoids) is the protection of the eggs by the females. In several species this behavior is elaborately developed: the adult female not only wards off potential predators but moves her body from side to side to inhibit egg laying by small parasitic wasps. After the eggs hatch, the females of some species guard the first-instar nymphs, which apparently do not feed. Egg guarding has arisen several times independently within the family. Eberhard (1975) stated that it is known in 12 genera and 14 species, but this figure is certainly too small because he mentions parental behavior in five species of *Antiteuchus* Dallas alone.

The most detailed work on maternal care is that of Eberhard (1975) on *Antiteuchus tripterus limbativentris* Ruckes (Discocephalinae) in Colombia. He found that these insects protected both the egg masses and first-instar and early second-instar nymphs. The protection is effective against general predators. But interestingly it appears to be detrimental in the case of two scelionid wasp egg parasites that are attracted by the females over the eggs; despite defensive behavior by the females a high proportion of the eggs are parasitized. Eberhard presented detailed information concerning differences in wasp attack behavior. This pentatomid, as well as several other species in the genus, is a serious pest of cacao and mango and is a vector of a fungus that attacks cacao. Interestingly, the bug also is a folk cure of intestinal parasites of humans. Oddly, Eberhard found the bugs to be more common in the city of Cali than in the countryside, and they were most common where heavy pedestrian and automobile traffic was present.

There are many striking cases of protective resemblance in the family, especially in those species that live on bark, some of which assume bizarre shapes as well as highly cryptic patterns. This adaptation reaches its apogee in the Cyrtocorinae, which cover themselves with a dense brownish white secretion that, together with the great expansions of the pronotum and abdomen, renders them nearly invisible on the bark of the principle host plants. Species of many other genera such as *Brochymena* Amyot and Serville are cryptically colored and also may resemble the bark of trees. Brightly colored species are thought to be aposematic, although the basis for the warning coloration has not been well established for most of them.

In *Cyrtocoris trigonus* (Germar), which feeds primarily on *Acalifa diversifolia* (Euphorbiaceae), nymphs form large aggregations of up to 700 individuals on a single plant. Some maternal care appears to occur (Brailovsky et al., 1988).

Distribution and faunistics. The family is worldwide in distribution and well represented in all of the major faunal regions. As with most of the large families of Heteroptera, the tropical and subtropical faunas are the most extensive.

Useful identification aids include Gross (1975–1976) for Australia, Cachan (1952b) for Madagascar; Rolston and McDonald (1979, 1981, 1984) and Rolston et al. (1980) for keys to Western Hemisphere subfamilies and for tribes of Pentatominae; McPherson (1982) for the eastern North American fauna; Schouteden (1903, 1905b) for world genera and African species of Podopinae; Gapud (1991) for world genera of Asopinae; and Thomas (1992) for the Asopinae of the Western Hemisphere. Schouteden's (1907) revision of the world genera of Asopinae contains excellent color plates. Other useful studies of smaller groups include Freeman's (1940) revision of *Nezara* Amyot and Serville, Rolston's (1974, 1982b, 1984) revisions of *Euschistus,* and Ruckes's (1947) revision of *Brochymena.*

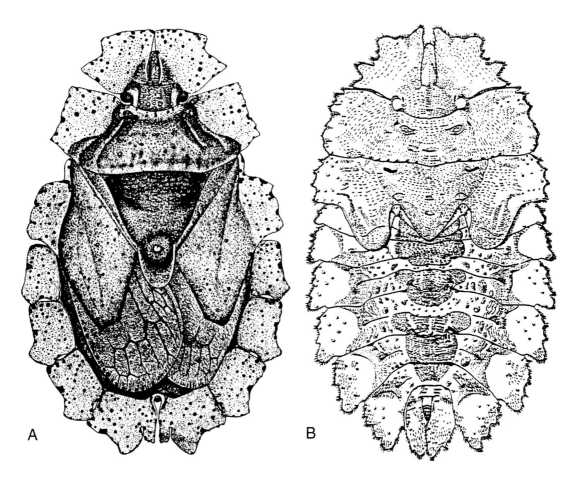

Fig. 74.1. Phloeidae. **A.** *Phloea corticata* (Drury) (from Lent and Jurberg, 1965). **B.** *Phloeophana* sp., nymph.

74
Phloeidae

General. These are bizarre, large, flattened, brownish insects with the lateral body margins strongly expanded (Fig. 74.1A, B) and with obvious protective shape and coloration associated with life on the bark of trees. They range in size from 20 to 30 mm and have no common name.

Diagnosis. Body extremely depressed, with outer margins of mandibular plates, pronotum, base of corium, and abdomen broadly foliate (Fig. 74.1A, B); eyes divided into a dorsal and ventral portion; bucculae long, low posteriorly, with labial channel very elongate; antennae 3-segmented, with segment 1 very long, segment 3 somewhat curved (Fig. 74.2A), antennal segments almost hidden below expanded mandibular plates; post-frenal portion of scutellum very elongate; scutellum not covering forewing; membrane extensively reticulate; hind wing with hamus; tarsi 3-segmented; abdominal sterna 3–7 with trichobothria arranged longitudinally mesad of spiracular line; metathoracic scent-gland opening near lateral margin of pleuron; dorsal abdominal scent-gland openings of nymphs present between terga 3/4, 4/5, 5/6 (Fig. 74.1B), those between 3/4 and 4/5 paired, those between 5/6 coalesced into a single opening (anterior gland sometimes lacking); abdominal connexivum with terga and sterna fused, no inner laterotergites; spiracle 2 present and partially exposed; ninth paratergites greatly elongated; genital capsule as in Fig. 74.2B; aedeagus with 3 pairs of conjunctival appendages (Fig. 74.2G; see Lent and Jurberg, 1965 for details); ovipositor plate-like, second valvifers fused medially (contra Lent and Jurberg [1965] and Gapud [1991]); spermatheca with well-defined pump region with flanges (Fig. 74.2F, H); parameres as in Fig. 74.2E; egg lacking a pseudoperculum.

Classification. This group was first recognized as a taxon above the generic level as the "Phleides" by Amyot and Serville (1843) and was treated as a sub-

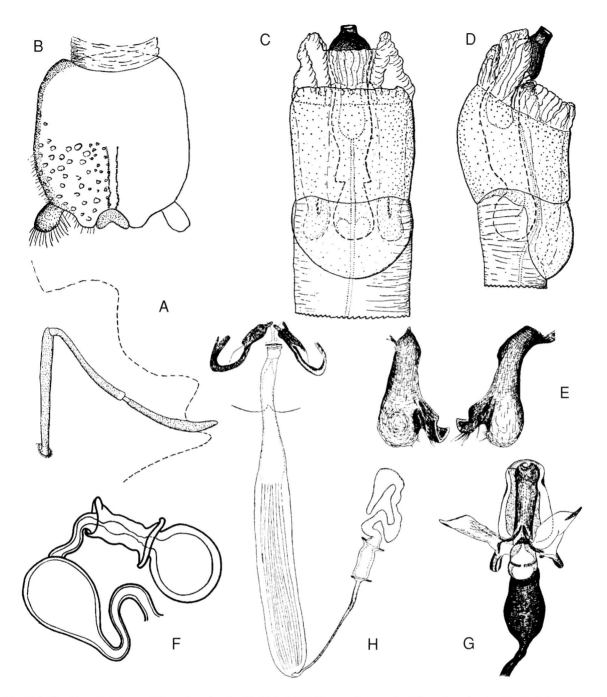

Fig. 74.2. Phoeidae. **A.** Antenna, *Phloeophana longirostris* (Spinola). **B.** Male genital capsule, *Phloeophana longirostris*. **C.** Aedeagus, sagittal view, *Phloeophana longirostris*. **D.** Aedeagus, lateral view, *Phloeophana longirostris* (A–D from Leston, 1953c). **E.** Parameres, *Phloea corticata* (from Lent and Jurberg, 1965). **F.** Spermatheca, *Phloeophana longirostris* (from Pendergrast, 1957). **G.** Conjunctiva, sagittal view, *Phloeophana longirostris*. **H.** Spermatheca, *Phloea subquadrata* Spinola (G, H from Lent and Jurberg, 1965).

family of Pentatomidae (the Phloeina) by Stål (1872) and others. Leston (1953c), Lent and Jurberg (1965), and Rolston and McDonald (1979) treated it as worthy of family status. *Phloeophana* Kirkaldy and *Phloea* Lepeletier and Serville—together containing three species—are recognized.

Specialized morphology. The 3-segmented antennae (Fig. 74.2A), strongly foliate body margins (Fig. 74.1A, B), and unique aedeagal morphology (Fig. 74.2C, D, G) are novel to the group (Lent and Jurberg, 1965).

Natural history. These curious insects are apparently phytophagous. They are wonderfully camouflaged to appear to be pieces of lichen on tree trunks or branches (Lent and Jurberg, 1966). They defend themselves when disturbed by "shooting" a liquid from the scent-gland ostioles a considerable distance from the body. Hussey (1934) summarized the early accounts for *Phloeophana longirostris* (Spinola). He quoted Magalhaes (who first reported maternal care in this species in a local newspaper) as stating that the females protect not only the eggs but also the first-instar nymphs, which attach themselves to the venter of the parent and are carried by her for many days. Magalhaes thought that the young nymphs were fed by the parent, but it is probable that this is merely another case of nonfeeding by the first-instar nymphs. Other accounts indicate maternal care through three instars (see Leston, 1953c, for a detailed discussion of conflicting biological observations).

Distribution and faunistics. The group is restricted to South America. Basic sources are those by Hussey (1934), Leston (1953c), and Lent and Jurberg (1965).

75
Plataspidae

Fig. 75.1. Plataspidae. *Ceratocoris* sp. (from Slater, 1982).

General. The members of this family are beetlelike in appearance, being ovoid or suborbicular and strongly convex, with the scutellum greatly enlarged to cover almost the entire abdomen (Fig. 75.1). Many species are as broad or broader than long—ranging in length from 2 to 20 mm—and some have the mandibular plates developed into horns (Fig. 75.1).

Diagnosis. Head flattened and keeled; antenniferous tubercles located below lateral margins of head, not visible from above; antennae appearing 4-segmented (division between segments 2 and 3 weakly developed); labial segment 2 sometimes much enlarged, flattened, and saclike, with stylets partially coiled (Fig. 75.2A); labium often swollen and sometimes also clypeus; scutellum completely covering abdomen (Fig. 75.1); wings complexly modified, forewings much longer than body (Fig. 75.2B); hind wings specialized to allow folding below scutellum by transverse constrictions (Fig. 75.2C); tarsi 2-segmented; abdominal sterna with a straight transverse sulcus on each side; nymphs with dorsal scent-gland openings between terga 3/4, 4/5 and 5/6, those between 3/4 sometimes much reduced; aedeagus as in Fig. 75.2D, E; parameres as in Fig. 75.2F; spermatheca with well-developed pump and 2 flanges as in Fig. 75.2G, H.

Classification. This taxon was first mentioned as a higher group by Dallas (1851–1852). Fieber (1861) used the name Arthropteridae, Kirkaldy (1909) the name Coptosominae, and Leston (1952) the name Brachyplatidae. Coloration, enlarged scutellum, and wing folding suggest relationships to the Aphylidae, Canopidae, Lestoniidae, Megarididae, and Scutelleridae. The origins and occurrences of these attributes, as well as other characteristics shared by at least some of these groups, have not been investigated using cladistic methods, and thus their status as synapomorphies remains unclear.

Key to Generic Groups of Plataspidae

1. Ocelli placed near eyes; ratio of distance between eyes and ocelli to interocellar distance less than 1:2; abdominal sterna usually convex; head usually narrow, approximately 0.3–0.5 times width of pronotum; base of scutellum (pseudoscutellum) usually raised, demarcated from rest of scutellum by an impressed line . . . . *Coptosoma* Group

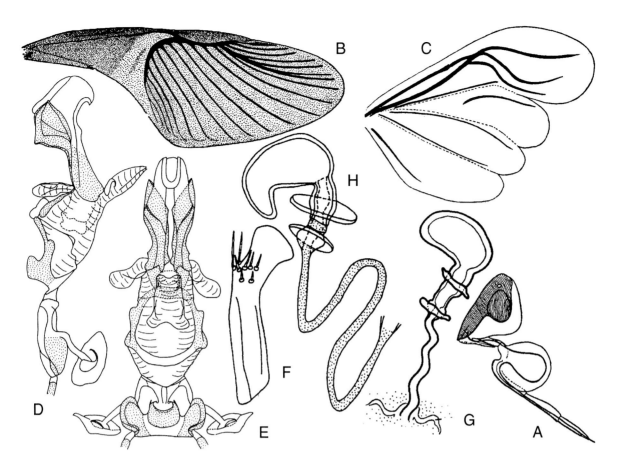

Fig. 75.2. Plataspidae. **A.** Mouthparts, *Bozius respersus* Distant (from China, 1931). **B.** Forewing, *Neocratoplatys salvazai* Miller. **C.** Hind wing, *N. salvazai* (B, C from Miller, 1955). **D.** Aedeagus, lateral view, *Coptosoma scutellatum* (Geoffroy). **E.** Aedeagus, sagittal view, *C. scutellatum*. **F.** Paramere, *C. scutellatum* (D–F from Davidova and Štys, 1980). **G.** Spermatheca, *C. inclusa* Stål (from Pendergrast, 1957). **H.** Spermatheca, *C. scutellatum* (from Davidova and Štys, 1980).

- Ocelli placed near one another, ratio of distance between eyes and ocelli to interocellar distance greater than 1:2; abdominal sterna not, or very slightly, convex; head transverse, usually 0.5–0.7 times width of pronotum; pseudoscutellum absent or weakly developed 2
2. Body flattened; color black, sometimes spotted with yellow, often with a yellow submarginal line on head, pronotum and scutellum .... *Brachyplatys* Group
- Body usually convex; color pattern red, yellow, or brown, maculated with dark brown to black punctures; dark areas sometimes extensive and spotted or flecked with yellow, without a yellow submarginal line on head, pronotum, and scutellum .... *Libyaspis* Group

Specialized morphology. In some species, the labial segments, and sometimes clypeus, are swollen, in which case the stylets extend backward through the thorax into the base of the abdomen (China, 1931).

The complex folding of the wings (Fig. 75.2B, C) and the greatly enlarged and subtruncate scutellum are obvious specializations of the family.

Several tropical African species reach a size of 15–20 mm, and many of these species are strongly sexually dimorphic. The males have enormously produced mandibular plates in the form of "horns" (Fig. 75.1), which may be bifid as in the genus *Ceratocoris* White or enormously prolonged as in *Severiniella* Montandon. Although the function of these horns is unknown, it seems likely that they are involved in territorial defense.

Natural history. China (1931) believed that some plataspids must be mycophagous because they possess very long stylets similar to those found in the Aradidae and Termitaphididae. But Maschwitz et al. (1987) showed a symbiotic relationship between two *Tropidoty-*

Key to generic groups of Plataspidae adapted from Jessop, 1983.

lus spp. and the ant *Meranoplus mucronatus*, wherein the ants tended and protected the bugs for their substantial honeydew secretions, indicating that the bugs are phloem feeders. These authors postulated that the long stylets function in reaching the phloem through the thick bark of the dicotyledonous host tree. Other species of the family are known to attack cowpeas and other legumes in Asia, the Pacific, and Australia.

Monteith (1982b) reported aestivation in large clusters by *Coptosoma lyncea* Stål in the monsoon forests of northern Australia. He described one aggregation on a tree 5 meters in height on which every leaf was covered with the insects. They assumed a regular spacing on the underside of the leaves and were arranged in rows along the petioles, giving the impression of galls or blemishes on the plants. When disturbed large numbers took flight instantly, buzzing loudly and producing a discharge of the stink glands. After several minutes all returned to their original roosting tree and became quiescent again. (See also Natural History, Chapters 88 and 89).

Distribution and faunistics. Plataspids are largely restricted to the tropics and subtropics of the Eastern Hemisphere, but a few species of *Coptosoma* Laporte occur in the temperate Palearctic.

A basic source on the group is that of Jessop (1983), which provides keys to generic groups and a detailed discussion of the *Libyaspis* Group. Davidova-Vilimova and Štys (1980) keyed the Western Palearctic *Coptosoma* spp.

76
Scutelleridae

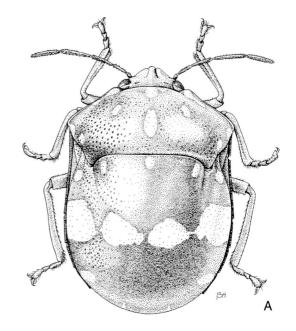

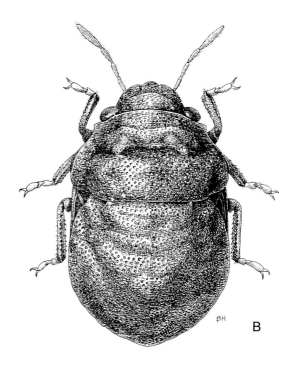

Fig. 76.1. Scutelleridae. **A.** *Pachycoris torridus* (Scopoli). **B.** *Acantholomidea* sp.

General. These insects, which range in length from 5 to 20 mm, are often known as shield bugs because of the greatly enlarged convex scutellum that usually entirely covers the abdomen (Fig. 76.1A, B). Some tropical species are vividly colored, even becoming iridescent, and others have a striking variety of strongly contrasting reds, blues, and yellows. Thus, some scutellerids are among the most spectacularly colored of all Heteroptera.

Diagnosis. Antennae 3- or 5-segmented (Fig. 76.1A); scutellum greatly enlarged and convex, covering or nearly covering entire abdomen (Fig. 76.1A, B); frena obsolete or lacking; strongly laminate propleural carinae present; forewing membrane with numerous veins; tarsi 3-segmented; prosternal sulcus present; sutures of abdominal venter extending to lateral margins; trichobothria paired;

aedeagus as in Fig. 76.2A, B; parameres as in Fig. 76.2C; second valvifers completely fused; first valvulae with reduced rami and largely membranous (Fig. 76.2D); base of spermatheca with a sclerotized groove and an enlarged bulb proximal to flanges (Fig. 76.2E, F).

Classification. The taxon was first established by

Leach (1815) as Scutellerida. Lattin (1964), Kumar (1965), and McDonald and Cassis (1984) reviewed the history of the group. They noted that Amyot and Serville (1843) were the first to subdivide the family (as a subfamily) and that they were followed by Dallas (1851–1852), Stål (1872), and Schouteden (1904–1906). Leston (1952) discussed the tribal arrangement, but treated the taxon as a subfamily, as did Lattin (1964). McDonald and Cassis (1984) stated that Van Duzee (1917) was the first to accord the group family status. But Fieber (1861) recognized a family level taxon as Tetyrae, and Stål (1867) used Tetyridae for the family name. Uhler (1863) recognized the group under the family name Pachycoridae. Dupuis (1947), Pendergrast (1957), and Kumar (1962) all treated the group as of family status.

The question of whether scutellerids should be considered a family or a subfamily of Pentatomidae has remained controversial. Kumar (1965) discovered that the internal male genitalia of scutellerids contain 6 "ecademe" tubules, four paired and two unpaired, opening into the "bulbus ejaculatorius." This feature is not found in any other pentatomoid, and at least suggests scutellerid monophyly. Gaffour-Bensebbane (1990) discovered that the spermatozoa of scutellerids are also distinctive and apparently derived within the Pentatomoidea. They possess a "plumed" structure, aiding the motility and thus providing additional evidence of the integrity of this family. A position outside the Pentatomidae is justified by the fact that scutellerids lack the distinctive spermathecal structure found in the Pentatomidae.

The family contains approximately 80 genera and at least 450 species (Lattin, 1964).

Key to Subfamilies of Scutelleridae

1. Antennae usually 5-segmented but sometimes 3-segmented; lateral portion of posterior margin of abdominal sterna emarginate, recurved anteriorly; metathoracic wing with trace of antevannal vein; 2 intervannal veins .... Scutellerinae
– Antennae 5-segmented (Fig. 76.1A); lateral portion of posterior margin of abdominal sterna not emarginate or recurved; metathoracic wing with antevannal vein absent; a single intervannal vein present 2
2. Abdominal venter with striated areas present on sterna 4, 5, and 6 .... Pachycorinae
– Abdominal venter lacking striated areas 3
3. Scutellum only moderately enlarged laterally, hemelytra exposed for entire length; connexivum broadly exposed; metathoracic wing with intervannal vein well developed .... Eurygastrinae
– Scutellum broadly developed, hemelytra exposed laterally only near base; connexivum at most narrowly exposed; metathoracic wing with intervannal vein greatly reduced ...... Odontotarsinae

EURYGASTRINAE. Moderate size; scutellum not strongly convex; dorsum punctate and glabrous; scutellum attaining end of abdomen, usually parallel-sided and leaving hemelytra and connexivum exposed laterally for almost entire length.

The subfamily is chiefly Old World, with a single Holarctic genus *Eurygaster* Laporte occurring in the Western Hemisphere.

ODONTOTARSINAE. Small to moderate size, length up to 11 mm; punctate, usually with considerable pubescence; head usually transverse and broadly rounded; hemelytra exposed only basally, laterad of large scutellum.

Unlike most other scutellerids the members of this subfamily are chiefly Holarctic with the largest fauna occurring in the Palearctic. However, representatives have been assigned to this subfamily from all major faunal regions. It is sparsely represented in Australia and the Indo-Pacific.

PACHYCORINAE (FIG. 76.1A). Size range small to large; strongly convex dorsally; abdominal sterna striated.

Key to subfamilies of Scutelleridae modified from Lattin, 1964.

The subfamily is primarily New World in distribution and constitutes the majority of New World Scutelleridae, with 27 genera and 125 species (Lattin, 1964). Two genera (*Hotea* Amyot and Serville and *Deroplax* Mayr) are African and are the only Old World representatives.

SCUTELLERINAE. Characterized by features in preceding key; antennae 5- (rarely 3-) segmented; many species large and colorful.

This subfamily is diverse in the Eastern Hemisphere, with only *Augocoris* Burmeister in the Western Hemisphere. Three tribes are recognized: the Elvisurini, Scutellerini, and Sphaerocorini, comprising five genera, all confined to the eastern Hemisphere, with three restricted to Australia. Their distribution extends far into Oceania and westward into Africa and the Middle East. The Elvisurini were recognized as a distinct subfamily by McDonald and Cassis (1984) and Gross (1975–1976). McDonald and Cassis (1984) recognized the Australian genus *Tectocoris* Hahn as a distinct subfamily. The genus contains only a single species, *Tectocoris diophthalmus* (Thunberg).

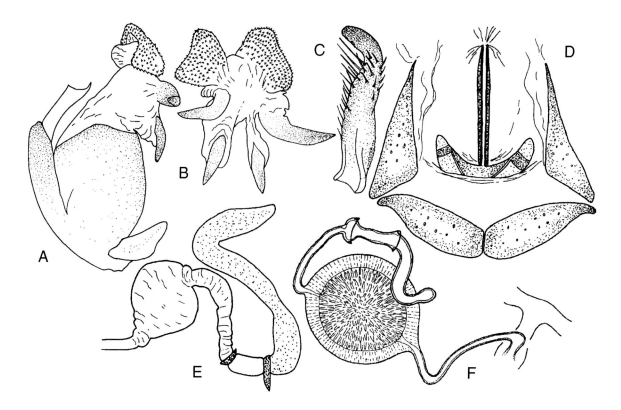

Fig. 76.2. Scutelleridae. **A.** Aedeagus, lateral view, *Eupychodera corrugata* (Van Duzee). **B.** Distal portion of aedeagus, sagittal view, *E. corrugata*. **C.** Paramere, *E. corrugata*. **D.** Terminal female abdominal segments, *Diolcus irroratus* (Fabricius). **E.** Spermatheca, *Pachycoris torridus* (A–E from McDonald, 1966). **F.** Spermatheca, *Sphaerocoris annulus* (Fabricius) (from Pendergrast, 1957).

Specialized morphology. The enlarged scutellum covering the entire dorsal surface of the abdomen, and often also the forewings, may have evolved independently more than once in the Pentatomoidea. Other specialized features of the Scutelleridae include the bizarre color patterns and unique characters of the male genitalia and sperm noted above. The 3-segmented antennae of the Scutellerinae are a derived feature.

Natural history. Despite the frequent large size and brilliant coloration of many scutellerids, biological information on the group is surprisingly sparse. All species are plant feeders and some are of economic importance, especially species of *Eurygaster* Laporte in the Middle East (see Chapter 8). Several species exhibit maternal care. Hussey (1934) discussed this phenomenon in the Neotropical bug *Pachycoris torridus* (Scopoli) in Paraguay. He noted lack of hatching of eggs at the periphery of the mass guarded by the female and suggested parasitism, which has subsequently been well established by Eberhard (1975) for a South American pentatomid (see Natural History, Chapter 73). Hussey also noted reports of egg guarding by *Tectocoris diophthalmus* (Thunberg) in Australia and *Cantao ocellatus* (Thunberg) in the Orient. Miller (1956a) referred to an observation of the latter species as being the only pollinator of the Moon Tree (*Macaranga roxburghi*) in India. *Tectocoris diophthalmus* often does serious damage to cotton bolls and other Malvaceae in Australia, where it is known as the cotton harlequin bug.

Shield bugs, because of the large, usually convex scutellum, often are mistaken for beetles. Several species have a longitudinal stripe down the middle of the scutellum, which increases the similarity of appearance to beetle elytra. This phenomenon merits careful field observations as it is not evident if, or why, this would be a selective advantage for the shield bugs.

The majority of scutellerids are not brightly colored, but the family does include some of the most strikingly colored of all heteropterans. Many are iridescent blue and green, others a rainbow of red, yellow, blue, and green markings. Some of the most bizarrely colored genera are *Callidea* Laporte, *Chrysocoris* Hahn, *Cryptacrus* Mayr, *Cosmocoris* Stål, *Poecilocoris* Dallas, and *Scutellera* Lamarck.

Distribution and faunistics. The family is represented in all major faunal regions, but it is most varied and numerous in the tropics and subtropics, where all of the spectacularly colored species occur.

McDonald and Cassis (1984) revised the Australian fauna and provided a key to the four recognized subfamilies. Lattin (1964) provided a key to subfamilies, detailed discussion of morphology, biological notes, and a treatment of the entire North American fauna. The last world key to genera was presented in the beautifully illustrated work of Schouteden (1904–1906).

77
Tessaratomidae

General. These are large to extremely large (often over 15 mm), robustly ovate or elongate-ovate bugs. They resemble large pentatomids (Fig. 77.1) in general habitus and as a group have no common name.

Diagnosis. Head small, triangular, much narrowed to apex, mandibular plates meeting mesally in front of clypeus; antennae usually 4-segmented, but if 5-segmented then segment 3 very short; head laterally keeled, antenniferous tubercles not visible from above; labium short, not exceeding forecoxae; pronotum extending over base of scutellum; scutellum triangular, not covering corium; hamus present on hind wings; strigil present on Pcu vein of hind wing and plectrum on abdominal tergum 1; metasternum produced laterad between coxae and anteriorly onto mesosternum, most strongly produced as an anterior wedge reaching nearly to front coxae with posterior margin truncate at its junction with abdomen; tarsi either 2- or 3-segmented; 6 pairs of abdominal spiracles usually visible, spiracle of segment 2 strongly exposed on an undifferentiated portion of sternum; abdominal trichobothria posterior to spiracles on sterna 3–7, arranged transversely mesad of spiracle; nymphs with dorsal abdominal scent-gland openings present between terga 3/4, 4/5, and 5/6 (that between 3/4 sometimes small); a small "scar" present between terga 6/7; ninth paratergites greatly enlarged; pygophore as in Fig. 77.2A; aedeagus as in Fig. 77.2B–D; parameres as in Fig. 77.2E, F; spermatheca as in Fig. 77.2G.

Classification. The group was first recognized as a higher taxon by Stål (1864–1865). Leston (1954b), while

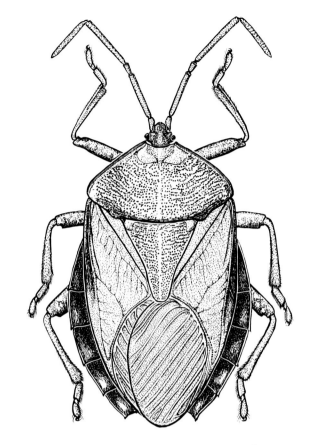

Fig. 77.1. Tessaratomidae. *Musgraveia sulciventris* (Stål) (drawn by S. Monteith; from CSIRO, 1991).

treating it as a subfamily of a very inclusive Pentatomidae, noted resemblances to the male genitalia of the Scutelleridae, such as third dorsal conjunctival appendages short and sclerotized, dorsal seminal duct, canal at junction of phallotheca and reservoir always somewhat undulating, reservoir and vesica merging imperceptibly into one another, and second conjunctival appendages ventral to vesica and membranous. He later (Leston, 1956) elevated the group to family status without comment, a status that has been maintained by Kumar (1969) and subsequent authors.

Three subfamilies, comprising 49 genera and about 235 species, are recognized (Rolston et al., 1993).

Key to Subfamilies of Tessaratomidae

1. Scutellum distinctly longer than wide; tarsi 3-segmented 2
 - Scutellum subequilateral; tarsi 2-segmented .... Natalicolinae
2. Membrane scarcely areolate basally; longitudinal veins arising from base of wing membrane ......
   .... Oncomerinae
 - Membrane areolate basally with longitudinal veins arising from this basal cell-like area ....
   .... Tessaratominae

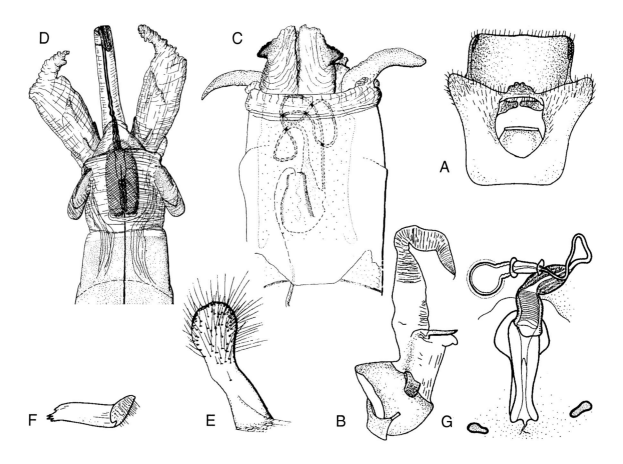

Fig. 77.2. Tessaratomidae. **A.** Genital capsule, *Piezosternum subulatum* (Thunberg). **B.** Aedeagus, lateral view, *P. subulatum* (A, B from McDonald, 1966). **C.** Aedeagus, sagittal view, *P. calidium* (Fabricius) (from Leston, 1954c). **D.** Aedeagus, sagittal view, *Tessaratoma papillosa* (Drury) (from Leston, 1954b). **E.** Paramere, *P. calidium* (from Leston, 1954c). **F.** Paramere, *P. subulatum* (from McDonald, 1966). **G.** Spermatheca, *T. javanica* (Thunberg) (from Pendergrast, 1957).

NATALICOLINAE. Tarsi 2-segmented; male genitalia of type found in Pentatominae (Leston, 1954b).

This is an almost exclusively Ethiopian subfamily.

ONCOMERINAE. Membrane lacking basal cells or with single elongate and narrow but feebly defined cell; hind wing with R+M and Cu parallel and contiguous on proximal two-thirds.

Leston (1955b) included 11 genera separated into two tribes, the Piezosternini and Oncomerini.

TESSARATOMINAE. Membrane with basal cells and longitudinal veins emanating posteriorly.

This taxon is widely distributed in the Old World tropics, with some of the species being among the largest of the Heteroptera outside the Belostomatidae and Coreidae.

Leston (1955b) treated this group as a tribe within his subfamily Tessaratominae. He reduced several species previously considered as tribes to subtribal level. The elevation of the tessaratomids to family rank returns the following to tribal status in the subfamily Tessaratominae: Prionogastrini Stål, Sepinini Horvath, Eusthenini Stål, Tessaratomini Stål, and Platytatini Horvath.

Specialized morphology. The extremely large size, the produced and enlarged wedgelike metasternum, and exposed second abdominal spiracle appear to be specialized features, as may be the brightly colored nymphs.

Natural history. All species so far as known are phytophagous (see also Chapter 8).

Distribution and faunistics. The family is primarily Old World tropical, with the widely distributed *Piezosternum* Amyot and Serville also having three species in the Neotropics.

Rolston et al. (1993) summarized the current classification. Kumar and Ghauri (1970) provided a key to the subfamilies of the world. Yang (1935) discussed the Chinese fauna. Kumar (1974b) keyed the genera of Natalicolinae. Leston (1955b) treated the genera of Oncomerini. Schouteden (1905a) provided keys to the African tribes, genera, and species with color plates. Horvath (1900) keyed family groups and many genera and species.

78
Thaumastellidae

General. These small, somewhat flattened, elongate, brown pentatomoids closely resemble some ground-living Lygaeidae (Fig. 78.1). They are never more than 3.5 mm long and occur as brachypterous and macropterous forms. They have no common name.

Diagnosis. Head porrect with rounded lateral margins; labium 4-segmented, segment 1 obscured in buccular groove; antennae 5-segmented, articulation between segments 2 and 3 not capable of flexure (Fig. 78.1); scutellum triangular, not enlarged (Fig. 78.1); scent-gland channel elongate, sometimes nearly reaching dorsal margin of metapleuron, adjacent evaporative area large, extending onto posterior area of mesopleuron; corium of macropterous morph divided into exo- and endocorium with furrow running along M vein, both parts separately rounded posteriorly, membrane penetrating a short distance between them; claval commissure present; membrane of forewing with reduced venation, no branching veins; flightless morphs with forewings reduced to short truncate (staphylinoid) pads, posterior margin straight, membrane absent (Fig. 78.1) (resembling lygaeids of tribe Plinthisini); coxal combs present, composed of flattened setae; foretibia spinose; tarsi 3-segmented; abdomen lacking inner laterotergites on segments 3–7; all abdominal spiracles ventral; nymphs with 3 pairs of dorsal abdominal scent glands, between terga 3/4, 4/5 and 5/6; abdominal sterna 3–6 with 2+2 trichobothria laterally in oblique rows, sternum 7 with 1+1 trichobothria in *Thaumastella aradoides and T. elizabethae* Jacobs (Fig. 78.2F) but 2+2 in *T. namaquensis* Schaefer and Wilcox (Jacobs, 1989); aedeagus as in Fig. 78.2A, B; parameres as in Fig. 78.2C; abdominal sternum 7 of female not divided; ovipositor valvulae platelike (Fig. 78.2D); second valvifers fused at midline; spermatheca with long coiled duct, pump flanges variable, both distal and proximal flanges present or absent (Fig. 78.2E).

Classification. The genus *Thaumastella* Horvath was originally described as a lygaeid and remained in that family until Seidenstücker (1960) noted that it lacked

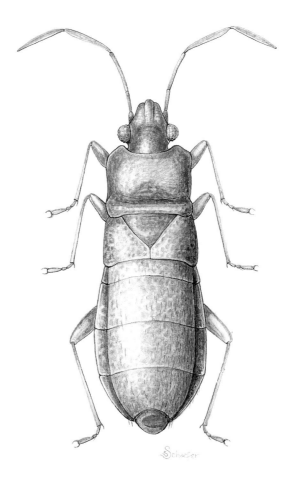

Fig. 78.1. Thaumastellidae. *Thaumastella namaquensis* Schaefer and Wilcox (from Schaefer and Wilcox, 1971).

the laciniate ovipositor characteristic of most Lygaeidae. Štys (1964a) emphasized its pentatomoid relationships and erected the family Thaumastellidae for the then single known species, *Thaumastella aradoides* Horvath, from North Africa and the Middle East. Dolling (1981) considered the group to be a subfamily of Cydnidae, whereas Jacobs (1989) treated it as a distinct family. Although we treat this group at the family level, it is clear that the coxal combs, spinose foretibia, and stridulatory structures ally it with the Cydnidae.

Jacobs (1989) found the chromosome number to be $2n = 20$ ($16XY + m + XY$) but also found conditions of 18 and 17 in different populations of *T. namaquensis*. The presence of an m-chromosome is particularly important, because it is unique in the Pentatomoidea, although it does occur frequently in the Lygaeidae. The presence of a claval commissure is also uncommon in the Pentatomoidea, occurring only in the Amnestinae (Cydnidae) and Urostylidae. Clearly, the Thaumastellidae is a family of great phylogenetic importance, but precise resolution of its relationships will require additional cladistic analysis.

Three species—all placed in the genus *Thaumastella*—

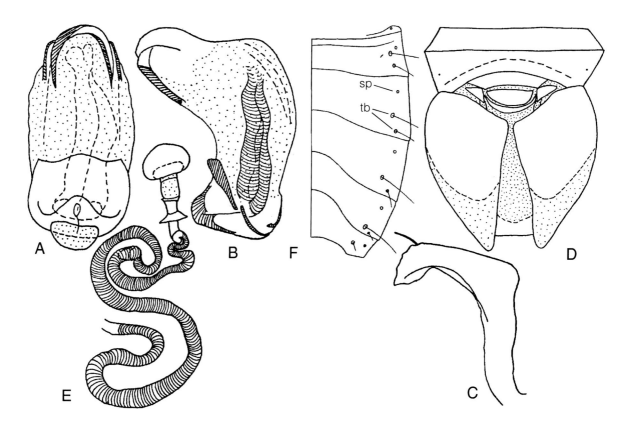

Fig. 78.2. Thaumastellidae. *Thaumastella aradoides* Horvath. **A.** Aedeagus, sagittal view. **B.** Aedeagus, lateral view. **C.** Paramere. **D.** Female terminal abdominal segments. **E.** Spermatheca (A–E from Štys, 1964a). **F.** Ventral view of abdomen with trichobothria (from Seidenstücker, 1964). Abbreviations: sp, spiracle; tb, trichobothria.

are known, the fully winged North African *T. aradoides* and the completely flightless *T. namaquensis* and *T. elizabethae* from Namaqualand, the Richterveld, and Namibia.

Specialized morphology. All species have a well-developed stridulatory mechanism (Jacobs, 1989). Schaefer (1980b) believed that the hind wings were absent in the flightless species from southern Africa and that the stridulitrum was situated on the hemelytra or possibly on the third abdominal tergum. Jacobs (1989), however, showed conclusively that very short, subtriangular hind wings have a stridulitrum formed of a well-sclerotized longitudinal ridge on the underside, situated above the abdominal tergum 1, which bears an extremely finely, transversely ridged, suboval stridulitrum situated near its anterolateral margin. Jacobs argued the "limae" were not involved in stridulation as was suggested by Seidenstucker (1960), Štys (1964a), and Schaefer (1980b).

Natural history. Jacobs (1989) found that the two Namaqualand species live chiefly in cavities under large stones. Despite having discovered a few individuals adjacent to these stones at dusk, he rarely found them leaving these sheltered cavities and believed that they may feed chiefly on seeds that accumulate in such areas by the force of the strong winds. Jacobs also gave an account, however, of *T. elizabethae* leaving its stone shelters to feed on the seeds of *Pharnaceum aurantium* (Aizoaceae), which it dragged along the ground, with the seeds attached to its stylets.

The eggs are large, ovoid, and half the size of the abdomen.

Distribution and faunistics. The family has a disjunct distribution including North Africa, the Near East, and the semidesert areas of southwestern Africa.

The major sources on the group are by Štys (1964a) and Jacobs (1989).

79
Urostylidae

General. These pentatomoids are usually relatively elongate, ranging in length from 3.5 to 14 mm, with elongate legs and a small head (Fig. 79.1). Many have a coreidlike habitus. They have no common name.

Diagnosis. Head small, peltoid, with lateral margins not keeled; antennae 5-segmented, segment 1 much longer than head; antenniferous tubercles broad, strongly exerted, often appearing annulate, placed on or slightly above line running through middle of eye; ocelli placed very close to one another (Fig. 79.2A) (closer than to the eyes, a unique condition in Pentatomoidea); bucculae small; claval commissure reduced (Fig. 79.1) or obsolete (Fig. 79.2A); radial and medial veins of forewings divergent from base; membrane usually with only 4 or 5 longitudinal veins (Figs. 79.1, 79.2A); frenum extending to apex of scutellum; metasternal scent-gland orifice frequently with an elevated spinose auricle; middle and hind coxae widely separated; tarsi 3-segmented; nymphal scent-gland openings present between abdominal terga 4/5 and 5/6; pygophore as in Fig. 79.2C; vesica short with conjunctival appendages (Fig. 79.2D); second valvifers fused to form an M-shaped or W-shaped sclerite (but said to be separate by Gapud, 1981); chromosome number $2n = 16$.

Classification. The systematic position of this family has long been ambiguous. Kumar (1971) reviewed the early literature, noting that Singh-Pruthi (1925) related it to the Acanthosomatidae, Yang (1938a, b, 1939) and Pendergrast (1957) to the Pyrrhocoridae, and Miyamoto (1961a) to the Pentatomidae. Kumar (1971) believed that the group represents an early divergence from the other pentatomomorphans, possibly together with the Pyrrhocoridae with which they share uniquely the fused second valvifers that form an M- or W-shaped sclerite. China and

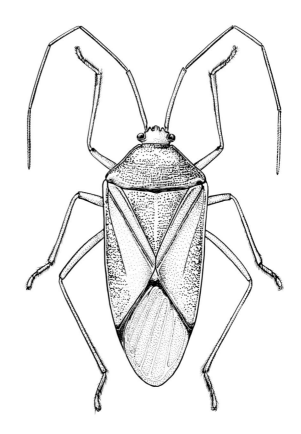

Fig. 79.1. Urostylidae. *Urolabida* sp. (drawn by S. Monteith; from CSIRO, 1991).

Slater (1956:411) indicated that the urostylids "must represent the Proto-Trichophora at the base of the Pentatomidae, Coreidae and Lygaeidae." Schaefer and Ashlock (1970:629) noted that, although primitive, they are "some distance from the origin of the Pentatomoidea." Since most of the above was based on study of only one or a very few species and the language mostly "precladistic," clearly a comprehensive study of the family and its relationships is badly needed. Whatever its more precise position, the trichobothrial number and pattern suggest placement in the Pentatomoidea.

Key to Subfamilies of Urostylidae

1. Body relatively large; well over 5 mm in length; spiracles ventral; scent-gland auricle with a spine; hamus present in hind wing .... Urostylinae
- Body very small, at most scarcely exceeding 4 mm in length; spiracles lateral (except segment 2 ventral when present); scent-gland auricle not spined; hamus absent .... Saileriolinae

SAILERIOLINAE. Very small, unlike Urostylinae in habitus; head strongly declivent, eyes close to base; antennal segment 3 very short; legs simple, mutic; meso- and metacoxae separated by distance nearly twice their length; tarsal segments 1 and 3 each longer than segment 2; scutellum swollen anteromesally; corium semihyaline, apex

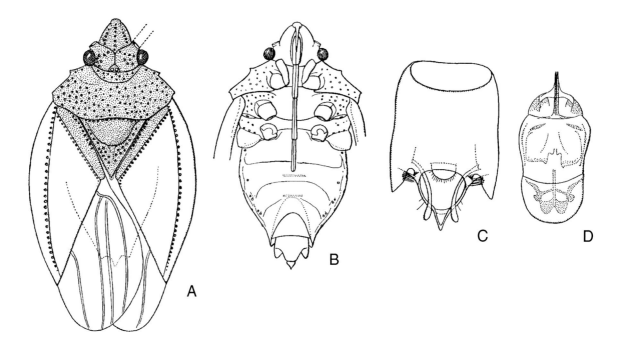

Fig. 79.2. Urostylidae. *Saileriola sandakanensis* China and Slater (from China and Slater, 1956). **A.** Habitus. **B.** Venter. **C.** Male genital capsule. **D.** Aedeagus, sagittal view.

extending beyond apex of abdomen; trichobothria apparently always absent on abdominal sterna 3 and 4; metathoracic scent-gland orifice slitlike, lacking a peritreme and evaporative area.

Three genera are known: *Saileriola* China and Slater (two species; Borneo and Vietnam) and the monotypic *Bannacoris* Hsiao (1964) from Yannan Province, China, and *Ruckesona* Schaefer and Ashlock from Thailand.

UROSTYLINAE (FIGS. 79.1, 79.2A). Relatively large; habitus somewhat coreoid; antennae usually long, sweeping; legs long; body more elongate than in most pentatomoids; abdominal trichobothria paired, transverse, present on sterna 3–7.

Approximately four genera and more than 80 species are known.

Specialized morphology. The reduced third antennal segment in the Saileriolinae and perhaps the fused second valvifers are certainly novel features, as presumably is the elongation of the corium in the Saileriolinae.

Natural history. Little is known of the biology of these pentatomoids other than a collection record of *Ruckesona vitrella* Schaefer and Ashlock on "palm at water edge." Nymphs of several instars were taken, suggesting that this unidentified palm is a true host plant. Schaefer and Ashlock (1970) noted that both adults and nymphs of this species had what appeared to be fragments of chloroplasts in the gut, suggesting that the insects do not feed exclusively upon sap.

Distribution and faunistics. Urostylids occur in southern and eastern Asia, reaching northward into the eastern Palearctic and southwest into New Guinea. A basic reference on the group is by Yang (1939).

Lygaeoidea

80
Berytidae

General. Commonly known as stilt bugs, these insects, which range in length from 2.5 to 11 mm, are usually elongate and slender, with threadlike legs and antennae (Fig. 80.1A, B). Many superficially resemble species of Hydrometridae or the reduviid subfamily Emesinae, but lack the raptorial forelegs of the latter. The majority are rather dull yellowish or reddish brown and mutic, but some species are bizarrely ornamented with spines and other protuberances (Fig. 80.1B).

Diagnosis. Head subspherical, often with clypeus

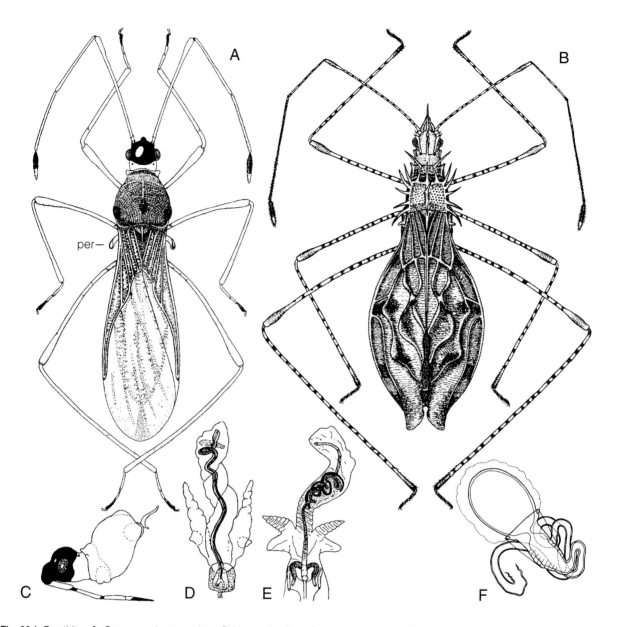

Fig. 80.1. Berytidae. **A.** *Gampsocoris panormimus* Seidenstücker (from Seidenstücker, 1965a). **B.** *Acanthoberytus wygodzinskyi* Stusak (from Stusak, 1968). **C.** Lateral view head, thorax, scutellum, *G. panormimus* (from Seidenstücker, 1965a). **D.** Aedeagus, sagittal view, *Berytinus hirticornis* (Brullé). **E.** Aedeagus, sagittal view, *G. culicinus* Seidenstücker (D, E from Péricart, 1984). **F.** Spermatheca, *B. minor* (Herrich-Schaeffer) (from Pendergrast, 1957). Abbreviation: per, peritreme.

produced anteriorly (Fig. 80.1C); antennae located above a line through middle of eye; antenniferous tubercles reduced; antennal segment 4 usually short and somewhat swollen; distal ends of femora often swollen (Fig. 80.1A, B); scutellum pointed posteriorly; peritreme of metathoracic scent gland usually uniquely produced, often as an elongate spine (Fig. 80.1A); corium usually in part desclerotized; all abdominal spiracles dorsal; adults usually with 3 trichobothria (sometimes 2) on abdominal sternum 3; abdominal mediotergites fused; nymphs with dorsal abdominal scent-gland openings between terga 3/4 and 4/5 or only between terga 3/4; aedeagus as in Fig. 80.1D, E; ovipositor reduced; sternum 7 of females entire; apical spermathecal bulb large, ovoid, and globular with distal pump flange well developed, proximal flange reduced or absent (Fig. 80.1F); nymphs usually with glandular setae.

Classification. The taxon was first established by Fieber (1851) as Berytidea. Costa (1853) (as Berytini) and Uhler (1876) (as Berytina) treated them as a subfamily of the Coreidae. Stål (1874) (as Berytina) considered them to be a subfamily of Lygaeidae. Kirkaldy (1902) recognized a subfamily Neidinae and treated the entire taxon at family rank. Because *Neides* Latreille, 1802 was considered to be a senior synonym of *Berytus* Fabricius, 1803 and because Kirkaldy believed that genera in synonymy could not serve as the basis for higher group names, he used the name Neididae and was followed by many subsequent authors.

Berytidae have long been considered to be closely related to some subgroups of Lygaeidae, especially the subfamily Cyminae. Southwood and Leston (1959) placed the Cyminae in the Berytidae, presumably on the basis of the very short claval commissure, similarities of the egg, female genitalia, and chromosome numbers. Hamid (1975) noted that Southwood and Leston gave no character evaluation, that the chromosome number was true for only one subfamily, and that some characters were obvious plesiomorphies. He also listed 15 character states in which the Cyminae and Berytidae differed. Hamid also disagreed with Štys's (1967c) discussion of the spiracle position in the two groups and concluded that the placement of the Cyminae in the Berytidae was unwarranted (see also Cyminae, Chapter 83). Péricart (1984) agreed, stating that some of the apparent similarities are the result of parallel evolution.

Two subfamilies are recognized, the Berytinae and Metacanthinae, the latter established by Douglas and Scott (1865) as a separate family. They comprise about 39 genera and 150 species (Péricart, 1984).

Key to Subfamilies of Berytidae

1. Frons usually produced above clypeus into a laterally compressed crest or cone; scutellum triangular without denticle or spine; abdomen ventrally punctate .... Berytinae
– Frons rounded anteriorly, without a crest or cone, but often spinose or tuberculate; scutellum more or less semicircular, usually with a long denticle or spine in middle, or with apex produced into a horizontal curved spine; abdomen ventrally impunctate .... Metacanthinae

BERYTINAE. Head truncate anteriorly or prolonged above postclypeus by an elongate process; pronotum elevated before the posterior margin in macropterous forms; scutellum lacking an upstanding spine; antennae and legs lacking extensive black annulations; frequently brachypterous or micropterous; hemelytral membrane with 5 distinct veins; metathoracic scent-gland auricle variable, but never prolonged into fingerlike process; abdomen densely punctate below; 2 pairs of trichobothria on sternum 3; female terga 8 and 9 divided completely by a longitudinal groove; first valifer fused with paratergite 8 (except in *Apoplymus* Fieber).

This group of 11 genera is broadly distributed in the Old World, with only *Neides muticus* (Say) occurring natively in North America.

METACANTHINAE. Head anteriorly annulate or subtruncate, but lacking a forward-projecting spinelike process; pronotum more or less trituberculate or trispinose posteriorly, sometimes spinose laterally; antennae and legs usually annulated with dark rings; usually macropterous; scutellum with an upstanding spine; metathoracic scent-gland auricle variable, frequently produced as an elongate spinelike process; abdomen impunctate ventrally or with at most scattered punctures; trichobothria on sterna 3–7; female terga 8 and 9 entire; first valifer not fused with paratergite 8.

The Metacanthinae is a group of worldwide distribution, with 28 genera currently recognized.

Specialized morphology. The extended metapleural scent-gland auricle, elongate slender appendages, and especially the glandular setae of the nymphs are specializations that are infrequently, or never, found in other Heteroptera.

Natural history. The majority of species are thought to be phytophagous and live above the ground on plants; others are geophilous. At least some species of *Jalysus* Stål, *Neides* Latreille, and *Berytinus* Kirkaldy are in part predatory (Poisson and Poisson, 1931; Wheeler and Henry, 1981). Many species are known to live on plants covered with a sticky glandular pilosity (Péricart, 1984). Péricart (1984) should be consulted for a summary of individual biological studies. Wheeler and Henry (1981) presented details of the biology of *Jalysus wickhami* Harris and *J. spinosus* (Say) (see Chapter 8). Wheeler and Schaefer (1982) presented a host list and discussed feeding trends.

Distribution and faunistics. The family is represented in all major zoogeographic regions.

Péricart's (1984) study of the Palearctic fauna is the most exhaustive recent work. Stusak has published many descriptive papers on the world fauna, for example,

Key to subfamilies of Berytidae adapted from Kerzhner and Jaczewski, 1964.

his recent paper on the Oriental fauna (Stusak, 1989). Froeschner (1981) provided a key to the South American fauna.

81
Colobathristidae

General. The members of this family, which range in length from 6 to 20 mm, are generally very elongate insects with slender legs and antennae and a punctate dorsal body surface (Fig. 81.1). They have no common name.

Diagnosis. Eyes substylate (Fig. 81.1; except in *Dayakiella* Štys); antennae elongate, 4-segmented (Fig. 81.1), located above a line running through middle of eyes; antenniferous tubercles reduced; ocelli present; infraocular ridges sometimes strongly developed; distal end of scutellum narrowed, elongate, sometimes armed with an erect spine; forewing narrowed, lateral corial margin concave; corium at least in part transparent or translucent, usually with a triangular distal cell; membrane of forewing with veins reduced (Fig. 81.2A) or absent; clavi usually overlapping, no claval commissure; abdomen constricted at base; abdominal spiracles 2, 3, and 4 dorsal, those on segments 5, 6, and 7 ventral; abdominal trichobothria as in Fig. 81.2B; nymphal scent-gland openings between terga 3/4, 4/5, 5/6, those between 3/4 and 4/5 reduced; aedeagus with elliptical phallobase and elongate desclerotized tubular vesica (Fig. 81.2C, E); parameres as in Fig. 81.2D, F; ovipositor platelike, sternum 7 not split mesally; spermathecal duct elongate, coiled, bulb globular, with reduced flanges (Fig. 81.2H); second valifers fused; eggs spindle-shaped with an obliquely set pseudoperculum.

Classification. The taxon was first recognized by Stål (1864–1865) as a subfamily of Lygaeidae and was ele-

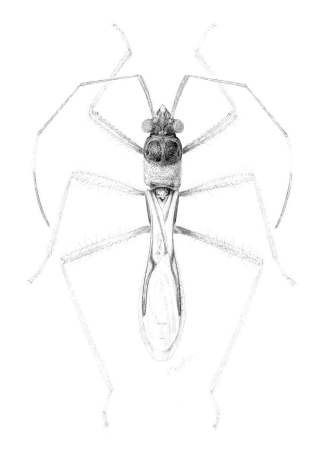

Fig. 81.1. Colobathristidae. *Colobathristes chalcocephalus* Burmeister (from Slater, 1982).

vated to family status by Bergroth (1910). Štys (1966a) considered it as belonging to the "malcid evolutionary line," which included the cymine lygaeids, the Malcidae, and Berytidae. Kumar (1968) discussed the close relationship to the Lygaeidae and believed that the family might well be included in the Lygaeidae as a subfamily, and Štys (1966a) commented on intrafamilial relationships in detail. The relationships of the family are in need of further clarification.

The group contains 23 genera and 83 species.

Key to Subfamilies of Colobathristidae

1. Antennae relatively short, subequal to half length of body; antennal segment 1 shorter than width of head across eyes .... Dayakiellinae
– Antennae elongate, longer than or subequal in length to body length; antennal segment 1 longer than width of head across eyes .... Colobathristinae

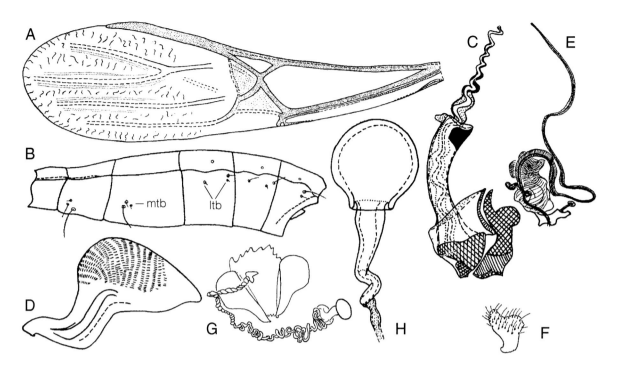

Fig. 81.2. Colobathristidae. **A.** Forewing, *Dayakiella brevicornis* Horvath. **B.** Abdomen, with trichobothria, *D. brevicornis* (A, B from Štys, 1966b). **C.** Aedeagus, lateral view, *Phaenacantha australiae* Kirkaldy. **D.** Paramere, *P. australiae*. **E.** Aedeagus, *Symphylax musiphthora* Ghauri. **F.** Paramere, *S. musiphthora*. **G.** Spermatheca, *S. musiphthora* (E–G from Ghauri, 1968). **H.** Spermatheca, *P. australiae* (C, D, H from Štys, 1966a). Abbreviations: ltb, lateral trichobothria; mtb, mesial trichobothria.

COLOBATHRISTINAE (FIG. 81.1). Antennal length greater than or subequal to body length; legs very long; tarsal segment 1 at least 1.66 times as long as (usually longer than) combined length of segments 2 and 3; hind wings lacking Cu vein; laterotergites of segment 7 "normally" developed, posteriorly free; male genitalia terminal (often telescoped).

The group contains 11 genera and 44 species from the Oriental and Australian regions and 11 genera and 37 species from the Neotropics (Kormilev 1949, 1951).

DAYAKIELLINAE. Antennae about one-half length of body; facies lygaeid-like; eyes sessile; head lacking infraocular ridges; scutellum with long horizontal spine; tarsal segment 1 at most 1.17 times as long as combined length of segments 2 and 3; hind wings with free distal part of Cu and with recurrent glochis; both dorsal and ventral laterotergites of segment 7 meeting posteriorly and fusing to form a genitoanal chamber.

This subfamily contains only *Dayakiella* Horvath, with two species (*D. brevicornis* Horvath and *D. sumatrensis* Štys) known solely from Indonesia. Štys (1966b) believed that it possesses several plesiomorphic characters.

SPECIALIZED MORPHOLOGY. The aedeagus is similar in form to the apparently derived condition found in the lygaeid subfamilies Pachygronthinae and Heterogastrinae, where the sperm reservoir is reduced and located near the distal end of the elongate membranous tubular phallus and in which holding sclerites are lacking.

Several Neotropical genera possess densely striate infraocular ridges representing the stridulitrum of the stridulatory apparatus. Fine spines present on the inner side of the anterior femora presumably function as the plectrum (Štys, 1966a).

Natural history. All colobathristids whose habits are known feed exclusively on grasses. *Phaenacantha saccharicida* (Karsch) is destructive to sugar cane in Indonesia (Miller, 1956a). Many species are myrmecomorphic; species of *Tricentrus* Horvath are strikingly so (Kormilev, 1949).

Distribution and faunistics. This family has extensive Neotropical and Oriental faunas, but not a single species occurs in Africa. Štys (1966a) believed that the Neotropical and Oriental faunas are quite distinct and may well be found to represent distinct subfamilies. *Phaenacantha*, however, is represented by a single Neotropical species (*P. saileri* Kormilev), whereas 27 other species of the genus are Oriental-Australian. Basic works on the group include those of Horvath (1904) as well as those of Kormilev and Štys cited above. The key of Carvalho and Costa (1989) allows for identification of Neotropical genera.

82
Idiostolidae

General. Members of this family range from 5 to 7 mm in length and resemble broad-bodied species of the lygaeid genus *Ozophora* Uhler (Fig. 82.1). They are of moderate size and have no common name.

Diagnosis. Head porrect; antennae inserted at level below ventral margin of eye; ocelli present; membrane of forewing with 5 longitudinal veins or numerous cells (Figs. 82.1, 82.2A); legs slender and mutic; tarsi 3-segmented; abdomen with numerous trichobothria as follows: adults with 7 on sterna 3 and 4 (positioned both medially and laterally) and 4 laterally on segments 5–7 (Fig. 82.2B); nymphs with 3 on segments 3 and 4, 1 on segment 5, 2 on segment 6, and 3 on segment 7 (Schaefer, 1966a); all spiracles ventral; dorsal inner laterotergites present (Fig. 82.2B); nymphal body shape ovoid; nymphal abdominal scent glands between abdominal terga 4/5 and 5/6 and a scar representing a small gland between terga 3/4; abdominal terga 1 and 2 and terga and sterna 3–6 fused; aedeagus as in Fig. 82.2D; parameres as in Fig. 82.2E; females with sternum 7 completely divided medially; ovipositor laciniate; spermatheca absent.

Classification. This taxon was originally described as a subfamily of the Lygaeidae by Scudder (1962a). Subsequently it was elevated to family and then superfamily status by Štys (1965) (see also Schaefer, 1966a; Štys and Kerzhner, 1975). Although we include the Idiostolidae within the Lygaeoidea, this may not correctly represent its phylogenetic position within the Pentatomomorpha, judging from the unusual combination of characters possessed by the group.

Three genera comprising four species are known: *Idiostolus insularis* Berg from southern South America, *Trisecus pictus* Bergroth from Tasmania, *T. armatus* Woodward from New South Wales, and *Monteithocoris hirsutus* Woodward from Tasmania.

Specialized morphology. The novel trichobothrial patterns (Fig. 82.2B), lack of spermatheca and vesica, and the broadly ovoid shape of the nymphs are all specialized conditions (Schaefer, 1966b; Schaefer and Wilcox, 1969).

Natural history. Idiostolids live in moss and litter in *Nothofagus* forests and are almost certainly phytophagous. They are most easily collected by processing in a Berlese funnel samples of litter and moss from the far southern forests.

Distribution and faunistics. The family possesses a strikingly disjunct transantarctic distribution.

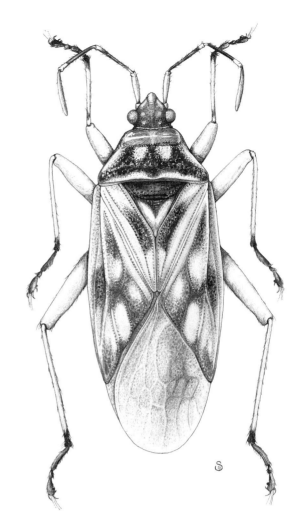

Fig. 82.1. Idiostolidae. *Trisecus pictus* Bergroth (from Schaefer and Wilcox, 1969).

Basic references in the group are by Scudder (1962a), Woodward (1968b), and Schaefer and Wilcox (1969).

83
Lygaeidae

General. Members of this large and diverse family are extremely varied in size (1.2–12 mm) and form (Figs. 83.3–83.6). Most species are rather small and obscurely brown or black, but many are brightly colored red or yellow and black. Some have conspicuously enlarged forefemora. They are often referred to as seed bugs.

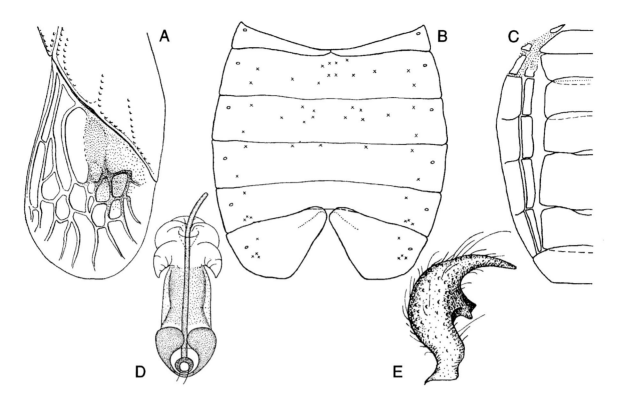

Fig. 82.2. Idiostolidae. **A.** Hemelytron, *Trisecus pictus* (from Schaefer and Wilcox, 1969). **B.** Abdomen, ventral view, *Idiostolus insularis* Berg, trichobothrial positions marked by "x." **C.** Abdomen, dorsal view, *I. insularis*. **D.** Aedeagus, sagittal view, *I. insularis* (B–D from Schaefer, 1966a). **E.** Paramere, *I. insularis* (from Schaefer and Wilcox, 1969).

Diagnosis. Ocelli present except in brachyterous forms; bucculae well developed; antennae located below a line drawn through middle of eye; forewing with 4–5 veins in membrane; abdominal spiracle position extremely variable; usually with 3 trichobothria submesally or laterally on abdominal sterna 3 and 4, and laterally on sterna 5 and 6, 2 trichobothria laterally on sternum 7 (Fig. 83.1A, B); aedeagus usually with conjunctival lobes and processes and a distinct vesica (Fig. 83.2A–F); sperm reservoir present (Fig. 83.1C); parameres variable in form, often elongate and slender or broadened (Fig. 83.1D); testes as in Fig. 83.2G; ovipositor usually laciniate (Fig. 83.1E); spermatheca usually with distinct bulb and flange (Figs. 83.1F, 83.2H, I); egg with 3 to 15 micropyles.

Classification. The Lygaeidae are probably paraphyletic, with some of the subfamilies presumably being the sister taxa of members of other groups such as Berytidae, Colobathristidae, and Malcidae (Southwood and Leston, 1959; Štys, 1967c). Consequently, the family is difficult to characterize, and the complex relationships have not yet been worked out. Many subfamilies will probably be elevated to family status in the future. Nonetheless, there has been a great deal of systematic work on the higher classification of the Lygaeidae in recent years, and the majority of subfamily and tribal taxa appear to be reasonably well established as monophyletic.

The family possesses great diversity of some features found to be relatively constant in other lygaeoid (and pentatomomorphan) families. For example, the spermatheca is extremely variable and useful taxonomically chiefly at subfamilial levels, as are the position of the abdominal spiracles, the dorsal abdominal scent-gland openings, and the number and placement of the abdominal trichobothria.

The taxon was first recognized by Schilling (1829), and the most complete early synthesis was by Stål (1872). Slater (1964b) provided a modern world catalog. Kumar (1968) discussed the relationships relative to related superfamilies.

At least 500 genera and 4000 species are known.

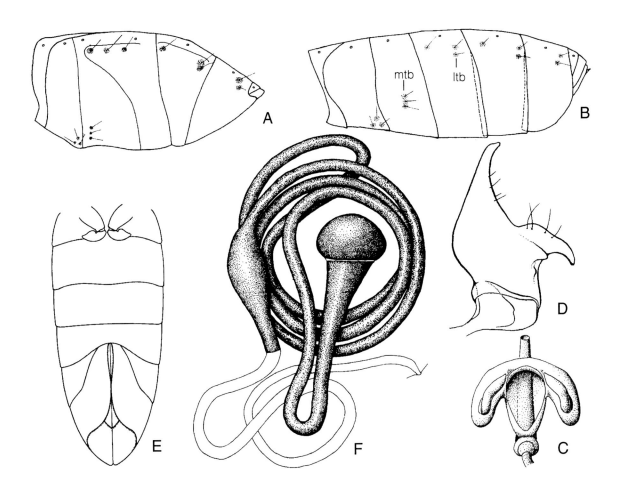

Fig. 83.1. Lygaeidae. **A.** Female abdomen and trichobothria, lateral view, *Targarema stali* (White). **B.** Female abdomen and trichobothria, lateral view, *Phasmosomus araxis* Kiritshenko (A, B from Sweet, 1967). **C.** Sperm reservoir, sagittal view, *Xenoblissus lutzi* Barber (from Slater, 1979). **D.** Paramere, *Ozophora singularis* Slater (from Slater, 1983). **E.** Female abdomen, ventral view, *Ischnodemus* sp. (from Slater, 1979). **F.** Spermatheca, *Plinthisus flindersi* Slater and Sweet (from Slater and Sweet, 1977). Abbreviations: ltb, lateral trichobothria; mtb, mesial trichobothria.

Key to Subfamilies of Lygaeidae

1. Suture between abdominal sterna 4 and 5 usually curving forward laterally and rarely (e.g., *Gastrodes* Westwood, *Phasmosomus* Kiritshenko, *Caenusia* Strand) attaining lateral margins of abdomen (Fig. 83.1A); if 4/5 suture complete, trichobothria usually present on head (Fig. 83.6A, B) .... Rhyparochrominae
– Suture between abdominal sterna 4 and 5 not curving forward, attaining lateral margins of abdomen; head without trichobothria 2
2. Spiracles on abdominal segments 2–7 all located dorsally[1] 3
– At least one pair (and often more) of spiracles on abdominal segments 2–7 located ventrally (Fig. 83.1A, B) 6
3. Clavus at least in part punctate; posterior margin of pronotum not depressed laterad of base of scutellum, or clavus and corium fused into a hard convex, coarsely punctate shell 4
– Clavus impunctate; posterior margin of pronotum depressed between scutellum and humeral angles 5
4. Forewings forming a convex, beetle-like shell, wings meeting evenly down midline (Fig. 83.5); tarsi 2-segmented; abdominal segment 5 with a single trichobothrium; no trichobothria present on abdominal segment 4 .... Psamminae

[1] A few Cyminae key out at this point but are readily recognizable by the coarsely punctate body surface.

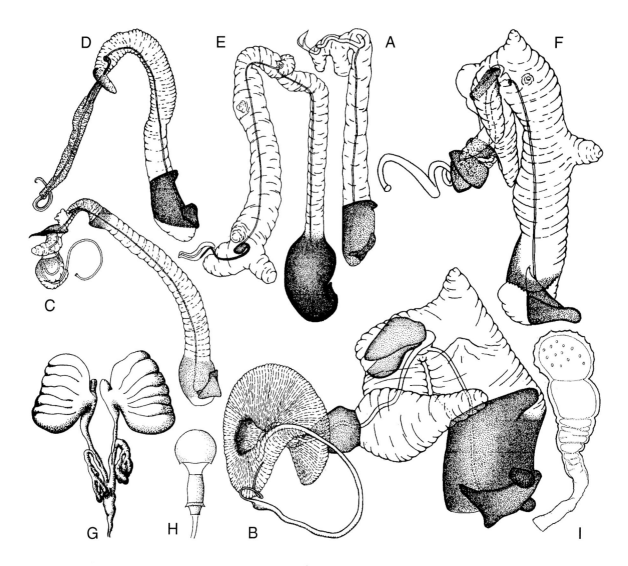

Fig. 83.2. Lygaeidae. **A.** Aedeagus, lateral view, *Cymus angustatus* Stål. **B.** Aedeagus, lateral view, *Oncopeltus fasciatus* (Dallas). **C.** Aedeagus, lateral view, *Orsillus depressus* Dallas. **D.** Aedeagus, lateral view, *Oxycarenus hyalinipennis* (Costa). **E.** Aedeagus, lateral view, *Oedancala dorsalis* (Say). **F.** Aedeagus, lateral view, *Myodocha serripes* Olivier (A–F from Ashlock, 1957). **G.** Testes, *Nysius thymi* (Wolff) (from Pendergrast, 1957). **H.** Spermatheca, *Macropes femoralis* Distant (from Slater, 1979). **I.** Spermatheca, *Oxycarenus hyalinipennis* (from Carayon, 1964b).

- Forewings with a distinct clavus, corium, and membrane, membranes overlapping one another; tarsi 3-segmented; 2 or 3 trichobothria present on both abdominal segments 4 and 5 .... Ischnorhynchinae
5. Apical corial margin straight; hind wing with a subcosta but lacking intervannals; often brightly colored with red, yellow, orange, and black .... Lygaeinae
- Apical corial margin sinuate on mesal half; hind wing lacking a subcosta but with intervannals present; usually dull brownish yellow with hemelytra partially hyaline .... Orsillinae
6. Spiracles of abdominal segment 7 ventral, all others dorsal 7
- At least spiracles of abdominal segments 6 and 7 ventral 9
7. Hemelytra impunctate or at most with only weak, scattered punctures (Fig. 83.3A) .... Blissinae
- Hemelytra coarsely punctate 8
8. Bucculae short not extending caudad of base of antenniferous tubercles; trichobothria present laterally on abdominal sterna 3–7 .... Cyminae
- Bucculae elongate, extending to base of head; trichobothria present laterally only on abdominal sterna 5 and 6 .... Cryptorhamphinae

9. Spiracles on abdominal segments 3 and 4 dorsal 10
– Spiracles on abdominal segments 3–7 ventral 12
10. Abdominal sterna 2–4 with sutures fused and obliterated; no lateral trichobothria on segments 3 and 4; body appearing myrmecomorphic .... Bledionotinae
– Abdominal sterna 2–4 with distinct sutures present, trichobothria on sterna 3 and 4 lateral; body generally short and stout, not myrmecomorphic 11
11. Spiracles on abdominal segment 5 dorsal; spiracles on segment 2 ventral .... Henestarinae
– Spiracles on abdominal segment 5 ventral; spiracles on segment 2 dorsal .... Geocorinae
12. Abdominal sterna 3–5 lacking lateral trichobothria; females without a spermatheca .... Henicocorinae
– Abdominal sterna 3 and 4 each with 2 or 3 lateral trichobothria; females with a spermatheca 13
13. Spiracles of abdominal segment 2 dorsal 14
– Spiracles of abdominal segment 2 ventral 15
14. Lateral pronotal margins explanate or laminate .... Artheneinae
– Lateral pronotal margins rounded or at most slightly carinate, never conspicuously explanate .... Oxycareninae
15. Cross vein present in membrane of forewing creating a closed basal cell; forefemora at most weakly incrassate and with few spines; hamus of hind wing arising distad of point on discal cell where cubitus diverges as a free vein .... Heterogastrinae
– No cross vein or closed cell basally in membrane of forewing; forefemora strongly incrassate and heavily spinose (Fig. 83.4); hamus of hind wing arising in discal cell basad of divergence of cubitus as a free vein .... Pachygronthinae

ARTHENEINAE. Pronotal margins explanate; hamus lacking; abdominal spiracles 3 through 7 located ventrally on sternal "shelf"; complete, unfused suture between abdominal sterna 4 and 5 in female; inner laterotergites lacking; trichobothrial pattern conventional, with those on sternum 5 located one above the other on a single elevation.

Four tribes are recognized: Polychismini Slater and Brailovsky (1986), containing only *Polychisme ferruginosus* (Stål) from northern South America; Artheneini, containing five genera and 16 species; Dilompini, containing only *Dilompus* Scudder, with two known species from southeastern Australia and New Zealand; and Nothochromini Slater, Woodward, and Sweet, containing a single species, *Nothochromus maoricus* Slater, Woodward, and Sweet from New Zealand.

BLEDIONOTINAE. Strikingly myrmecomorphic; pronotum sometimes with bizarre ornamentation, lateral margins rounded; anterior abdominal segments fused, sutures obliterated; spiracles dorsal on abdominal segments 2, 3 and 4, ventral on segments 5–7; dorsal abdominal scent-gland openings present between terga 4/5 and 5/6, these segments curving strongly posteriorly from lateral margins to meson.

Two tribes are recognized: Bledionotini, which includes only the strongly myrmecomorphic *Bledionotus systellonotoides* Reuter from the Near East, and Pamphantini, which contains seven genera and 20 species. The Pamphantini are chiefly Neotropical; most species are known from Cuba and Hispaniola, but there are also a few South American species. *Austropamphantus woodwardi* Slater occurs in northern Australia.

There is considerable doubt about the monophyly of the subfamily. Most of the characters used to associate the two tribes are the result of myrmecomorphic modifications.

BLISSINAE (FIG. 83.3A). Hemelytra not, or only weakly, punctate; body frequently covered with a pruinose layer formed of minute spicules; spiracles dorsal on abdominal segments 2–6, ventral on segment 7; nymphal scent-gland openings present between terga 4/5 and 5/6; body shape ranging from elongate and extremely slender to short and stout.

This is the so-called chinch bug subfamily. Approximately 50 genera and 385 species are currently recognized. The group is worldwide in distribution but most abundant in the tropics. *Ischnodemus* Fieber has a large element that shows close South American–African relationships, but the group is absent from the rest of the Old World tropics. Slater (1979) monographed the subfamily, including keys to genera and species and cladograms of the genera and some species groups. All members of the subfamily breed only on monocotyledonous plants. Poaceae is the most frequent host group, although Cyper-

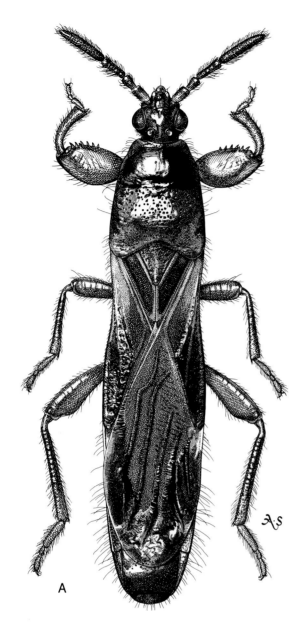
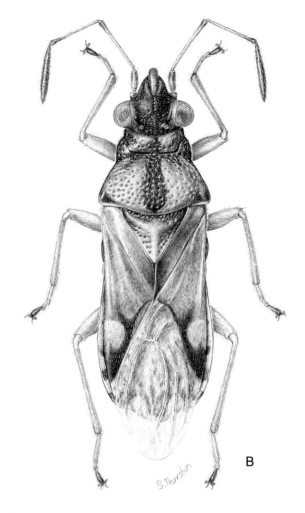

Fig. 83.3. Lygaeidae. **A.** Blissinae. *Lucerocoris brunneus* Slater (from Slater, 1968). **B.** Orsillinae. *Hyalonysius ashlocki* Slater.

aceae are common hosts and groups such as the Restionaceae are less common (Slater, 1976).

CRYPTORHAMPHINAE. Elongate, brownish yellow; body coarsely punctate; bucculae long, extending to base of head; pronotum lacking a median carina; corium with a distinct subcosta, radius, media, and cubitus; hind wing with well-developed hamus and anterior and posterior vannals; dorsal abdominal scent-gland scars present between terga 4/5 and 5/6; ninth paratergite secondarily cleft; trichobothria present only on sterna 5 and 6 (Hamid, 1971a).

Two genera and four species are currently known. *Cryptorhamphus* Stål (two species) from Australia and Tasmania and *Gonystus* Stål (two species) from Australia, with *G. nasutus* Stål occurring on Fiji.

Nothing appears to be known of the biology.

CYMINAE. Small, usually brownish yellow, body coarsely punctate; body shape elliptical; spiracles dorsal on abdominal segments 2–6 but usually ventral on segment 7 (sometimes all spiracles dorsal); nymphal scent glands usually present between abdominal terga 3/4, and 4/5, sometimes also between terga 5/6 and occasionally only between terga 3/4 (Hamid, 1975).

Fourteen genera and 76 species are currently recognized, segregated into three tribes—Ontiscini, Ninini, and Cymini. The group occurs worldwide, with the Ontiscini being restricted to the Australian and Oriental regions.

The high chromosome number (20–28 autosomes + XY), the short uncoiled gonoporal process, and the side-by-side method of copulation suggest that the sister taxon of the Cyminae may be the Berytidae, which currently are treated as a separate family (see Southwood and Leston, 1959; Štys, 1967c).

GEOCORINAE. Eyes large, reniform, prominent, often projecting backward and usually overlapping or nearly overlapping anterolateral pronotal angles; body usually stout and ovoid; pronotum broad with a transverse furrow; spiracles dorsal on abdominal segments 2–4, ventral on segments 5–7; nymphal scent glands present dorsally between abdominal terga 4/5 and 5/6, sutures between segments usually curving strongly backward from lateral margins to mesal scent-gland openings.

The subfamily is distributed worldwide and contains 14 genera and approximately 219 species. Readio and Sweet (1982) revised the eastern Nearctic fauna. These insects are unusual in the Lygaeidae in being chiefly predaceous on other small arthropods. The "big-eyed bugs" have been studied intensively in recent years as possible biological control agents against several destructive insects.

HENESTARINAE. Body strongly punctate; eyes stalked; pronotal margin sinuately convex posteriorly; hind wing with hamus and intervannals, latter basally fused; nonspinous forefemora; spiracles ventral on abdominal segments 2, 6, and 7, dorsal on segments 3–5.

This group contains three genera and 19 species. The distribution is chiefly southern Palearctic and African. *Henestaris* Spinola and *Engistus* Fieber are widespread in the Palearctic.

HENICOCORINAE. Forewing with reduced membrane and tendency to coleoptery; hind wings absent; setae with strongly raised granular bases; metathoracic scent-gland auricle with a blocklike spout; lateral trichobothria lacking on sterna 3–5, sterna 6–7 each with 2 lateral trichobothria placed transversely behind level of spiracle; all abdominal spiracles ventral; inner laterotergites present on segments 2–4; no spermatheca; nymphal dorsal abdominal scent-gland openings present between terga 4/5 and 5/6; roof of genital chamber of female with a large membranous sac.

This monotypic subfamily was established by Woodward (1968a) for *Henicocoris monteithi* Woodward from Victoria, Australia.

HETEROGASTRINAE. Membrane of forewing with 1 or 2 closed cells at base; hamus and intervannals present in hind wing; all abdominal spiracles ventral; spermatheca elongate, coiled, nonflanged, in common with Pachygronthinae; nymphal abdominal scent-gland openings between terga 3/4, 4/5, and 5/6.

Twenty-two genera and 92 species are recognized. The group is widespread in the Old World tropics and has several Palearctic genera. The New World fauna consists of two Nearctic species of *Heterogaster* Schilling.

ISCHNORHYNCHINAE. Small, usually dull brownish or reddish brown, frequently ovoid, sometimes shining or subshining; membrane often hyaline; corium frequently translucent; clavus punctate; posterior margin of pronotum nondepressed; nymphs with dorsal abdominal scent-gland openings between terga 4/5 and 5/6 (except in *Kleidocerys* Stephens, with scent glands between terga 3/4).

Acanthocrompus Scudder is unique within the subfamily in having the forefemora strongly incrassate and heavily spined below, although the condition is widespread in the Rhyparochrominae.

Fifteen genera and 75 species are currently recognized. *Kleidocerys* Stephens is widespread in the Northern Hemisphere, but the majority of Ischnorhynchinae are tropical or south temperate in distribution. *Pylorgus* Stål is widespread in mountainous areas of Africa and the Orient.

LYGAEINAE. Hemelytra impunctate; membrane of forewing usually possessing a distinct cell, subcostal vein, and hamus; all abdominal spiracles dorsal; nymphs with dorsal abdominal scent-gland openings between terga 4/5 and 5/6.

This large subfamily is found worldwide. Fifty-eight genera and about 500 species are currently recognized. As with most lygaeid taxa, the greatest diversity is in the tropics and subtropics. A few genera occur in both the Old World and New World.

Many species are large, with showy red and black or orange and black aposematic coloration, although other color combinations are also present. Most lygaeines feed above the ground, although many will live among ground litter when seeds from the host plant are abundant there. A small component appears secondarily geophilous; some of these species are flightless and cryptically colored. Examples include *Apterola* Mulsant and Rey (five species; Africa, southern Palearctic), *Melanerythrus* Stål (three species; Australia), *Stenaptula* Seidenstucker (two species; Africa and India), and *Lygaeospilus* Barber (four species; Nearctic).

ORSILLINAE (FIG. 83.3B). Relatively small, dull, gray brown; hemelytra in large part impunctate; hind wing with hamus, lacking subcostal vein; abdominal spiracles all dorsal; inner laterotergites present on abdominal terga 2–6; nymphal dorsal abdominal scent glands present between terga 4/5 and 5/6.

Four tribes are currently recognized, comprising a total of 28 genera and 250 species: Lepionysiini, Metrargini, Orsillini, and Nysiini. Ashlock (1967) monographed the world genera.

Although the orsillines occur worldwide, in temperate as well as tropical areas and on many oceanic islands, a striking feature of this group is its tremendous radiation on the Hawaiian Islands, with at least half of the world species occurring there (Usinger, 1942b). Species of the

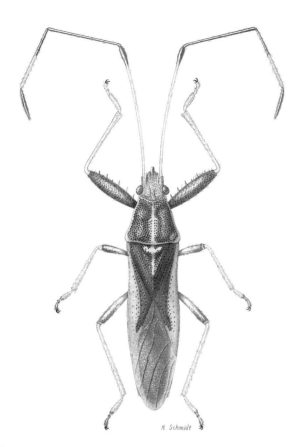

Fig. 83.4. Lygaeidae: Pachygronthinae. *Pachygrontha* sp.

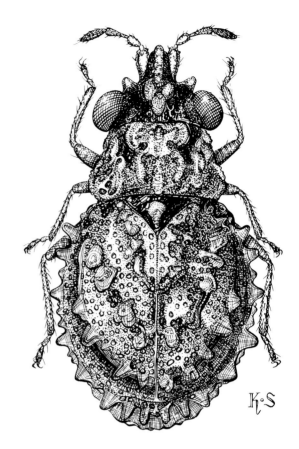

Fig. 83.5. Lygaeidae: Psamminae. *Saxicoris verrucosus* Slater (from Slater, 1970).

endemic genera *Oceanides* Kirkaldy and *Neseis* Kirkaldy tend to be restricted to one island and are often host-specific. By contrast, the 19 species of *Nysius* Dallas endemic to Hawaii occur on all of the major islands and are generalist feeders.

OXYCARENINAE. Usually small; often flattened; sometimes strongly myrmecomorphic; head commonly porrect; hemelytra with explanate margins; pronotal margins rounded, nonexplanate; hamus absent; abdominal spiracles 3–7 ventral; inner laterotergites absent; abdominal sternum 5 with at most a single posterior trichobothrium.

Twenty-three genera and 144 species are known. The group is most diverse in the Palearctic and is poorly represented in the Western Hemisphere. Samy (1969) revised the extensive African *Oxycarenus* Fieber fauna.

PACHYGRONTHINAE (FIG. 83.4). Body coarsely punctate; forefemora armed; abdominal spiracles ventral; spermatheca peculiar, elongate, nonflanged; nymphal scent glands present between terga 4/5 and 5/6.

Slater (1955) treated the world fauna of this primarily tropical and subtropical group, providing keys to genera and species. The Pachygronthini, comprising elongate, slender, bright brown bugs with long sweeping antennae, contain four genera and 50 species. The Teracriini are usually relatively short and stout, although elongate species occur in the genera *Cymophyes* Fieber and *Stenophyella* Horvath, but like most other members of the tribe they have short, stout antennae. Nine genera and 28 species are known.

PSAMMINAE (FIG. 83.5). Small, beetle-like; length not over 3–4 mm; body surface bearing waxy scalelike setae (Slater and Sweet, 1965); eyes large, reniform; ocelli lacking; antennae short and subclavate; hemelytra highly convex, completely covering abdomen and meeting along midline; membrane of forewing and entire hind wing lacking, clavus and corium indistinguishably fused; tarsi 2-segmented; spiracles dorsal on all abdominal segments; nymphal dorsal abdominal scent-gland openings between terga 4/5 and 5/6; trichobothria reduced, none on segments 4 and 7, a single one located mesally on segment 5; inner laterotergites present.

Three genera—all monotypic—are known. *Psammium* Breddin and *Saxicoris* Slater are found in the xeric regions of South Africa and Namibia, and *Sympeplus* Bergroth is known from India.

RHYPAROCHROMINAE. Usually dull brown or mottled brown, black and white; frequently myrmecomorphic;

cephalic trichobothria usually present; forefemora usually incrassate, strongly armed below with stout spines; suture between sterna 4 and 5 fused, usually curving forward anterolaterally from midline of sternum, not reaching dorsal margin of abdomen.

The subfamily was established by Amyot and Serville (1843) as Rhyparochromides. It was called Myodochina by Stål (1872), Pachymeriidae by Uhler (1860), Aphanini by Puton (1887), and Megalonotinae by Slater (1957). The basic separation into tribes was by Stål (1872), who recognized the Lethaeini, Drymini, Myodochini, Rhyparochromini, and Gonianotini. Stål (1874) later added the Cleradini, and Gulde (1936) the Stygnocorini. Scudder (1957b) discussed tribal relationships, placing some of Stål's tribes as subtribes, and expanded the concept of the Stygnocorini. Slater (1957) separated the Megalonotini from the Rhyparochromini. Slater and Sweet (1961) established the Plinthisini. Ashlock (1964) recognized the polyphyletic nature of the Lethaeini and removed the Antillocorini and Targaremini as distinct tribes. Sweet (1964) recognized the Ozophorini and later (Sweet, 1967) reviewed the higher classification in detail and removed the Udeocorini from the Myodochini. Slater and Woodward (1982) established the Lilliputocorini.

Because this subfamily is so large and diverse, and because the tribal classification is well established, we present a key to the tribes and briefly discuss each of them.

Key to Tribes of Rhyparochrominae

1. All abdominal spiracles located on sternum 2
- At least spiracle of abdominal segment 4 located dorsally on outer laterotergite 12
2. Posterior pair of trichobothria on abdominal sternum 5 located one above the other 3
- Posterior pair of trichobothria on abdominal sternum 5 located one in front of the other so that the 3 abdominal trichobothria of segments 4 and 5 occur as a linear series 10
3. Females with a conjunctiva present between abdominal sterna 4 and 5; males with a stridulatory mechanism involving abdominal segment 1 (plectrum) and hind wing (stridulitrum); pronotum wider across anterior one-third than across humeral angles .... Plinthisini
- Both sexes with abdominal sterna 4 and 5 fused, lacking a conjunctiva between them; males lacking an abdominal and hind wing stridulatory mechanism; pronotum variable, usually wider across humeral angles than across anterior lobe 4
4. Ocelli lateral, behind eyes; suture between abdominal sterna 4 and 5 attaining lateral connexival margin; abdominal tergum 3 usually desclerotized; labial segment 2 usually not attaining base of head; nymphs lacking a Y-shaped suture but with a lateral suture along length of abdomen .... Cleradini
- Ocelli located between and slightly posterior to eyes; suture between abdominal sterna 4 and 5 usually not attaining lateral connexival margin and usually markedly curving anteriorly from venter dorsally; labium variable, but usually with segment 2 reaching or exceeding base of head 5
5. Nymphs lacking a Y-suture 6
- Nymphs with a Y-suture 7
6. Nymphal abdominal scent-gland openings only between terga 3/4 and 4/5; tarsi 2-segmented; inner laterotergites absent; metathoracic scent-gland auricle strongly curving anteriorly .... Lilliputocorini
- Nymphal abdominal scent-gland openings between terga 3/4, 4/5 and 5/6; tarsi 3-segmented; inner laterotergites present; metathoracic scent-gland auricle straight or curving posteriorly .... Antillocorini
7. Abdominal trichobothria on sternum 5 closer to spiracle than to posterior margin of segment 5; pronotum lacking a distinct anterior collar; hind wing usually with both hamus and secondary veins 8
- Abdominal trichobothria on sternum 5 located closer to posterior margin of sternum 5 than to spiracle; distinct pronotal collar present; hind wing veins reduced, either hamus or secondary veins absent (usually both) 9
8. All trichobothria on sterna 4 and 5 located anterior to spiracle of sternum 5; spiracle 5 located in central third of segment; pores present near spiracles of segments 3 and 4 .... Drymini
- Posterior trichobothria of sternum 5 located posterior to spiracle 5; spiracle 5 located in posterior third of segment; no pores present near spiracles 3 and 4 .... Stygnocorini
9. Inner laterotergites present; suture between abdominal sterna 4 and 5 incised, straight and attaining

connexival margin (Fig. 83.1B); spermathecal duct short, only 3 times as long as bulb; abdominal scent-gland scars small .... Phasmosomini

– Inner laterotergites absent; suture between abdominal sterna 4 and 5 sometimes attaining connexival margin, usually remote from margin and strongly curving dorsoanteriorly (Fig. 83.1A); spermathecal duct very elongate and coiled; abdominal scent-gland scars broad ..... Ozophorini

10. All trichobothria on segment 5 located anterior to spiracle and usually equidistant from each other; nymphs with a Y-suture .... Targaremini

– Usually with one trichobothrium on segment 5 posterior to spiracle; middle trichobothrium not equally distant from the other 2; nymphs lacking a Y-suture 11

11. Apical corial margin deeply concave; inner laterotergites present; head lacking iridescent areas; abdominal scent-gland scars present between terga 3/4, 4/5, and 5/6 .... Antillocorini

– Apical corial margin straight; no inner laterotergites present; head frequently with iridescent areas present basally; abdominal scent-gland scar between terga 5/6 minute or absent ...... Lethaeini

12. Abdominal spiracles of segments 2–4 located dorsally 13

– Abdominal spiracles located dorsally only on either segments 3 and 4 or 4 only, always ventral on segment 2 14

13. Inner laterotergites absent; lateral pronotal margins almost always rounded; nymphs lacking large black sclerotized areas around dorsal abdominal scent-gland openings .... Myodochini

– Inner laterotergites present; lateral pronotal margins variable from rounded to carinate; nymphs frequently with black sclerotized areas around dorsal abdominal scent-gland openings .... Udeocorini

14. Spiracles located dorsally only on abdominal segment 4 .... Gonianotini

– Spiracles located dorsally on abdominal segments 3 and 4 15

15. Nymphs with a distinct Y-suture present between abdominal terga 3/4 .... Rhyparochromini

– Nymphs lacking a Y-suture between abdominal terga 3/4 .... Megalonotini

ANTILLOCORINI (FIG. 83.6A). Very small to minute; bucculae joined by a carina well behind labium; anterolateral pronotal trichobothria absent; apical corial margin deeply concave; inner laterotergites present; abdomen laterally with evaporative areas; well-developed nymphal scent glands between terga 3/4, 4/5, and 5/6; trichobothria linearly arranged on all segments.

Twenty-nine genera and at least 93 species are known. This group is primarily tropical and subtropical, with a few species extending into the temperate Northern Hemisphere, and is well represented on oceanic islands in the Pacific. The group may not be monophyletic, because some Neotropical species lack the linear arrangement of abdominal trichobothria and do not have the deeply concave apical corial margin. Unfortunately, nymphs of these species are unknown, precluding determination of whether they possess the abdominal evaporative areas that are apparently a synapomorphy for the Antillocorini and Lethaeini.

CLERADINI. Ocelli located behind rather than between eyes; antennal segment 3 short; labium relatively short, segment 2 usually not exceeding base of head; forefemora usually slender, unarmed below; no inner laterotergites; connexival membrane greatly expanded; parameres bifurcate.

The body form ranges from short and stout to large with an elongate neck. Species of several genera possess stridulatory mechanisms incorporating an abdominal stridulitrum and a femoral plectrum. Nymphs uniquely possess an impressed lateral suture running the length of the abdomen. Most remarkable is the presence of a distinct pseudoperculum otherwise not present in the eggs of Lygaeidae.

The Cleradini are confined to the tropics of the Eastern Hemisphere (except for the introduction of *Clerada apicicornis* Signoret into the Western Hemisphere). Twenty genera and 50 species are currently recognized. Malipatil (1981) revised the extensive Australian fauna, and Malipatil (1983) revised the world fauna and analyzed the cladistic relationships.

DRYMINI. Usually small to medium-sized; lateral pronotal margins usually carinate or narrowly explanate; apical corial margin usually straight; posterior pair of trichobothria on sternum 5 located dorsoventrad of one another, placed anterior to spiracle 5; pore present near spiracles 3 and 4; usually with 9 chromosomes.

Fifty-two genera and 269 species are currently known. The group is speciose in the Old World. There is a limited Nearctic fauna closely related to that of the Palearctic, but the group is absent from the Neotropics.

GONIANOTINI. Body usually broad, elliptical, with strongly explanate lateral pronotal margins; spiracles of abdominal segment 4 dorsal; nymphal abdomen usually dark, heavily sclerotized; no Y-suture; nymphs with dorsal abdominal scent-gland openings between terga 4/5 and 5/6.

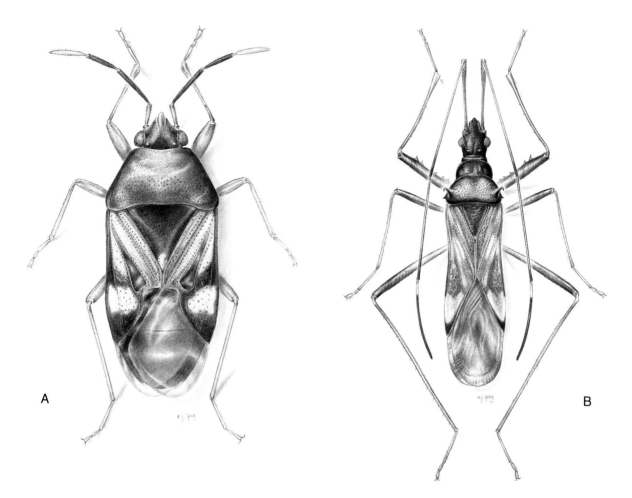

Fig. 83.6. Lygaeidae: Rhyparochrominae. **A.** *Botocudo cavernicola* Slater (Antillocorini) (from Slater, 1984). **B.** *Primierus quadrispinosus* Slater and Zheng (Ozophorini) (from Slater and Zheng, 1985).

Twenty-two genera and approximately 123 species are currently recognized. Gonianotines are chiefly Palearctic and Ethiopian and, although present in other faunal regions, are absent from the Neotropical and Australian regions.

LETHAEINI. Usually shining to subshining; head with single or double iridescent area(s) at base; bucculae joined posteriorly by carina immediately behind labial base; pronotal margins usually carinate or explanate; anterolateral angle of pronotum frequently with an elongate trichobothrium; nymphs lacking Y-suture but with evaporative areas laterally; Y chromosome lacking (synapomorphy); sperm reservoir complex, highly modified (O'Donnell, 1991).

Lethaeines comprise 33 genera and 150 species. They are chiefly tropical, although a few species extend northward into either the Palearctic or the Nearctic.

LILLIPUTOCORINI. Minute, 2 mm long; body yellowish brown, somewhat flattened; antennae rather clavate; forefemora mutic; tarsi 2-segmented; abdominal spiracles ventral; abdominal trichobothria variably placed, posterior pair on sternum 5 sometimes lacking, when present placed in dorsoventral position; inner laterotergites lacking; nymphal scent glands between terga 3/4 and 4/5, troughed groove present between terga 3/4 and 4/5 leading to evaporative areas laterally on abdomen; ovipositor reduced.

One genus and 10 species are known. The Lilliputocorini are known from northern Australia, New Guinea, Borneo, Ceylon, Nepal, Seychelles, Ghana, South Africa, and Brazil (Slater and Woodward, 1982). Nothing is known of the biology of these minute insects other than that they live in ground litter. Wing reduction is common with both staphylinoid and micropterous forms known. Wing reduction is sexually dimorphic, a condition common in some families of Heteroptera, but almost unknown in the Lygaeidae.

MEGALONOTINI. Usually resembling Rhyparochromini, with no character yet discovered to differentiate these two tribes in the adult stage. Both have carinate or explanate pronotal margins, spiracles on abdominal segments 3 and 4 dorsal, inner laterotergites present, and 3 well-developed nymphal abdominal scent-gland openings. Megalonotine nymphs have dark, heavily sclerotized abdomens and lack a Y-suture, a character that is well developed in nymphs of the Rhyparochromini.

Eighteen genera and approximately 87 species are known. The Megalonotini, most diverse in the Old World tropics and the Palearctic, are represented by a few taxa in the Nearctic and are absent from the Neotropics.

MYODOCHINI. Body form variable, ranging from short and stout to elongate and slender, from rather small to very large; often myrmecomorphic; myodochines share the following features with the Udeocorini: anterior pronotal lobe almost always rounded laterally, abdominal spiracles 2–4 located dorsally, nymphs with Y-suture, 3 pairs of abdominal scent glands, inner laterotergites absent.

Sixty-seven genera and approximately 307 species are known. Myodochini are abundant in all major zoogeographic regions, sometimes reaching high latitudes in temperate areas. Some species also appear to be extremely vagile, and there has been colonization of many oceanic islands. The Neotropical fauna is especially diverse. Harrington (1980) provided a cladistic analysis of the world genera.

OZOPHORINI (FIG. 83.6B). Medium-sized, rather slender; head generally porrect, grooved; pronotal collar usually well developed; legs and antennae elongate; hind wing lacking hamus and secondary veins; spiracles ventral; nymphal dorsal abdominal scent glands between terga 3/4, 4/5, and 5/6, strongly developed Y-suture in nymphs; inner laterotergites absent.

Twenty-four genera and 173 species are currently recognized. Some species are myrmecomorphic, and some show body shape development parallel to that of myodochines. Ozophorines are worldwide in distribution. They are most abundant in the Neotropics and in the New Guinea island arc.

PHASMOSOMINI. Ocelli absent; cubital furrow absent; Y-suture present; inner laterotergites present; deeply incised suture between sterna 4 and 5; 4/5 abdominal suture complete although fused.

This tribe contains two south-central Palearctic species. They are similar in habitus to many species of Ozophorini.

PLINTHISINI. Small; shining or subshining; usually with pronotum expanded across anterior lobe (especially in flightless forms); forefemora heavily incrassate, spined; wings often greatly reduced, usually in a staphylinoid manner, posterior half of abdominal dorsum exposed; stridulatory mechanism novel (as noted in key); sperm reservoir continuous with body when not reduced.

Uniquely within Rhyparochrominae, the Plinthisini have a conjunctival membrane present between abdominal sterna 4 and 5 in the female, the lack of which feature has been used as a synapomorphy for the Rhyparochrominae (Sweet, 1967). Thus, the Plinthisini may merit subfamily status. The presence of head trichobothria, however, suggests that this may be the sister group of all other rhyparochromines.

There are nearly 100 described species, with large numbers of undescribed species present in collections. The group occurs worldwide.

RHYPAROCHROMINI. Often large and robust; lateral pronotal margins carinate or explanate; nymphs frequently with variegated, lightly sclerotized abdomen and well developed Y-suture; inner laterotergites present; 3 nymphal dorsal abdominal scent-gland openings between terga 3/4, 4/5, 5/6; spiracles of abdominal segments 3 and 4 located dorsally.

Forty-one genera and approximately 350 species are currently recognized. The Rhyparochromini are most diverse in the Old World tropics and subtropics, with an extensive Palearctic fauna. The group is poorly developed in the Nearctic and natively absent from the Neotropics. Eyles's (1973) revision of the genus *Dieuches* Dohrn is the single most comprehensive work on this large and complex group.

STYGNOCORINI. Short, rather stout-bodied; vertex without longitudinal grooves; hamus and secondary veins strongly developed; abdominal spiracles ventral; inner laterotergites present; no pronotal collar; nymphs with Y-suture; nymphal dorsal abdominal scent-gland openings between terga 3/4, 4/5, and 5/6.

Sixteen genera and 168 species are currently recognized. This group includes several Palearctic genera, a montane element in Africa and Madagascar, and a south-temperate group in Africa, Tasmania, and New Zealand (see Slater and Sweet, 1970; O'Rourke, 1974, 1975).

TARGAREMINI. Small to medium-sized; all abdominal spiracles ventral; trichobothria on abdominal sternum 5 located anterior to spiracle, usually equidistant from one another, in linear sequence as in Lethaeini and many Antillocoriini; nymphs with a Y-suture.

Twenty-three genera and 59 species are known. The Targaremini appear to be an ancient group, being found in Australia, New Guinea, New Caledonia, the New Hebrides, and New Zealand. On the last islands and associated islets there has been extensive radiation (Malipatil, 1977).

UDEOCORINI. Medium-sized to large; body form varying from elongate, slender, and long-legged, with rounded pronotal margins, to short, stout, and subflattened, with carinate, or even explanate, pronotal margins; spiracles dorsal on abdominal segments 2–4; Y-suture well developed in nymphs; nymphal dorsal abdominal scent-gland openings between terga 3/4, 4/5, 5/6; inner laterotergites present.

Seventeen genera and 33 species are currently recognized. Udeocorines are found chiefly in Australia, where they have radiated. They have the appearance of other tribes in other regions of the world and (presumably) occupy the same niches. There are also three Neotropical taxa.

Bathycles amarali Correa apparently is haematophagous and has converged remarkably in general habitus to species of *Clerada* Signoret.

Specialized morphology. Nowhere else in the Heteroptera do we observe such marked differences in the position of the spiracles or in the Trichophora such variation in the number and position of the abdominal trichobothria (Sweet, 1967). Many lygaeids possess stridulatory structures. These often occur as stridulitra along the lateral margins of the hemelytra, as crescent-shaped areas on the abdominal sternum, along the sides of the head and the prothorax, or on the hind wings.

The forefemora are sometimes fossorial, but much more frequently are enlarged and spinose, appearing raptorial, but actually serving either to hold and manipulate seeds or to aid in dragging the body in enclosed spaces.

Many species are myremcomorphic, with constricted anterior abdominal segments and expanded posterior segments as well as conical antlike protrusions on both the pronotum and scutellum.

Brachyptery, coleoptery, and microptery are widespread (see Chapter 6).

Natural history. Although the great majority of Lygaeidae feed on mature seeds (Sweet, 1964), there is substantial diversity in the feeding habits for the family. The Blissinae are sap suckers, most Geocorinae are largely predaceous on other small arthropods, and the Cleradini feed on the blood of vertebrate hosts. There is a very large ground-living component. In some species, especially those from stable habitats, the flying morph appears to have been completely eliminated.

Although many members of the genus *Nysius* are generalists, some orsillines are specialists. These often live on the seeds of trees, such as *Belonochilus numenius* (Say) on *Platanus* spp. (sycamores) in North America and *Hyalonysius pallidomaculatus* Slater on *Buddleia* spp. in South Africa.

Species of the Holarctic genus *Kleidocerys* often are found in large numbers on catkins and seed heads of *Betula, Rhododendron, Spiraea, Typha,* and other plants. They produce sound by means of a stridulitrum on a vein of the hind wing that contacts a plectrum on the front wing (Southwood and Leston, 1959). Such stridulation can be heard by the human ear by placing an insect in a small vial and shaking it vigorously.

The various species of the genus *Stilbocoris* Bergroth (Rhyparochrominae: Drymini) are remarkable in being ovoviviparous. Males have a complex mating ritual in which they secrete salivary fluid into fig seeds, which they then offer to females to facilitate copulation. The males frequently fly about with the small fig seeds impaled upon their beaks (Carayon, 1964a).

Aposematic coloration is widespread in the Lygaeinae, many species of which feed on toxic or unpalatable plants. They are involved in complex Müllerian and Batesian mimicry rings with beetles and moths, as well as other heteropterans (see Chapter 7).

There is an interesting case of interspecific displacement of other insects from a host-plant by the aposematically colored red and black *Neacoryphus bicrucis* (Say) in North America. This species feeds on *Senecio* spp. (Asteraceae), from which it sequesters pyrrolizidine alkaloids and apparently is protected against attack by potential predators. It co-occurs with at least five other insect species, including members of the Miridae, Rhopalidae, and Coreidae. Males are extremely aggressive in attempting to copulate with other species as well as with females of their own species. This behavior drives the other species from the host plants. When *Neacoryphus* males are removed, populations of the other species increase and vice versa (McLain and Shure, 1987). This interspecific displacement by pseudocompetition may be an important component in the success of several of the aposematically colored Lygaeinae.

Oncopeltus fasciatus (Dallas), known as the large milkweed bug, has been used for many years as a laboratory animal, and there is an enormous literature dealing with various aspects of its physiology and biochemistry. This species has a flight cycle and a reproductive cycle during which energy may be directed to flight or to egg production with concomitant loss of flight musculature in the latter case and marked dispersal in the former (Dingle, 1978, 1979, 1981, 1985). Work in the Western Hemisphere tropics has shown interesting aspects of habitat segregation even on the same host plants by closely related species of *Oncopeltus* (Blakley, 1980; Dingle and Baldwin, 1983). For example, some species are dependent for reproductive success upon mature seeds and, in order to find new plants with mature seeds, have increased dispersal ability. Other species are able to breed on the host plant whether or not seeds are present, and such species are relatively more sedentary.

Some lygaeids feed above ground in the seed heads of plants. Many of these, especially the nymphs of species of Cyminae and Pachygronthini, resemble the seed heads of the sedges and rushes upon which they feed. Others, however, live in such habitats but have the body flattened rather than resembling seeds. One of the best known of these is the Palearctic *Chilacis typhae* Perrin, all stages of which live on and overwinter in the catkins of *Typha* species. This species recently was introduced into the eastern United States, where it is now established (Wheeler and Fetter, 1987). In South Africa, several *Oxycarenus* spp. have adapted themselves for feeding on the deeply set seeds of *Protea* (Proteaceae) and have an enormously elongated labium that in nymphs may be nearly one-third longer than the body length of the insect.

Many species of Heterogastrinae, especially those that feed on *Ficus*, have very elongate ovipositors. Two rather distinct groups occur within the subfamily. One occurs on the bark and trunks of *Ficus* spp. and feeds by inserting the very elongate rostral stylets through the synconium of the fig to reach the seeds within. These species, many of which belong to the genus *Dinomachus* Distant, seem to be gregarious, with large numbers of adults and nymphs occurring together on the host plants. The second is short and stout and occurs primarily on Labiatae (Slater, 1971).

In Trinidad and Peru, *Cligenes subcavicola* Scudder lives in caves inhabited by fruit-eating bats, where it sometimes occurs in tremendous numbers, feeding on the seeds present in the guano. Scudder et al. (1967) reported numbers from 1000 to 100,000 per square meter in the Tamana Caves in Trinidad. We saw thousands of the bugs on the floor of an abandoned building at the Simla Tropical Research Station on Trinidad, where there roosted a large bat population that was feeding chiefly on fruits of a *Piper* sp. At the Cueva de las Lechusas near Tingo Maria, Peru, the same lygaeid apparently fed primarily on fig seeds.

Sweet (1964), in a major study of the biology of the Rhyparochrominae of northeastern North America, discovered that the fauna could be divided into two ecological groups. The first group was found in temporary habitats, had bivoltine life cycles, produced large numbers of eggs, usually had weak diapause as adults, and was entirely macropterous. The second group was found in more permanent habitats, had univoltine life cycles, produced more limited numbers of eggs, had strong diapause, often in the egg stage, and most species were largely brachypterous.

Distribution and faunistics. The family is worldwide in distribution. A number of taxa show distribution patterns that imply past continental connections, as between Africa and South America, transantarctic distributions between Australia and southern South America, transsiberian distributions between boreal Asia and North America, and a few that appear to represent more widespread Gondwanaland relationships.

Putchkov (1969) presented keys to the fauna of the Ukraine. This extensive treatment also includes detailed host plant and ecological data and figures of many nymphal stages. Kerzhner and Jaczewski (1964) presented keys to the entire European fauna of the former USSR, including almost all of the Palearctic genera. Slater and Baranowski (1990) treated the fauna of Florida and consequently provided access to much of the eastern North American fauna. Slater (1964a) treated the South African fauna in detail.

Many other papers should be consulted for identification of various groups within the family. Among them are those by Harrington (1980) and Malipatil (1978) for the Myodochini, Ashlock (1967) for the Orsillinae, A. Slater (1985, 1992) for the Lygaeinae, Gross (1965) for Australian Drymini, Gross and Scudder (1963) for Australian Rhyparochromini, Scudder (1962b) for the Heterogastrinae, and Scudder (1962c) for the Ischnorhynchinae.

84
Malcidae

General. Members of this family are small (3–4 mm), thick-bodied (Fig. 84.1A), and coarsely punctate (Fig. 84.1B), with flattened scalelike or curved glandular setae on the body surface. They have no common name.

Diagnosis. Head strongly declivent; ocelli present; bucculae large; antennae placed above a line drawn through middle of eye; membrane of forewing with five nonbranching veins (Fig. 84.1A, B); tarsi 3-segmented; abdominal sterna 2–5 fused; inner laterotergites absent; trichobothria located on "loaflike" tubercles, submedial on sternum 3 as well as laterally where 2 arranged diagonally, missing on sternum 4 (Fig. 84.1D); lateral margins of abdominal segments 5–7 usually expanded into distinct flanges; spiracles dorsal on abdominal segments 2–6; nymphal dorsal abdominal scent glands small, situated between terga 3/4, 4/5, and 5/6 or between terga 4/5 and 5/6; aedeagus as in Fig. 84.1E; parameres as in Fig. 84.1F; ovipositor laciniate; sternum 7 entire; spermatheca often with elongate ductus (Fig. 84.1G); eggs quadrate in cross section.

Classification. This taxon was established by Stål (1864–1865) as Malcida in a key to subfamilies of Lygaeidae. Despite the lack of included taxa, the name was obviously based on *Malcus* Stål (1859c). Štys (1967c) reviewed the history of the group in detail, noting that many authors credited authorship to Horvath (1904), who first used the term Malcinae. Subsequent authors usually treated the taxon as a lygaeid subfamily, although Lethierry and Severin (1893–1896) and Distant (1904) placed it in the Colobathristidae.

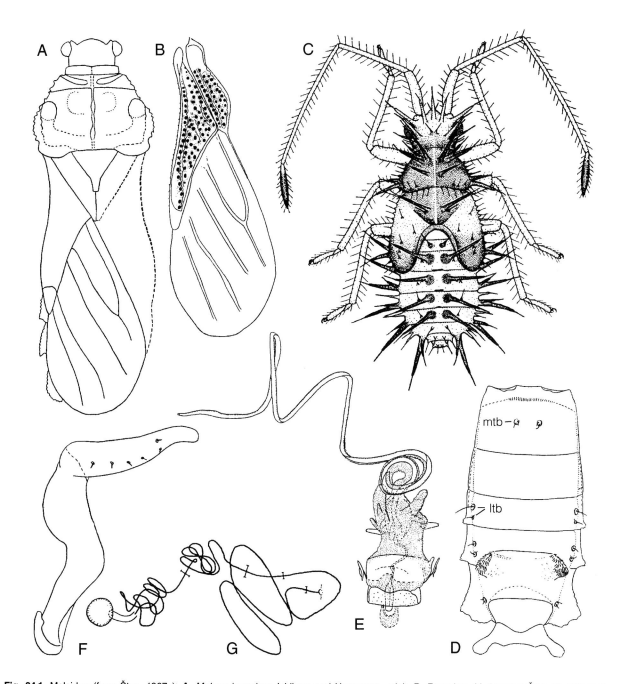

Fig. 84.1. Malcidae (from Štys, 1967c). **A.** *Malcus japonicus* Ishihara and Hasegawa, adult. **B.** Forewing, *M. furcatus* Štys. **C.** Fifth-instar nymph, *M. flavidipes* Stål. **D.** Ventral view, male abdomen, *M. furcatus*. **E.** Aedeagus, sagittal view, *M. furcatus*. **F.** Paramere, *M. furcatus*. **G.** Spermatheca, *M. furcatus*. Abbreviations ltb, lateral trichobothria; mtb, mesial trichobothria.

Štys (1967c) discussed the relationships of the Malcidae—and other lygaeoid groups—exhaustively. He concluded that the Chauliopinae and Malcinae were closely related and together merited family status.

Key to Subfamilies of Malcidae

1. Eyes stylate; ocelli widely separated from one another; body bearing waxlike scales; claval commissure much shorter than scutellum . . . . Chauliopinae
– Eyes sessile (Fig. 84.1A); ocelli close together, situated on a common tubercle; body lacking waxlike scales; claval commissure subequal in length to scutellum (Fig. 84.1A, B) . . . . Malcinae

CHAULIOPINAE. Body short and stout; antenniferous tubercles large, spinous, produced; metathoracic scent-gland auricle not produced; no tubercle at apex of corium; abdominal terga 3–6 separate; nymphs lacking elongate spines; nymphal dorsal abdominal scent glands paired between terga 4/5 and 5/6.

The Chauliopinae was erected as a subfamily of Lygaeidae by Breddin (1907) but was placed in the Heterogastrinae by Distant (1910), an action followed much later by Miller (1956a). Other authors, including Štys (1963), treated the group as a subfamily of Lygaeidae. Two genera are recognized, *Chauliops* Scott (seven species) and *Neochauliops* Štys (two species from Africa) (Štys, 1963). The Chauliopinae occur in both the Oriental and Ethiopian regions.

MALCINAE (FIG. 84.1A). Body relatively elongate; antenniferous tubercles reduced; metathoracic scent-gland auricle protruding, dorsally perpendicular to metapleuron; corium with a conspicuous tubercle at apical angle; abdominal terga 3–6 fused; nymphs with numerous long spines over body surface (Fig. 84.1C); dorsal abdominal scent glands paired between terga 3/4 and 4/5, unpaired between terga 5/6; eggs with 3 club-shaped micropylar processes, lacking pseudoperculum.

One genus, *Malcus* Stål containing 19 species from the Oriental region, is known. The subfamily was monographed by Štys (1967c).

Specialized morphology. The elongate nymphal spines of the Malcinae (Fig. 84.1C), fused abdominal sterna 2–5, tuberculate trichobothria (Fig. 84.1D), flanged abdominal segments, stylate eyes, and frequent occurrence of glandular setae are all specialized features.

Natural history. The biology of *Malcus* spp. is essentially unknown. Ishihara and Hasegawa (1941) reported on *Malcus japonicus* Ishihara and Hasegawa on *Morus bombycis*. Štys (1967c) recorded *Malcus flavidipes* Stål as abundant on leaves of banana in Cambodia. Species of *Chauliops* apparently feed chiefly on Solanaceae but are sometimes destructive to beans (Sweet and Schaefer, 1985; see Chapter 8).

Distribution and faunistics. Štys (1963:213) suggested "that Burma and the surrounding mountains on the west must be considered a center of speciation and probably also the original country of both subfamilies."

The basic reference on the group is Štys (1967c).

85
Piesmatidae

General. These small insects, not over 5 mm in length, have reticulate cell-like forewings and prothorax (Fig. 85.1A). Superficially they resemble small grayish or yellowish lace bugs. They are sometimes referred to as ash-gray leaf bugs.

Diagnosis. Body surface reticulate (Fig. 85.1A); ocelli present; mandibular plates elongate, projecting strongly forward (Fig. 85.1A); scutellum exposed; metathoracic scent-gland openings obsolete; hind wing with stridulitrum on Cu vein and plectrum on first tergum; tarsi 2-segmented; nymphs with scent glands on abdominal terga 3/4 and 4/5 (small, sometimes visible only between terga 3/4), these apparently functional in adults; one prespiracular trichobothrium present on abdominal sternum 5 and one on 6 in *Piesma* (sometimes one pair or none); some, or all, abdominal spiracles dorsal; aedeagus as in Fig. 85.1B, vesica greatly elongate; parameres as in Fig. 85.1C; spermatheca as in Fig. 85.1D, E; 2 pairs of Malphigian tubules opening into anterior end of rectum (apparent apomorphy for family).

Classification. This taxon was first recognized as a higher group by Amyot and Serville (1843) as the "Piesmides." Early classifications placed the Piesmatidae with the Tingidae because of the reticulate wings. Reuter (1910) considered them a distinct family. Tullgren (1918) and Leston et al. (1954) removed the family from a position related to the Tingidae to the Pentatomomorpha. Drake and Davis (1958) further showed that the presence of trichobothria and details of genitalic morphology indicated that the resemblance to some Tingidae was only superficial. Both Štys (1961b) and Kumar (1968) considered the family to be isolated and believed it merited superfamily status. Two subfamilies have been recognized.

PIESMATINAE (FIG. 85.1A). *Piesma* Lepeletier and Serville, with 31 species, is widespread and has been segregated into three subgenera by Péricart (1974). Of these, *Parapiesma* Péricart is Holarctic, *Piesma* sensu stricto is Holarctic and African, and the monotypic *Afropiesma* Péricart is African. Two other genera have restricted distributions: *Miespa* Drake (one species; Chile) and *Mcateella* Drake (four species; Australia).

THAIOCORINAE. Kormilev (1969) erected this subfamily for *Thaicoris* Kormilev, with one species from southeast Asia. If Schaefer (1972a, 1981b) is correct, however, *Thaicoris* is actually the sister taxon of *Miespa*

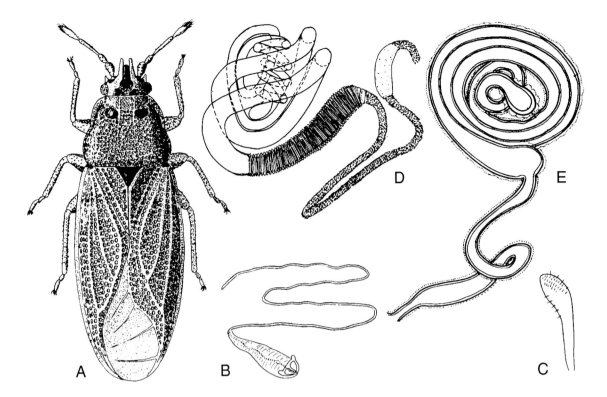

Fig. 85.1. Piesmatidae. **A.** *Piesma capitatum* (Wolff). **B.** Phallus, sagittal view, *P. cinereum* (Say). **C.** Paramere, *P. quadratum* (Fieber). **D.** Spermatheca, *P. maculatum* (A–D from Heiss and Péricart, 1983). **E.** Spermatheca, *P. quadratum* (from Pendergrast, 1957).

and *Mcateella,* and therefore not deserving of subfamily status.

Specialized morphology. The unique position of the Malpighian tubules, functional abdominal scent glands in adults, and greatly reduced trichobothrial numbers are all specialized features.

Natural history. All species are phytophagous. Schaefer (1981b) summarized known host-plant associations of this strictly phytophagous group, listing 10 genera of Chenopodiaceae as being utilized by various species of *Piesma*. Species of the Australian genus *Mcateella* have been reported feeding on *Acacia* and *Beyeria*. There are also records of feeding on species of Caryophyllaceae, Amaranthaceae, and Cistaceae.

Distribution and faunistics. This small family occurs in all major zoogeographic regions. Schaefer (1981b) believed that the distribution of the family "is not inconsistent" with a Gondwana origin. The works of Drake and Davis (1958) and Heiss and Péricart (1983) serve as basic references on the group.

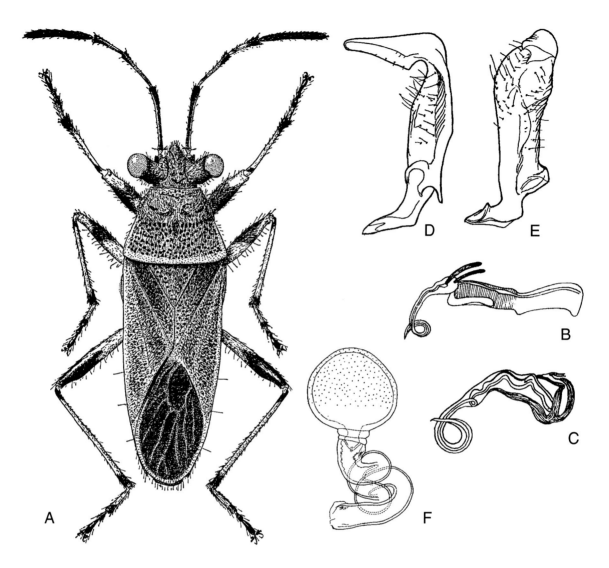

Fig. 86.1. Largidae. **A.** *Acinocoris stehliki* van Doesburg (from van Doesburg, 1966). **B.** Aedeagus, lateral view, *Physopelta famelica* Stål. **C.** Aedeagus, detail, *P. famelica* (B, C from Kumar, 1967). **D.** Paramere, *A. calidus* (Fabricius). **E.** Paramere, rotated 90°, *A. calidus* (D, E from van Doesburg, 1966). **F.** Spermatheca, *Largus rufipennis* Laporte (from Pluot, 1970).

Pyrrhocoroidea

86 Largidae

General. These are moderately small to large insects (up to 55 mm), which are frequently brightly colored. The body is ovoid, oblong, or elongate (Fig. 86.1), except in myrmecomorphic species (Fig. 86.2), often with relatively short antennae and legs. The group appears to have no common name.

Diagnosis. Antennae inserted below a line through middle of eye; ocelli absent; membrane of forewing with basal cells and at least 7 distally radiating veins (Fig. 86.1A); metathoracic scent-gland openings reduced; sometimes an incomplete suture between abdominal sterna 4 and 5; 3 trichobothria on sterna 3–6, 2 on sternum 7, those on segments 5–7 lateral and dispersed; nymphal dorsal abdominal scent-gland openings between terga 3/4, 4/5, and 5/6, 2 anterior gland openings reduced; aedeagus as in Fig. 86.1B, C; parameres as in Fig. 86.1D, E; ovipositor laciniate; abdominal sternum 7 cleft mesally; spermatheca as in Fig. 86.1F.

Classification. The taxon was first recognized as a higher group, the Largides, by Amyot and Serville (1843). Van Duzee (1916) treated it as a subfamily of the Pyrrhocoridae, calling it Euryophthalminae, but separate family status has been accepted by most recent workers on the basis of great differences in the female genitalia (Štys and Kerzhner, 1975; Henry, 1988).

Hussey (1929), in a world catalog, treated the largids as a subfamily, but stated that the differences were of such a magnitude that family status was warranted. He recognized two tribes: the Largini (his Euryophthalmini), confined to the Western Hemisphere, and the Physopeltini (Old World tropics).

Bliven (1973), apparently unable to distinguish myrmecomorphy from fundamental morphological features, considered the largids to belong to the Alydidae. In so doing he erected a second subfamily, the Arhaphinae, which may have some validity as a higher taxon although it represents only one (or more) myrmecomorphic genera and certainly does not reflect a fundamental reorganization.

The corial vein connecting M and Cu is lacking and is considered by some to be a synapomorphy with the Hyocephalidae. Reduction of the anterior nymphal scent-gland orifices suggests a possible relationship with the Colobathristidae. Schaefer (1964) suggested that the Pyrrhocoridae evolved from the Largidae, basing the idea in large part upon the laciniate ovipositor of the latter. Kumar (1968) disagreed, noting a number of specialized features in the internal anatomy of the Largidae, and believed that the laciniate ovipositor was a secondary development.

Approximately 15 genera and over 100 species are known.

Specialized morphology. The lack of ocelli (shared with Pyrrhocoridae), complex venation of the membrane of the forewing, and frequent fusion and obliteration of abdominal sutures are all specialized conditions.

Natural history. All known species feed on seeds and plant juices. Two rather distinct elements are present: one group of genera lives on the ground and resembles ground-living Lygaeidae; a second group lives on forbs, shrubs, and trees and resembles species of Pyrrhocoridae. The species of *Euryophthalmus* Laporte show bewildering variability in color, with yellow and black or red and black morphs and with strikingly different metallic blue nymphs, which are presumably aposematic.

Several genera of largids are striking, not only with

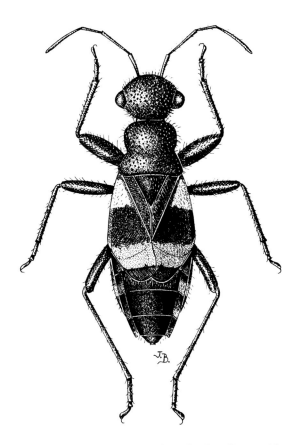

Fig. 86.2. *Arhaphe carolina* Herrich-Schaeffer (from Slater and Baranowski, 1978).

regard to body shape, but because of their movements in the field. Hussey (1927) illustrated the remarkably myrmecomorphic (Neotropical) *Thaumastaneis montandoni* Kirkaldy and Edwards, which has a swollen, antlike head, abdomen strongly constricted at the base, wings reduced to short pads, and a strong spine projecting from the posterolateral areas of the pronotum. Species of the genera *Arhaphe* Herrich-Schaeffer (Fig. 86.2), *Japetus* Stål, and *Theraneis* Spinola also are strongly myrmecomorphic, presumably either Müllerian or Batesian mimics of species of velvet ants (Mutillidae).

Distribution and faunistics. The family is represented in all major zoogeographic regions, but is most abundant and diverse in the tropics and subtropics.

Hussey's (1929) catalog remains the most important reference. Froeschner (1981) provided a key to South American genera. The confusing literature of Bliven is summarized by Henry (1988).

87
Pyrrhocoridae

General. This family includes species often referred to as cotton stainers. Most are medium-sized to large (8–30 mm), frequently colored with red or yellow and black. Many closely resemble in size and shape species of the lygaeid subfamily Lygaeinae (Fig. 87.1).

Diagnosis. Ocelli lacking; metathoracic scent-gland openings reduced; membrane of forewing with 2 basal cells and a series of 7–8 anastomosing veins distally (Fig. 87.1); 3 trichobothria on abdominal segments 3–6 and 2 on segment 7; sometimes a curved suture between abdominal sterna 4 and 5 not attaining dorsal margin (similar to condition in most rhyparochromine lygaeids); inner laterotergites absent; abdominal spiracles ventral; nymphs with dorsal abdominal scent-gland openings between terga 3/4, 4/5 and 5/6, posterior openings reduced; aedeagus as in Fig. 87.2B; parameres as in Fig. 87.2C–E; abdominal sternum 7 of female complete; ovipositor platelike rather than laciniate; spermatheca lacking a distal pump flange (Fig. 87.2F–H); alimentary canal as in Fig. 87.2A.

Classification. Although the recognition of this family as a higher taxon is usually attributed to Fieber (1861), the group was actually first recognized by Amyot and Serville (1843) as the "Pyrrhocorides." For many years, the Largidae were included in the Pyrrhocoridae as a subfamily, but recent authors have believed these two groups to be distinct (see Chapter 86). Approximately 30 genera and 300 species are known.

Specialized morphology. The platelike ovipositor valvulae, reduced scent glands in the adult, and loss of ocelli (shared with Largidae) are specialized conditions.

Natural history. Most species whose biology is known feed chiefly on seeds and fruits, particularly of the Malvales. While the most conspicuous species are arboreal as adults, there is a significant Old World ground-litter fauna that presumably feeds on mature seeds.

In West Africa, studies have shown a yearly succession of species, with large colonies frequently being composed of more than one species (Fuseini and Kumar, 1975). These colonies have associated with them a group of reduviids of the genus *Phonoctonus* that feed on *Dysdercus*. Each reduviid species mimics the coloration of its pyrrhocorid prey species (Stride, 1956) (see Chapter 7).

Derr et al. (1981) demonstrated that larger species of *Dysdercus* had a greater migratory capacity and a greater degree of survival of diapause than smaller species. The larger species in an area were entirely or largely confined

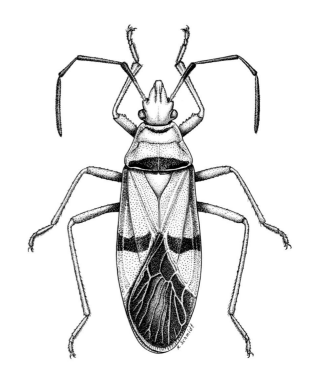

Fig. 87.1. Pyrrhocoridae. *Dysdercus fasciatus* Signoret.

to woody Malvales, whereas smaller species utilized herbaceous plants. Thus larger size appeared to confine such insects to large oil-rich fruits of large trees that were widely separated, and the greater migratory ability of such species was selected for as a consequence.

In most of the colonial *Dysdercus* species, the gravid females absorb the flight muscles after a migration flight and at the onset of oogenesis (Davis, 1975). They then are no longer capable of migrating from one host to another.

Carroll and Loye (1990) studied dimorphism in wing muscle histolysis in *Dysdercus bimaculatus* Stål. Females lose flight ability prior to egg laying, but not until they have fed and mated. Thus the ability to use muscle nutrients and energy for increased egg productivity must be balanced against the danger of low reproduction if a poor habitat is utilized. Several *Dysdercus* spp. feed on malvaceous seed crops that are ephemeral because of feeding pressure by the bugs, and eggs are laid in large numbers and in a very short period of time. Nymphs thus presumably have a chance to mature before the seed crop is exhausted. Females live for a much shorter period than do males, presumably because of exhaustion from such massive egg production, which may account for the male sex bias so often found in these insects.

Pyrrhocoris apterus (Linnaeus) is a widespread Palearctic species. It is a ground-living insect, and most populations are predominantly brachypterous. This species has been studied with respect to the function of juvenile hor-

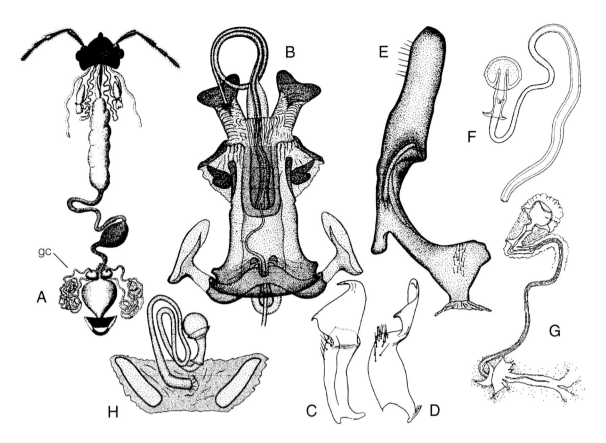

Fig. 87.2. Pyrrhocoridae. **A.** Alimentary canal, *Pyrrhocoris apterus* (Linnaeus) (from Dufour, 1833). **B.** Aedeagus, sagittal view, *Paradindymus madagascariensis* (Blanchard) (from Stehlik, 1966). **C.** Paramere, *Dysdercus blotei* van Doesburg. **D.** Paramere, *D. blotei* (C, D from van Doesburg, 1968). **E.** Paramere, *Paradindymus madagascariensis* (from Stehlik, 1966). **F.** Spermatheca, *D. discolor* Walker (from Pluot, 1970). **G.** Spermatheca, *D. fasciatus* (from Pendergrast, 1957). **H.** Gynatrial complex, showing spermatheca and ring sclerites, *Paradindymus madagascariensis* (from Stehlik, 1966). Abbreviation: gc, gastric caecum.

mone as well as inheritance of wing polymorphism. Several species of Pyrrhocoridae have been used as laboratory animals. An extensive physiological and biochemical literature exists for such species as *D. koenigii* (Fabricius).

Distribution and faunistics. Members of the family are chiefly tropical and subtropical, with only a very few species reaching into the temperate Holarctic. They are found, however, in all major zoogeographic regions.

Freeman (1947) should be consulted for identification of the Old World *Dysdercus* fauna, and van Doesburg (1968) for the New World; Stehlik (1965) erected subgenera for this large genus. Cachan (1952c) revised the Malagasy fauna. Hussey's (1929) catalog remains the basic literature source. There is no general identification work for the other taxa.

Coreoidea

88
Alydidae

General. Members of the group are usually elongate and slender with disproportionately large heads (Fig. 88.1). They range in length from 8 to 20 mm. Many species are myrmecomorphic, particularly as nymphs. These insects are sometimes referred to as broad-headed bugs.

Diagnosis. Bucculae very short, not extending posteriorly beyond antennal insertion; antennae inserted dorsally, segment 1 not constricted at base; ocelli not placed on sclerotized elevations; corium elongated on costal margin; membrane of forewing with numerous veins (Fig.

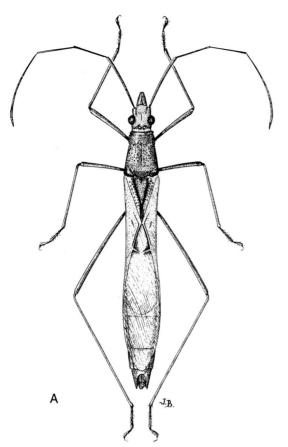
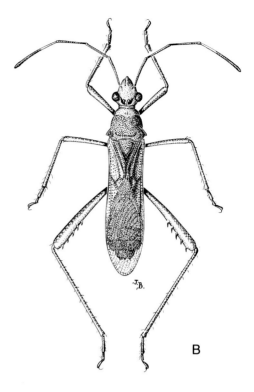

Fig. 88.1. Alydidae (from Slater and Baranowski, 1978). **A.** *Stenocoris tipuloides* (De Geer). **B.** *Megalotomus quinquespinosus* (Say).

88.1A, B); metathoracic scent-gland auricle well developed; tibiae nonsulcate; abdominal trichobothria placed laterally or sublaterally on segments 5–7, submedially on sterna 3–4, either clustered or dispersed; spiracle present on abdominal segment 8, all spiracles ventral; nymphal dorsal abdominal scent-gland openings between terga 4/5 and 5/6; aedeagus as in Fig. 88.2C; pygophore and parameres as in Fig. 88.2B; spermatheca lacking proximal flange (Fig. 88.2D–F); alimentary canal as in Fig. 88.2A.

Classification. Amyot and Serville (1843) first characterized the group as "Alydides." Stål (1867), who treated the group as a subfamily "Alydida" of the Coreidae, was followed by many authors because of the multiveined membrane of the forewing and the flattened plate-like ovipositor. Schaefer (1965b) and subsequent authors have treated the group as a family within the Coreoidea and elevated the tribes to subfamily status, but have combined the Micrelytrinae and Leptocorisinae into a single subfamily with two tribes. We follow Ahmad (1965) in recognizing three taxa at the subfamily level, a position accepted by Schaefer (1972b).

The family contains approximately 42 genera and 250 species.

Key to Subfamilies of Alydidae

1. Maximum width of pronotum at least 1.5 times maximum width of head and distinctly longer than head (Fig. 88.1A) .... Leptocorisinae
– Pronotum at most only slightly wider and longer than head 2
2. Hind femora swollen, with a series of ventral spines (Fig. 88.1B) (except in *Euthetus* Dallas); labial segment 2 distinctly shorter than combined length of segments 3 and 4; posterior margin of abdominal sternum 7 in females without a median split .... Alydinae
– Hind femora never with ventral spines; labial segment 2 usually distinctly longer than combined length of segments 3 and 4; posterior margin of abdominal sternum 7 in females usually with a median split .... Micrelytrinae

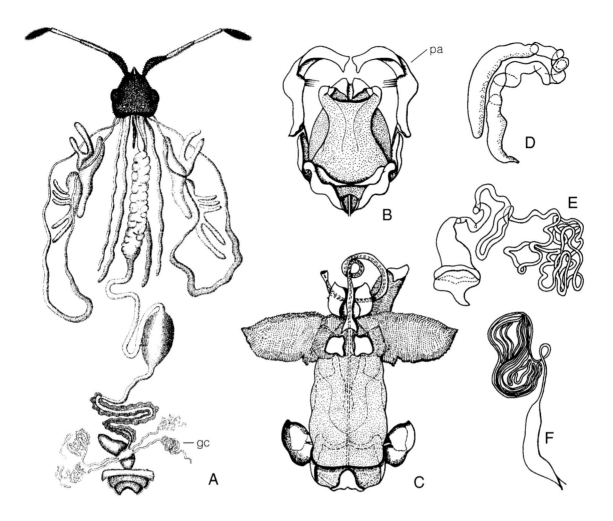

Fig. 88.2. Alydidae. **A.** Alimentary canal, *Camptopus lateralis* (Germar) (from Dufour, 1833). **B.** Male genitalia, including parameres, *Leptocorisa acuta* (Thunberg). **C.** Aedeagus, sagittal view, *Acestra malayana* (Dallas). **D.** Spermatheca, *Daclera punctata* (Signoret). **E.** Spermatheca, *L. chinensis* Dallas. **F.** Spermatheca, *Micrelytra fossularum* (Rossi) (B–F from Ahmad and Southwood, 1964). Abbreviations: gc, gastric caecum; pa, paramere.

ALYDINAE (FIG. 88.1B). Head transverse, broader than thorax (Fig. 88.1B); hind femora with ventral spines (except in *Euthetus*); metathoracic processes of scent-gland peritreme fused and evaporative area not lateral to opening, smooth; abdominal sternum 7 with well-developed anterior spur; second inner laterotergite absent; subcosta present and distinct on forewing.

Some of the better known genera are *Alydus* Fabricius, *Hyalymenus* Amyot and Serville, *Camptopus* Amyot and Serville, and *Riptortus* Stål. The subfamily is found in all major zoogeographic regions. It contains many myrmecomorphic species and others that appear to be wasp mimics.

LEPTOCORISINAE (FIG. 88.1A). Head elongate and relatively slender; labial segment 1 extending well beyond posterior margin of compound eyes, segment 2 usually distinctly shorter than segments 3 and 4 combined, segment 4 usually subequal in length to segment 3; posterior angles of metasternum acutely produced; legs lacking spines, usually very long and slender; hamus of hind wing at most only slightly separated from base of wing; posterior margin of abdominal tergum 2 truncate; spermatheca usually balloon-shaped with a coiled tube.

Ahmad (1965) recognized two tribes. The Leptocorisini, with four genera and a worldwide distribution, contains *Leptocorisa* Berthold from the Orient and Australia and *Stenocoris* Burmeister from the southern Nearctic, Neotropical, and Ethiopian regions. The Noliphini include the genus *Lyrnessus* Stål from South America, whereas the remaining two genera occur in Australia and in the Orient as far west as Sumatra.

MICRELYTRINAE. Antennal segments 2 and 3 sometimes triquetrous; ratio length of head to antennal segment 1 not less than 1:1; labial segment 3 very short, less than one-half length of segment 4, both together shorter than segment 2; posterior margins of metathorax more or less produced, often acutely so; pronotal collar lacking; metathoracic scent-gland evaporative area well developed, occupying nearly one-half of metapleuron; moderately slender or strongly myrmecomorphic.

Micrelytra Laporte is widely distributed in the southern Palearctic, and *Protenor* Stål in the Nearctic.

Specialized morphology. Several genera possess a forewing stridulitrum and a hind femoral plectrum and also long spines on the dorsal surface of the male genital capsule (*Alydus* Fabricius, *Megalotomus* Fieber, *Burtinus* Stål, *Tollius* Stål) (Schaefer et al., 1989). The short bucculae, which do not extend beyond the antennal insertion, appear to be unique to the group.

Natural history. Most species of Alydinae live on legumes, whereas the Leptocorisinae and Micrelytrinae feed primarily on grasses (Schaefer, 1980a; Schaefer and Mitchell, 1983). There are reports of feeding on carrion, but this is not a general food source.

Nymphs of Alydinae are strikingly antlike. The early instars are almost indistinguishable from ants in the field, not only in body form but in their jerky movements as well. Adults with reduced wings are also often strikingly antlike, whereas macropterous forms are remarkably like pompilid wasps (Southwood and Leston, 1959). This likeness may not be evident in museum specimens, but in the field the display of the brightly colored abdominal dorsum and the wasplike twisting and turning are remarkably convincing. Some micrelytrines such as *Trachelium* Herrich-Schaeffer and *Cydamus* Stål are also strikingly antlike.

Other members of the family for which observations are available live either on plant stems or on the ground among stem litter and are cryptic both in body shape and color. Some of these are associated with monocots, as for example three species of the Neotropical genus *Bactrophyamixia* Brailovsky living on bamboos (*Guadua* spp.) in Mexico (Brailovsky, 1991).

Monteith (1982) reported several species of Heteroptera, including *Leptocorisa acuta,* aggregated in large clusters during the dry season in monsoon forests in northern Australia. Apparently this is a phenomenon for protection during nonfeeding periods. They were found about a meter above the ground beneath leaves or in rows along twigs of low shrubs and ferns. When disturbed, the clusters burst into buzzing brief flight and discharged their repellent scents before settling again in the original site. This aestivation probably covers a period of up to six months. A similar phenomenon has been observed in New Guinea, where with the onset of the dry season migration from open grassland to shaded sites is triggered. Gregarious aestivation takes place for up to two months (Sands, 1978). Brown (1965) noted that in the Near East *Camptopus lateralis* Germar migrates from wheat fields, where it feeds chiefly on weeds, to montane overwintering areas many kilometers away.

Distribution and faunistics. The family is represented in all major zoogeographic areas. Štys and Riha (1977) indicated Oligocene-Miocene records of all three subfamilies from sites in the Holarctic. The extinct Monstrocoreinae Popov are recognized from the Upper Jurassic of Kazakhstan.

Schaffner (1964) provided the most comprehensive taxonomic study of the group. Ahmad (1965) monographed the Leptocorisinae; Froeschner (1981) keyed the South American subfamilies and genera; Gross (1963) keyed the fauna of Micronesia; Linnavuori (1987) treated the West and Central African fauna; and Putchkov (1962) treated the fauna of the Ukraine.

89
Coreidae

General. Most coreids are relatively heavy-bodied insects usually robustly elongate or broadly elliptical (Fig. 89.1). The family includes some of the largest of living heteropterans, as well as other species that are delicate (Fig. 89.2) or slender. Many have bizarre dilations and expansions of the hind femora or tibiae and antennal segment 3. Body length ranges from 7 to 45 mm. Members of the group are sometimes referred to as leaf-footed bugs.

Diagnosis. Very diverse in size and shape; head usually small relative to body size; antennae inserted above a line running through center of eye; membrane of forewing multiveined; femora and tibiae of hind legs frequently incrassate or dilated (Fig. 89.1); inner laterotergites usually present; abdominal spiracles all ventral; 3 trichobothria on abdominal segments 3–6, 2 on segment 7, those on segments 5–7 sublateral, clustered; abdominal spiracles ventral (but see Agriopocorinae); nymphal dorsal abdominal scent-gland openings between terga 4/

5 and 5/6; aedeagus as in Fig. 89.3A, B; parameres as in Fig. 89.3C; ovipositor valvulae flattened, platelike (Fig. 89.3D); sternum 7 usually cleft for about half its length (Fig. 89.3D), very rarely cleft throughout, sometimes not cleft; spermatheca usually with proximal pump flange but no distal flange, duct usually short (Fig. 89.3E, F); egg usually with pseudoperculum, usually as a well-defined circular cap, although nonoperculate in Pseudophloeinae and *Hydara*.

Classification. The family was established by Leach (1815) and included Rhopalidae and Alydidae as subfamilies. These groups were retained as subfamilies for many years. The basic suprageneric classification was established by Stål (1867, 1870).

Although there has been considerable recent work on the higher classification of this great family, the infrafamilial relationships remain surprisingly obscure, and relationships within the taxon are obviously in need of a modern synthesis. Schaefer (1964, 1965) provided the most comprehensive treatment. The phylogenetic scheme of Ahmad (1970) recognized the Coreinae, Colpurinae, Phyllomorphinae, Hydarinae, Pseudophloeinae, and Procamptinae. Schaefer (1982) discussed the status of several of these subfamilies. Other authors have placed some of them at the tribal level. As with the Reduviidae, we have adopted a conservative course with regard to the subfamily classification, treating all other proposed higher taxa at the tribal level. The family contains at least 250 genera and 1800 species.

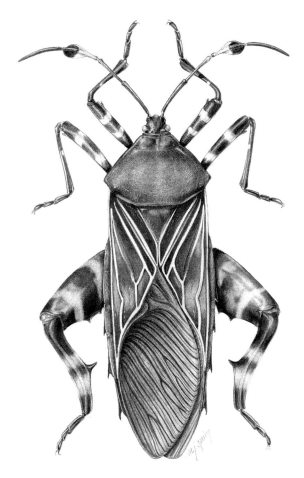

Fig. 89.1. Coreidae. *Thasus acutangulus* (Stål) (Coreinae) (from Slater, 1982).

Key to Subfamilies of Coreidae

1. Spiracles near lateral margin of abdomen, those on segments 2 and 3 visible from above; wings rarely present .... Agriopocorinae
– All spiracles ventral and located away from lateral abdominal margins, not visible from above; usually macropterous, sometimes wings reduced but always at least micropterous 2
2. Hind tibia with distal end produced into a tooth or spine .... Meropachydinae
– Hind tibia lacking a distal spine, or if spine present then head large and only slightly shorter and narrower than thorax 3
3. Median sulcus present on head before eyes; tibiae sulcate on outer surface; hind wing usually lacking an antevannal spur of cubitus .... Coreinae
– Head lacking a median sulcus in front of eyes; tibiae not sulcate on outer surface; antevannal spur of cubitus usually present on hind wing .... Pseudophloeinae

AGRIOPOCORINAE. Body flattened and aradid-like or elongate and sticklike; antenniferous tubercles occupying entire anterodorsal head surface; usually apterous, rarely macropterous; macropterous forms with membrane of forewing reticulately multicellular; metathoracic scent-gland auricles bilobate; tibiae indistinctly sulcate; abdominal segment 8 of males with spiracles; some abdominal spiracles marginal, those on segments 2 and 3 visible from above; valvulae of ovipositor intermediate between laciniate and platelike.

Two genera are known from Australia. *Agriopocoris* Miller (five species) are flattened and broad-bodied and

Fig. 89.2. Coreidae. *Pephricus paradoxus* (Sparrman) (Coreinae) (used with permission of Kathleen Schmidt).

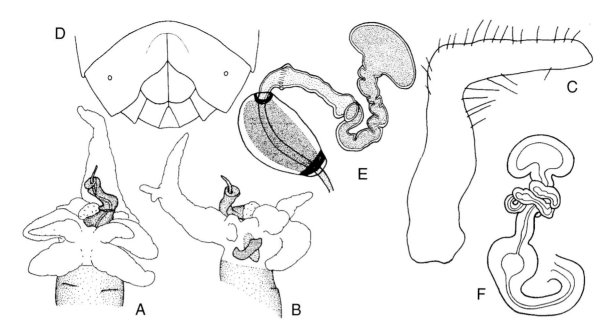

Fig. 89.3. Coreidae. **A.** Aedeagus, lateral view, *Agathyrna ceramica* Dolling. **B.** Aedeagus, sagittal view, *Agathyrna ceramica* (A, B from Dolling, 1987a). **C.** Paramere, *Anasa alfaroi* Brailovsky (from Brailovsky, 1985). **D.** Female terminal abdominal segments, *Catorhintha siblica* Brailovsky and Garcia (from Brailovsky, 1987). **E.** Spermatheca, *Anasa guayaquila* Brailovsky (from Brailovsky, 1985). **F.** Spermatheca, *Anoplocnemis* sp. (from Pendergrast, 1957).

have been taken on *Acacia*. The slender parallel-sided *Tylocryptus egenus* Horvath is the only other genus. It feeds on *Casuarina* branches, on which it is very cryptic.

COREINAE (FIGS. 89.1, 89.2). Usually medium-sized to very large; interocellar distance greater than that from eye to ocellus; anterolateral opening of the metathoracic scent gland well developed, peritreme with well-developed projections, size of evaporative area usually twice that of scent-gland auricle; corial margins straight or slightly sinuate; membranal veins of forewing arising from a transverse vein near, or touching, corial margin; abdominal terga 1–2 and 3–7 fused in both sexes; genital capsule of male without lateral prolongations; articulation of first valvifer and valvulae usually membranous; gonangulum usually flat and folded.

This subfamily contains the vast majority of coreid bugs. It is worldwide in distribution, but most species occur in the tropics. Many of the currently recognized tribes, some of which may be polyphyletic, are either Old or New World in distribution. Only the Coreini (38 genera) and Hydarini (six genera) occur worldwide.

Tribes occurring only in the Eastern Hemisphere are Acanthocorini (= Physomeraria) (eight genera); Amorbini (seven genera); Anhomoeini (one genus); Cloresmini (four genera); Colpurini (19 genera); Cyllarini (one genus); Daladerini (= Brachytini) (nine genera); Dasynini (16 genera); Gonocerini (seven genera); Homoeocerini (12 genera); Latimbini (two genera); Manocoreoini (one genus); Mecocnemini (one genus); Mictini (47 genera); Petascelidini (12 genera); Phyllomorphini (four genera); Prionotylini (three genera); Procamptini (one genus); Sinotagini (one genus).

Tribes occurring only in the Western Hemisphere are Acanthocephalini (19 genera); Acanthocerini (16 genera); Anisoscelidini (nine genera); Barreratalpini (one genus); Chariesterini (four genera); Chelinideini (one genus); Discogastrini (seven genera); Leptoscelidini (nine genera); Nematopodini (14 genera); Spartocerini (five genera). There are six described genera of uncertain tribal affiliation.

Of the above-mentioned tribes, the following six may prove to merit subfamily status.

COLPURINI. This taxon was raised to subfamily status from a tribe of the Coreinae by Ahmad (1970). Nineteen genera are recognized. The genus *Brachylybas* Stål, which is widespread in the Pacific, is of special interest because 40% of its species have reduced wings (including coleopteroid types), a situation that, while widespread in the Heteroptera, is uncommon in the Coreidae.

HYDARINI. This taxon has until recently always been considered a tribe within the Coreinae. Ahmad (1970) raised it to subfamily status. Schaefer (1982) suggested a close relationship with the Alydidae and indicated that in his view it is a relatively primitive taxon. Eight genera are currently recognized.

PHYLLOMORPHINI (FIG. 89.2). This group was raised to

subfamily status from a tribe of the Coreinae by Ahmad (1970). The insects are remarkably cryptic, with wide, flattened, often spinose marginal plates giving them somewhat the appearance of large Tingidae. Four genera are recognized.

PROCAMPTINI. This tribe was established by Ahmad (1964) for *Procamptus segrex* Bergroth from the Philippines and considered by him to probably merit subfamily status.

MEROPACHYDINAE. Head small; thorax narrow; posterior tibiae toothed or spined; hind femora large, clavate; metathoracic scent-gland opening located deep between coxae, opening anteriorly, projections of peritreme fused, evaporative area lacking ridges; abdomen with inner laterotergites fused to connexivum; sterna 2–5 fused in both sexes.

Kormilev (1954) recognized three tribes: Merocorini (two genera), Meropachydini (six genera), and Spathophorini (three genera). The Meropachydinae are chiefly Neotropical, with *Merocoris* Hahn extending northward into the central and northern United States.

PSEUDOPHLOEINAE. Small to moderate-sized coreids; antennae inserted at sides of head, antenniferous tubercles provided with porrect or deflexed processes at outer apical angles; metathoracic scent-gland peritreme with dorsal ridge entire or shortly bilobed, not produced into a Y-shaped auricle; membrane of hemelytron with a compound vein near base almost parallel with apical margin of corium; posterior coxae separated from one another by about width of a coxa; femora moderately to strongly clavate, hind femora typically with 2 or more large subdistal spines ventrally on outer side; tibiae terete, never sulcate; female paratergite 8 without functional spiracle; spermatheca with lunate bulb and without a prominent flange; egg not operculate or pseudoperculate, opening by a transverse eclosion rent.

The Pseudophloeinae, which comprise 28 genera and 166 species, are predominately Old World in distribution but are absent from temperate Australia. Only *Vilga* Stål occurs in the Neotropics, and two genera are found in the Nearctic, both of which also occur in the Palearctic. Dolling (1986) noted the apparently long isolation of the Neotropical fauna and believed it possible that the separation occurred prior to the opening of the Atlantic Ocean before the end of the Cretaceous.

Specialized morphology. Many coreids have strongly expanded, often leaflike hind tibiae (Fig. 89.1). Antennal segments 2 and 3 are also frequently dilated and flattened. Many species are ornamented with spines and tubercles, this being particularly true of the humeral angles of the pronotum, which are frequently produced into acute processes (Fig. 89.2). The eggs frequently have a large number of micropyles, up to a maximum of about 60.

Natural history. Coreids are all phytophagous, and the majority live on plants above the ground. Most appear to feed in the plant vascular system (Mitchell, 1980b). Some are of considerable economic importance (see Chapter 8).

Brown (1965, 1966) noted that in the Near East *Anoplocerus elevatus* (Fieber) and *Ceraleptus obtusus* (Brullé) show a migration similar to that found in several destructive scutellerids and pentatomids—that is, from wheat fields to montane areas many kilometers away. These coreids, however, appear to feed mainly on weeds in the fields rather than on the wheat itself. All species of Pseudophloeinae whose food habits are known feed on herbaceous legumes.

Schaefer and Mitchell (1983) gave a useful summary of the food plants of the family. They concluded that many groups of coreid bugs show definite associations with particular plant groups; others, by contrast, contain members that feed on unrelated plant taxa. Within a given genus, some species may be host-specific, whereas others feed on a variety of plants. Mitchell (1980b) studied species of *Leptoglossus* in detail. In this genus, species such as *L. phyllopus* (Linnaeus) and *L. gonagra* (Fabricius) are very general feeders, whereas *L. fulvicornis* (Westwood), *L. ashmeadi* Heidemann, and *L. corculus* (Say) are extremely restricted in their host associations. Polyphagous species tend to feed on annuals and have a labium of intermediate length, whereas specialists have an elongated labium. Mitchell (1980b) discussed host selection, survival, and parasitism and provided an extensive analysis of the literature on feeding by phytophagous Hemiptera. She demonstrated that in *Leptoglossus* the specialist feeders did not feed on related plants, indicating that the plant-insect coevolutionary theory so strongly advocated by Ehrlich and Raven (1965) is not applicable to many piercing-sucking insects.

Despite those observations, Schaefer and O'Shea (1979) noted that three coreine tribes (Mictini, Acanthocerini, and Nematopodini) feed chiefly on plants of the family Leguminosae. They raised the possibility that such feeding habits could be used in phylogenetic reconstruction—that is, that this habit constitutes a synapomorphy. Schaefer and Mitchell (1983), however, noted that legume feeding may be primitive in the Coreoidea, being found in the Alydinae, Pseudophloeinae, Hydarini, and Hyocephalidae.

It has recently been shown that some of the large species of coreids defend territories on flower heads and fight vigorously with other males who attempt to enter their territories. This phenomenon may be widespread and could account for the sexually dimorphic hind legs of many coreids, in which the male hind femora are often much larger than those of the females and are provided with wicked series of sharp spines (Mitchell, 1980a).

Although the majority of Coreidae are dull-colored, many display strikingly bright coloration, some of which is presumably aposematic. Other coloration is probably deflective because it includes metallic hues and occurs chiefly in species found in tropical forest habitats.

Members of the genus *Thasus* Stål are among the largest of the terrestrial Heteroptera. *Thasus acutangulus* Stål (Fig. 89.1) has aposematically colored bright orange, yellow, and black nymphs that form feeding aggregations. Apparently the aggregation pheromone is perceived by receptors on the distal segment of each antenna; thus nymphs displaced from the aggregation are able to re-aggregate. If these feeding aggregations are disturbed, the members pulsate, spray jets of anal fluid into the air, and exude scent-gland secretions over the abdominal terga (Aldrich and Blum, 1978).

Distribution and faunistics. Coreid bugs are worldwide in distribution but are most abundant in the tropics and subtropics, where they also attain their largest size and most bizarre appearance.

There is no modern key to the major groups. The works of Stål (1867, 1870) are still important. For special groups see those of O'Shea (1980a; Acanthocerini), O'Shea (1980b; Nematopodini), O'Shea and Schaefer (1978; Mictini), Yonke (1972; Chariesterini), Kormilev (1954; Meropachydinae), Herring (1980; Chelinideini), Dolling (1978, 1979; Clavigrallini), Breddin (1900; Colpurini), Osuna (1984; Anisoscelidini), Gross (1963; Micronesian fauna), Putchkov (1962; fauna of Ukraine), Allen (1969; *Leptoglossus*), Brailovsky (1985; *Anasa* Amyot and Serville), and Brailovsky and Garcia (1987; *Catorhintha* Stål).

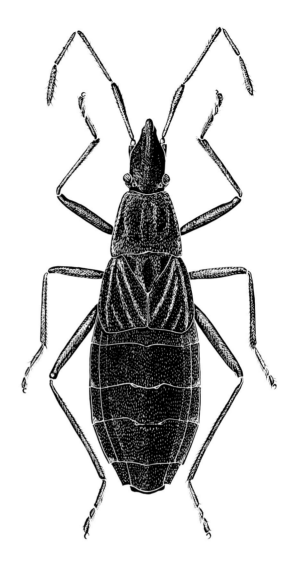

Fig. 90.1. Hyocephalidae. *Maevius indecorus* Stål (drawn by S. Monteith; from CSIRO, 1991).

90
Hyocephalidae

General. These insects are rather large (length up to 15 mm), reddish brown or black, moderately elongate and parallel-sided (Fig. 90.1). They are dorsally flattened and resemble some pseudophloeine coreids and certain lygaeids.

Diagnosis. Head very elongate, strongly tapered (Figs. 90.1, 90.2B, C), surface tuberculate; ocelli very small, placed near posterolateral margins of eyes; clypeus elevated, compressed in middle; antenniferous tubercles placed below a line drawn through middle of eye; bucculae large, elongate, contiguous anteriorly, extending posteriorly to anterior margin of eyes (Fig. 90.2B, C); mandibular plates short; gula with a labial groove; membrane of forewing with basal cells formed by cross veins connecting 4 primary longitudinal veins, and with several distal veins (Fig. 90.2A); vein connecting M and Cu on corium lacking; metathoracic scent-gland ostiole with bristlelike processes; tibiae sulcate; base of abdomen ventrally with an ovoid pore-bearing organ (sieve plate) on each side (Fig. 90.2D), comparable to porous area on sternum of some rhyparochromine lygaeids; trichobothria of abdominal sterna 3 and 4 mesal, 5 and 6 lateral, clustered (Fig. 90.2D); all spiracles ventral (Fig. 90.2D); inner laterotergites present; nymphs with 2 pairs of dorsal abdominal scent-gland openings between terga 4/5 and 5/6; ovipositor laciniate; spermathecal bulb lack-

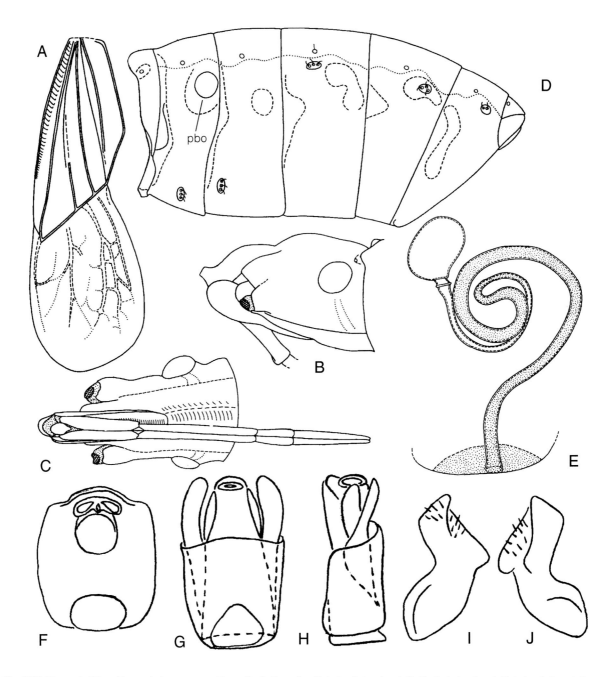

Fig. 90.2. Hyocephalidae. *Hyocephalus aprugnus* Bergroth. **A.** Forewing. **B.** Lateral view head. **C.** Ventral view head. **D.** Lateral view abdomen, showing trichobothria and pore-bearing organ. **E.** Spermatheca (A–E from Štys, 1964b). *Maevius indecorus* Stål. **F.** Genital capsule. **G.** Aedeagus, sagittal view. **H.** Aedeagus, lateral view. **I.** Paramere. **J.** Paramere, obverse view. (F–J modified from Schaefer, 1981a.) Abbreviation: pbo, pore-bearing organ.

ing distinct flanges (Fig. 90.2E); eggs elongate, with 3 micropylar processes; nymphs with a U-shaped ecdysial line removed from eyes.

Classification. The taxon was first established by Bergroth (1906) as a subfamily of Coreidae but later was reduced to tribal status (Bergroth, 1912). Reuter (1912a) raised the group to family rank. Štys (1964b) reviewed the position of the family, noting that it had been variously placed in the Coreoidea and Lygaeoidea, a factor that influenced him (Štys, 1961b) to combine the two superfamilies.

Two genera and two species—*Hyocephalus aprugnus* Bergroth and *Maevius indecorus* Stål—both from Australia, are known. Schaefer (1981a) gave evidence for placing the Hyocephalidae and Stenocephalidae in the Coreoidea.

Specialized morphology. The bristlelike processes projecting from the scent-gland orifice appear to be an apomorphy for the family. The ovoid pore-bearing plate on the abdomen has been referred to as a "strainer" (Štys, 1964b).

Natural history. Hyocephalids live on the underside of stones in sandy gravelly areas, where they are very cryptic. They feed on the ripe seeds of *Acacia* and *Eucalyptus*.

Distribution and faunistics. The family occurs only in Australia. The main references on the group are by Štys (1964b) and Schaefer (1981a).

91
Rhopalidae

General. The members of this family range in length from 4 to 15 mm. They vary greatly in shape and color. The majority are dull brownish and resemble species of Orsillinae (Lygaeidae), with which one often finds them confused in collections. The remainder are much larger and similar in shape, body form, and bright coloration (Fig. 91.1) to species of Lygaeinae (Lygaeidae) and many species of Pyrrhocoridae and Largidae. They are frequently called the scentless plant bugs.

Diagnosis. Clypeus surpassing mandibular plates; ocelli situated on low tubercles; antennae never dilated, first segment constricted basally; metathoracic scent-gland openings usually obsolete or obsolescent; corium frequently with large hyaline areas; membrane of forewing always with numerous veins; trichobothria on abdominal sterna 3 and 4 mediolateral, those of 5, 6, and 7 lateral; abdominal spiracles ventral; inner laterotergites present; nymphs with dorsal abdominal scent-gland openings between terga 4/5 and 5/6, the latter displaced forward, a unique and universally occurring character in the family; pygophore with lateral, median, and paralateral lobes (Fig. 91.2A); aedeagus as in Fig. 91.2B; parameres as in Fig. 91.2C; ovipositor platelike, abdominal sternum 7 of females entire; spermatheca consisting of a round bulb, small pump, and long, generally coiled duct.

Classification. The taxon was first recognized as a higher group by Amyot and Serville (1843) as the "Rhopalides." This name was used subsequently by many

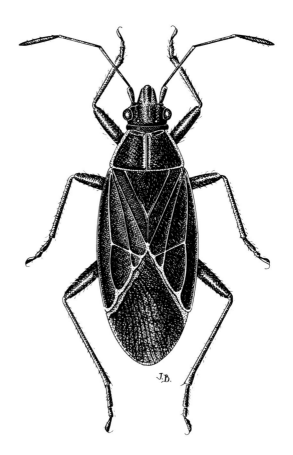

Fig. 91.1. Rhopalidae. *Boisea trivittata* (Say) (from Slater and Baranowski, 1978).

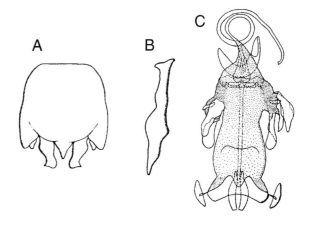

Fig. 91.2. Rhopalidae (from Chopra, 1967). **A.** Genital capsule, *Myrmus miriformis* (Fallén). **B.** Aedeagus, sagittal view, *Liorhyssus hyalinus* (Fabricius). **C.** Paramere, *Ithamar hawaiiensis* Kirkaldy.

authors including Dallas (1851–1852) and Stål (1862). Costa (1853) used Corizini as a tribe, and Douglas and Scott (1865) used Corizidae as a family. Mayr (1866) used "Corizida," and much of the literature is under this name. The family has often been considered to be a subfamily of an inclusive Coreidae, but modern workers such as Chopra (1967) and Göllner-Scheiding (1983) treated it as a distinct family. The last-mentioned work, a world catalog, serves as a summary of the classification of the group and an exhaustive introduction to the literature. Schaefer and Chopra (1982) provided a cladistic analysis to the tribal level. Two subfamilies, comprising 18 genera and 209 species are recognized.

Key to Subfamilies of Rhopalidae

1. Lateral margins of pronotum straight or slightly sinuate, lacking a distinct notch immediately behind collar .... Rhopalinae
– Lateral margins of pronotum with a distinct notch behind collar .... Serinethinae

RHOPALINAE. Generally of small to moderate size; lateral pronotal margin lacking a distinct notch immediately behind collar; usually dull-colored; abdominal sterna 3 and 4 fused in both sexes; male and female genitalia also distinctive (see Chopra, 1967).

Six tribes are generally recognized: Chorosomini, Corizomorphini, Harmostini, Maccevethini, Niesthreini, and Rhopalini. The European *Corizus hyoscyami* (Linnaeus), a presumed mimic of *Lygaeus equestris* (Linnaeus), is unusual in the subfamily, along with a few other species, for its bright coloration.

SERINETHINAE (FIG. 91.1). Relatively large, elongate, elliptical; usually brightly colored; head broader than long; clypeus slightly raised; antenniferous tubercles without lateral projections; pronotum trapezoidal, lateral margins notched just behind collar; metapleuron not, or only indistinctly, divided; a well-developed metathoracic third axillary sclerite spur; metathoracic scent-gland openings in coxal cavities; dorsal abdominal scent glands retained and functional in adults (Ribeiro, 1989); hind femora not incrassate or spined; abdominal sterna 3 and 4 not fused; phallus with sclerotized ventral conjunctival appendages.

Three genera are currently recognized. *Leptocoris* Hahn, with 30 species from the Old World tropics, is particularly speciose in Africa. *Jadera* Stål, with 17 species, is chiefly tropical and subtropical in the New World, with several species reaching into the southern United States. *Boisea* Kirkaldy, with two Nearctic, one central and west African, and one South Indian species, was until recently considered to be a junior synonym of *Leptocoris*. It was elevated to generic status by Göllner-Scheiding (1980) for the Western Hemisphere species previously placed in *Leptocoris*, a decision questioned by Schaefer and Chopra (1982). It includes the common box elder bug, *Boisea trivittata* (Say).

Specialized morphology. The functional abdominal scent glands in the adults are a specialized feature in the Serinethinae. They may represent a neotenic condition (see Chapter 85 for a similar situation).

Natural history. All species are phytophagous and feed on a variety of herbs and woody plants; host use was reviewed by Schaefer and Chopra (1982). None are of major economic importance, although *Boisea trivittata* often becomes a nuisance when it forms aggregations in the fall and enters houses in large numbers to hibernate. It is an arboreal species almost monophagous on box elder trees (*Acer negundo*).

The soapberry bug, *Jadera haematoloma* Herrich-Schaeffer, feeds on three native species of Sapindaceae: *Sapindus saponaria* var. *drummondii* (the soapberry tree) in the south-central United States, *Serjania brachycarpa* (serjania vine) in southernmost Texas, and the perennial *Cardiospermum corindum* (balloon vine) in southern Florida (Carroll and Boyd, 1992). The insect has colonized three introduced species of Sapindaceae: *Koelreuteria paniculata* (golden rain tree), *K. elegans* (flat-podded golden rain tree), and *Cardiospermum halicacabum* (heartseed vine) (Carroll and Loye, 1987). Carroll and Boyd (1992) discussed the establishment of recognizable populations upon these various hosts, distinguishable primarily by large differences in the length of the rostrum. Large fruits with deeply embedded seeds have populations with very long rostra, whereas small fruits have populations with significantly shorter rostra. These authors noted that much of this differentiation must have occurred within the past 30 to 50 years because the introduced host plants were planted commonly only within that time period. This rapid establishment of host-plant races directly attributable to a selective trait has great evolutionary significance.

Carroll (1988, 1991) reported that *Jadera haematoloma* has developed two different mating systems in different populations. Where females are abundant, promiscuous mating occurs, but where males are much more abundant than females a guarding system has developed due to the increased cost of male searching in such populations. Nymphs are reddish and aposematic and tend to congregate on the trunks and seeds, whereas the adults are cryptically colored and disperse widely when on tree hosts. The nymphs that form aggregations, especially during the time of molting, have lower mortality in early instars than do isolated nymphs, and they tend to molt earlier and with greater synchrony. Older nymphs do not show this advantage, possibly because of increased cannibalism. Experiments on toads and blue jays show that all nymphs are distasteful and indeed appear also to be distasteful to praying mantids. Interestingly, however, the adults also are distasteful. So the aposematic nymphal coloration appears to be associated with gregarious behavior, serving especially as protection during periods of molting and being most significant in early-instar nymphs (Ribeiro, 1989). A similar phenomenon appears to obtain for the related box elder bug (see Chapter 8).

Distribution and faunistics. Rhopalids are found in all major faunal regions. Some of their distributions are unusual. The Niesthreini contains three genera; two are confined to the Western Hemisphere, and the third is found in South Africa and India. The Chorosomini contains six genera; three are Palearctic (one also occurs in the Old World tropics), *Xenogenus* Berg is Neotropical, and *Ithamar* Kirkaldy is endemic to the Hawaiian Islands. *Boisea* contains two Nearctic species, one species from west and central Africa and one from southern India. Additional cladistic analyses should improve our understanding of origins and relationships in rhopalid higher taxa.

Chopra (1967) discussed the higher classification and provided valuable generic revisions (1968, 1973). Putchkov (1986) monographed the fauna of the former Soviet Union. Gross (1960) revised the Australian and Pacific *Leptocoris,* and Gross (1963) provided keys to the Micronesian genera and species. Göllner-Scheiding (1979) revised the genus *Jadera,* and Göllner-Scheiding (1980) treated the African species of *Leptocoris* and *Boisea.* The Göllner-Scheiding (1983) world catalog is a basic resource.

92
Stenocephalidae

General. These bugs are relatively slender, elongate, parallel-sided, brown and yellow, and have a distinct coreidlike appearance (Fig. 92.1A). They range in length from 8 to 15 mm. They have no common name.

Diagnosis. Mandibular plates surpassing and usually contiguous anterior to apex of clypeus (Fig. 92.1A); membrane of forewing opaque with a large and a small basal cell from which numerous radiating and anastamosing veins arise (Fig. 92.1A); abdominal spiracles ventral; inner laterotergites present; trichobothria of abdominal segments 5 and 6 placed laterally and clustered posterior to spiracle (Fig. 92.1H); nymphs with dorsal abdominal scent-gland openings between terga 4/5 and 5/6; genital capsule and abdominal organs as in Fig. 92.1B (Lansbury, 1965); aedeagus as in Fig. 92.1C; parameres as in Fig. 92.1D, E; ovipositor laciniate, completely dividing sternum 7 (Fig. 92.1F); spermatheca as in Fig. 92.1G; m-chromosome present; egg oblong with 4–9 micropylar processes.

Classification. This family is of special interest in that it shows characteristics that are transitional between the Coreidae and the Lygaeidae. Its relationships have been discussed in detail by Scudder (1957a) and Schaefer (1965a, b).

The above characteristics as well as such specialized features as the four-lobed principal salivary gland, the structure of the phallus, and the coreidlike appearance of the nymphs all suggest close relationships to the Coreidae. The ovipositor, however, is distinctly laciniate (Fig. 92.1F); the spermatheca has a spherical apical bulb and a differentiated pump and duct (Fig. 92.1G), and the egg is oblong with 4–9 micropylar processes grouped at the anterior pole. These features are definitely of the lygaeid type and represent conditions not generally found in the Coreoidea.

Two genera, *Dicranocephalus* Hahn and *Psotilnus* Stål, and over 30 species are known (Lansbury, 1965–1966). The majority are from the tropics and subtropics of the Eastern Hemisphere, including Australia, but some species occur in the temperate Palearctic.

Specialized morphology. The unusual combination of features possessed by the group is discussed above under Classification.

Natural history. Stenocephalids are frequently swept from various species of Euphorbiaceae, upon which they breed. The eggs are laid on the surface of the stems rather

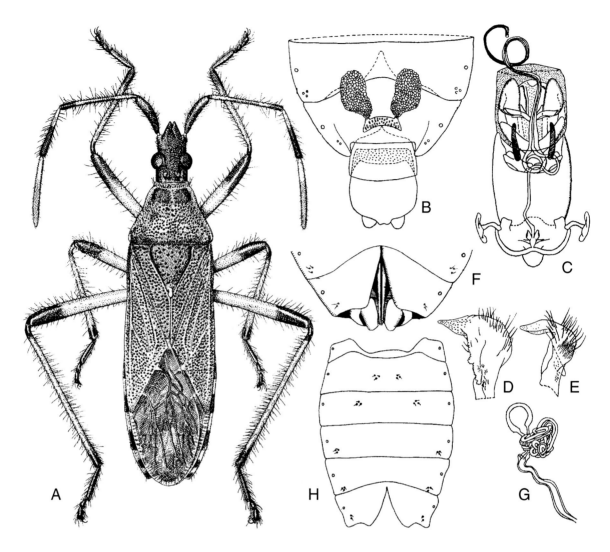

Fig. 92.1. Stenocephalidae. **A.** *Dicranocephalus setulosus* (Ferrari) (from Lansbury, 1965–1966). **B.** Genital capsule, showing abdominal organs, *D. pallidus* (Signoret) (from Lansbury, 1965, with permission from *Nature,* copyright 1965 Macmillan Magazines Limited). **C.** Aedeagus, sagittal view, *D. albipes* (Fabricius). **D.** Paramere, *D. agilis* (Scopoli). **E.** Paramere, *D. albipes.* **F.** Female terminal abdominal segments, *D. agilis.* **G.** Spermatheca, *D. agilis.* **H.** Ventral view, abdomen with trichobothria, *D. albipes* (C–H from Lansbury, 1965–1966).

than inserted in the plant tissues, despite the fact that the ovipositor is laciniate.

Distribution and faunistics. An endemic species has been reported in the Galápagos Islands and was treated as distinct in Lansbury's (1965–1966) revisional study, the major work on the group. Usinger and Ashlock, the first heteropterists to do careful collecting in the islands, were not able to obtain this species, suggesting that this species is thus very scarce or has become extinct. Although originally described from the Galápagos, it is now thought to be conspecific with an African species. It was certainly introduced, probably by sailing ships, not later than the early nineteenth century.

LITERATURE CITED

Abreu, J. M. de. 1977. Mirideos neotrópicais associados ao cacaueiro. *In:* E. M. Lavabre (ed.), Les mirides du cacaoyer, pp. 85–106. Institut Français du Café et du Cacao, Paris.

Afzelius, B. A., R. Dallai, and P. Lindskog. 1985. Spermatozoa of saldid bugs (Insecta, Hemiptera, Leptopodomorpha). J. Ultrastruct. Res. 90:304–312.

Ahmad, I. 1964. Systematic position of *Dicrorymbus, Xenoceraea* and *Procamptus* (Hemiptera: Coreidae), three of Bergroth's genera from the Philippine Islands with description of two new species. Ann. Entomol. Fenn. 30:17–34.

Ahmad, I. 1965. The Leptocorisinae (Heteroptera: Alydidae) of the world. Bull. Br. Mus. (Nat. Hist.) Entomol. Suppl. 5:1–156.

Ahmad, I. 1970. Some aspects of the female genitalia of *Hygia* Uhler 1861 (Coreidae: Colpurinae) and their bearing on classification. Pak. J. Zool. 2:235–243.

Ahmad, I., and J. E. McPherson. 1990. Male genitalia of the type species of *Corimelaena* White, *Galgupha* Amyot and Serville, and *Cydnoides* Malloch (Hemiptera: Cydnidae: Corimelaenidae) and their bearing on classification. Ann. Entomol. Soc. Am. 83:162–170.

Ahmad, I., and T. R. E. Southwood. 1964. The morphology of the alydid abdomen with special reference to the genitalia and its bearing on classification (Heteroptera). Tijdschr. Entomol. 107:365–378.

Akingbohunbe, A. E. 1979. A new genus and four new species of Hyaliodinae (Heteroptera: Miridae) from Africa with comments on the status of the subfamily. Rev. Zool. Afr. 93:500–522.

Aldrich, J. R. 1988. Chemical ecology of the Heteroptera. Ann. Rev. Entomol. 33:211–238.

Aldrich, J. R., and M. S. Blum. 1978. Aposematic aggregation of a bug (Hemiptera: Coreidae): the defensive display and formation of aggregations. Biotropica 10:58–61.

Aldrich, J. R., W. R. Lusby, J. P. Kochansky, M. P. Hoffmann, L. T. Wilson, and F. G. Zalom. 1988. *Lygus* bug pheromones vis-à-vis stink bugs. Proc. Beltwide Cotton Prod. Conf., pp. 213–216.

Aldrich, J. R., S. P. Carroll, J. E. Oliver, W. R. Lusby, A. A. Rudmann, and R. M. Waters. 1990. Exocrine secretions of scentless plant bugs: *Jadera, Boisea* and *Niesthrea* (Hemiptera: Heteroptera: Rhopalidae). Biochem. Syst. Ecol. 18:369–376.

Allen, R. C. 1969. A revision of the genus *Leptoglossus* Guerin (Hemiptera: Coreidae). Entomol. Am. 45:35–140.

Ambrose, D. P., and S. J. Vennison. 1990. Diversity of spermatophore capsules in reduviids (Insecta, Heteroptera, Reduviidae) of South India. Mitt. Zool. Mus. Berlin 66:309–317.

Amyot, C. J.-B., and A. Serville. 1843. Histoire naturelle des insectes Hémiptères. Librairie Encyclopédique de Roret, Paris. 675 pp.

Andersen, N. M. 1973. Seasonal polymorphism and developmental changes in organs of flight and reproduction in bivoltine pondskaters (Hem. Gerridae). Entomol. Scand. 4:1–20.

Andersen, N. M. 1975. The *Limnogonus* and *Neogerris* of the Old World with character analysis and a reclassification of the Gerrinae (Hemiptera: Gerridae). Entomol. Scand. Suppl. 7:1–96.

Andersen, N. M. 1976. A comparative study of locomotion on the water surface in semiaquatic bugs (Insecta, Hemiptera, Gerromorpha). Vidensk. Medd. Dan. Naturhist. Foren. 139:337–396.

Andersen, N. M. 1977a. A new and primitive genus and species of Hydrometridae (Hemiptera, Gerromorpha) with a cladistic analysis of relationships within the family. Entomol. Scand. 8:301–316.

Andersen, N. M. 1977b. Fine structure of the body hair layers and morphology of the spiracles of semiaquatic bugs (Insecta, Hemiptera, Gerromorpha) in relation to life on the water surface. Vidensk. Medd. Dan. Naturhist. Foren. 140:7–37.

Andersen, N. M. 1977c. On the taxonomy of the subfamily Microveliinae (Heteroptera: Veliidae) of West and Central Africa. Ann. Entomol. Fenn. 43(2):41–61.

Andersen, N. M. 1978. A new family of semiaquatic bugs for *Paraphrynovelia* Poisson with a cladistic analysis of relationships (Insecta, Hemiptera, Gerromorpha). Steenstrupia 4:211–225.

Andersen, N. M. 1980. Hygropetric water striders of the genus *Onychotrechus* Kirkaldy with description of a related genus (Insecta, Hemiptera, Gerridae). Steenstrupia 6:113–146.

Andersen, N. M. 1981a. A new genus of Veliinae and descriptions of new Oriental species of the subfamily (Hemiptera: Veliidae). Entomol. Scand. 12:339–356.

Andersen, N. M. 1981b. Semiaquatic bugs: phylogeny and classification of the Hebridae (Heteroptera: Gerromorpha) with revisions of *Timasius, Neotimasius* and *Hyrcanus*. Syst. Entomol. 6:377–412.

Andersen, N. M. 1982a. The Semiaquatic Bugs (Hemiptera, Gerromorpha): Phylogeny, Adaptations, Biogeography, and Classification. Entomonograph 13. Scandinavian Science Press, Klampenborg, Denmark. 455 pp.

Andersen, N. M. 1982b. The first species of *Heterocleptes* Villiers from the Oriental region (Hemiptera: Hydrometridae). Entomol. Scand. 13:105–108.

Andersen, N. M. 1982c. Semiterrestrial water striders of the genera *Eotrechus* Kirkaldy and *Chimarrhometra* Bianchi (Insecta, Hemiptera, Gerridae). Steenstrupia 9:1–25.

Andersen, N. M. 1989a. The coral bugs, genus *Halovelia* Bergroth (Hemiptera, Veliidae). I. History, classification, and taxonomy of species except the *H. malaya*–group. Entomol. Scand. 20:75–120.

Andersen, N. M. 1989b. The Old World Microveliinae (Hemiptera: Veliidae). II. Three new species of *Baptista* Distant and a new genus from the Oriental region. Entomol. Scand. 19:363–380.

Andersen, N. M. 1989c. The coral bugs, genus *Halovelia* Bergroth (Hemiptera, Veliidae). II. Taxonomy of the *N. malaya*–group, cladistics, ecology, biology, and biogeography. Entomol. Scand. 20:179–227.

Andersen, N. M. 1990. Phylogeny and taxonomy of water striders,

Andersen, N. M. 1991a. Cladistic biogeography of marine water striders (Hemiptera, Gerromorpha) in the Indo-Pacific. Aust. Syst. Bot. 4:151–163.

Andersen, N. M. 1991b. Marine insects: genital morphology, phylogeny, and evolution of sea skaters, genus *Halobates* (Hemiptera, Gerridae). Zool. J. Linn. Soc. 103:21–60.

Andersen, N. M. 1991c. A new genus of marine water striders (Hemiptera, Veliidae) with five new species from Malesia. Entomol. Scand. 22:389–404.

Andersen, N. M. 1993. The evolution of wing polymorphism in water striders (Gerridae): a phylogenetic approach. Oikos 67:433–443.

Andersen, N. M., and W. A. Foster. 1992. Sea skaters of India, Sri Lanka, and the Maldives, with a new species and a revised key to Indian Ocean species of *Halobates* and *Asclepios* (Hemiptera, Gerridae). J. Nat. Hist. 26:533–553.

Andersen, N. M., and J. T. Polhemus. 1976. Water striders (Hemiptera: Gerridae, Veliidae, etc.). *In:* L. Cheng (ed.), Marine Insects, pp. 187–224. North Holland Publishing, Amsterdam.

Andersen, N. M., and J. T. Polhemus. 1980. Four new genera of Mesoveliidae (Hemiptera, Gerromorpha) and the phylogeny and classification of the family. Entomol. Scand. 11:369–392.

Andersen, N. M., and J. R. Spence. 1992. Classification and phylogeny of the Holarctic water strider genus *Limnoporus* Stål (Hemiptera, Gerridae). Can. J. Zool. 70:753–785.

Anderson, A. B. 1963. Morphology and biology of *Macrovelia hornii* Uhler (Heteroptera, Mesoveliidae). M.S. thesis, Oregon State University, Corvallis. 99 pp.

Anderson, L. D. 1932. A monograph of the genus *Metrobates* (Hemiptera, Gerridae). Univ. Kans. Sci. Bull. 20(16):297–311.

Ando, H. 1988. Obituary. Ryuichi Matsuda. Int. J. Insect Morphol. Embryol. 17:91–94.

Ashlock, P. D. 1957. An investigation of the taxonomic value of the phallus in the Lygaeidae (Hemiptera-Heteroptera). Ann. Entomol. Soc. Am. 50:407–426.

Ashlock, P. D. 1960. H. G. Barber: bibliography and list of names proposed. Proc. Entomol. Soc. Wash. 62:129–138.

Ashlock, P. D. 1964. Two new tribes of Rhyparochrominae: a re-evaluation of the tribe Lethaeini (Hemiptera- Heteroptera: Lygaeidae). Ann. Entomol. Soc. Am. 57:414–422.

Ashlock, P. D. 1967. A generic classification of the Orsillinae of the World. (Hemiptera-Heteroptera-Lygaeidae). Univ. Calif. Publ. Entomol. 48:1–82.

Ashlock, P. D. 1969. Robert L. Usinger bibliography and list of proposed names. Pan-Pac. Entomol. 45:185–203.

Ashlock, P. D., and J. D. Lattin. 1963. Stridulatory mechanisms in the Lygaeidae, with a new American genus of Orsillinae (Hemiptera: Heteroptera). Ann. Entomol. Soc. Am. 56:693–703.

Asquith, A. 1991. Revision of the genus *Lopidea* in America North of Mexico (Heteroptera: Miridae). Koeltz Scientific Books, Königstein, Germany. 280 pp.

Asquith, A., and J. D. Lattin. 1990. *Nabicula* (*Limnonabis*) *propinqua* (Reuter) (Heteroptera: Nabidae): dimorphism, phylogenetic relationships and biogeography. Tijdschr. Entomol. 133:3–16.

Bachmann, A. O. 1966. Catálogo sistemático y clave para la determinación de las subfamilias, géneros y especies de las Gerridae de la República Argentina (Insecta: Hemiptera). Physis 26(71):207–218.

Bacon, J. A. 1956. A taxonomic study of the genus *Rhagovelia* (Hemiptera, Veliidae) of the Western Hemisphere. Univ. Kans. Sci. Bull. 38:695–914.

Baehr, M. 1989. Revision of the genus *Ochterus* Latreille in the Australian region (Heteroptera: Ochteridae). Entomol. Scand. 20:449–477.

Bailey, N. S. 1951. The Tingoidea of New England and their biology. Entomol. Am., n.s., 31:1–140.

Balduf, W. V. 1941. Life history of *Phymata pennsylvanica* americana Melin (Phymatidae, Hemiptera). Ann. Entomol. Soc. Am. 34:204–214.

Balduf, W. V. 1964. Numbers of ovarioles in the Heteroptera (Insecta). Proc. Entomol. Soc. Wash. 66:2–5.

Balwin, W. F., and T. N. Salthouse. 1959. Dermal glands and mucin in the moulting cycle of *Rhodnius prolixus* Stål. J. Insect Physiol. 3:345–348.

Baptist, B. A. 1941. The morphology and physiology of the salivary glands of Hemiptera-Heteroptera. Q. J. Microsc. Sci. 82:91–139.

Baranowski, R. M. 1958. Notes on the biology of the royal palm bug, *Xylastodoris luteolus* Barber (Hemiptera, Thaumastocoridae). Ann. Entomol. Soc. Am. 51:547–551.

Barber, H. G. 1920. A new member of the family Thaumastocoridae. Bull. Brooklyn Entomol. Soc. 15:97–104.

Barber, H. G. 1924. Corrections and comments Hemiptera-Heteroptera. J. N.Y. Entomol. Soc. 32:133–137.

Barber, H. G. 1930. Essay on the subfamily Stenopodinae of the world. Entomol. Am., n.s., 10:149–238.

Barber, H. G. 1939. Scientific survey of Porto Rico and the Virgin Islands: insects of Porto Rico and the Virgin Islands: Hemiptera-Heteroptera (excepting Miridae and Corixidae). Sci. Surv. Porto Rico 14(3):263–441.

Bare, C. O. 1928. Haemoglobin cells and other studies of the genus *Buenoa* (Hemiptera Notonectidae). Univ. Kans. Sci. Bull. 18:265–349.

Barth, R. 1954. Estudos anatômicos e histológicos sôbre a subfamília Triatominae (Heteroptera, Reduviidae). IV. parte: O complexo das glândulas salivares de *Triatoma infestans*. Mem. Inst. Oswaldo Cruz 52:517–585.

Barth, R. 1961. Sôbre o orgão abdominal glandular de *Arilus carinatus* (Forster, 1771) (Heteroptera: Reduviidae). Mem. Inst. Oswaldo Cruz 59:37–43.

Baudoin, R. 1946. Contribution à l'éthologie d'*Aepophilus bonnairei* Signoret et à celle de queleques autres arthropodes à respiration aérienne de la zone intercôtidale. Bull. Soc. Zool. Fr. 71:109–113.

Baudoin, R. 1955. La physico-chimie des surfaces dans la view des arthropodes aériens des miroirs d'eau, des rivages marins et lacustres et de la zone intercôtidale. Bull. Biol. Fr. Belg. 89:16–164.

Baunacke, W. 1912. Statische Sinnesorgane bei Nepiden. Zool. Jahrb., Abt. Anat. Ontog. 34:179–346.

Beier, M. 1935. [Biography and bibliography of] Anton Handlirsch. Konowia 14:340–347.

Beier, M. 1938. 28. Ordnung der Pterygogenea: Heteroptera = Wanzen. *In:* Handbuch der Zoologie, vol. 4, Insecta 2, pp. 2041–2204. Walter de Gruyter, Berlin.

Bennett, D. V., and E. F. Cook. 1981. The semiaquatic Hemiptera of Minnesota (Hemiptera: Heteroptera). Univ. Minn. Agric. Exp. Sta. Tech. Bull. 332. 58 pp.

Bequaert, J. 1935. Presocial behavior among the Hemiptera. Bull. Brooklyn Entomol. Soc. 30:177–191.

Berenbaum, M. R., and E. Miliczky. 1984. Mantids and milkweed bugs: efficacy of aposematic coloration against invertebrate predators. Am. Midl. Nat. 111:64–68.

Berg, C. 1879. Hemiptera Argentina enumeravit speciesque novas. Pauli E. Coni, Bonariae. 316 pp.

Berg, C. 1884. Addenda et emendanda ad Hemiptera Argentinae. Pauli E. Coni, Bonariae. 213 pp.

Bergroth, E. 1891. Eine neue Saldiden-Gattung. Wiener Entomol. Ztg. 10:263–267.

Bergroth, E. 1898. Sur la place systématique du genre *Joppeicus* Put. Rev. Entomol., Caen 17:188.

Bergroth, E. 1906. Aphylinae und Hyocephalinae, zwei neue Hemipteren-Subfamilien. Zool. Anz. 29:644–649.

Bergroth, E. 1907. Über die systematische Stellung der Gattung *Eumenotes* West. Dtsch. Entomol. Z. 1907:498–500.

Bergroth, E. 1910. Remarks on Colobathristidae with descriptions of two new genera. Ann. Soc. Entomol. Belg. 54:297–305.

Bergroth, E. 1912. New or little known Hemiptera, chiefly from Australia, in the American Museum of Natural History. Bull. Am. Mus. Nat. Hist. 31:343–348.

Bergroth, E. E. 1920. List of the Cylapinae (Hem., Miridae) with descriptions of new Philippine forms. Ann. Soc. Entomol. Belg. 55:67–83.

Billberg, G. J. 1820. Enumeratio Insectorum in Museo Gust. Joh. Billberg. Stockholm.

Blakley, N. 1980. Divergence in seed resource use among Neotropical milkweed bugs, *Oncopeltus*. Oikos 25:8–15.

Blakley, N., and H. Dingle. 1978. Competition: butterflies eliminate milkweed bugs from a Caribbean Island. Oecologia 37:133–136.

Blatchley, W. S. 1926. Heteroptera, or True Bugs, of Eastern North America. Nature Publishing, Indianapolis. 1116 pp.

Blatchley, W. S. 1928. "Quit-claim" specialists vs. the making of manuals. Bull. Brooklyn Entomol. Soc. 23:10–18.

Blatchley, W. S. 1930. Blatchleyana. Nature Publishing, Indianapolis. 77 pp.

Blatchley, W. S. 1939. Blatchleyana. II. Nature Publishing, Indianapolis. 46 pp.

Bliven, B. P. 1973. A third paper on Hemiptera associated with the Pyrrhocoridae. Occident. Entomol. 1:10:123–133.

Blöte, H. C. 1945. On the systematic position of *Scotomedes* (Heteroptera, Nabidae). Zool. Meded. 25:321–324.

Boas, G., L. Villas, and A. R. Panizzi. 1980. Biologia de *Euschistus heros* (Fabricius, 1798) em soja (*Glycine max* (L.) Merrill). An. Soc. Entomol. Brasil 9:105–113.

Bobb, M. L. 1951. Life history of *Ochterus banksi* Barber. Bull. Brooklyn Entomol. Soc. 46:92–100.

Bobb, M. L. 1974. The Insects of Virginia, no. 7: The Aquatic and Semi-aquatic Hemiptera of Virginia. Res. Div. Bull. 87. Virginia Polytechnic and State University, Blacksburg. 195 pp.

Boer, M. H. den. 1965. Revisionary notes on the genus *Metrocoris* Mayr (Heteroptera: Gerridae), with descriptions of four new species. Zool. Verh., Leiden 74:1–38.

Bonhag, P. F., and J. R. Wick. 1953. The function anatomy of the male and female reproductive systems of the milkweed bug, *Oncopeltus fasciatus* (Dallas) (Heteroptera: Lygaeidae). J. Morphol. 93:177–284.

Börner, C. 1934. Über System und Stammesgeschichte der Schabelkerfe. Entomol. Beihefte Berlin-Dahlem. 1:138–144.

Borror, D. J., C. A. Triplehorn, and N. F. Johnson. 1989. An Introduction to the Study of Insects, fifth ed. Sauders College Publishing, Philadelphia. 875 pp.

Bradley, G. A., and J. D. Hinks. 1968. Ants, aphids, and jack pine in Manitoba. Can. Entomol. 100:40–50.

Brailovsky, H. 1985. Revisión del género *Anasa* Amyot-Serville (Hemiptera-Heteroptera-Coreidae-Coreinae-Coreini). Monogr. Inst. Biol. Univ. Natl. Autón. México 2:1–266.

Brailovsky, H. 1989. Un género y dos especies nuevas de Hemípteros (Lygaeidae, Bledionotinae, Pamphantini) del Brasil. An. Inst. Biol. Univ. Natl. Autón. México, Ser. Zool. 59:193–202.

Brailovsky, H. 1991. Hemiptera-Heteroptera from Mexico. XLIII. A new genus and three new species of Neotropical Micrelytrinae (Alydidae) collected on bamboos. J. N.Y. Entomol. Soc. 99:487–495.

Brailovsky, H., and M. Garcia. 1987. Revisión del género *Catorhintha* Stål (Hemiptera-Heteroptera-Coreidae-Coreinae-Coreini). Monogr. Inst. Biol. Univ. Natl. Autón. México 4:1–148.

Brailovsky, H., L. Cervantes, and C. Mayorga. 1988. Hemípteros-Heterópteros de México. XL. La familia Cyrtocoridae Distant en la estación de Biología Tropical "Los Tuxtlas" (Pentatomoidea). An. Inst. Biol. Univ. Natl. Autón. México, Ser. Zool. 58(2):537–560.

Brandt, E. 1878. Vergleichend-anatomische Untersuchungen über das Nervensystem der Hemipteren. Horae Soc. Entomol. Rossicae 14:496–505.

Breddin, G. 1897. Hemipteren. *In:* Hamburger Megasaensisehen Sammelreise. Friedrichsen & Co., Hamburg.

Breddin, G. 1900. Materiae ad conitionem subfamiliae Pachycephalini (Lybantini olim), ex Hemipteris-Heteropteris, Fam. Coreidae. Rev. Entomol. Fr. xix:194–217.

Breddin, G. 1907. Berytiden und Myodochiden von Ceylon aus der Sammelausbeute von Dr. W. Horn (Rhynch. Het.). Dtsch. Entomol. Z. 1907(1):34–47.

Brindley, M. D. 1930. On the metasternal scent glands of certain Heteroptera. Trans. Entomol. Soc. Lond. 78:199–207.

Brinkhurst, R. O. 1959. Alary polymorphism in the Gerroidea (Hemiptera-Heteroptera). J. Anim. Ecol. 28:211–230.

Brinkhurst, R. O. 1961. Alary polymorphism in the Gerroidea. Verh. Inst. Verein. Limnol. 14:978–982.

Brinkhurst, R. O. 1963. Observations on wing-polymorphism in the Heteroptera. Proc. R. Entomol. Soc. Lond. (A) 38:15–22.

Brooks, A. R., and L. A. Kelton. 1967. Aquatic and semiaquatic Heteroptera of Alberta, Saskatchewan, and Manitoba (Hemiptera). Mem. Entomol. Soc. Can. 51. 92 pp.

Brooks, G. T. 1951. A revision of the genus *Anisops* (Notonectidae, Hemiptera). Univ. Kans. Sci. Bull. 34:301–519.

Brown, E. S. 1948. *Poissonia longifemorata*, a new genus and species of giant water-bug (Hemiptera, Belostomatidae) and a new variety in the allied genus *Hydrocyrius* Spinola. Proc. R. Entomol. Soc. Lond. 17:109–144.

Brown, E. S. 1951. The relation between migration rate and type of habitat in aquatic insects, with special reference to certain species of Corixidae. Proc. Zool. Soc. Lond. 121:539–545.

Brown, E. S. 1955. *Pseudotheraptus wayi*, a new genus and species of coreid (Hemiptera) injurious to coconuts in East Africa. Bull. Entomol. Res. 46:240.

Brown, E. S. 1962a. The distribution of the species of *Eurygaster* Lap. (Hemiptera, Scutelleridae) in Middle East countries. Ann. Mag. Nat. Hist., ser. 13, (5):65–81.

Brown, E. S. 1962b. Researches on the ecology and biology of *Eurygaster integriceps* Put. (Hemiptera, Scutelleridae) in Middle East countries, with special reference to the overwintering period. Bull. Entomol. Res. 53:445–514.

Brown, E. S. 1963. Report on research on the soun pest (*Eurygster integriceps* Put.) and other wheat pentatomids in Middle East countries, 1958–1961. Misc. Rept. No. 4. Dept. Technical Co-operation, Technical Assistance Prog., Central Treaty Organization, London. 46 pp.

Brown, E. S. 1965. Notes on the migration and direction of flight of *Eurygster* and *Aelia* species (Hemiptera, Pentatomoidea) and their possible bearing on invasions of cereal crops. J. Anim. Ecol. 34:93–107.

Brown, E. S. 1966. An account of the fauna associated with *Eurygaster integriceps* Put. and *Aelia* species (Hem. Pentatomoidea) in their overwintering area in the Middle East. Entomol. Monthly Mag. 102:29–46.

Brundin, L. 1966. Transantarctic relationships and their significance, as evidenced by chironomid midges, with a monograph of the subfamilies Podonominae and Afroteniinae and the Austral Heptagyiae. Kongl. Svenska Vet.-Akad. Handl. 11(1):1–472.

Büchner, P. 1921. Über eine neues, symbiotisches Organ der Bettwanze. Biol. Zentralbl. 41:570–574.

Büchner, P. 1965. Endosymbiosis of Animals with Plant Microorganisms. Interscience Publishers, New York. 909 pp.

Burmeister, H. 1832–1839. Handbuch der Entomologie, vols. 1 and 2. G. Reimer, Berlin. 696 pp.; 1050 pp.

Butler, E. A. 1923. A Biology of the British Hemiptera-Heteroptera. H. F. & G. Witherby, London. 682 pp.

Cachan, P. 1952a. Etude de la prédation chez les réduvides de la région éthiopienne. 1. La prédation en groupe chez *Ectrichodia gigas* H.-Sch. Physiol. Comp. 2:378–385.

Cachan, P. 1952b. Les Pentatomidae de Madagascar (Hémiptères, Hétéroptères). Mém. Inst. Sci. Madagascar, ser. E, 1:231–462.

Cachan, P. 1952c. Pyrrhocoridae de Madagascar. Mém. Inst. Sci. Madagascar, ser. E, 1:71–92.

Calam, D. H., and A. Youdeowei. 1968. Identification and functions of secretion from the posterior scent gland of fifth instar larva of the bug *Dysdercus intermedius*. J. Insect Physiol. 14:1147–1158.

Callahan, P. S. 1975. Insect antennae with special reference to the mechanism of scent detection and the evolution of sensilla. Int. J. Insect Morphol. Embryol. 4:381–430.

Carayon, J. 1948a. Les organes parastigmatiques des Hémiptères Nabidae. C. R. Acad. Sci. Fr. 227:864–866.

Carayon, J. 1948b. Dimorphisme sexuel des glandes odorantes métathoraciques chez quelques Hémiptères. C. R. Acad. Sci. Fr. 227:303–305.

Carayon, J. 1949. Observations sur la biologie des Hémiptères Microphysides. Bull. Mus. Natl. Hist. Nat., Paris, ser. 2, 21:710–716.

Carayon, J. 1950a. Nombre et disposition des ovarioles dans les ovaires des Hémiptères-Hétéroptères. Bull. Mus. Natl. Hist. Nat., Paris, ser. 2, 22:470–475.

Carayon, J. 1950b. Caractères anatomiques et position systématique des Hémiptères Nabidae (note préliminaire). Bull. Mus. Natl. Hist. Nat., Paris, ser. 2, 22:95–101.

Carayon, J. 1951. Les mécanismes de transmission héréditaire des endosymbiontes chez les insectes. Tijdschr. Entomol. 49:111–142.

Carayon, J. 1952. Les fécondations hémocoeliennes chez les Hémiptères Nabidés du genre *Alloeorhynchus*. C. R. Acad. Sci., Paris 234:751–753.

Carayon, J. 1954. Un type nouveau d'appareil glandulaire propre aux mâles de certains Hémiptères Anthocoridae. Bull. Mus. Natl. Hist. Nat., Paris, ser. 2, 26:602–606.

Carayon, J. 1957. Introduction à l'étude de Anthocoridae omphalophores (Hemiptera Heteroptera). Ann. Soc. Entomol. Fr. 126:159–196.

Carayon, J. 1958. Etudes sur les Hémiptères Cimicoidea. 1. Position des genres *Bilia*, *Bilioa*, *Bilianella* et *Wollastoniella* dans une tribe nouvelle (Oriini) des Anthocoridae; differences entre derniers et les Miridae Isometopinae (Heteroptera). Mém. Mus. Natl. Hist. Nat., ser. A, Zool. 16:141–172.

Carayon, J. 1960. *Stethoconus frapai* n. sp., miridé prédateur du tingidé du caféier, *Dulinius unicolor* (Sign.), à Madagascar. J. Agric. Trop. Bot. Appl. 7:110–120.

Carayon, J. 1961a. Hemiptera (Heteroptera) Anthocoridae. South Africa Animal Life 8:533–557.

Carayon, J. 1961b. Valeur systématique de voies ectodermique de l'appareil génital femelle chez les Hémiptères Nabidae. Bull. Mus. Natl. Hist. Nat., Paris, ser. 2, 33:183–196.

Carayon, J. 1961c. La viviparite chez les Hétéroptères. Verh. Int. Kongr. Entomol. Wien 1:711–714.

Carayon, J. 1962. Observations sur l'appareil odorifique de Hétéroptères. Particulièrement celui de Tingidae, Vianaidae et Piesmatidae. Cah. Nat., n.s., 18:1–16.

Carayon, J. 1963. La transmission héréditaire des bactéries symbiotiques chez les Lygaeidae vivipares (Heteroptera). Proc. 16th Int. Congr. Zool., vol. 1. Washington, D.C.

Carayon, J. 1964a. Ethologie: Un cas d'offrande nuptiale chez les Hétéroptères. C. R. Acad. Sci., Paris 259:4815–4818.

Carayon, J. 1964b. La spermathèque et les voies génitales femelles des lygaeides Oxycareninae. Rev. Fr. Entomol. 31:196–218.

Carayon, J. 1966. Traumatic insemination and the paragenital system. *In:* R. L. Usinger (ed.), Monograph of Cimicidae (Hemiptera-Heteroptera), vol. 7, pp. 81–166. The Thomas Say Foundation. Entomological Society of America, Lanham, Md.

Carayon, J. 1969. Emploi du noir chlorazol en anatomie microscopique des insectes. Ann. Entomol. Soc. Fr., n.s., 5:179–193.

Carayon, J. 1970. Etude de *Alloeorhynchus* d'Afrique Centrale, avec quelques remarques sur la classification des Nabidae [Hemiptera]. Ann. Soc. Entomol. Fr., n.s., 6:899–931.

Carayon, J. 1971a. Notes et documents sur l'appareil odorant métathoracique des Hémiptères. Ann. Soc. Entomol. Fr., n.s., 7:737–770.

Carayon, J. 1971b. *Lyctocoris* (*Paralyctocoris*) *menieri*, Anthocoridae nouveau des Iles Canaries. Bull. Soc. Entomol. Fr. 76:161–165.

Carayon, J. 1972a. Caractères systématiques et classification des Anthocoridae (Hemipt.). Ann. Soc. Entomol. Fr., n.s., 8:309–349.

Carayon, J. 1972b. Le genre *Xylocoris:* Subdivision des espèces nouvelles [Hem. Anthocoridae]. Ann. Soc. Entomol. Fr., n.s., 8:579–606.

Carayon, J. 1974. Etude sur les Hémiptères Plokiophilidae. Ann. Soc. Entomol. Fr., n.s., 10:409–525.

Carayon, J. 1977. Insemination extra-génitale traumatique. *In:* P.-P. Grassé (ed.), Anatomie, systématique, biologie, insectes. Gamètogenèses, fécondation, métamorphoses pp. 351–390. Traité de zoologie. Masson, Paris.

Carayon, J. 1981. Dimorphisme sexuel des glandes tégumentaires et production de pheromones chez les Hémiptères Pentatomoidea. C. R. Acad. Sci., ser. 3, 292:867–870.

Carayon, J. 1984. Les androconies de certains Hémiptères Scutelleridae. Ann. Entomol. Soc. Fr., n.s., 20:113–134.

Carayon, J., and J.-R. Steffan. 1959. Observations sur le régime alimentaire des *Orius* et particulièrement d'*Orius pallidicornis* (Reuter) (Heteroptera: Anthocoridae). Cahiers Nat., n.s., 15:53–63.

Carayon, J., and A. Villiers. 1968. Etude sur les Hémiptères Pachynomidae. Ann. Entomol. Soc. Fr., n.s., 4:703–739.

Carayon, J., R. L. Usinger, and P. Wygodzinsky. 1958. Notes on the higher classification of the Reduviidae, with the description of a new tribe of the Phymatinae (Hemiptera-Heteroptera). Rev. Zool. Bot. Afr. 57:256–281.

Carpenter, G. H. 1892. Rhynchota from Murray Island and

Mabuiag. Reports on the zoological collections made in Torres Straights by Professor A. C. Haddon, 1888–1889. Sci. Proc. R. Dublin Soc., n.s., 7:137–146.

Carroll, S. P. 1988. Contrasts in the reproductive ecology of temperate and tropical populations of *Jadera haematoloma* (Rhopalidae), a mate-guarding hemipteran. Ann. Entomol. Soc. Am. 81:54–63.

Carroll, S. P. 1991. The adaptive significance of mate guarding in the soapberry bug, *Jadera haematoloma* (Hemiptera: Rhopalidae). J. Insect Behav. 4:509–530.

Carroll, S. P., and C. Boyd. 1992. Host race radiation in the soapberry bug: natural history with the history. Evolution 46:1052–1069.

Carroll, S. P., and J. E. Loye. 1987. Specialization of *Jadera* species (Hemiptera: Rhopalidae) on seeds of the Sapindaceae, and coevolution of defense and attack. Ann. Entomol. Soc. Am. 80:373–378.

Carroll, S. P., and J. E. Loye. 1990. Male-biased sex ratios, female promiscuity, and copulatory mate guarding in an aggregating tropical bug, *Dysdercus bimaculatus*. J. Insect Behav. 3:33–48.

Carvalho, J. C. M. 1951. New genera and species of Isometopidae in the collection of the British Museum of Natural History (Hemiptera). An. Acad. Bras. Cienc. 23:381–391.

Carvalho, J. C. M. 1952. On the major classification of the Miridae (Hemiptera). (With keys to the subfamilies and tribes and a catalogue of the world genera). An. Acad. Brasil. Cien. 24:31–110.

Carvalho, J. C. M. 1955. Keys to the genera of Miridae of the World (Hemiptera). Bol. Mus. Paraense Emilio Goeldi, Belem 11: 1–151.

Carvalho, J. C. M. 1956. Heteroptera: miridae. *In:* Insects of Micronesia, vol. 7, no. 1, pp. 1–100. Bernice P. Bishop Museum, Honolulu.

Carvalho, J. C. M. 1957. A catalogue of the Miridae of the world, pt. 1. Arq. Mus. Nac., Rio de Janeiro 44:158 pp.

Carvalho, J. C. M. 1958a. A catalogue of the Miridae of the world, pt. 2. Arq. Mus. Nac., Rio de Janeiro 45:216 pp.

Carvalho, J. C. M. 1958b. A catalogue of the Miridae of the world, pt. 3. Arq. Mus. Nac., Rio de Janeiro 47:161 pp.

Carvalho, J. C. M. 1959. A catalogue of the Miridae of the world, pt. 4. Arq. Mus. Nac., Rio de Janeiro 48:384 pp.

Carvalho, J. C. M. 1960. A catalogue of the Miridae of the world, pt. 5. Arq. Mus. Nac., Rio de Janeiro 51:194.

Carvalho, J. C. M. 1973. Neotropical Miridae, CLXXVIII: studies on the tribe Herdoniini Distant. XVI: Key to the world genera (Hemiptera). Rev. Brazil. Biol. 33 (suppl.):197–200.

Carvalho, J. C. M. 1976. Mirideos neotrópicais, CC: Revisão do gênero *Horicias* Distant, com descrições de espécies novas (Hemiptera). Rev. Brasil. Biol. 36:429–472.

Carvalho, J. C. M. 1981. The Bryocorinae of Papua New Guinea (Hemiptera, Miridae). Arq. Mus. Nac., Rio de Janeiro 56:35–89.

Carvalho, J. C. M. 1984a. On a new species of intertidal water strider from Brazil (Hemiptera, Gerromorpha, Mesoveliidae). Amazoniana 8:519–523.

Carvalho, J. C. M. 1984b. On the subfamily Palaucorinae Carvalho (Hemiptera, Miridae). Rev. Brasil. Biol. 44:81–86.

Carvalho, J. C. M., and W. E. China. 1952. The "Cyrtopeltis-Engytatus" complex (Hemiptera, Miridae, Dicyphini). Ann. Mag. Nat. Hist., ser. 12, 5:158–166.

Carvalho, J. C. M., and L. A. A. Costa. 1989. Chave para identifição dos gêneros neotrópicos da família Colobathristidae (Hemiptera). Rev. Brasil. Biol. 49:271–277.

Carvalho, J. C. M., and A. V. Fontes. 1968. Mirideos neotrópicais, CI: Revisão do complexo "Cylapus" Say, com descrições de gêneros and espécies novos (Hemiptera). Rev. Brasil. Biol. 28:273–282.

Carvalho, J. C. M., and A. V. Fontes. 1971. Mirideos neotrópicais, CXXIX: Chave sistemática para os gêneros da tribo Resthenini Reuter (Hemiptera). Rev. Brasil. Biol. 31:141–144.

Carvalho, J. C. M., and G. F. Gross. 1979. The tribe Hyalopeplini of the World (Hemiptera: Miridae). Rec. South Aust. Mus. 17:429–531.

Carvalho, J. C. M., and G. F. Gross. 1980. The distribution in Australia of the grass bugs of the tribe Stenodemini (Heteroptera-Miridae-Mirinae). Rec. S. Aust. Mus. 18(2):75–82.

Carvalho, J. C. M., and G. F. Gross. 1982. Taxonomy of S. Australian ant-mimetic Miridae (Hemiptera: Heteroptera). I. The *Leucophoroptera* Group of the subfamily Phylinae. Aust. J. Zool., suppl. ser., 86:1–75.

Carvalho, J. C. M., and L. M. Lorenzato. 1978. The Cylapinae of Papua New Guinea (Hemiptera: Miridae). Rev. Brasil. Biol. 38:121–149.

Carvalho, J. C. M., A. V. Fontes, and T. J. Henry. 1983. Taxonomy of South American species of *Ceratocapsus*, with descriptions of 45 new species (Hemiptera: Miridae). U.S. Dep. Agric. Tech. Bull. 1676. 58 pp.

Carver, M. 1990. Integumental morphology of the ventral thoracic scent gland system of *Poecilometis longicornis* (Dallas) (Hemiptera: Pentatomidae). Int. J. Insect Morphol. Embryol. 19:319–321.

Cassis, G. 1984. A systematic study of the subfamily Dicyphinae (Heteroptera: Miridae). Ph.D. dissertation, Oregon State University, Corvallis. 389 pp.

Caudell, A. N. 1924. Some insects from the Chilibrillo bat caves of Panama. Insecutor Inscitiae Menstruus 12:133–136. (Cited in Froeschner, 1960:661).

Champion, G. C. 1897–1901. Biologia Centrali Americana. Insecta. Rhynchota. Hemiptera-Heteroptera, vol. 2. xvi + 416 pp., 22 pls.

Chapman, H. C. 1962. The Saldidae of Nevada (Hemiptera). Pan-Pac. Entomol. 38(3):147–159.

Chen, P., and N. Nieser. 1993. A taxonomic revision of the Oriental water strider genus *Metrocoris* Mayr (Hemiptera, Gerridae), pts. 1 and 2. Steenstrupia 19:1–43, 44–82.

Cheng, L. 1965. The genus *Ventidius* Distant (Heteroptera: Gerridae) in Malaya, with a description of four new species. Proc. R. Entomol. Soc. Lond. 34:153–163.

Cheng, L. 1966. Three new species of *Esakia* Lundblad (Heteroptera: Gerridae) from Malaya. Proc. R. Entomol. Soc. Lond. 35:16–22.

Cheng, L. 1977. The elusive sea bug *Hermatobates* (Heteroptera). Pan-Pac. Entomol. 53:87–97.

Cheng, L., and C. H. Fernando. 1969. A taxonomic study of the Malayan Gerridae (Hemiptera: Heteroptera) with notes on their biology and distribution. Orient. Insects 3:97–160.

China, W. E. 1931. Morphological parallelisms in the structure of the labium in the hemipterous genera *Coptosomoides*, gen. new and *Bozius* Dist. (Fam. Plataspidae) in connection with mycetophagous habits. Ann. Mag. Nat. Hist., ser. 10, 7:281–286.

China, W. E. 1933. A new family of Hemiptera-Heteroptera with notes on the phylogeny of the suborder. Ann. Mag. Nat. Hist., 10, 12:180–196.

China, W. E. 1935. New and little known Helotrephidae (Hemiptera, Helotrephidae). Ann. Mag. Nat. Hist., ser. 10, 15:593–614.

China, W. E. 1940. New South American Helotrephidae (Hemiptera-Heteroptera). Ann. Mag. Nat Hist., ser. 11, 5:106–126.

China, W. E. 1953. A new subfamily of Microphysidae (Hemiptera-Heteroptera). Ann. Mag. Nat. Hist., ser. 12, 6:97–125.

China, W. E. 1955a. A new genus and species representing a new subfamily of Plataspidae, with notes on the Aphylidae (Hemiptera, Heteroptera). Ann. Mag. Nat. Hist., ser. 12, 8:204-210.

China, W. E. 1955b. A reconsideration of the systematic position of the family Joppeicidae Reuter (Hemiptera-Heteroptera), with notes on the phylogeny of the suborder. Ann. Mag. Nat. Hist., ser. 12, 8:353–370.

China, W. E. 1955c. The evolution of the water bugs. In: Symposium on Organic Evolution, pp. 91–103. Natl. Inst. Sci. India, Bull. 7.

China, W. E. 1956. A new species of the genus *Hermatobates* from the Hawaiian Islands (Hemiptera: Heteroptera, Gerridae, Halobatinae). Ann. Mag. Nat. Hist., ser. 12, 9:353–357.

China, W. E. 1962. Hemiptera-Heteroptera collected by the Royal Society Expedition to south Chile 1958-1959. Ann. Mag. Nat. Hist., ser. 13, 5:705–723.

China, W. E. 1963. *Lestonia haustorifera* China (Hemiptera: Lestoniidae)—a correction. J. Entomol. Soc. Queensland 2:67–68.

China, W. E., and N. C. E. Miller. 1959. Check-list and keys to the families and subfamilies of the Hemiptera-Heteroptera. Bull. Br. Mus. (Nat. Hist.), Entomol. 8(1):1–45.

China, W. E., and J. G. Myers. 1929. A reconsideration of the classification of the cimioid families (Heteroptera), with the description of two new spider-web bugs. Ann. Mag. Nat. Hist., 10, 3:97–126.

China, W. E., and J. A. Slater. 1956. A new subfamily of Urostylidae from Borneo (Hemiptera: Heteroptera). Pac. Sci. 10:410–414.

China, W. E., and R. L. Usinger. 1949a. Classification of the Veliidae (Hemiptera) with a new genus from South Africa. Ann. Mag. Nat. Hist., ser. 12, 2:343–354.

China, W. E., and R. L. Usinger. 1949b. A new genus of Hydrometridae from the Belgian Congo, with a new subfamily and a key to the genera. Rev. Zool. Bot. Afr. 41:314–319.

China, W. E., R. L. Usinger, and A. Villiers. 1950. On the identity of *Heterocleptes* Villiers 1948 and *Hydrobatodes* China and Usinger 1949. Rev. Zool. Bot. Afr. 43:336–344.

Chopra, N. P. 1967. The higher classification of the family Rhopalidae (Hemiptera). Trans. R. Entomol. Soc. Lond. 119:363–399

Chopra, N. P. 1968. A revision of the genus *Arhyssus* Stål. Ann. Entomol. Soc. Am. 61:629–655.

Chopra, N. P. 1973. A revision of the genus *Niesthrea* Spinola (Rhopalidae: Hemiptera). J. Nat. Hist. 7:441–459.

Chu, Y.I. 1969. On the bionomic of *Lyctocoris beneficus* (Hiura) and *Xylocoris glactinus* (Fieber) (Anthocoridae, Heteroptera). J. Fac. Agric. Kyushu Univ. 15:1–136.

Cloarec, A. 1976. Interactions between different receptors involved in prey capture in *Ranatra linearis*. Biol. Behav. 1:251–266.

Cobben, R. H. 1957. Beitrag zur Kenntnis der Uferwanzen (Hem. Het. Fam. Saldidae). Entomol. Berichten 17:245–257.

Cobben, R. H. 1959. Notes on the classification of Saldidae with the description of a new species from Spain. Zool. Meded. 36(22):303–316.

Cobben, R. H. 1960a. Die Uferwanzen Europas. Hemiptera-Heteroptera Saldidae. In: W. Stichel (ed.), Illustrierte Bestimmungstabellen der Wanzen. II. Europa (Hemiptera-Heteroptera Europae), vol. 3, pp. 209–263.Hermsdorf, Berlin.

Cobben, R. H. 1960b. The Heteroptera of the Netherlands Antilles. I. Foreword. Gerridae, Veliidae, Mesoveliidae (Water Striders), no. 11, pp. 1–34. Wageningen, Netherlands.

Cobben, R. H. 1960c. The Heteroptera of the Netherlands Antilles. II. Hebridae, no. 11, pp. 35–43. Wageningen, Netherlands.

Cobben, R. H. 1960d. The Heteroptera of the Netherlands Antilles. III. Saldidae (shore bugs). Studies on the Fauna of Curaçao and Other Caribbean Islands, no. 52, pp. 44–61. Wageningen, Netherlands.

Cobben, R. H. 1961. A new genus and four new species of Saldidae (Heteroptera). Entomol. Ber. 21:96–107.

Cobben, R. H. 1965. Egg-life and symbiont transmission in a predatory bug, *Mesovelia furcata* Ms and Rey (Heteroptera, Mesoveliidae). Proc. 12th Int. Congr. Entomol., pp. 166–168. London.

Cobben, R. H. 1968a. Evolutionary Trends in Heteroptera. Part I. Eggs, Architecture of the Shell, Gross Embryology, and Eclosion. Centre for Agricultural Publishing and Documentation, Wageningen, Netherlands. 475 pp.

Cobben, R. H. 1968b. A new species of Leptopodidae from Thailand (Hemiptera-Heteroptera). Pac. Insects 10:529–533.

Cobben, R. H. 1970. Morphology and taxonomy of intertidal dwarfbugs (Heteroptera: Omaniidae fam. nov.). Tijdschr. Entomol. 113:61–90.

Cobben, R. H. 1971. A fossil shore bug from the tertiary amber of Chiapas, Mexico (Heteroptera, Saldidae). Univ. Calif. Pubs. Entomol. 63:49–56.

Cobben, R. H. 1976. La faune terrestre de l'ile de Sainte-Helene. Troisième partie. 2. Fam. Saldidae. Ann. Mus. R. Afr. Centr., Zool. 215:345–353.

Cobben, R. H. 1978. Evolutionary Trends in Heteroptera. Part 2. Mouthpart-structures and Feeding Strategies. Mededlingen Landbouwhogeschool 78-5. H. Veeman, Wageningen, Netherlands. 407 pp.

Cobben, R. H. 1980a. The Saldidae of the Hawaiian archipelago (Hemiptera: Heteroptera). Pac. Insects 22:1–34.

Cobben, R. H. 1980b. On some species of *Pentacora*, with the description of a new species from Australia (Heteroptera, Saldidae). Zool. Meded. 55:115–126.

Cobben, R. H. 1981a. Comments on some cladograms of major groups of Heteroptera. Rostria 33(suppl.):29–39.

Cobben, R. H. 1981b. The recognition of grades in the Heteroptera and comments on R. Schuh's cladograms. Syst. Zool. 30:181–191.

Cobben, R. H. 1982. The hebrid fauna of the Ethiopian Kaffa Province, with considerations on species grouping (Hebridae, Heteroptera). Tijdschr. Entomol. 125:1–24.

Cobben, R. H. 1985. Additions to the Eurasian saldid fauna, with a description of fourteen new species (Heteroptera, Saldidae). Tijdschr. Entomol. 128:215–270.

Cobben, R. H. 1987a (1986). New African Leptopodomorpha (Heteroptera: Saldidae, Omaniidae, Leptopodidae), with an annotated checklist of Saldidae from Africa. I. New species of the genus *Saldula* (Saldidae). Rev. Zool. Afr. 100:399–421.

Cobben, R. H. 1987b (1986). New African Leptopodomorpha (Heteroptera: Saldidae, Omaniidae, Leptopodidae), with an annotated checklist of Saldidae from Africa. II. New taxa of Saldidae (except *Saldula*), Omaniidae, Leptopodidae, and a checklist of African shorebugs. Rev. Zool. Afr. 100:3–30.

Cook, M. L. 1977. A key to the genera of Asian Ectrichodiinae (Hemiptera: Reduviidae) together with a checklist of genera and species. Orient. Insects 11:63–88.

Cooper, G. M. 1981. The Miridae (Hemiptera: Heteroptera) associated with noble fir, *Abies procera* Rehd. M.S. thesis, Oregon State University, Corvallis. 135 pp.

Costa, A. 1853. Cimicum regni Neapolitani centuria. Atti del Reale Istituto d'Incorraggiamento alle Scienze Naturali 3:239–279.

Costa Lima, A. da, C. R. Hathaway, and C. A. C. Seabra. 1948. Sôbre algumas espécies de Apiomerinae representadas nas nossas coleções. Mem. Inst. Oswaldo Cruz 45:761–772.

Costa Lima, A. da, C. A. C. Seabra, and C. R. Hathaway. 1951. Estudo dos Apiomeros. Mem. Inst. Oswaldo Cruz 49:273–442.

Coutière, H., and J. Martin. 1901. Sur un nouvel Hémiptère halophile, *Hermatobatodes marchei*, n. gen., n. sp. Bull. Mus. Hist. Nat. 5:214–226.

Crampton, G. C. 1923. Preliminary note on the terminology applied to the parts of an insect's leg. Can. Entomol. 55:126–132.

Csiki, E. 1944. Dr. Horváth Géza T. tag emlékezete 1847–1937. A Magyar Tudományos Akadémia e hunyt tagjai fölött tartott emlékbeszédek, Budapest. 48 pp.

CSIRO (Commonwealth of Scientific and Industrial Research Organization). 1991. Hemiptera. *In:* The Insects of Australia, 2nd ed., vol. 1, pp. 429–509. Cornell University Press, Ithaca, N.Y.

Cummings, C. 1933. The giant water bugs (Belostomatidae, Hemiptera). Univ. Kans. Sci. Bull. 21:197–219.

Dallai, R., and B. A. Afzelius. 1980. Characteristics of the sperm structure in Heteroptera (Hemiptera, Insecta). J. Morphol. 164:301–309.

Dallas, W. S. 1851–1852. List of the Specimens of Hemipterous Insects in the Collection of the British Museum, pts. 1 and 2. London. 368 pp., 11 pls.; 223 pp., 4 pls.

Darlington, P. J., Jr. 1971. Carabidae of tropical islands, especially the West Indies. *In:* W. L. Stern (ed.), Adaptive Aspects of Insular Evolution, pp. 7–15. Washington State University Press, Pullman.

Darlington, P. J., Jr. 1973. Carabidae of mountains and islands: data on the evolution of isolated faunas and on atrophy of wings. Ecol. Monogr. 13:37–61.

Darwin, C. 1859. On the Origin of Species. John Murray, London. 490 pp.

Dashman, T. 1953. Terminology of the pretarsus. Ann. Entomol. Soc. Am. 46:56–62.

Davidova-Vilimova, J., and P. Štys. 1980. Taxonomy and phylogeny of West Palearctic Plataspidae (Heteroptera). Stud. Czech. Akad. Ved. 4:1–155.

Davis, N. T. 1955. Morphology of the female organs of reproduction in the Miridae (Hemiptera). Ann. Entomol. Soc. Am. 48:132–150.

Davis, N. T. 1957. Contributions to the morphology and phylogeny of the Reduvioidea (Hemiptera: Heteroptera). Part I. The morphology of the abdomen and genitalia of Phymatidae. Ann. Entomol. Soc. Am. 50:432–443.

Davis, N. T. 1961. Morphology and phylogeny of the Reduvioidea (Hemiptera: Heteroptera). Part II. Wing venation. Ann. Entomol. Soc. Am. 54:340–354.

Davis, N. T. 1966. Contributions to the morphology and phylogeny of the Reduvioidea (Hemiptera: Heteroptera). The male and female genitalia. Ann. Entomol. Soc. Am. 59:911–924.

Davis, N. T. 1969. Contribution to the morphology and phylogeny of the Reduvioidea. Part IV. The harpactoroid complex. Ann. Entomol. Soc. Am. 62:74–94.

Davis, N. T. 1975. Hormonal control of flight muscle histolysis in *Dysdercus fulvoniger*. Ann. Entomol. Soc. Am. 68:710–714.

Davis, N. T., and R. L. Usinger. 1970. The biology and relationships of the Joppeicidae (Heteroptera). Ann. Entomol. Soc. Am. 63:577–586.

De Carlo, J. A. 1938. Los belostomátidos americanos. An. Mus. Arg. Cienc. Nat. 39:189–260.

De Carlo, J. A. 1958. Identificación de las especies del género *Horvathinia* Montandon. Descripción de tres especies nuevas (Hemiptera-Belostomatidae). Rev. Soc. Entomol. Arg. 20:45–52.

De Carlo, J. A. 1968. Tres especies nuevas del género *Coleopterocoris* y una especie nueva del género *Heleocoris* (Hemiptera, Naucoridae). Physis 28(76):193–197.

DeCoursey, R. M. 1971. Keys to the families and subfamilies of the nymphs of North American Hemiptera-Heteroptera. Proc. Entomol. Soc. Wash. 73:413–428.

De Geer, C. 1752–1771. Mémoires pour servir a l'histoire naturelle des insectes, Heteroptera, vol. 3. Stockholm.

Delamare Deboutteville, C., and R. Paulian. 1966. Le Professeur René Jeannel. Ann. Soc. Entomol. Fr., n.s., 2:3–37.

Derksen, W., and U. Göllner-Scheiding. 1963–1975. Index Literaturae Entomologicae, ser. 2: Die Welt-Literatur über die gesamte Entomologie von 1864–1900. Deutsche Akademie der Landwirtschaftswissenschaften, Berlin. 5 vols.

Derr, J. A., B. Alden, and H. Dingle. 1981. Insect life histories in relation to migration, body size and host plant array: a comparative study of *Dysdercus*. J. Anim. Ecol. 50:181–193.

Dethier, M. 1974. Les organes odoriférants métathoraciques des Cydnidae. Bull. Soc. Vaud. Sci. Nat. (Lausanne) 72:127–140.

de Vrijer, P. W. F. 1988. In memoriam R. H. Cobben (1925–1987), een biografie en een bibliografie. Entomol. Ber. 48:165–168.

Dingle, H. 1978. Migration and diapause in tropical, temperate, and island milkweed bugs. *In:* H. Dingle (ed.), Evolution of Insect Migration and Diapause, pp. 254–276. Springer-Verlag, New York.

Dingle, H. 1979. Adaptive variation in the evolution of insect migration. Movement of highly mobile insects. *In:* R. L. Rabb and G. G. Kennedy (eds.), Concepts and Methodology in Research, pp. 64–87. North Carolina State University, Raleigh.

Dingle, H. 1981. Geographical variation and behavioral flexibility in milkweed bug life histories *In:* R. F. Denno and H. Dingle (eds.), Insect Life History Patterns: Habitat and Geographic Variation, pp. 57–73. Springer-Verlag, New York. x + 225 pp.

Dingle, H. 1985. Migration. *In:* G. A. Kerkut and L. I. Gilbert (eds.), Comprehensive Insect Physiology, Biochemistry and Pharmacology, vol. 9, pp. 375–415. Pergamon Press, New York.

Dingle, H., and G. Arora. 1973. Experimental studies of migration in bugs of the genus *Dysdercus*. Oecologia 12:119–140.

Dingle, H., and J. D. Baldwin. 1983. Geographic variation in life histories: a comparison of tropical and temperate milkweed bugs (*Oncopeltus*). Series Entomol. 23:143–165.

Dispons, P., and W. Stichel. 1959. Fam. Reduviidae Lt. *In:* W. Stichel (ed.), Illustrierte Bestimmungstabellen der Wanzen. II. Europa (Hemiptera-Heteroptera Europae), vol. 3, pp. 81–185. Hermsdorf, Berlin.

Distant, W. L. 1880–1893. Biologia Centrali Americana. Insecta. Rhynchota. Hemiptera-Heteroptera, vol. 1. London. xx + 462 pp., 39 pls.

Distant, W. L. 1902–1918. The Fauna of British India including Ceylon and Burma. Rhynchota (Heteroptera, Homoptera), vols. 1–7. Taylor and Francis, London.

Distant, W. L. 1904. The Fauna of British India including Ceylon and Burma. Rhynchota, vol. 2: Heteroptera. Taylor and Francis, London. 503 pp.

Distant, W. L. 1910. The Fauna of British India including Ceylon and Burma. Rhynchota, vol. 5: Heteroptera; Appendix. Taylor and Francis, London. 362 pp., 1 map.

Dolling, W. R. 1978. A revision of the Oriental pod bugs of the tribe Clavigrallini (Hemiptera: Coreidae). Bull. Br. Mus. (Nat. Hist.), Entomol. 36(6):281–321.

Dolling, W. R. 1979. A revision of the African pod bugs of the tribe Clavigrallini (Hemiptera: Coreidae) with a checklist of the world species. Bull. Br. Mus. (Nat. Hist.), Entomol. 39(1):1–84.

Dolling, W. R. 1981. A rationalized classification of the burrower bugs (Cydnidae). Syst. Entomol. 6:61–76.

Dolling, W. R. 1984. Pentatomid bugs (Hemiptera) that transmit a flagellate disease of cultivated palms in South America. Bull. Entomol. Res. 74:473–476.

Dolling, W. R. 1986. The tribe Pseudophloeini (Hemiptera: Coreidae) in the Old World tropics with a discussion on the distribution of the Pseudophloeinae. Bull. Br. Mus. (Nat. Hist.), Entomol. 53(3):151–212.

Dolling, W. R. 1987a. A mimetic coreid bug and its relatives (Hemiptera: Coreidae). J. Nat. Hist. 21:1259–1271.

Dolling, W. R. 1987b. Obituary. Norman Cecil Egerton Miller. Entomol. Mon. Mag. 123:251–259.

Dolling, W. R. 1991a. Bibliographies of the works of W. L. Distant and G. W. Kirkaldy. Tymbal, suppl. no. 1. 60 pp.

Dolling, W. R. 1991b. The Hemiptera. Natural History Museum Publication, Oxford University Press. 274 pp.

Dolling, W. R., and J. M. Palmer. 1991. *Pameridea* (Hemiptera: Miridae): predaceous bugs specific to the highly viscid plant genus *Roridula*. Syst. Entomol. 16:319–328.

Dougherty, V. M. 1980. A systematic revision of the New World Ectrichodiinae (Hemiptera: Reduviidae). Ph.D. dissertation, University of Connecticut, Storrs.

Douglas, J. W., and J. Scott. 1865. The British Hemiptera, vol. 1: Hemiptera-Heteroptera. R. Hardwicke, London. xii + 627 pp.

Drake, C. J. 1949a. Concerning North American Saldidae. Ark. Zool. 42B(3):1–4.

Drake, C. J. 1949b. Some American Saldidae (Hemiptera). Psyche 56(4):187–193.

Drake, C. J. 1955. New South American Saldidae (Hemiptera). J. Kans. Entomol. Soc. 28(4):152–158.

Drake, C. J. 1961. A new subfamily, genus and two new species of Dipsocoridae (Hemiptera). Publ. Cult. Co. Diam. Angola 52:75–80.

Drake, C. J. 1962. Synonymic data and two new genera of shore bugs (Hemiptera: Saldidae). Proc. Biol. Soc. Wash. 75:115–124.

Drake, C. J., and H. C. Chapman. 1953. Preliminary report on the Pleidae (Hemiptera) of the Americas. Proc. Biol. Soc. Wash. 66:53–59.

Drake, C. J., and H. C. Chapman. 1958. New Neotropical Hebridae, including a catalogue of the American species (Hemiptera). J. Wash. Acad. Sci. 48:317–326.

Drake, C. J., and H. C. Chapman. 1963. A new genus and species of water striders from California. Proc. Biol. Soc. Wash. 76:227–234.

Drake, C. J., and N. T. Davis. 1958. The morphology and systematics of the Piesmatidae (Hemiptera), with keys to world genera and American species. Ann. Entomol. Soc. Am. 51:567–581.

Drake, C. J., and N. T. Davis. 1960. The morphology, phylogeny, and higher classification of the family Tingidae, including the description of a new genus and species of the subfamily Vianaidinae (Hemiptera: Heteroptera). Entomol. Am., n.s., 39:1–100.

Drake, C. J., and R. C. Froeschner. 1962. A new myrmecophilous lacebug from Panama (Hemiptera: Tingidae). Great Basin Nat. 22:8–11.

Drake, C. J., and H. M. Harris. 1932a. A synopsis of the genus *Metrobates* Uhler. Ann. Carnegie Mus. 21:83–89.

Drake, C. J., and H. M. Harris. 1932b. A survey of the species of *Trepobates* Uhler (Hemiptera, Gerridae). Bull. Brooklyn Entomol. Soc. 27:113–123.

Drake, C. J., and H. M. Harris. 1934. The Gerrinae of the Western Hemisphere (Hemiptera). Ann. Carnegie Mus. 23:179–240.

Drake, C. J., and H. M. Harris. 1964. The genus *Nidicola* (Hemiptera: Anthocoridae). Proc. Entomol. Soc. Wash. 77:53–64.

Drake, C. J., and L. Hoberlandt. 1965. A revision of the genus *Potamometra* (Hemiptera: Gerridae). Acta Entomol. Mus. Natl. Pragae 36:303–310.

Drake, C. J., and F. C. Hottes. 1951. Stridulatory organs in Saldidae (Hemiptera). Great Basin Nat. 11:43–47.

Drake, C. J., and R. F. Hussey. 1955. Concerning the genus *Microvelia* Westwood, with descriptions of two new species and a check list of the American forms (Hemiptera: Veliidae). Florida Entomol. 38:95–115.

Drake, C. J., and J. D. Lattin. 1963. American species of the lacebug genus *Acalypta* (Hemiptera: Tingidae). Proc. U.S. Natl. Mus. 115(3486):331–345.

Drake, C. J., and D. R. Lauck. 1959. Descriptions, synonymy, and check-list of American Hydrometridae (Hemiptera: Heteroptera). Great Basin Nat. 19:43–52.

Drake, C. J., and F. A. Ruhoff. 1965. Lacebugs of the world: a catalog. Smithsonian Institution, U.S. Natl. Mus. Bull. 243. 634 pp.

Drake, C. J., and J. A. Slater. 1957. The phylogeny and systematics of the family Thaumastocoridae (Hemiptera: Heteroptera). Ann. Entomol. Soc. Am. 50:353–370.

Dufour, L. 1833. Recherches anatomiques et physiologiques sur les Hémiptères accompagnées de considérations relatives à l'histoire naturelle et à la classification de ces insectes. Mém. Savants Etrang. Acad. Sci., Paris. 4:123–432, 19 pls.

Dupuis, C. 1947. Données sur la morphologie des glandes dorsoabdominales des Hémiptères-Hétéroptères. Feuill Nat., n.s., 2:13–21.

Dupuis, C. 1970. Heteroptera. *In:* S. L. Tuxex (ed.), Taxonomist's Glossary of Genitalia of Insects, pp. 158–169. Ejnar Munksgaard, Copenhagen. 284 pp.

Durai, P. S. S. 1987. A revision of the Dinidoridae of the World (Heteroptera: Pentatomoidea). Orient. Insects 21:163–360.

Eberhard, W. G. 1975. The ecology and behavior of a subsocial pentatomid bug and two scelionid wasps: strategy and counterstrategy in a host and its parasites. Smithsonian Contrib. Zool. 205:1–39.

Eberhard, W. G., N. I. Platnick, and R. T. Schuh. 1993. Natural history and systematics of arthropod symbionts (Araneae; Hemiptera; Diptera) inhabiting webs of the spider *Tengella radiata* (Araneae, Tengellidae). Am. Mus. Novit. 3065:17 pp.

Edwards, F. J. 1969. Environmental control of flight muscle histolysis in the bug *Dysdercus intermedius*. J. Insect Physiol. 15:2013–2020.

Edwards, J. S. 1962. Observations on the development and predatory habit of two reduviid Heteroptera, *Rhinocoris carmelita* Stål and *Platymeris rhadamanthus* Gerst. Proc. R. Entomol. Soc. Lond. (A) 37:89–98.

Ehrlich, P. R., and P. H. Raven. 1965. Butterflies and plants: a study in coevolution. Evolution 18:586–608.

Ekblom, T. 1926. Morphological and biological studies of the Swedish families of Hemiptera-Heteroptera. Part I. The families Saldidae, Nabidae, Lygaeidae, Hydrometridae, Veliidae, and Gerridae. Zool. Bidr. 10:31–180.

Ekblom, T. 1930. Morphological and biological studies of the Swedish families of Hemiptera-Heteroptera. Part II. The families Mesoveliidae, Corizidae, and Corixidae. Zool. Bidr. 12:113–150.

Elsey, K. D., and R. E. Stinner. 1971. Biology of *Jalysus spinosus*, an insect predator found on tobacco. Ann. Entomol. Soc. Am. 64:779–783.

Emel'yanov, A. F. 1987. The phylogeny of the Cicadina (Homoptera, Cicadina) based on comparative morphological data. Trudy Vsesoyuznogo Entomol. Obsh. 69:19–109. [In Russian].

Emsley, M. G. 1969. The Schizopteridae (Hemiptera: Heteroptera) with the description of new species from Trinidad. Mem. Am. Entomol. Soc. 25:154 pp.

Esaki, T. 1924. On the curious halophilous water strider *Halovelia maritima* Bergroth (Hemiptera, Gerridae). Bull. Brooklyn Entomol. Soc. 19:29–34.

Esaki, T. 1926. The water striders of the subfamily Halobatinae in the Hungarian National Museum. Ann. Mus. Natl. Hung. 23:117–164.

Esaki, T. 1927. An interesting new genus and species of Hydrometridae (Hem.) from South America. Entomol. 60:181–184.

Esaki, T. 1929. A remarkable speo-halophilous water strider (Heteroptera, Mesoveliidae). Ann. Mag. Nat. Hist., ser. 10, 4:341–346.

Esaki, T. 1930. New and little known Gerridae from the Malay Peninsula. J. Fed. Malay. St. Mus. 16:13–24.

Esaki, T., and W. E. China. 1927. A new family of aquatic Heteroptera. Trans. Entomol. Soc. Lond. 1927:279–295.

Esaki, T., and W. E. China. 1928. A monograph of the Helotrephidae, subfamily Helotrephinae (Hem. Heteroptera). EOS 4:129–172.

Esaki, T., and S. Miyamoto. 1955. Veliidae of Japan and adjacent territory (Hemiptera: Heteroptera). I. *Microvelia* Westwood and *Pseudovelia* Hoberlandt of Japan. Sieboldia 1:169–215.

Esaki, T., and S. Miyamoto. 1959a. A new genus and species of Helotrephidae (Hemiptera-Heteroptera). Sieboldia 2:83–89, pls. 7–9.

Esaki, T., and S. Miyamoto. 1959b. Veliidae of Japan and its adjacent territory (Hemiptera: Heteroptera). II. *Xiphovelia* Lundblad. Sieboldia 2:91–115.

Esaki, T., and S. Miyamoto. 1959c. A new or little known *Hypselosoma* from Amami-Oshima and Japan, with the proposal of a new tribe for the genus (Hemiptera). Sieboldia 2:109–120, pls. 14–17.

Eyles, A. C. 1971. List of Isometopidae (Heteroptera: Cimicoidea). New Zealand J. Sci. 14:940–944.

Eyles, A. C. 1972. Supplement to list of Isometopidae (Heteroptera: Cimicoidea). New Zealand J. Sci. 15:463–464.

Eyles, A. C. 1973. Monograph of *Dieuches* Dohrn (Heteroptera: Lygaeidae). Otago Daily Times, Dunedin, New Zealand. 465 pp.

Fabricius, J. C. 1803. Systema Rhyngotorum. Apud Carolum Reichard, Brunsvigiae. vi + 314 pp.

Fairmaire, M. L. 1889. Notice nécrologique sur Victor-Antoine Signoret. Ann. Soc. Entomol. Fr., ser. 6, 9:505–512.

Falkenstein, R. B. 1931. A general biological study of the lychee stink bug, *Tessaratoma papillosa* Drur. (Heteroptera, Pentatomidae). Lignan Sci. J. 10:29–80, pls. 11, 12.

Fallén, C. F. 1807. Monographic Cimicum Sueciae. Hafniae. 123 pp.

Fauvel, G. 1974. Sur l'alimentation pollinique d'un anthocoride prédateur *Orius (Heterorius) vicinus* Rib. (Hémiptère). Ann. Zool.-Ecol. Anim. 6:245–258.

Fedotov, D. M. (ed.). 1947–1960. The noxious pentatomid, *Eurygaster integriceps* Puton, vols. 1–4. Moscow. Akad. Nauk SSSR. (Vol. 1, 1947, 272 pp.; vol. 2, 1947, 271 pp.; vol. 3, 1955, 278 pp.; vol. 4, 1960, 239 pp.)

Ferris, G. F., and R. L. Usinger. 1939. The family Polyctenidae (Hemiptera; Heteroptera). Microentomology. 4:1–50.

Fieber, F. X. 1844. Entomologische Monographien. Leipzig. 138 pp., 10 pls.

Fieber, F. X. 1851. Genera Hydrocoridum secundum ordinem naturalem in familias disposita. Abh. Konigl. Bohemischen Gesellsch. Wissenschaften Prague, ser. 5, 7:181–212. (Separate Act. Reg. Bohemicae Soc. Sci. Pragae 1:1–30).

Fieber, F. X. 1861. Die europäischen Hemiptera. Halbflügler. (Rhynchota Heteroptera). Carl Gerold's Sohn, Vienna. 444 pp., 2 pls.

Filshie, B. K., and D. F. Waterhouse. 1969. The structure and development of a surface pattern on the cuticle of the green vegetable bug *Nezara viridula*. Tissue and Cell 1:367–385.

Finke, C. 1968. Lautäusserungen und Verhalten von *Sigara striata* und *Callicorixa praeusta* (Corixidae Leach., Hydrocorisae Latr.). Z. Vergl. Physiol. 58:398–422.

Forbes, A. R. 1976. The stylets of the large milkweed bug, *Oncopeltus fasciatus* (Hemiptera: Lygaeidae), and their innervation. J. Entomol. Soc. Br. Columbia 73:29–32.

Foster, W. A. 1989. Zonation, behaviour and morphology of the intertidal coral-treader *Hermatobates* (Hemiptera: Hermatobatidae) in the south-west Pacific. Zool. J. Linn. Soc. 96:87–105.

Fracker, S. B., and S. C. Bruner. 1924. Notes on some Neotropical Reduviidae. Ann. Entomol. Soc. Am. 17:163–174.

Freeman, P. 1940. A contribution to the study of the genus *Nezara* Amyot & Serville (Hemiptera: Pentatomidae). Trans. R. Entomol. Soc. Lond. 90:351–374.

Freeman, P. 1947. A revision of the genus *Dysdercus* Boisduval (Hemiptera, Pyrrhocoridae), excluding the American species. Trans. R. Entomol. Soc. Lond. 98:373–424.

Friend, W. G., and J. J. B. Smith. 1971. Feeding in *Rhodnius prolixus:* mouthpart activity and salivation, and their correlation with changes of electrical resistance. J. Insect Physiol. 17:233–242.

Froeschner, R. C. 1960. Cydnidae of the Western Hemisphere. Proc. U.S. Natl. Mus. 111:337–680.

Froeschner, R. C. 1969. Zoogeographic and systematic notes on the lace bug tribe Litadeini, with the description of the new genus *Strongulotingis* (Hemiptera: Tingidae). Great Basin Nat. 29:129–132.

Froeschner, R. C. 1981. Heteroptera or true bugs of Ecuador: a partial catalog. Smithsonian Contrib. Zool. 322:1–147.

Froeschner, R. C., and Q. L. Chapman. 1963. A South American cydnid, *Scaptocoris castaneus* Perty, established in the United States (Hemiptera: Cydnidae). Entomol. News 74:95–98.

Froeschner, R. C., and N. A. Kormilev. 1989. Phymatidae or ambush bugs of the world: a synonymic list with keys to species, except *Lophoscutus* and *Phymata* (Hemiptera). Entomography 6:1–76.

Froeschner, R. C., and W. E. Steiner, Jr. 1983. Second record of South American burrowing bug, *Scaptocoris castaenus* Perty (Hemiptera: Cydnidae), in the United States. Entomol. News 94:276.

Fujita, K. 1977. Wing form composition in the field population of two species of lygaeid bugs, *Dimorphopterus pallipes* and *D. japonicus,* and its relation to environmental conditions. Jpn. J. Ecol. 27:263–267.

Furth, D. G., and K. Suzuki. 1990. Comparative morphology of the tibial flexor and extensor tendons in insects. Syst. Entomol. 15:433–441.

Fuseini, B. A., and R. Kumar. 1975. Ecology of cotton stainers (Hemiptera: Pyrrhocoridae) in southern Ghana. Biol. J. Linnaean Soc. 7:113–146.

Gaedike, R., and O. Smetana. 1978–1984. Ergänzungen und Berichtigungen zu Walter Horn und Sigmund Schenkling: *Index Literaturae Entomologicae,* ser. 1: Die Welt-Literatur über die gesamte Entomologie bis inklusiv 1863. Beitr. Entomol. 28:329–436; 34:167–291.

Gaffour-Bensebbane, C. 1990. Importance systématique et phylogenetique des spermatozoïdes chez les Scutelleridae (Insecta, Heteroptera). Bull. Soc. Zool. Fr. 115:271–276.

Gagné, W. C., and F. G. Howarth. 1975. The cavernicolous fauna of Hawaiian lava tubes. 6. Mesoveliidae or water treaders (Heteroptera). Pac. Insects 16:399–413.

Galbreath, J. E. 1975. Thoracic polymorphism in *Mesovelia mulsanti* (Hemiptera: Mesoveliidae). Univ. Kans. Sci. Bull. 50:457–482.

Gallardo, A. 1902. El Doctor Carlos Berg. Apuntes biográficos. Bibliografía del Doctor Carlos Berg. An. Soc. Cienc. Argentina 53:98–128.

Galliard, H. 1935. Recherches sur les Réduviidés hématophages *Rhodnius* et *Triatoma*. Ann. Parsit. Humaine Comp. 8:401–423.

Gapud, V.P. 1981. A generic revision of the subfamily Asopinae, with consideration of its phylogenetic position in the family Pentatomidae and superfamily Pentatomoidea (Hemiptera-Heteroptera). Ph.D. dissertation, University of Kansas. 2 vols.

Gapud, V. 1991. A generic revision of the subfamily Asopinae with consideration of its phylogenetic position in the family Pentatomidae and superfamily Pentatomoidea (Hemiptera-Heteroptera). Philippine Entomol. 8(3):865-961.

Ghauri, M. S. K. 1964. A remarkable phenomenon amongst the males of Piratinae (Reduviidae, Heteroptera). Ann. Mag. Nat. Hist., ser. 13, 7:733–737.

Ghauri, M. S. K. 1968. Notes on Colobathristidae (Heteroptera), including descriptions of new species and a supsected virus vector of Musa in Sabah. Proc. R. Entomol. Soc. Lond. (B) 37:80–88.

Ghauri, M. S. K. 1975. Anomalous Miridae (Heteroptera) from Australasia. J. Nat. Hist. 9:611–618.

Ghazi-Bayat, A., and I. Hasenfuss. 1979. Zur Oberflächenstruktur der tarsalen Haftlappen von *Coreus marginatus* (L.) (Coreidae, Heteroptera). Zool. Anz. 203:345–347.

Ghazi-Bayat, A., and I. Hasenfuss. 1980a. Die Oberflächenstrukturen der Prätarsus on *Elasmucha ferrugata* (Fabricius) (Acanthosomatidae, Heteroptera). Zool. Anz. 205:76–80.

Ghazi-Bayat, A., and I. Hasenfuss. 1980b. Zur Herkunft der Adhäsionsflüssigkeit der tarsalen Haftlappen bei den Pentatomidae (Heteroptera). Zool. Anz. 204:13–18.

Giacchi, J. C. 1984. Revisión de los Stenopodainos Americanos. VI. Las especies americanas del género *Oncocephalus* Klug, 1830 (Heteroptera-Reduviidae). Physis, Buenos Aires, C 42:39–62.

Giglioli, H. 1864. On some parasitical insects from China. Q. J. Microscop. Sci., n.s., 4:18–26.

Gilbert, P. 1977. A Compendium of the Biographical Literature on Deceased Entomologists. Br. Mus. (Nat. Hist.), London. 455 pp.

Gittelman, S. H. 1974. The habitat preference and immature stages of *Neoplea striola* (Hemiptera: Pleidae). J. Kans. Entomol. Soc. 47:491–503.

Gittelman, S. H. 1975. Physical gill efficiency and winter dormancy in the pygmy backswimmer, *Neoplea striola* (Hemiptera: Pleidae). Ann. Entomol. Soc. Am. 68:1011–1017.

Goel, S. C. 1972. Notes on the structure of the unguitractor plate in Heteroptera (Hemiptera). J. Entomol. (A) 46:167–173.

Goel, S. C., and C. W. Schaefer. 1970. The structure of the pulvillus and its taxonomic value in the land Heteroptera (Hemiptera). Ann. Entomol. Soc. Am. 63:307–313.

Gogala, M. 1984. Vibration producing structures and songs of terrestrial Heteroptera as systematic character. Biol. Vestn. 1:19–36.

Gogala, M., and A. Cokl. 1983. The acoustic behavior of the bug *Phmyata crassipes* (F.) (Heteroptera). Rev. Can. Biol. Exper. 42:249–256.

Gogala, M., M. Virant, and A. Blejec. 1984. Mocking bug, *Phymata crassipes* (Heteroptera). Acoustic Letters 8:44–51.

Göllner-Scheiding, U. 1979. Die Gattung *Jadera* Stål, 1862. Dtsch. Entomol. Z., n.f., 26:47–75.

Göllner-Scheiding, U. 1980. Revision der afrikanischen Arten sowie Bemerkungen zu weiteren Arten der Gattungen *Leptocoris* Hahn, 1833 und *Boisea* Kirkaldy, 1910. Dtsch. Entomol. Z., n.f., 27:103–148.

Göllner-Scheiding, U. 1983. General-Katalog der Familie Rhopalidae (Heteroptera). Mitt. Zool. Mus. Berlin 59:37–189.

Göllner-Scheiding, U. 1991. Die Entwicklung des Systems der Heteroptera (*Wanzen*). Entomol. Nachr. Berichte 34:1–8.

Golub, V. B. 1988. Tingidae. *In:* P. A. Lehr (ed.), Keys to the Insects of the Far East of the USSR, vol. 2. Nauka, Leningrad. [In Russian].

Goodchild, A. J. P. 1963. Studies on the functional anatomy of the intestines of Heteroptera. Proc. Zool. Soc. Lond. 141:851–910.

Graham, H. M., A. A. Negm, and L. R. Ertle. 1984. Worldwide literature of the *Lygus* complex (Hemiptera: Miridae), 1900–1980. U.S. Dep. Agric., Bibliography and Literature of Agriculture 30:1–205.

Grassé, P.-P. 1974. Raymond A. Poisson (1895–1973). Sa vie et son oeuvre. Bull. Biol. Fr. Belg. 108:191–204.

Grassé, P.-P. 1975. Mécanorécepteurs et mécanoréception. Les organes chimiorécepteurs: Olfaction, gustation. *In:* P.-P. Grassé (ed.), Traité de zoologie, Insects, vol. 8, no. 3, pp. 541–662. Masson, Paris.

Greathead, D. J. 1966. A taxonomic study of the species of *Antestiopsis* (Hemiptera, Pentatomidae) associated with *Coffea arabica* in Africa. Bull. Entomol. Res. 56:515–554.

Grimaldi, D., C. Michalski, and K. Schmidt. 1993. Amber fossil Enicocephalidae (Heteroptera) from the Lower Cretaceous of Lebanon and Oligo-Miocene of the Dominican Republic, with biogeographic analysis of *Enicocephalus*. Am. Mus. Novit. 3971:30 pp.

Gross, G. F. 1954–1957. A revision of the flower bugs (Heteroptera Anthocoridae) of the Australian and adjacent Pacific regions. Rec. S. Aust. Mus. (Pt. 1, 1954, 11:129–164; pt. 2, 1955, 11:409–433; pt. 3, 1957, 13:131–142.)

Gross, G. F. 1960. A revision of the genus *Leptocoris* Hahn (Heteroptera: Coreidae: Rhopalinae) from the Indo-Pacific and Australian regions. Rec. S. Aust. Mus. 13:403–451.

Gross, G. F. 1963. Heteroptera: Coreidae (Alydini by J. C. Schaffner), Neididae and Nabidae. *In:* Insects of Micronesia, vol. 7, pp. 7:357–390. Bernice P. Bishop Museum, Honolulu.

Gross, G. F. 1965. A revision of the Australian and New Guinea Drymini (Het.-Lyg.). Rec. S. Aust. Mus. 15:39–77.

Gross, G. F. 1972. A revision of the species of Australian and New Guinea shield bugs formerly placed in the genera *Poecilometis* Dallas and *Eumecopus* Dallas (Heteroptera: Pentatomidae), with description of new species and selection of lectotypes. Aust. J. Zool., suppl. ser., 15:1–192.

Gross, G. F. 1975–1976. Plant feeding and other bugs (Hemiptera) of South Australia–Heteroptera, pts. 1 and 2. Handbook Flora Fauna South Australia. A. B. James, Adelaide, South Australia.

Gross, G. F., and G. G. E. Scudder. 1963. The Australian Rhyparochromini (Hemiptera: Lygaeidae). Rec. S. Aust. Mus. 14:427–469.

Grozeva, S. M., and I. M. Kerzhner. 1992. On the phylogeny of aradid subfamilies (Heteroptera, Aradidae). Acta Zool. Hung. 38:199–205.

Gulde, J. 1902. Die Dorsalrüsen der Larven der Hemiptera-Heteroptera. Ber. Senkenb. Nat. Ges. 1902:85–136.

Gulde, J. 1936. Die Wanzen Mitteleuropas. Hemiptera Heteroptera Mitteleuropas. Teil 5, 1:1–4.

Guthrie, D. M. 1961. The anatomy of the nervous system in the genus *Gerris* (Hemiptera: Heteroptera). Phil. Trans. R. Soc. Lond. 708(244):65–102.

Hagan, H. R. 1931. The embryogeny of the polyctenid, *Hesperoctenes fumarius* Westwood, with reference to viviparity in insects. J. Morph. Physiol. 51:1–92, 12 pls.

Hagemann, J. 1910. Beiträge zur Kenntnis von *Corixa*. Zool. Jahrb. Abt. Anat. Ontog. Tiere 36c:189–191.

Hahn, C. W. 1831–1835. Die wanzenartigen Insecten, vols. 1–3. Nuremberg, Germany.

Hale, H. M. 1926. Studies in Australian aquatic Hemiptera. Rec. Aust. Mus. 3:195–217.

Hamid, A. 1971a. A revision of Cryptorhamphinae (Heteroptera: Lygaeidae) including the description of the two new species from Australia. J. Aust. Entomol. Soc. 10:163–174.

Hamid, A. 1971b. The life cycles of three species of *Cymus* (Hemiptera: Lygaeidae) in Connecticut. Univ. Conn. Occ. Pap., Biol. Sci. Ser. 2:21–28.

Hamid, A. 1975. A systematic revision of the Cyminae (Heteroptera: Lygaeidae) of the world with a discussion of the morphology, biology, phylogeny and zoogeography. Occ. Papers Entomol. Soc. Nigeria 14:1–179.

Hamilton, M. A. 1931. Morphology of the water scorpion, *Nepa cinerea*. Proc. Zool. Soc. Lond. 104:1067–1136.

Handlirsch, A. 1897. Monographie der Phymatiden. Ann. Kaiserlich-königlichen Naturhist. Ges. Hofmuseum 12:127–230.

Handlirsch, A. 1900a. Zur Kenntniss der Stridulationsorgane bei den Rhynchoten. Ein morphologish-biologischer Beitrag. Ann. Hofmus. Wien 15:127–141, pl. xii.

Handlirsch, A. 1900b. Neue Beiträge zur Kenntniss der Stridulationsorgane bei den Rhynchoten. Verh. Ges. Wien 1:555–560.

Haridass, E. T. 1985. Ultrastructure of the eggs of Reduviidae. I. Eggs of Piratinae (Insecta–Heteroptera). Proc. Indian Acad. Sci. (Anim. Sci.) 94:533–545.

Haridass, E. T. 1986a. Ultrastructure of the eggs of Reduviidae. II. Eggs of Harpactorinae (Insecta–Heteroptera). Proc. Indian Acad. Sci. (Anim. Sci.) 95:237–246.

Haridass, E. T. 1986b. Ultrastructure of the eggs of Reduviidae. III. Eggs of Triatominae and Ectrichodiinae (Insecta–Heteroptera). Proc. Indian Acad. Sci. (Anim. Sci.) 95:447–456.

Haridass, E. T. 1988. Ultrastructure of the eggs of Reduviidae. IV. Eggs of Rhaphidosomatinae (Insecta–Heteroptera). Proc. Indian Acad. Sci. (Anim. Sci.) 97:49–54.

Haridass, E. T., and T. N. Ananthakrishnan. 1981. Functional morphology of the salivary system in some Reduviidae (Insecta–Heteroptera). Proc. Indian Acad. Sci. (Anim. Sci.) 90:145–160.

Harrington, B. J. 1980. A generic level revision and cladistic analysis of the Myodochini of the world. (Hemiptera, Lygaeidae, Rhyparochrominae). Bull. Am. Mus. Nat. Hist. 167:45–116.

Harris, H. M. 1928. A monographic study of the hemipterous family Nabidae as it occurs in North America. Entomol. Am. 9:1–90, pls. I–IV.

Harris, H. M. 1930a. Notes on some South American Nabidae, with descriptions of new species (Hemiptera). Ann. Carnegie Mus. 19:241–248.

Harris, H. M. 1930b. Notes on Philippine Nabidae, with a catalogue of the species of *Gorpis* (Hemiptera). Philippine J. Sci. 43:415–423.

Harris, H. M. 1931a. Nabidae from the state of Paraná. Ann. Mus. Zool. Polonici 9:179–185.

Harris, H. M. 1931b. The genus *Aphelonotus* (Hemiptera: Nabidae). Bull. Brooklyn Entomol. Soc. 26:13–20.

Harris, H. M. 1939a. A contribution to our knowledge of *Gorpis* Stål (Hemiptera). Philippine J. Sci. 69:147–155.

Harris, H. M. 1939b. Miscelánea sobre Nabidae sudamericanos (Hemiptera). Notas Mus. La Plata 4:367–377.

Hart, E. 1986. The genus *Zelus* Fabricius in the United States, Canada, and northern Mexico. Ann. Entomol. Soc. Am. 79:535–548.

Hart, E. 1987. The genus *Zelus* Fabricius in the West Indies (Hemiptera: Reduviidae). Ann. Entomol. Soc. Am. 80:293–305.

Harvey, G. W. 1907. A ferocious water bug. Proc. Entomol. Soc. Wash. 8:72–75.

Hasan, S. A., and I. J. Kitching. 1993. A cladistic analysis of the tribes of the Pentatomidae (Heteroptera). Jpn. J. Entomol. 61:651–669.

Hasegawa, H. 1967. [Bibliography of T. Esaki; in Japanese]. Kontyû 35(suppl.):17–19.

Heiss, E. 1990. In memoriam Gustav Seidenstücker, einem bedeutendem deutschen Entomologen (1912–1989). Nachrichtenbl. Bayer. Entomol. 39:65–70.

Heiss, E., and J. Péricart. 1983. Revision of Palaearctic Piesmatidae (Heteroptera). Mitt. Münchner Entomol. Ges. 73:61–171.

Heming–Van Battum, K. E., and B. S. Heming. 1986. Structure, function and evolution of the reproductive system in females of *Hebrus pusillus* and *H. ruficeps* (Hemiptera, Gerromorpha, Hebridae). J. Morph. 190:121–167.

Heming–Van Battum, K. E., and B. S. Heming. 1989. Structure, function, and evolutionary significance of the reproductive system in males of *Hebrus ruficeps* and *H. pusillus* (Heteroptera, Gerromorpha, Hebridae). J. Morph. 202:281–323.

Henry, T. J. 1977. *Teratodia* Bergroth, new synonym of *Diphleps* Bergroth with descriptions of two new species (Heteroptera: Miridae: Isometopinae). Florida Entomol. 60:201–210.

Henry, T. J. 1979. Review of the New World species of *Myiomma* with descriptions of eight new species (Hemiptera: Miridae: Isometopinae). Proc. Entomol. Soc. Wash. 81:551–569.

Henry, T. J. 1984. New species of Isometopinae (Hemiptera: Miridae) from Mexico, with new records for previously described species. Proc. Entomol. Soc. Wash. 86:337–345.

Henry, T. J. 1988. Family Largidae Amyot and Serville, 1843. The largid bugs. *In:* T. J. Henry and R. C. Froeschner (eds.), Catalog of the Heteroptera, or True Bugs, of Canada and the Continental United States, pp. 159–165. E. J. Brill, Leiden.

Henry, T. J. 1991. Revision of *Keltonia* and the cotton fleahopper genus *Pseudatomoscelis*, with the description of a new genus and an analysis of their relationships (Heteroptera: Miridae: Phylinae). J. N.Y. Entomol. Soc. 99:351–404.

Henry, T. J., and R. C. Froeschner (eds.). 1988. Catalog of the Heteroptera, or True Bugs, of Canada and the Continental United States. E. J. Brill, Leiden. 958 pp.

Henry, T. J., and R. T. Schuh. 1979. Redescription of *Beamerella* Knight and *Hambletoniola* Carvalho and included species (Hemiptera, Miridae), with a review of their relationships. Am. Mus. Novit. 2689:13 pp.

Henry, T. J., J. W. Neal, Jr., and K. M. Gott. 1986. *Stethoconus japonicus* (Heteroptera: Miridae): a predator of *Stephanitis* lace bugs newly discovered in the United States, promising in the biocontrol of azalea lace bugs (Heteroptera: Tingidae). Proc. Entomol. Soc. Wash. 88:722–730.

Herrich-Schaeffer, G. A. W. 1839–1853. Die wanzenartigen Insecten, vols. 4–9. Nuremberg, Germany.

Herring, J. L. 1961. The genus *Halobates* (Hemiptera: Gerridae). Pac. Insects 3(2–3):223–305.

Herring, J. L. 1965. *Hermatobates*, a new generic record for the Atlantic ocean, with descriptions of new species (Hemiptera: Gerridae). Proc. U.S. Natl. Mus. 117:123–129.

Herring, J. L. 1967. Heteroptera: Anthocoridae. *In:* Insects of Micronesia, vol. 7, no. 8, pp. 391–414. Bernice P. Bishop Museum, Honolulu.

Herring, J. L. 1976a. Keys to the genera of Anthocoridae of America north of Mexico, with description of a new genus (Hemiptera: Heteroptera). Florida Entomol. 59:143–150.

Herring, J. L. 1976b. A new genus and species of Cylapinae from Panama (Hemiptera: Miridae). Proc. Entomol. Soc. Wash. 78:91–94.

Herring, J. L. 1980. A review of the cactus bugs of the genus *Chelinidea* with the description of a new species (Hemiptera: Coreidae). Proc. Entomol. Soc. Wash. 82:237–251.

Herring, J. L., and P. D. Ashlock. 1971. A key to the nymphs of the families of Hemiptera (Heteroptera) of America north of Mexico. Florida Entomol. 54:207–212.

Hesse, A. J. 1947. A remarkable new dimorphic isometopid and two other species of Hemiptera predaceous upon the red scale of citrus. J. Entomol. Soc. S. Afr. 10:31–45.

Hickman, V. V., and J. L. Hickman. 1981. Observations on the biology of *Oncylocotis tasmanicus* (Westwood) with descriptions of the immature stages (Hemiptera, Enicocephalidae). J. Nat. Hist. 15:703–715.

Hill, L. 1980. Tasmanian Dipsocoroidea (Hemiptera: Heteroptera). J. Aust. Entomol. Soc. 19:107–127.

Hill, L. 1984. New genera of Hypselosomatinae (Heteroptera: Schizopteridae) from Australia. Aust. J. Zool., suppl. ser., 103:1–55.

Hill, L. 1987. First record of Dipsocoridae (Hemiptera) from Australia with the description of four new species of *Cryptostemma* Herrich-Schaeffer. J. Aust. Entomol. Soc. 26:129–139.

Hill, L. 1988. The identity and biology of *Baclozygum depressum* Bergroth (Hemiptera: Thaumastocoridae). J. Aust. Entomol. Soc. 27:37–42.

Hill, L. 1990. A revision of Australian *Pachyplagia* Gross (Heteroptera: Schizopteridae). Invert. Taxon. 3:605–617.

Hinks, C. F. 1966. The dorsal vessel and associated structures in some Heteroptera. Trans. R. Entomol. Soc. Lond. 118:375–392.

Hoberlandt, L., and P. Štys. 1979. *Tamopocoris asiaticus* gen. and sp. n.—a new aphelocheirine from Vietnam and further studies on Naucoridae (Heteroptera). Acta Mus. Natl. Pragae 33(B):1–20.

Hoke, S. 1926. Preliminary paper on the wing-venation of the Hemiptera (Heteroptera). Ann. Entomol. Soc. Am. 19:13–34.

Horn, W., and I. Kahle. 1935–1937. Über entomologische Sammlungen, Entomologen and Entomo-Museologie. Entomol. Beihefte 2–4:536 pp., 38 pls., 3 figs.

Horn, W., and S. Schenkling. 1928–1929. Index Literaturae Entomologicae, ser. 1: Die Welt-Literatur über die gesamte Entomologie bis inklusiv 1863. Selbstverlag W. Horn, Berlin-Dahlem. 4 vols., 1486 pp.

Horn, W., I. Kahle, G. Friese, and R. Gaedike. 1990. Collectiones entomologicae: Ein Kompendium über den Verbleib entomologischer Sammlungen der Welt bis 1960. Akad. Landwirtsch., Berlin. 2 vols.

Horvath, G. 1900. Analecta ad cognitionem Tessaratominorum. Termez. Fuzetek 23:339–374.

Horvath, G. 1904. Monographia Colobathristinarum. Ann. Mus. Natl. Hung. 2:117–172.

Horvath, G. 1911. Révision de Leptopodides. Ann. Mus. Natl. Hung. 9:358–370.

Horvath, G. 1929. Mesoveliidae. General Catalogue of the Hemiptera. Smith College, Northampton, Mass. 2:1–15.

Howard, L. O., E. A. Schwarz, and A. Busck. 1916. A biographical and bibliographical sketch of Otto Heidemann. Proc. Entomol. Soc. Wash. 18:203–205.

Hsiao, T.Y. 1964. New species and new record of Hemiptera-Heteroptera from China. Acta Zootaxonomica Sin. 1:283–292.

Hsiao, T.-Y. (ed.). 1977. A Handbook for the Determination of the Chinese Hemiptera-Heteroptera, vol. 1. Science Press, Beijing. 330 pp., 52 pls. [In Chinese].

Hsiao, T.-Y., S.-Z. Ren, L.-Y. Zheng, H.-L. Jing, H.-G. Zou, and S.-L. Liu. 1981. A Handbook for the Determination of the Chinese Hemiptera-Heteroptera, vol. 2. Science Press, Beijing. 654 pp., 85 pls. [In Chinese].

Hungerford, H. B. 1917. The life-history of *Mesovelia mulsanti* White. Psyche 24:73–84.

Hungerford, H. B. 1919. The biology and ecology of aquatic and semiaquatic Hemiptera. Univ. Kans. Sci. Bull. 11:1–341, pls. 1–30.

Hungerford, H. B. 1922a. Oxyhaemoglobin present in backswimmer *Buenoa margaritacea* Bueno. Can. Entomol. 54:262–263.

Hungerford, H. B. 1922b. The life history of the toad bug. Univ. Kans. Sci. Bull. 24:145–171.

Hungerford, H. B. 1933. The genus *Notonecta* of the world. Univ. Kans. Sci. Bull. 21(9):5–195.

Hungerford, H. B. 1941. A remarkable new naucorid water bug (Hemiptera). Ann. Entomol. Soc. Am. 34:1–4.

Hungerford, H. B. 1942. *Coleopterocoris,* an interesting new genus of the subfamily Potamocorinae. Ann. Entomol. Soc. Am. 35:135–139.

Hungerford, H. B. 1948. The Corixidae of the Western Hemisphere (Hemiptera). Univ. Kans. Sci. Bull. 32:827 pp.

Hungerford, H. B. 1954. The genus *Rheumatobates* Bergroth (Hemiptera: Gerridae). Univ. Kans. Sci. Bull. 36:529–588.

Hungerford, H. B., and N. E. Evans. 1934. The Hydrometridae of the Hungarian National Museum and other studies in the family (Hemiptera). Ann. Mus. Natl. Hung. 28:31–112.

Hungerford, H. B., and R. Matsuda. 1958a. The genus *Esakia* Lundblad with two new species. J. Kans. Entomol. Soc. 31:193–197.

Hungerford, H. B., and R. Matsuda. 1958b. The *Tenagogonus-Limnometra* complex of the Gerridae. Univ. Kans. Sci. Bull. 39(9):371–457.

Hungerford, H. B., and R. Matsuda. 1960a. Concerning the genus *Ventidius* and five new species (Heteroptera, Gerridae). Univ. Kans. Sci. Bull. 40:323–343.

Hungerford, H. B., and R. Matsuda. 1960b. Keys to the subfamilies, tribes, genera and subgenera of the Gerridae of the World. Univ. Kans. Sci. Bull. 41:3–23.

Hungerford, H. B., and R. Matsuda. 1961. Some new species of *Rhagovelia* from the Philippines (Veliidae, Heteroptera). Univ. Kans. Sci. Bull. 42:257–279.

Hungerford, H. B., and R. Matsuda. 1962. The genus *Cylindrostethus* Feiber from the Eastern Hemisphere. Univ. Kans. Sci. Bull. 43:83–111.

Hungerford, H. B., and R. Matsuda. 1965. The genus *Ptilomera* Amyot and Serville (Gerridae: Hemiptera). Univ. Kans. Sci. Bull. 45:397–515.

Hurd, M. P. 1946. Generic classification of North American Tingoidea (Hemiptera-Heteroptera). Iowa State Coll. J. Sci. 20(4):429–492.

Hussey, R. F. 1925. A new hydrometrid genus from Honduras (Hemiptera). Bull. Brooklyn Entomol. Soc. 20:115–119.

Hussey, R. F. 1927. On some American Pyrrhocoridae (Hemiptera). Bull. Brooklyn Entomol. Soc. 22:227–235.

Hussey, R. F. 1929. General Catalogue of the Hemiptera. Fascicle III. Pyrrhocoridae. Smith College, Northampton, Mass. 144 pp.

Hussey, R. F. 1934. Observations on *Pachycoris torridus* (Scop.), with remarks on parental care in other Hemiptera. Bull. Brooklyn Entomol. Soc. 29:133–145.

Hutchinson, G. E. 1929. A revision of the Notonectidae and Corixidae of South Africa. Ann. S. Afr. Mus. 25:359–474.

International Commission on Zoological Nomenclature. 1943. Opinion 143. Opinions rendered by the International Commission on Zoological Nomenclature 2:81–99.

Ishihara, T., and H. Hasegawa. 1941. A new *Malcus* species from Japan, with a list of Malcinae of the World (Hemiptera: Lygaeidae). Mushi 13:105–107.

Ishiwatari, T. 1974. Studies on the scent of stink bugs (Hemiptera: Pentatomidae). I. Alarm pheromone activity. Appl. Entomol. Zool. 9:153–158.

Jacobs, D. H. 1986. Morphology and taxonomy of sub-Saharan *Aneurus* species with notes on their phylogeny, biology and cytogenetics (Heteroptera: Aradidae: Aneurinae). Entomol. Mem. Dept. Agric. Wat. Supply Repub. S. Afr. 64:1–64.

Jacobs, D. H. 1989. A new species of *Thaumastella* with notes on the morphology, biology and distribution of the two southern African species (Heteroptera: Thaumastellidae). J. Entomol. Soc. S. Afr. 52:301–316.

Jacobson, E. 1911. Biological notes on the hemipteran *Ptilocerus ochraceus*. Tijdschr. Entomol. 54:175–179.

Jamieson, B. G. M. 1987. The ultrastructure and phylogeny of insect spermatozoa. Cambridge University Press, Cambridge. 320 pp.

Jansson, A. 1972. Mechanisms of sound production and morphology of the stridulatory apparatus in the genus *Cenocorixa* (Hemiptera, Corixidae). Acta Zool. Fenn. 9:120–129.

Jansson, A. 1976. Audiospectrographic analysis of stridulatory signals of some North American Corixidae (Hemiptera). Ann. Zool. Fenn. 13:48–62.

Jansson, A. 1986. The Corixidae (Heteroptera) of Europe and some adjacent regions. Acta Entomol. Fenn. 47:1–94.

Jeannel, R. 1919. Voyage de Ch. Alluaud et R. Jeannel en Afrique Orientale (1911–1912). Résultats scientifiques. Insectes Hémiptères. III. Henicocephalidae et Reduviidae, pp. 133–313, pls. 5–12. L. Lhomme, Paris.

Jeannel, R. 1941. Les Hénicocéphalides. Monographie d'un groupe d'Hémiptères hématophages. Ann. Soc. Entomol. Fr. 110:273–368.

Jessop, L. 1983. A review of the genera of Plataspidae (Hemiptera) related to *Libyaspis*, with a revision of *Cantharodes*. J. Nat. Hist. 17:31–62

Johansson, A. S. 1957a. On the functional anatomy of the metathoracic scent glands of the milkweed bug, *Oncopeltus fasciatus* (Dallas) (Heteroptera: Lygaeidae). Norsk. Entomol. Tidssk. 10:95–109.

Johansson, A. S. 1957b. The nervous system of the milk weed bug, *Oncopeltus fasciatus* (Dallas) (Heteroptera: Lygaeidae). Trans. Am. Entomol. Soc. 83:119–183.

Johansson, A. S., and T. Braten. 1970. Cuticular morphology of the scent gland areas of some heteropterans. Entomol. Scand. 1:158–162.

Jordan, K. 1951. Autotomie bei *Mesovelia furcata* Mls. R. (Hem. Het. Mesoveliidae). Zool. Anz. 147:205–209.

Josifov, M. 1967. Zur Systematik der Gattung *Cryptostemma* H.-S (Heteroptera). Ann. Zool. Warsz. 25:215–226.

Josifov, M., and I. M. Kerzhner. 1984. Zur Systematik der Gattung *Dryophilocoris* Reuter 1875 (Heteroptera, Miridae). Reichenbachia 22(31)215–226.

Kalin, M., and F. M. Barrett. 1975. Observations on the anatomy, histology, release site, and function of Brindley's glands in the blood-sucking bug, *Rhodnius prolixus* (Heteroptera: Reduviidae). Ann. Entomol. Soc. Am. 68:127–134.

Kanyukova, E. V. 1974. Water bugs of the family Aphelocheiridae (Heteroptera) in the fauna of the USSR. Zool Zh. 53:1726–1731. [In Russian].

Kanyukova, E. V. 1982 (1981). Water striders (Heteroptera, Gerridae) of the fauna of the USSR. Trudy Zool. Inst. AN SSSR. 105:62–94. [In Russian].

Kanyukova, E. V. 1988. Gerromorpha. *In:* P. A. Ler (ed.), Keys to the Insects of the Far East of the USSR, vol. 2, pp. 755–760. Nauka, Leningrad. [In Russian].

Kellen, W. R. 1959. Notes on the biology of *Halovelia marianarum* Usinger in Samoa (Veliidae: Heteroptera). Ann. Entomol Soc. Am. 52:53–62.

Kellen, W. R. 1960. A new species of *Omania* from Samoa, with notes on its biology (Heteroptera: Saldidae). Ann. Entomol. Soc. Am. 53:494–499.

Kelton, L. A. 1959. Male genitalia as taxonomic characters in the Miridae (Hemiptera). Can. Entomol., suppl., 11:72 pp.

Kelton, L. A. 1964. Revision of the genus *Reuteroscopus* Kirkaldy 1905 with descriptions of eleven new species (Hemiptera: Miridae). Can. Entomol. 96:1421–1433.

Kelton, L. A. 1965. *Chlamydatus* Curtis in North America (Hemiptera: Miridae). Can. Entomol. 97:1132–1144.

Kelton, L. A. 1967. Synopsis of the genus *Lyctocoris* in North America and description of a new species from Quebec (Heterptera: Anthocoridae). Can. Entomol. 99:807–814.

Kelton, L. A. 1968. Revision of the North American species of *Slaterocoris* (Heteroptera: Miridae). Can. Entomol. 100:1121–1137.

Kelton, L. A. 1971. Review of *Lygocoris* species found in Canada and Alaska (Heteroptera: Miridae). Mem. Entomol. Soc. Canada 83:53 pp.

Kelton, L. A. 1975. The lygus bugs (genus *Lygus* Hahn) of North America (Heteroptera: Miridae). Mem. Entomol. Soc. Canada 95:101 pp.

Kelton, L. A. 1978. The Insects and Arachnids of Canada. Part 4. The Anthocoridae of Canada and Alaska. Heteroptera: Anthocoridae. Biosystematics Research Institute, Agriculture Canada, Ottawa. 101 pp.

Kelton, L. A. 1980a. First record of a European bug, *Loricula pselaphiformis*, in the Nearctic region (Heteroptera: Microphysidae). Can. Entomol. 112:1085–1087.

Kelton, L. A. 1980b. The insects and arachnids of Canada. Part 8. The plant bugs of the prairie provinces of Canada (Heteroptera: Miridae). Agriculture Canada, Publ. 1703, 408 pp.

Kelton, L. A. 1981. First record of a European bug, *Myrmedobia exilis* (Heteroptera: Microphysidae), in the Nearctic region. Can. Entomol. 113:1125–1127.

Kelton, L. A., and H. H. Knight. 1970. Revision of the genus *Platylygus*, with descriptions of 26 new species (Hemiptera: Miridae). Can. Entomol. 102:1429–1460.

Kenaga, E. E. 1941. The genus *Telmatometra* Bergroth (Hemiptera: Gerridae). Univ. Kans. Sci. Bull. 27:169–183.

Kerzhner, I. M. 1963. Beitrag zur Kenntnis der Unterfamilie Nabinae (Heteroptera: Nabidae). Acta Entomol. Mus. Natl. Pragae 35:5–61.

Kerzhner, I. M. 1968. New and little known Palearctic bugs of the family Nabidae (Heteroptera). Entomol. Obozr. 47:848–863. [In Russian; English translation *in:* Entomol. Rev. 47:517–525].

Kerzhner, I. M. 1970a. Neue und wenig bekannte Nabidae (Heteroptera) aus den tropischen Gebieten der Alten Welt. Acta Entomol. Mus. Natl. Pragae 38:279–359.

Kerzhner, I. M. 1970b. Some Heteroptera Nabidae (Hemiptera) from the southern Philippines and the Bismarek Islands. Entomol. Medd. 38:177–194.

Kerzhner, I. M. 1971. Classification and Phylogeny of the True Bugs of the Family Nabidae (Heteroptera). Special Scientific Session on Results of Studies during 1970, pp. 23–24. Zool. Inst. Acad. Nauk SSSR. [In Russian].

Kerzhner, I. M. 1974. New and little-known Heteroptera from Mongolia and adjacent regions of the USSR. 2. Dipsocoridae, Reduviidae. Nasekomye Mongol. 4(2):72–79. [In Russian].

Kerzhner, I. M. 1981. Fauna of the USSR. Bugs. Vol. 13, no. 2. Heteroptera of the Family Nabidae. Acad. Sci. USSR, Zool. Inst. Nauka, Leningrad. 326 pp. [In Russian].

Kerzhner, I. M. 1986a. African species of the genus *Arbela* Stål (Heteroptera, Nabidae). Ann. Soc. Entomol. Fr. 22:235–240.

Kerzhner, I. M. 1986b. Neotropical Nabidae (Heteroptera). 1. A new genus, some new species, and notes on synonymy. J. N.Y. Entomol. Soc. 94:180–193.

Kerzhner, I. M. 1988. Miridae. *In:* P. A. Lehr (ed.), Keys to the Insects of the Far East of the USSR, vol. 2. Nauka, Leningrad. [In Russian].

Kerzhner, I. M. 1989. On taxonomy and habits of the genus *Medocostes* (Heteroptera: Nabidae). Zool. Zhurnal 68:150–151. [In Russian].

Kerzhner, I. M. 1990. Neotropical Nabidae (Heteroptera). 3. Species of the genus *Arachnocoris* from Costa Rica. J. N.Y. Entomol. Soc. 98:133–138.

Kerzhner, I. M. 1992. Nomenclatural and bibliographic corrections to J. Maldonado Capriles (1990) "Systematic Catalogue of the Reduviidae of the World (Insecta: Heteroptera)." Zoosyst. Rossica 1:46–60.

Kerzhner, I. M., and T. L. Jaczewski. 1964. 19. Order Hemiptera. *In:* G. Ya. Bei Bienko. (ed.), Keys to the Insects of the European USSR, vol. 1: Apterygota, Palaeoptera, Hemimetabola, pp. 851–1118. Academy of Sciences of the USSR, Zoological Institute. Translated from Russian, 1967. Israel Program for Scientific Translations, Jerusalem.

Kerzhner, I. M., and A. A. Stackelberg. 1971. [Portrait and bibliography of A. N. Kiritshenko]. Entomol. Obozr. 50:719–729.

King, P. E., and M. R. Fordy. 1984. Observations on *Aepophilus bonnairei* (Signoret) (Saldidae: Hemiptera), an intertidal insect of rocky shores. Zool. J. Linn. Soc. 80:231–238.

King, P. E., and N. A. Ratcliffe. 1970. The surface structure of the cuticle of an intertidal hemipteran, *Aepophilus bonnairei* (Signoret). Entomol. Mon. Mag. 106:1–2, pl. I.

Kiritani, K., N. Hokyo, and J. Yukawa. 1963. Co-existence of the two related stink bugs *Nezara viridula* and *N. antennata* under natural conditions. Res. Popul. Ecol. 5:11–12.

Kiritshenko, A. N. 1940. [Bibliography and portraits of V. F. Oshanin]. Isp'talii prirod., hist. ser., 5:1–30.

Kirkaldy, G. W. 1897. Synonymic notes on aquatic Rhynchota. Entomologist 30:258.

Kirkaldy, G. W. 1902. Memoirs on Oriental Rhynchota. Bombay Nat. Hist. Soc. 14:294–309.

Kirkaldy, G. W. 1906. List of the genera of the pagiopodous Hemiptera-Heteroptera, with their type species, from 1758 to 1901 (and also of the aquatic and semiaquatic Trochalopoda). Trans. Am. Entomol. Soc. 32:117–156.

Kirkaldy, G. W. 1908. Memoir on a few heteropterous Hemiptera from eastern Australia. Proc. Linn. Soc. NSW 32:768–788.

Kirkaldy, G. W. 1909. Catalogue of the Hemiptera (Heteroptera) with Biological and Anatomical References, List of Food Plants and Parasites, etc., vol. 1. Cimicidae [= Pentatomidae]. Berlin. xl + 392 pp.

Klausner, E., E. R. Miller, and H. Dingle. 1981. Genetics of brachyptery in a lygaeid bug island population. J. Heredity 72:288–289.

Klug, J. C. F. 1830. Symbolae physicae, seu icones et descriptiones insectorum, quae ex itinere per African borealem et Asiam F. G. Hemprich et C. H. Ehrenberg studo novae aut illustratae redierunt. Berolini. 2.

Knight, H. H. 1922. Studies on the life history and biology of *Perillus bioculatus* Fabricius, including observations on the nature of the color pattern. pp. 50–96. 19th Rep. State Entomologist Minnesota.

Knight, H. H. 1923. Guide to the insects of Connecticut. Part IV. The Hemiptera or sucking insects of Connecticut. Family Miridae (Capsidae). Conn. Geol. Nat. Surv. Bull. 34:422–658.

Knight, H. H. 1924. On the nature of the color patterns in Heteroptera with data on the effects produced by temperature and humidity. Ann. Entomol. Soc. Am. 17:258–272

Knight, H. H. 1927. On the Miridae in Blatchley's "Heteroptera of Eastern North America." Bull. Brooklyn Entomol. Soc. 22:98–105.

Knight, H. H. 1941. The plant bugs, or Miridae, of Illinois. Bull. Illinois Nat. Hist. Surv. 22:234 pp.

Knight, H. H. 1968. Taxonomic review: Miridae of the Nevada Test Site and the western United States. Brigham Young Univ. Sci. Bull., ser. 9. 282 pp.

Knight, W. J. 1980. Obituary and bibliography. William Edward China. Entomol. Mon. Mag. 115:164–175.

Kormilev, N. A. 1949. La familia "Colobathristidae" Stål en la Argentina con la descripción de tres especies nuevas neotropicales. Acta Zool. Lilloana 7:369–383.

Kormilev, N. A. 1951. Notas sobre "Colobathristidae" neotropicales (Hemiptera) con la descripción de tres géneros y siete especies nuevas. Rev. Brasil Biol. 11:60–84.

Kormilev, N. A. 1954. Notas sobre Coreidae neotropicales II: (Hemiptera) Merocorinae de la Argentina y países limítrofes. Rev. Ecuatoriana Entomol. Parasit. 2:153–186.

Kormilev, N. A. 1955a. Una curiosa familia de Hemípteros nueva para la fauna Argentina. Thaumastotheriidae (Kirkaldy), 1907. Rev. Soc. Entomol. Argentina 17:5–10.

Kormilev, N. A. 1955b. A new myrmecophil family of Hemiptera from the delta of Rio Parana, Argentina. Rev. Ecuatoriana Entomol. Parasit. 2:465–477, 1 pl.

Kormilev, N. A. 1962. Revision of Phymatinae (Hemiptera, Phymatidae). Philippine J. Sci. 89:287–486, pls. 1–19.

Kormilev, N. A. 1969. Thaicorinae, n. subfam. from Thailand (Hemiptera: Heteroptera: Piesmatidae). Pac. Insects 11:645–648.

Kormilev, N. A. 1971a. Mezirinae of the Oriental region and South Pacific. Pacific Insects Monographs 26:1–165.

Kormilev, N. A. 1971b. Ochteridae from the Oriental and Australia regions. Pac. Insects 12:429–444.

Kormilev, N. A., and R. C. Froeschner. 1987. Flat bugs of the world. A synonymic list (Heteroptera: Aradidae). Entomography 5:1–246.

Krugman, S. L., and T. W. Koerber. 1969. Effect of cone feeding by *Leptoglossus occidentalis* on ponderosa pine seed development. Forest Sci. 15:104–111.

Kudo, S. 1990. Brooding behavior in *Elasmucha putoni* (Heteroptera: Acanthosomatidae) and a possible nymphal alarm substance triggering guarding responses. Appl. Entomol. Zool. 25:431–437.

Kudo, S., M. Sato, and M. Ohara. 1989. Prolonged maternal care in *Elasmucha dorsalis* (Heteroptera: Acanthosomatidae). J. Ecol. 7:75–81.

Kullenberg, B. 1944. Studien über die Biologie der Capsiden. Zool. Bidr., Uppsala 23:522 pp.

Kullenberg, B. 1947. Über Morphologie und Function des Kopulationsapparats der Capsiden und Nabiden. Zool. Bidr. 24:217–418, pls. 2–23.

Kumar, R. 1961. Studies on the genitalia of some aquatic and semi-aquatic Heteroptera. Entomol. Tidskr. 82:163–179.

Kumar, R. 1962. Morpho-taxonomical studies on the genitalia and salivary glands of some Pentatomoidea. Entomol. Tidskr. 83:44–88.

Kumar, R. 1964. Anatomy and relationships of Thaumastocoridae (Hemiptera: Cimicoidea). J. Entomol. Soc. Queensland 3:48–51.

Kumar, R. 1965. Contributions in the morphology and relationships of Pentatomoidea (Hemiptera: Heteroptera). Part 1. Scutelleridae. J. Entomol. Soc. Queensland 4:41–55.

Kumar, R. 1967. Morphology of the reproductive and alimentary systems of the Aradoidea (Hemiptera), with comments on relationships with the superfamily. Ann. Entomol. Soc. Am. 60:17–25.

Kumar, R. 1968. Aspects of the morphology and relationships of the superfamilies Lygaeoidea, Piesmatoidea and Pyrrhocoroidea (Hemiptera: Heteroptera). Entomol. Mon. Mag. 103:251–261.

Kumar, R. 1969. Morphology and relationships of the Pentatomoidea (Heteroptera). III. Natalicolinae and some Tessaratomidae of uncertain position. Ann. Entomol. Soc. Am. 62:681–695.

Kumar, R. 1971. Morphology and relationships of the Pentatomoidea (Heteroptera). 5: Urostylidae. Am. Midl. Nat. 85:63–73.

Kumar, R. 1974a. A revision of world Acanthosomatidae (Heteroptera: Pentatomoidea): keys to and descriptions of subfamilies, tribes and genera with designation of types. Aust. J. Zool., suppl. ser. N, 34:1–60.

Kumar, R. 1974b. A key to the genera of Natalicolinae Horvath, with the description of a new species of Tessaratominae Stål and with new synonymy (Pentatomoidea: Heteroptera). J. Nat. Hist. 8:675–679.

Kumar, R., and M. S. K. Ghauri. 1970. Morphology and relationships of the Pentatomoidea (Heteroptera). 2. World genera of Tessaratomini. Dtsch. Entomol. Z. 17:1–32.

Laboulbène, A. 1865. Paroles d'adieu adressées à M. Léon Dufour. Liste des travaux d'entomologie publiés de 1811 à 1864. Ann. Soc. Entomol. Fr., ser. 4, 5:214–252.

Lai-Fook, J. 1970. The fine structure of developing type "B" dermal glands in *Rhodnius prolixus*. Tissue and Cell 2:119–138.

Lansbury, I. 1962. Notes on the genus *Anisops* in Bishop Museum (Hem.: Notonectidae). Pac. Insects 4:141–151.

Lansbury, I. 1964. A revision of the genus *Paranisops* Hale (Heteroptera: Notonectidae). Proc. R. Entomol. Soc. Lond. (B) 33:181–188.

Lansbury, I. 1965. New organ in Stenocephalidae (Hemiptera-Heteroptera). Nature 205:106.

Lansbury, I. 1965–1966. A revision of the Stenocephalidae Dallas 1852. (Hemiptera: Heteroptera). Entomol. Mon. Mag. 101:52–92, 145–160

Lansbury, I. 1968. The *Enithares* (Hemiptera-Heteroptera: Notonectidae) of the Oriental region. Pac. Insects 10:353–442.

Lansbury, I. 1972. A review of the Oriental species of *Ranatra* Fabricius (Hemiptera-Heteroptera: Nepidae). Trans. R. Entomol. Soc. Lond. 124:387–341.

Lansbury, I. 1974a. A new genus of Nepidae from Australia with a revised classification of the family (Hemiptera: Heteroptera). J. Aust. Entomol. Soc. 13:219–227.

Lansbury, I. 1974b. Notes on *Ranatra* (*Amphischizops*) *compressicollis* Montandon with a review of the its systematic position within the American *Ranatra* (Hemiptera-Heteroptera, Nepidae). Zool. Scripta 3:23–30.

Laporte, F. 1833. Essai d'une classification systématique de l'ordre de Hémiptères (Hémiptères Hétéroptères, Latr.). Guerin Magasin de Zoologie 2:1–88, 4 pls.

La Rivers, I. 1950. A new species of the genus *Potamocoris* from Honduras. Proc. Entomol. Soc. Wash. 52:301–304.

La Rivers, I. 1969. New naucorid taxa. Occ. Pap. Biol. Soc. Nevada 20:1–12.

La Rivers, I. 1971. Studies on Naucoridae. Biol. Surv. Nevada Memoire II. iii + 120 pp.

La Rivers, I. 1974. Catalogue of taxa described in the family Naucoridae (Hemiptera). Supplement no. 1: Corrections, emendations and additions, with descriptions of new species. Occ. Pap. Biol. Soc. Nevada 38:1–17.

La Rivers, I. 1976. Supplement no. 2 to the catalogue of taxa described in the family Naucoridae (Hemiptera), with descriptions of new species. Occ. Pap. Biol. Soc. Nevada No. 41:1–18, 6 figs.

Larsén, O. 1938. Untersuchungen über den Geschlechtsapparat der aquatilen Wanzen. Opusc. Entomol., Suppl. 1, pp. 1–388.

Larsén, O. 1957. Truncale Scolopalorgane in den Pterothoakalen und den beiden ersten abdominalen Segmenten der aquatilen Heteropteren. Acta Univ. Lund Avd. 2. 53(1):1–68.

Latreille, P. A. 1802. Histoire naturelle, générale et particulière des crustaces et des insectes, vol. 3. F. Dufart, Paris. xii + 467 pp.

Latreille, P. A. 1809. Genera crustaceorum et insectorum secundum ordinem naturalem in familias disposit, iconibus exemplisque plurimis explicata, vol. 4. A. Konig, Parisiis et Argentorati. 399 pp.

Latreille, P. A. 1810. Considérations générales sur l'ordre naturel des animaux. F. Schoell, Paris.

Lattin, J. D. 1964. The Scutellerinae of America north of Mexico (Hemiptera: Heteroptera: Pentatomidae). Ph.D. dissertation, University of California, Berkeley. 346 pp.

Lattin, J. D. 1989. Bionomics of the Nabidae. Ann. Rev. Entomol. 34:383–400.

Lattin, J. D., and N. L. Stanton. 1992. A review of the species of Anthocoridae (Hemiptera: Heteroptera) found on *Pinus contorta*. J. N.Y. Entomol. Soc. 100:424–479.

Lattin, J. D., and G. M. Stonedahl. 1984. *Campyloneura virgula*, a predaceous Miridae not previously recorded from the United States (Hemiptera). Pan-Pac. Entomol. 60:4–7.

Lauck, D. R. 1962. A monograph of the genus *Belostoma*. Part I. Introduction and *B. dentatum* and *subspinosum* groups. Bull. Chicago Acad. Sci. 11:34–81.

Lauck, D. R. 1963. A monograph of the genus *Belostoma*. Part II. *B. aurivillianum, stollii, testaceopalidum, dilatatum,* and *discretum* groups. Bull. Chicago Acad. Sci. 11:82–101.

Lauck, D. R. 1964. A monograph of the genus *Belostoma*. Part III. *B. triangulum, bergi, minor, bifoveolatum* and *flumineum* groups. Bull. Chicago Acad. Sci. 11:102–154.

Lauck, D. R. 1979. Family Corixidae/water boatmen. *In:* A. S. Menke (ed.), The Semiaquatic and Aquatic Hemiptera of California (Heteroptera: Hemiptera), pp. 87–123. Univ. Calif. Publ., Bull. Calif. Insect Surv., vol. 21.

Lauck, D. R., and A. S. Menke. 1961. The higher classification of the Belostomatidae (Hemiptera). Ann. Entomol. Soc. Am. 54:644–657.

Lavabre, E. M. (ed.). 1977. Les mirides du cacaoyer. Institut Français du Café et du Cacao, Paris. 366 pp.

Lawrence, P. A., and B. W. Staddon. 1975. Peculiarities of the epidermal gland system of the cotton stainer *Dysdercus fasciatus* Singoret (Heteroptera: Phyrhocoridae). J. Entomol. (A), 49:121–136.

Lawry, J. V., Jr. 1973. A scanning electron microscopic study of mechanoreceptors in the walking legs of *Gerris remigis*. J. Anat. 116:25–30.

Leach, W. E. 1815. Hemiptera. *In:* Brewster's Edinburgh Encyclopedia, vol. 9, pp. 57–192. Edinburgh, Scotland.

LeConte, J. L. (ed.). 1859. American Entomology. A description of the insects of North America, by Thomas Say, with illustrations drawn and colored after nature. With a memoir of the author by George Ord. Boston. 2 vols.

Lee, C. E. 1969. Morphological and phylogenetic studies on the larvae and male genitalia of the East Asiatic Tingidae (Heteroptera). J. Fac. Agric. Kyushu Univ. 15:137–256, pls. 1–16.

Lent, H., and J. Jurberg. 1965. Contribuição ao conhecimento dos Phloeidae Dallas, 1851, com un estudo sôbre genitalia (Hemiptera, Pentatomoidea). Rev. Brasil. Biol. 25:123–144.

Lent, H., and J. Jurberg. 1966. Os estádios larvares de "Phloeophana longirostris" (Spinola, 1837) (Hemiptera, Pentatomoidea). Rev. Brasil. Biol. 26:1–4.

Lent, H., and P. Wygodzinsky. 1944. Sôbre uma nova espécies do gênero "Chryxus" Champion, 1898. Rev. Brasil. Biol. 4:167–171.

Lent, H., and P. W. Wygodzinsky. 1979. Revision of the Triatominae (Hemiptera, Reduviidae), and their significance as vectors of Chagas' disease. Bull. Am. Mus. Nat. Hist. 163:123–520.

Leston, D. 1952. Notes on the Ethiopian Pentatomoidea (Hemiptera). V. On the specimens collected by Mr. A. L. Capener, mainly in Natal. Ann. Mag. Nat. Hist., ser. 12, 5:512–520.

Leston, D. 1953a. On the wing venation, male genitalia and spermatheca of *Podops inuncta* (F.) with a note on the diagnosis of the subfamily Podopinae Dallas (Hem., Pentatomidae). J. Soc. Br. Entomol. 4:129–135.

Leston, D. 1953b. Notes on the Ethiopian Pentatomoidea (Hemiptera). 14. On acanthosomid from Angola, with remarks upon the status and morphology of Acanthosomidae Stål. Publ. Cult. Cia Diam. Angola 16:123–132.

Leston, D. 1953c. "Ploeidae" Dallas: systematics and morphology, with remarks on the phylogeny of "Pentatomoidea" Leach and upon the position of "Serbana" Distant (Hemiptera). Rev. Brasil. Biol. 13:121–140.

Leston, D. 1954a. Strigils and stridulation in Pentatomoidea (Hem.): some new data and a review. Entomol. Mon. Mag. 90:49–56.

Leston, D. 1954b. Wing venation and male genitalia of *Tessaratoma* Berthold, with remarks on Tessaratominae Stål (Hemiptera: Pentatomidae). Proc. R. Entomol. Soc. Lond. (A) 29:9–16.

Leston, D. 1954c. Notes on the Ethiopian Pentatomoidea (Hemiptera). XVII. Tessaratominae, Dinidorinae and Phyllocephalinae of Angola. Publ. Cult. Comp. Diam. Angola, Lisboa 24:11–22.

Leston, D. 1955a. The aedeagus of Dinidorinae (Hem., Pentatomidae). Entomol. Mon. Mag. 91:214–215.

Leston, D. 1955b. A key to the genera of Oncomerini Stål (Heteroptera: Pentatomidae, Tessaratominae), with the description of a new genus and species from Australia and new synonymy. Proc. R. Entomol. Soc. Lond. (B) 24:62–68.

Leston, D. 1955c. Remarks on the male and female genitalia and abdomen of Aradidae. Proc. R. Entomol. Soc. Lond. 30:63–69.

Leston, D. 1956. Results from the Danish Expedition to the French Cameroons 1919–50. IX. Hemiptera, Pentatomoidea. Bull. Inst. Fr. Afr. Noire, ser. A, 18:618–626.

Leston, D. 1957a. Systematics of the marine-bug. Nature 178:427–428.

Leston, D. 1957b. The stridulatory mechanisms in terrestrial species of Hemiptera Heteroptera. Proc. Zool. Soc. Lond. 128(3):369–386.

Leston, D. 1961. Testis follicle number and the higher systematics of the Miridae (Hemiptera-Heteroptera). Proc. Zool. Soc. Lond. 137:89–106.

Leston, D. 1962. Tracheal capture in ontogenetic and phylogenetic phases of insect wing development. Proc. R. Entomol. Soc. Lond. (A) 37:135–144.

Leston, D., J. G. Pendergrast, and T. R. E. Southwood. 1954. Classification of the terrestrial Heteroptera (Geocorisae). Nature 174:91.

Lethierry, L., and G. Severin. 1893–1896. Catalogue général des Hémiptères. Musée Royal d'Histoire Naturelle de Belgique. (Vol. 1, 1893, x + 286 pp.; vol. 2, 1894, 277 pp.; vol. 3, 1886, 275 pp.)

Leuckhart, R. 1855. Über die Mikropyle und den feineren Bau der Schalenhaut bei den Insekteneiern. Arch. Anat. Physiol. Wiss. Med. 1855:90–264.

Lindberg, H. 1928. Sketch in commemoration of Ernst Evald Bergroth. Mem. Soc. Fauna Flora Fenn. 4:292–317.

Lindberg, H. 1953. Hemiptera Insularum Canariensium. Soc. Sci. Fenn., Comm. Biol. 14:1–304.

Lindberg, H. 1958. Hemiptera Insularum Caboverdensium. Soc. Sci. Fenn., Comm. Biol. 19:1–246.

Linder, H. J. 1956. Structure and histochemistry of the maxillary glands in the milkweed bug, *Oncopeltus fasciatus* (Hem.). J. Morphol. 99:575–611.

Lindskog, P., and J. T. Polhemus. 1992. Taxonomy of *Saldula*: revised genus and species group definitions, and a new species of the *pallipes* group from Tunisia (Heteroptera: Saldidae). Entomol. Scand. 23:63–88.

Linnaeus, C. 1758. Systema naturae, 10th ed.

Linnavuori, R. 1971. Hemiptera of the Sudan, with remarks on some species of the adjacent countries. 1. The aquatic and subaquatic families. Ann. Zool Fenn. 8:340–366.

Linnavuori, R. 1974. Hemiptera of the Sudan, with remarks on some species of the adjacent countries. 3. Families Cryptostemmatidae, Cimicidae, Polyctenidae, Joppeicidae, Reduviidae, Pachynomidae, Nabidae, Leptopodidae, Saldidae, Henicocephalidae and Berytidae. Ann. Entomol. Fenn. 42(3):116–138.

Linnavuori, R. 1975. Hemiptera of the Sudan, with remarks on some species of the adjacent countries. 4. Miridae and Isometopidae. Ann. Zool. Fenn. 12:1–118.

Linnavuori, R. 1977. On the taxonomy of the subfamily Microveliinae (Heteroptera: Veliidae) of West and Central Africa. Ann. Entomol. Fenn. 43:41–61.

Linnavuori, R. 1981. Hemiptera of Nigeria, with remarks on some species of the adjacent countries. 1. The aquatic and subaquatic families, Saldidae and Leptopodidae. Acta Entomol. Fenn. 37:1–39.

Linnavuori, R. 1986. Heteroptera of Saudi Arabia. Fauna of Saudi Arabia 8:31–197.

Linnavuori, R. 1987. Alydidae, Stenocephalidae and Rhopalidae of West and Central Africa. Acta Entomol. Fenn. 49:1–36.

Linnavuori, R. 1993. Cydnidae of west, central and north-east Africa (Heteroptera). Acta Zool. Fenn. 192:1–148.

Linsenmair, K. E., and R. Jander. 1963. Das Entspannungsschwimmen von *Velia* und *Stenus*. Naturwissenschaften 50:231.

Liquido, N. J., and T. Nishida. 1985. Variation in number of instars, longevity, and fecundity of *Cyrtorhinus lividipennis* Reuter (Hemiptera: Miridae). Ann. Entomol. Soc. Am. 78:459–463.

Lis, J. A. 1990. New genera, new species, new records and checklist of the Old World Dinidoridae (Heteroptera, Pentatomoidea). Ann. Upper Silesian Mus., Entomol. 1:103–147.

Livingstone, D. 1967. On the morphology of *Tingis buddleiae* Drake (Heteroptera: Tingidae). Part XIII. The wings—homology of the areas, venation and pteralia. Bull. Entomol. 8:1–11.

Livingstone, D. 1968. The morphology and biology of *Tingis buddleiae* Drake (Heteroptera: Tingidae). Part V. The nervous system, endocrine glands and sense organs. J. Anim. Morph. Physiol. 15:1–25.

Livingstone, D. 1978. On the body outgrowths and the phenomenon of "sweating" in the nymphal instars of Tingidae (Hemiptera: Heteroptera). J. Nat. Hist. 12:377–394.

Louis, D. 1974. Biology of Reduviidae of cocoa farms in Ghana. Am. Midl. Nat. 91:68–89.

Lundblad, O. 1933. Zur Kenntnis der aquatilen und semiaquatilen Hemipteren von Sumatra, Java und Bali. Arch. Hydrobiol., suppl., 12:1–105, 263–489.

Lundblad, O. 1936. Die altweltlichen Arten der Veliidengattungen *Rhagovelia* und *Tetraripis*. Ark. Zool. 28(21):1–76.

Maa, T. C. 1964. A review of the Old World Polyctenidae. Pac. Insects 6:494–516.

Maa, T. C., and K.-S. Lin. 1956. A synopsis of the Old World Phymatidae (Hem.). Q. J. Taiwan Mus. 9:109–154, pls. I–IV.

Mahner, M. 1993. Systema cryptoceratorum phylogeneticum (Insecta, Heteroptera). Zoologica 143:ix + 302 pp.

Maldonado Capriles, J. 1981. A new *Ghilianella* and a new saicine genus, *Buninotus* (Hemiptera: Reduviidae), from Panama. J. Agric. Univ. Puerto Rico 65:401–407.

Maldonado Capriles, J. 1990. Systematic Catalogue of the Reduviidae of the World (Insecta: Heteroptera). Caribbean J. Sci. (special ed.). 694 pp.

Malipatil, M. B. 1977. Distribution, origin and speciation, wing development, and host-plant relationships of New Zealand Targaremini (Hemiptera: Lygaeidae). New Zealand J. Zool. 4:369–381.

Malipatil, M. B. 1978. Revision of the Myodochini (Hemiptera: Lygaeidae: Rhyparochrominae) of the Australian region. Aust. J. Zool., suppl. ser., 56:1–178.

Malipatil, M. B. 1981. Revision of Australian Cleradini (Heteroptera: Lygaeidae). Aust. J. Zool. 29:773–819.

Malipatil, M. B. 1983. Revision of world Cleradini (Heteroptera: Lygaeidae) with a cladistic analysis of relationships within the tribe. Aust. J. Zool. 31:205–225.

Malipatil, M. B. 1985. Revision of Australian Holoptilinae (Reduviidae: Heteroptera). Aust. J. Zool. 33:283–299.

Malouf, N. S. R. 1933. Studies of the internal anatomy of the stink-bug *Nezara viridula* L. Bull. Soc. Entomol. Egypte 17:96–117.

Marshall, A. G. 1982. The ecology of the bat ectoparasite *Eoctenes spasmae* (Hemiptera: Polyctenidae) in Malaysia. Biotropica 14:50–55.

Maschwitz, U., B. Fiala, and W. R. Dolling. 1987. New trophobiotic symbioses of ants with South East Asian bugs. J. Nat. Hist. 21:1097–1107.

Matsuda, R. 1956. A supplementary taxonomic study of the genus *Rhagovelia* (Hemiptera, Veliidae) of the Western Hemisphere. A deductive method. Univ. Kans. Sci. Bull. 38:915-1017.

Matsuda, R. 1960. Morphology, evolution, and classification of the Gerridae (Hemiptera-Heteroptera). Univ. Kans. Sci. Bull. 41:25–632.

Mayr, E. E., G. Linsley, and R. L. Usinger. 1953. Methods and Principles of Systematic Zoology. McGraw-Hill, New York. 328 pp.

Mayr, G. 1866 (1868). Hemiptera. *In:* Reise der österreichschen Fregatte Novara um die Erde in den Jahren 1857, 1858, 1859. Zool. Pt. 2. Abt. 1, B. 2. Hemiptera. Hof und Staatsdruckerei. Karl Gerold's Sohn, Vienna. 204 pp.

McAtee, W. L., and J. R. Malloch. 1923. Notes on American Bacrodinae and Saicinae (Heteroptera: Reduviidae). Ann. Entomol. Soc. Am. 16:247–254, pl. 16.

McAtee, W. L., and J. R. Malloch. 1924. Some annectant bugs of the superfamily Cimicoidea (Heteroptera). Bull. Brooklyn Entomol. Soc. 19:69–82.

McAtee, W. L., and J. R. Malloch. 1925. Revision of bugs of the family Cryptostemmatidae in the collection of the United States National Museum. Proc. U.S. Natl. Mus. 67(13):1–42, pls. 1–4.

McAtee, W. L., and J. R. Malloch. 1928. Synopsis of pentatomid bugs of the subfamilies Megaridinae and Canopinae. Proc. U.S. Natl. Mus. 72:1–21.

McAtee, W. L., and J. R. Malloch. 1932. Notes on the genera of Isometopinae (Heteroptera). Entomol. Mon. Mag. 118:79–86.

McAtee, W. L., and J. R. Malloch. 1933. Revision of the subfamily Thyreocorinae of the Pentatomidae (Hemiptera-Heteroptera). Ann. Carnegie Mus. 21:191–411.

McClure, H. E. 1932. Incubation of bark-bug eggs (Aradidae). Entomol. News 43:188–189.

McDonald, F. J. D. 1966. The genitalia of North American Pentatomoidea (Hemiptera: Heteroptera). Quest. Entomol. 2:7–150.

McDonald, F. J. D. 1969. A new species of Lestoniidae (Hemiptera). Pac. Insects 11:187–190.

McDonald, F. J. D. 1970. The morphology of *Lestonia haustorifera* China (Het. Lestoniidae). J. Nat. Hist. 4:413–417.

McDonald, F. J. D. 1979. A new species of *Megaris* and the status of the Megarididae McAtee and Malloch and Canopidae Amyot and Serville (Hemiptera: Pentatomoidea). J. N.Y. Entomol. Soc. 87:42–54.

McDonald, F. J. D., and G. Cassis. 1984. Revision of the Australian Scutelleridae Leach (Hemiptera). Aust. J. Zool. 32:537–572.

McDonald, F. J. D., and J. Grigg. 1981. Life cycle of *Biprorulus bibax* Breddin (Hemiptera: Pentatomidae). Gen. Appl. Entomol. 13:54–58.

McHugh, J. V. 1994. On the natural history of Canopidae (Heteroptera: Pentatomoidea). J. N.Y. Entomol. Soc. 102:112–114.

McIver, J. D., and J. D. Lattin. 1990. Evidence for aposematism in the plant bug *Lopidea nigridea* Uhler (Hemiptera: Miridae: Orthotylinae). Biol. J. Linn. Soc. 40:99–112.

McIver, J. D., and G. M. Stonedahl. 1987a. Biology of the myrmecomorphic plant bug *Coquillettia insignis* Uhler (Heteroptera: Miridae: Phylinae). J. N.Y. Entomol. Soc. 95:258–277.

McIver, J. D., and G. M. Stonedahl. 1987b. Biology of the myrmecomorphic plant bug *Orectoderes obliquus* Uhler (Heteroptera: Miridae: Phylinae). J. N.Y. Entomol. Soc. 95:278–289.

McIver, J. D., and G. M. Stonedahl. 1993. Myrmecomorphy: morphological and behavior mimicry of ants. Ann. Rev. Entomol. 38:351–379.

McKinstry, A. P. 1942. A new family of Hemiptera-Heteroptera proposed for *Macrovelia hornii* Uhler. Pan-Pac. Entomol. 18(2):90–96.

McLain, D. K. 1984. Coevolution: Müllerian mimicry between a plant bug (Miridae) and a seed bug (Lygaeidae) and the relationship between host plant choice and unpalatability. Oikos 43:143–148.

McLain, D. K., and D. J. Shure. 1987. Pseudocompetition: interspecific displacement of insect species through misdirected courtship. Oikos 49:291–296.

McMahan, E. 1982. Bait-and-capture strategy of a termite-eating assassin bug. Insectes Sociaux 29:346–351.

McPherson, J. E. 1977. Effects of developmental photoperiod on adult color and pubescence in *Thyanta calceata* (Hemiptera: Pentatomidae), with information on ability of adults to change color. Ann. Entomol. Soc. Am. 70:373–376.

McPherson, J. E. 1982. The Pentatomoidea (Hemiptera) of Northeastern North America with Emphasis on the Fauna of Illinois. Southern Illinois University Press, Carbondale. 240 pp.

Menke, A. S. 1960a. A review of the genus *Lethocerus* (Hemiptera: Belostomatidae) in the Eastern Hemisphere with the description of a new species from Australia. Aust. J. Zool. 8:285–288.

Menke, A. S. 1960b. A taxonomic study of the genus *Abedus* Stål. Univ. Calif. Publ. Entomol. 16:393–440.

Menke, A. S. (ed.). 1979a. The Semiaquatic and Aquatic Hemiptera of California (Heteroptera: Hemiptera). Univ. Calif. Publ., Bull. Calif. Insect Surv., vol. 21. 166 pp., 277 figs.

Menke, A. S. 1979b. Family Ochteridae/velvety shore bugs. In: A. S. Menke (ed.), The Semiaquatic and Aquatic Hemiptera of California (Heteroptera: Hemiptera), pp. 124–125. Univ. Calif. Publ., Bull. Calif. Insect Surv., vol. 21.

Menke, A. S., and L. A. Stange. 1964. A new genus of Nepidae from Australia with notes on the higher classification of the family. Proc. R. Soc. Queensland 75:67–72.

Miles, P. W. 1972. The saliva of Hemiptera. Adv. Insect Physiol. 9:183–255.

Miller, N. C. E. 1938. A new subfamily of Malyasian Dysodiidae (Rhynchota). Ann. Mag. Nat. Hist., ser. 11, 2:498–510.

Miller, N. C. E. 1940. New genera and species of Malaysian Reduviidae. J. Fed. Malay St. Mus. 18:415–599.

Miller, N. C. E. 1941. New genera and species of Malaysian Reduviidae. J. Fed. Malay St. Mus. 18:601–804.

Miller, N. C. E. 1952. Three new subfamilies of Reduviidae (Hemiptera-Heteroptera). EOS 28:85–90.

Miller, N. C. E. 1953. Notes on the biology of the Reduviidae of Southern Rhodesia. Trans. Zool. Soc. Lond. 27:541–672, pls. 1–12.

Miller, N. C. E. 1954a. A new subfamily, new genera and species of Malaysian Reduviidae (Hem., Heteroptera). Idea 10:1–8.

Miller, N. C. E. 1954b. New genera and species of Reduviidae from Indonesia and the description of a new subfamily (Hemiptera-Heteroptera). Tijdschr. Entomol. 97:76–114.

Miller, N. C. E. 1955. New genera and species of Plataspidae Dallas, 1851 (Hemiptera-Heteroptera). Ann. Mag. Nat. Hist., ser. 12, 8:576–596.

Miller, N. C. E. 1956a. The Biology of the Heteroptera. Leonard Hill Books, London. 162 pp.

Miller, N. C. E. 1956b. Centrocneminae, a new sub-family of the Reduviidae (Hemiptera-Heterotera). Bull. Br. Mus. (Nat. Hist.), Entomology 4(6):219–283.

Miller, N. C. E. 1958. The presence of a stridulatory organ in *Rhytocoris* (Hemiptera-Heteroptera-Coreidae-Coreinae). Entomol. Mon. Mag. 94:238.

Miller, N. C. E. 1971. The Biology of the Heteroptera, 2nd ed., rev. Hampton, Classey. xiii + 206 pp.

Miller, P. L. 1961. Some features of the respiratory system of *Hyrdrocyrius columbiae* Spin. (Belostomatidae, Hemiptera). J. Insect Physiol. 6:243–271.

Miller, P. L. 1966. The function of haemoglobin in relation to the maintenance of neutral buoyancy in *Anisops pellucens*. J. Exp. Biol. 44:529–543.

Miller, R., and R. T. Schuh. 1995. On the feeding habits of *Clivinema*. J. N.Y. Entomol. Soc. 102:383–384.

Mitchell, P. L. 1980a. Combat and territorial defense of *Acanthocephala femorata* (Hemiptera: Coreidae). Ann. Entomol. Soc. Am. 73:404–408.

Mitchell, P. L. 1980b. Host plant utilization by leaf-footed bugs: an investigation of generalist feeding strategy. Ph.D. dissertation, University of Texas, Austin. 226 pp.

Miura, T., and R. Takahashi. 1987. Predation of *Microvelia pulchella* (Hemiptera: Veliidae) on mosquito larvae. J. Am. Mosquito Contr. Assoc. 3:91–93.

Miura, T., and R. Takahashi. 1988. Augmentation of *Notonecta unifasciata* eggs for suppressing *Culex tarsalis* larval population densities in rice fields. Proc. Calif. Mosquito Vec. Contr. Assoc. 55:45–49.

Miyamoto, S. 1952. Biology of *Helotrephes formosanus* Esaki et Miyamoto, with descriptions of larval stages. Sieboldia 1:1–10, pls. 1–3.

Miyamoto, S. 1953. Biology of *Microvelia diluta* Distant, with descriptions of its brachypterous form and larval stages. Sieboldia 1:113–133.

Miyamoto, S. 1955. On a special mode of locomotion utilizing surface tension at the water-edge in some semi-aquatic insects. Kontyû 23:45–52, pl. 7. [In Japanese with English summary].

Miyamoto, S. 1957. List of ovariole numbers in Japanese Heteroptera. Sieboldia 2:69–82.

Miyamoto, S. 1959. Additions and correction to my "List of Ovariole Numbers in Japanese Heteroptera." Sieboldia 2:121–123.

Miyamoto, S. 1961a. Comparative morphology of alimentary organs of Heteroptera, with the phylogenetic consideration. Sieboldia 2:197–259, pls. 20–49.

Miyamoto, S. 1961b. Insecta Japonica (Hemiptera: Gerridae). 55(1):1–40. Hokuryukan Publ. Co., Tokyo.

Miyamoto, S. 1965a. Hebridae in Formosa. Sieboldia 3:281–294.

Miyamoto, S. 1965b. Three new species of Cimicomorpha from Japan (Hemiptera). Sieboldia 3:271–280.

Miyamoto, S. 1967. Gerridae of Thailand and North Borneo taken by the Joint Thai-Japanese Biological Expedition 1961–62. Nature and Life in Southeast Asia 5:217–257.

Miyamoto, S. 1981. An estimation of the gross structure of heart for the higher classification of the Heteroptera. Rostria 33(suppl.):53–66.

Möller, H. 1921. Über *Lethocerus uhleri* Mont. Zool. Jahrb. Abt. Anat. Ontog. 42:43–90.

Monod, T., and J. Carayon. 1958. Observations sur les *Copium* (Hemipt. Tingidae) et leur action cécidogène sur les fleurs de *Teucrium* (Labiées). Arch. Zool. Exp. Gen. 95, Notes et Revue 1:1–31.

Monteith, G. B. 1969. A remarkable case of alary dimorphism in the Aradidae (Hemiptera) with a generic synonymy and a new species. J. Aust. Entomol. Soc. 8:87–94.

Monteith, G. B. 1980. Relationships of the genera of Chinamyersiinae, with description of a relict species from mountains of North Queensland (Hemiptera: Heteroptera: Aradidae). Pac. Insects 21:275–285.

Monteith, G. B. 1982a. Biogeography of the New Guinea Aradidae (Heteroptera). Monographiae Biologicae 42, pp. 645–657. Ed. J. L. Gressitt. Dr. W. Junk. The Hague.

Monteith, G. B. 1982b. Dry season aggregations of insects in Australian monsoon forests. Mem. Queensland Mus. 20:533–543.

Monteith, G. B. 1986. Obituary. Thomas Emmanuel Woodward. Aust. Entomol. Soc. News Bull. 22:55–57.

Mukhamedov, K. K. 1962. On the biology of the lucerne pest *Camptopus lateralis* Germ. (Heteroptera, Coreidae) in Turkmenia. Rev. Entomol. 41:310–312. [From Entomol. Obozr. 41:505–509].

Mulsant, E., and C. Rey. 1866. Histoire naturelle des punaises de France, vol. 2: Pentatomides. Paris.

Muraji, M., and F. Nakasuji. 1988. Comparative studies on life history traits of three-wing dimorphic water bugs, *Microvelia* spp. Westwood (Heteroptera: Veliidae). Res. Pop. Ecol. 30:315–327

Muraji, M., T. Miura, and F. Nakasuji. 1989. Phenological studies on the wing dimorphism of a semi-aquatic bug, *Microvelia douglasi* (Heteroptera: Veliidae). Res. Pop. Ecol. 31:129–138

Murphey, R. K. 1971. Sensory aspects of the control of orientation to prey by the waterstrider, *Gerris remigis*. Z. Vergl. Physiol. 72:168–185.

Myers, J. G. 1924. On the systematic position of the family Termitaphididae (Hemiptera, Heteroptera), with a description of a new genus and species from Panama. Psyche 31:259–278.

Myers, J. G. 1932. Observations on the family Termitaphididae (Hemiptera-Heteroptera) with the description of a new species from Jamaica. Ann. Mag. Nat. Hist., ser. 10, 9:366–372.

Myers, J. G., and G. Salt. 1926. The phenomenon of myrmecoidy, with new examples from Cuba. Trans. Entomol. Soc. Lond. 74:427–436, 1 pl.

Nelson, G., and N. Platnick. 1981. Systematics and Biogeography: Cladistics and Vicariance. Columbia University Press, New York. 567 pp.

Nieser, N. 1975. The water bugs of the Guyana region. Studies on the Fauna of Suriname and Other Guyanas, no. 59. M. Nijhoff, The Hague. 310 pp.

Nuorteva, P. 1956. Studies on the comparative anatomy of the salivary glands in four families of Heteroptera. Ann. Entomol. Fenn. 22:45–54.

Odhiambo, T. R. 1958a. The camouflaging habits of *Acanthaspis petax* Stål (Hem., Reduviidae) in Uganda. Entomol. Mon. Mag. 94:47.

Odhiambo, T. R. 1958b. Some observations on the natural history of *Acanthaspis petax* Stål (Hemiptera: Reduviidae) living in termite mounds in Uganda. Proc. R. Entomol. Soc. Lond. (A) 33:167–175.

Odhiambo, T. R. 1961. A study of some African species of the "Cyrtopeltis" complex (Hemiptera: Miridae). Rev. Entomol. Moçambique 4:1–36.

Odhiambo, T. R. 1962. Review of some genera of the subfamily Bryocorinae (Hemiptera: Miridae). Bull. Br. Mus. (Nat. Hist.), Entomol. 2(6):247–331.

O'Donnell, J. 1991. A survey of male genitalia in lethaeine genera (Heteroptera: Lygaeidae: Rhyparochrominae). J. N.Y. Entomol. Soc. 99:441–470.

O'Rourke, F. 1974. *Sweetocoris*, a new genus of Stygnocorini from South Africa with the description of fourteen new species (Hemiptera: Lygaeidae). Entomol. Soc. S. Afr. 37:215–250.

O'Rourke, F. 1975. A revision of the genus *Lasiosomus* and its relationship to the Stygnocorini (Hemiptera: Lygaeidae). Rev. Zool. Afr. 89:1–36.

Oshanin, B. 1906–1909. Verzeichnis der paläarktischen Hemipteren mit besonderer Berücksichtigung ihrer Verteilung im russischen Reiche. Annu. Mus. Zool. Acad. Imp. Sci., St. Petersburg, suppl., 11:i–lxxiv + 1–393 (1906); 13:395–586 (1908); 14:587–1087 (1909).

Oshanin, B. 1912. Katalog der paläarktischen Hemipteren (Heteroptera, Homoptera-Auchenorrhyncha und Psylloidea). Berlin, R. Friedlander & Sohn. 187 pp.

Oshanin, B. 1916. Vade mecum destiné à faciliter la détermination des Hémiptères. Catalogue systématique des faunes, des monographies et des synopsis traitant les Hétéroptères, les Cicadines et les Psyllides. Horae Soc. Entomol. Rossicae 42(2):1–106 + i–iv + 1.

Oshanin, B. 1922. Sur les genres de la tribu des *Stracharia* Put. (Heteroptera, Pentatomidae). Ezhegodnik Zool. Mus. Imperat. Akad. Nauk (Leningrad). 23:143–148.

O'Shea, R. 1980a. A generic revision of the Acanthocerini (Hemiptera: Coreidae: Coreinae). Stud. Neotrop. Fauna Environ. 15:57–80.

O'Shea, R. 1980b. A generic revision of the Nematopodini (Heteroptera: Coreidae: Coreinae). Stud. Neotrop. Fauna Environ. 15:197–225.

O'Shea, R., and C. W. Schaefer. 1978. The Mictini are not monophyletic (Hemiptera: Coreidae: Coreinae). Ann. Entomol. Soc. Am. 71:776–784.

Osuna, E. 1984. Monografía de la tribu Anisoscelidini (Hemiptera, Heteroptera, Coreidae) I. Revisión genérica. Bol. Entomol. Venezolana 3:77–148.

Otten, E. 1956. Heteroptera, Wanzen Halbflügler. *In:* P. Sorauer (ed.), Handbuch der Pflanzenkrankheiten, vol. 5 (5th ed., ed. H. Blunck), pp. 1–149, figs. 1–49. Paul Marey, Berlin. viii + 399 pp.

Owusu-Manu, E. 1977. Distribution and abundance of the cocoa shield bug, *Bathycoelia thalassina* (Hemiptera: Pentatomidae) in Ghana. J. Appl. Ecol. 14:331–341.

Paddock, F. B. 1918. Studies on the harlequin bug. Texas Agr. Expt. Stn. Bull. 227. 65 pp.

Palmén, J. A. 1914. Odo Moranal Reuter [Obituary and bibliography]. Acta Soc. Sci. Fenn. 46:1–44, 1 pl.

Palmer, L. S., and H. H. Knight. 1924. Carotin–the principal cause of the red and yellow colors in *Perillus bioculatus* (Fab.), and its biological origin from the lymph of *Leptinotarsa decemlineata* (Say). J. Biol. Chem. 59:443–449.

Papáček, M., P. Štys, and M. Tonner. 1988. A new subfamily of Helotrephidae (Heteroptera, Nepomorpha) from Southeast Asia. Acta Entomol. Bohemoslav. 85:120–152.

Papáček, M., P. Štys, and M. Tonner. 1989. A new genus and species of Helotrephidae from Afghanistan and Iran (Heteroptera: Nepomorpha). Vest. Cs. Spolec. Zool. 53:107–122.

Parker, A. H. 1965. The predatory behaviour and life history of *Pisilus tipuliformis* Fabricius (Hemiptera: Reduviidae). Entomol. Exp. Appl. 8:1–12.

Parker, A. H. 1969. The predatory and reproductive behaviour of *Rhinocoris bicolor* and *R. tropicus* (Hemiptera: Reduviidae). Entomol. Exp. Appl. 12:107–117.

Parshley, H. M. 1925. A Bibliography of the North American Hemiptera-Heteroptera. Smith College, Northampton, Mass. 252 pp.

Parsons, M. C. 1959. Skeleton and musculature of the head of *Gelastocoris oculatus* (Fabricius) (Hemiptera-Heteroptera). Bull. Mus. Comp. Zool. 122:3–53.

Parsons, M. C. 1960a. The nervous system of *Gelastocoris oculatus* (Fabricius) (Hemiptera: Heteroptera). Bull. Mus. Comp. Zool. 123:131–199.

Parsons, M. C. 1960b. Skeleton and musculature of the thorax of *Galastocoris oculatus* (Fabricius) (Hemiptera-Heteroptera). Bull. Mus. Comp. Zool. 122:299–357.

Parsons, M. C. 1962. Scolopophorous organs in the pterothorax and abdomen of *Gelastocoris oculatus* (Fabricius) (Hemiptera-Heteroptera). Bull. Mus. Comp. Zool. 127:207–236.

Parsons, M. C. 1969. The labium of *Aphelocheirus aestivalis* F. as compared to that of typical Naucoridae (Heteroptera). Can. J. Zool. 47:295–306.

Parsons, M. C. 1972. Morphology of the three anterior pairs of spiracles of *Belostoma* and *Ranatra* (aquatic Heteroptera: Belostomatidae, Nepidae). Can. J. Zool. 50:865–876.

Parsons, M. C. 1973. Morphology of the eighth abdominal spiracles of *Belostoma* and *Ranatra* (aquatic Heteroptera: Belostomatidae, Nepidae). J. Nat. Hist. 7:255–265.

Parsons, M. C., and R. J. Hewson. 1975. Plastral respiratory devices in adult *Cryphocricos* (Naucoridae: Heteroptera). Psyche 81:510–527.

Peet, W. B. 1979. Description and biology of *Nidicola jaegeri*, n. sp., from southern California (Hemiptera: Anthocoridae). Ann. Entomol. Soc. Am. 72:430–437.

Pendergrast, J. G. 1957. Studies on the reproductive organs of Heteroptera with a consideration of their bearing on classification. Trans. R. Entomol. Soc. Lond. 109:1-63.

Péricart, J. 1969. Description de quelques Hétéroptères Anthocoridae et Microphysidae nouveaux d'Union Soviétique. Ann. Soc. Entomol. Fr., n.s., 5:569–583.

Péricart, J. 1972. Hémiptères. Anthocoridae, Cimicidae, Microphysidae de l'Ouest-Paléarctique. Faune de l'Europe et du Bassin Méditerranéen, 7. Masson et Cie, Paris. 402 pp.

Péricart, J. 1974. Subdivision du genre *Piesma* (Hem. Piesmatidae) et remarques diverses. Ann. Soc. Entomol. Fr. 10:51–58.

Péricart, J. 1983. Hémiptères Tingidae Euro-Méditerranéens. Faune de France, 69. Fédération Française des Sociétés de Sciences Naturelles, Paris.

Péricart, J. 1984. Hémiptères Berytidae Euro-Méditerranéens. Faune de France 70. Fédération Française des Sociétés de Sciences Naturelles, Paris. 162 pp. + appendixes.

Péricart, J. 1987. Hémiptères Nabidae d'Europe occidentale et du Maghreb. Faune de France, 71. Fédération Française des Sociétés de Sciences Naturelles, Paris. 185 pp.

Péricart, J. 1990. Hémiptères Saldidae et Leptopodidae d'Europe et du Maghreb. Faune de France, 77. Fédération Française des Sociétés de Sciences Naturelles, Paris.

Péricart, J., and J. T. Polhemus. 1990. Un appareil stridulatoire chez les Leptopodidae de l'Ancien Monde (Heteroptera). Ann. Soc. Entomol. Fr. 26:9–17.

Pflugfelder, O. 1937. Vergleichend-anatomische, experimentelle und embryologische Untersuchungen über das Nervensystem und die Sinnesorgane der Rhynchoten. Zoologica, Stuttgart 34(93): 102 pp.

Phillips, D. M. 1970. Insect sperm: their structure and morphogenesis. J. Cell Biol. 44:243–277.

Picchi, V. D. 1977. A systematic review of the genus *Aneurus* Curtis of North and Middle America and the West Indies (Hemiptera: Aradidae). Quaest. Entomol. 13:155–308.

Pinto, C. 1927. Sphaeridopidae, nova família de Hemiptero Reduvioideae, com a descrição de um gênero e espécie nova. Bol. Biol., São Paulo 6:43–51.

Pinto, J. D. 1978. The parasitism of blister beetles by species of Miridae (Coleoptera: Meloidae; Hemiptera: Miridae). Pan-Pac. Entomol. 54:57–60.

Pluot, D. 1970. La spermathèque et les voies génitales femelles des Pyrrhocorides [Hemiptera]. Ann. Soc. Entomol. Fr., n.s., 6:777–807.

Poinar, G., Jr., and J. T. Doyen. 1992. A fossil termite bug, *Termitaradus protera* sp. n. (Hemiptera: Termitaphididae), from Mexican amber. Entomol. Scand. 23:89–93.

Poisson, R. 1924. Contribution à l'étude des Hémiptères aquatiques. Bull Biol., Paris 58:49–305.

Poisson, R. 1940. Contribution à la connaissance des espèces africaines du genre *Microvelia* Westwood. Rev. Fr. Entomol. 8:161–188.

Poisson, R. 1944. Contribution à la connaissance des espèces africaines du genre *Hebrus* Curtis 1833. Rev. Fr. Entomol. 10:89–112.

Poisson, R. 1949. Hémiptères aquatiques. Explor. Parc. Natl. Albert 58:1–94.

Poisson, R. 1951. Ordre de Hétéroptères. *In:* P. P. Grassé (ed.), Traité de zoologie vol. 10, pp. 1657–1803. Mason, Paris.

Poisson, R. 1957a. Chapter 8. Hemiptera Heteroptera: Hydrocorisae and Geocorisae-Gerroidea. South African Animal Life 4:327–373.

Poisson, R. 1957b. Hémiptères aquatique. Faune de France, 61. Fédération Française des Sociétés de Sciences Naturelles, Paris. 263 pp.

Poisson, R. 1959. Sur un nouveau représentant africain de la faune terrestre commensale des biotopes hyropétriques: *Madeovelia guineensis* nov. gen., n. sp. (Insectes, Hétéroptères). Bull. Inst. Fr. Afr. Noire (A) 21:658–663.

Poisson, R. A. 1965. Catalogue des insectes Hétéroptères Gerridae Leach, 1807, Africano-malgaches. Bull. Inst. Fr. Afr. Noire, ser. A, 27:1466–1503.

Poisson, R., and A. Poisson. 1931. Les Hémiptères de Normandie. Géocorises (4e liste des espèces et observations diverses). Bull. Soc. Linn. Norm. 3(1932):1–19.

Polhemus, D. A. 1990a. Heteroptera of Aldabra Atoll and nearby islands, western Indian Ocean. Part 1. Marine Heteroptera (Insecta): Gerridae, Hermatobatidae, Saldidae and Omaniidae, with notes on ecology and insular biogeography. Atoll Res. Bull., no. 345. 16 pp., 1 map, 9 figs.

Polhemus, D. A. 1990b. A revision of the genus *Metrocoris* Mayr (Heteroptera: Gerridae) in the Malay Archipelago and the Philippines. Entomol. Scand. 21:1–28.

Polhemus, D. A. 1992. The first records of the families Ochteridae and Hebridae (Heteroptera) from the granitic Seychelles, with descriptions of two new species. J. N.Y. Entomol. Soc. 100:418–423.

Polhemus, D. A., and J. T. Polhemus. 1988. The Aphelocheirinae of Tropical Asia (Heteroptera: Naucoridae). Raffles Bull. Zool. 36:167–300.

Polhemus, J. T. 1974. The *austrina* group of the genus *Microvelia* (Hemiptera: Veliidae). Great Basin Nat. 34:207–217.

Polhemus, J. T. 1981. African Leptopodomorpha (Hemiptera: Heteroptera): a checklist and descriptions of new taxa. Ann. Natal Mus. 24:603–619.

Polhemus, J. T. 1985. Shore Bugs (Heteroptera, Hemiptera; Saldidae). A World Overview and Taxonomy of Middle American Forms. The Different Drummer, Englewood, Colo. 252 pp.

Polhemus, J. T. 1990a. Miscellaneous studies on the genus *Rhagovelia* Mayr (Heteroptera: Veliidae) in Southeast Asia and the Seychelles Islands, with keys and descriptions of new species. Raffles Bull. Zool. 38:65–75.

Polhemus, J. T. 1990b. Surface wave communication in water striders: field observations of unreported taxa (Gerridae, Veliidae: Heteroptera). J. N.Y. Entomol. Soc. 98:383–384.

Polhemus, J. T. 1990c. A new tribe, a new genus and three new species of Helotrephidae (Heteroptera) from Southeast Asia, with a world checklist. Acta Entomol. Bohemoslov. 87:45–63.

Polhemus, J. T. 1991a. A new and primitive genus of Cryphocricinae (Heteroptera: Naucoridae). Pan-Pac. Entomol. 67:119–123.

Polhemus, J. T. 1991b. Three new species of *Saldoncula* Brown from the Malay Archipelago, with a key to the known species (Heteroptera: Saldidae). Raffles Bull. Zool. 39:153–160.

Polhemus, J. T. 1994. Stridulatory mechanisms in aquatic and semiaquatic Heteroptera. J. N.Y. Entomol. Soc. 102:270–274.

Polhemus, J. T., and N. M. Andersen. 1984. A revision of *Amemboa* Esaki with notes on the phylogeny and ecological evolution of eotrechine water striders (Insecta, Hemiptera, Gerridae). Steenstrupia 10:65–111.

Polhemus, J. T., and H. C. Chapman. 1979a. Family Saldidae. *In:* A. S. Menke (ed.), The Semiaquatic and Aquatic Hemiptera of California (Heteroptera: Hemiptera), pp. 16–33. Univ. Calif. Publ., Bull. Calif. Insect Surv., vol. 21.

Polhemus, J. T., and H. C. Chapman. 1979b. Family Macroveliidae. *In:* A. S. Menke (ed.), The Semiaquatic and Aquatic Hemiptera of California (Heteroptera: Hemiptera), pp. 46–48. Univ. Calif. Publ., Bull. Calif. Insect Surv., vol. 21.

Polhemus, J. T., and H. C. Chapman. 1979c. Family Mesoveliidae. *In:* A. S. Menke (ed.), The Semiaquatic and Aquatic Hemiptera of California (Heteroptera: Hemiptera), pp. 39–42. Univ. Calif. Publ., Bull. Calif. Insect Surv., vol. 21.

Polhemus, J. T., and H. C. Chapman. 1979d. Family Veliidae. *In:* A. S. Menke (ed.), The Semiaquatic and Aquatic Hemiptera of California (Heteroptera: Hemiptera), pp. 49–57. Univ. Calif. Publ., Bull. Calif. Insect Surv., 21.

Polhemus, J. T., and H. C. Chapman. 1979e. Family Gerridae. *In:* A. S. Menke (ed.), The Semiaquatic and Aquatic Hemiptera of California (Heteroptera: Hemiptera), pp. 58–69. Univ. Calif. Publ., Bull. Calif. Insect, Surv., vol. 21.

Polhemus, J. T., and J. L. Herring. 1979. A further description of *Hermatobates breddini* Herring, and a new record for Cuba (Hemiptera: Hermatobatidae). Proc. Entomol. Soc. Wash. 81:253–254.

Polhemus, J. T., and P. B. Karunaratne. 1993. A review of the genus *Rhagadotarsus*, with descriptions of three new species (Heteroptera: Miridae). Raffles Bull. Zool. 41:95–112.

Polhemus, J. T., and D. A. Polhemus. 1982. Notes on Neotropical Naucoridae. II. A new species of *Ambrysus* and review of the genus *Potamocoris* (Hemiptera). Pan-Pac. Entomol. 58:326–329.

Polhemus, J. T., and D. A. Polhemus. 1984. Notes on Neotropical Veliidae (Hemiptera). VI. Revision of the genus *Euvelia* Drake. Pan-Pac. Entomol. 60:55–62.

Polhemus, J. T., and D. A. Polhemus. 1988. A new genus of foam-inhabiting Veliidae (Heteroptera) from western Madagascar. J. N.Y. Entomol. Soc. 96:274–280.

Polhemus, J. T., and D. A. Polhemus. 1989. Zoogeography, ecology, and systematics of the genus *Rhagovelia* (Heteroptera: Veliidae) in Borneo, Celebes, and the Moluccas. Insecta Mundi 2:161–230.

Polhemus, J. T., and D. A. Polhemus. 1991a. A review of the veliid fauna of bromeliads, with a key and description of a new species (Heteroptera: Veliidae). J. N.Y. Entomol. Soc. 99:204–216.

Polhemus, J. T., and D. A. Polhemus. 1991b. Distributional data and new synonymy for species of *Halobates* Escholtz (Heteroptera: Gerridae) occurring on Aldabra and nearby atolls, western Indian Ocean. J. N.Y. Entomol. Soc. 99:217–223.

Polhemus, J. T., and D. A. Polhemus. 1991c. A revision of the Leptopodomorpha (Heteroptera) of Madagascar and nearby Indian Ocean islands. J. N.Y. Entomol. Soc. 99:496–526.

Polhemus, J. T., and D. A. Polhemus. 1993. The Trepobatinae (Heteroptera: Gerridae) of New Guinea and surrounding regions, with a review of the world fauna. Part 1. Tribe Metrobatini. Entomol. Scand. 24:241–284.

Polhemus, J. T., and R. T. Schuh. 1995. A new species of *Leotichius* from Bali, with notes on immature states and habitat (Heteroptera, Leptopodidae). J. N.Y. Entomol. Soc. 102:367–373.

Polhemus, J. T., and P. J. Spangler. 1989. A new species of *Rheumatobates* Bergroth from Ecuador, and disribution of the genus (Hemiptera: Gerridae). Proc. Entomol. Soc. Wash. 91:421–428.

Popham, E. J., M. T. Bryant, and A. A. Savage. 1984. The function of the abdominal strigil in male corixid bugs. J. Nat. Hist. 118:441–444.

Popov, Y. 1971. Historical development of Hemiptera infraorder Nepomorpha (Heteroptera). Trudy Paleo. Inst. Acad. Sci. USSR 129:230 pp., pls. I–IX. [In Russian].

Popov, Y. 1985. Jurrasic Heteroptera and Peloridiina of southern Siberia and western Mongolia. *In:* A. P. Rasnitsyn (ed.), Jurassic Insects of Siberia and Mongolia, pp. 28–47. Trudy Paleontol. Inst., Moscow. 211:192 pp. [In Russian].

Poppius, B. 1912. Die Miriden der Aethiopischen Region. I. Mirina, Cylapina, Bryocorina. Acta Soc. Sci. Fenn. 41:203 pp.

Poppius, B. 1914. Die Miriden der Aethiopischen Region. II. Macrolophinae, Heterotominae, Phylinae. Acta Soc. Sci. Fenn. 44:136 pp.

Poppius, B., and E. Bergroth. 1921. Beiträge zur Kenntnis der myrmecoiden Heteropteren. Ann. Hist. Natl. Mus. Hung. 18:31–88.

Prager, J. 1973. Die Horschwelle des mesothorakalen Tympanalorgans von *Corixa punctata* Ill. (Heteroptera, Corixidae). J. Comp. Physiol. 86:55–58.

Prager, J. 1976. Das mesothorakale Tympanalorgan von *Corixa punctata* Ill. (Heteroptera, Corixidae). J. Comp. Physiol. 110:33–50.

Prager, J., and R. Streng. 1982. The resonance properties of the physical gill of *Corixa punctata* and their significance in sound reception. J. Comp. Physiol. 148:323–335.

Pupedis, R. J., C. W. Schaefer, and P. Duarte Rodrigues. 1985. Postembryonic changes in some sensory structures of the Tingidae (Hemiptera: Heteroptera). J. Kans. Entomol. Soc. 58:277–289.

Putchkov, P. 1987. Assassin bugs. *In:* Fauna of the Ukraine, vol. 21, p. 5. Naukova Dumka, Kiev. 248 pp. [In Ukrainian].

Putchkov, V. G. 1962. Coreoidea. *In:* Fauna of the Ukraine, vol. 21, p. 2. Zool. Inst. Acad. Sci., Kiev. 338 pp. [In Ukrainian].

Putchkov, V. G. 1969. Lygaeidae. *In:* Fauna of the Ukraine, vol. 21, p. 3. Akad. Nauk., Ukraine RsR. 388 pp. [In Ukrainian].

Putchkov, V. G. 1974. Berytidae, Pyrrhocoridae, Piesmatidae, Aradidae and Tingidae. *In:* Fauna of the Ukraine, vol. 21, p. 4. Zool. Instit. Acad. Sci., Kiev. 332 pp. [In Ukrainian].

Putchkov, V. G. 1986. Opredeliteli po faune SSSR. 146. Bugs of the family Rhopalidae (Heteroptera) of the fauna of USSR. Nauka, Leningrad. 132 pp. [In Russian].

Putchkov, V. G., and P. V. Putchkov. 1985. A Catalog of Assassin-Bug Genera of the World (Heteroptera, Reduviidae). [Published by the authors], Kiev. 137 pp.

Putchkov, V. G., and P. V. Putchkov. 1986–1989. A Catalog of the Reduviidae (Heteroptera) of the World. Vinity, Lyubertsy. 6 vols.

Putchkova, L. V. 1979. Adaptive features of leg structures in the Heteroptera. Entomol. Rev. 58:189–196.

Puton, J.-P. A. 1881. Enumération des Hémiptères récoltés en Syrie par M. Abeille de Perrin avec la description de espèces nouvelles. Mitt. Schweiz. Entomol. Ges. 6:119–129.

Puton, J.-P. A. 1887. Hémiptères nouveaux ou peu connus de la faune Paléarctique. Rev. Entomol., Caen 6:96–105.

Quentin, R. M. 1983. L'oeuvre scientifique d'André Villiers. Entomologiste 39:167–208.

Radinovsky, S. 1964. Cannibal of the pond. Natural History 73:16–25.

Rastogi, S. C., and K. Kumari. 1962. Observations on the life history of the red pumpkin bug, *Coridius (Aspongopus) janus* (Fabr.) (Heteroptera: Dinidorinae). Zool. Poloniae 12(1):69–77.

Rawat, B. L. 1939. Notes on the anatomy of *Naucoris cimicoides* L. (Hemiptera: Heteroptera). Zool. Jahrb. (Abt. Anat.) 65:535–600.

Readio, J., and M. H. Sweet. 1982. A review of the Geocorinae of the United States East of the 100th Meridian (Hemiptera: Lygaeidae). Misc. Publ. Entomol. Soc. Am. 12:1–91.

Readio, P. A. 1927. Studies on the biology of the Reduviidae of America north of Mexico. Univ. Kans. Sci. Bull. 17(1):5–291.

Reissig, W. H., E. A. Heinrichs, J. A. Litsinger, K. Moody, L. Faedler, T. W. Maw, and A. T. Bergen. 1985. Illustrated guide of integrated pest management in rice in tropical Asia. Int. Rice Res. Inst., Los Baños, Laguna, Philippines. xi + 441 pp.

Remane, R. 1953. Zur Systematik der Untergattung *Reduviolus* (Hem. Het. Nabidae). Zool. Anz. 150:190–199.

Remane, R. 1962. Kenntnis der Gattung *Nabis* Latr. (Hem. Het. Nabidae). Mem. Soc. Entomol. Ital. 41:5–14.

Remane, R. 1964. Weitere Beiträge zur Kenntnis der Gattung *Nabis* Latr. (Hemiptera-Heteroptera, Nabidae). Zool. Beitr. 10:253–314.

Remold, H. 1962. Über die biologische Bedeutung der Duftdrüsen bei den Landwanzen (Geocorisae). Z. Vgl. Physiol. 45:636–694.

Remold, H. 1963. Scent-glands of land bugs, their physiology and biological function. Nature 198:764–768.

Ren, S.-Z. 1992. An Iconograph of Hemiptera-Heteroptera Eggs in China. Science Press, Beijing. 118 pp., 80 pls. [In Chinese].

Reuter, O. M. 1875. Hemiptera Gymnocerata Scandinaviae et Fenniae diposuit et descripsit. Pars I. Cimicidae (Capsina). Acta Faun. Flora Fenn. 206 pp.

Reuter, O. M. 1878–1896. Hemiptera Gymnocerata Europae. Acta Soc. Sci. Fenn. Part 1, 1878 (as separate) [8:1–188, 8 pls., 1885]; Part 2, 1879 (as separate) [8:193–312, 5 pls., 1885]; Part 3, 1883 (as separate) [8:313–496, 5 pls., 1885]; Part 4, 1891, 23:1–179, 6 pls.; Part 5, 1896, 33(2):1–392, 10 pls.

Reuter, O. M. 1884. Monographia Anthocoridarum orbis terrestris. Acta Soc. Sci. Fenn. 14:555–758.

Reuter, O. M. 1891. Monographia Ceratocombidarum orbis terrestris. Acta Soc. Sci. Fenn. 19(6):27 pp., 1 pl.

Reuter, O. M. 1895. Species palaearcticae generis *Acanthia* Fabr., Latr. Acta Soc. Sci. Fenn. 21(2):1–58, 1 pl.

Reuter, O. M. 1905. Hemipterologische Spekulationen. I. Die Klassifikation der Capsiden. Festschr. Palmen 1:58 pp.

Reuter, O. M. 1908. Bemerkungen über Nabiden nebst Beschreibung neuer Arten. Mem. Soc. Entomol. Belg. 15:87–130.

Reuter, O. M. 1910. Neue Beiträge zur Phylogenie und Systematik der Miriden nebst einleitenden Bemerkungen über die Phylogenie der Heteropteren-Familien. Acta Soc. Sci. Fenn. 37:169 pp., 1 pl.

Reuter, O. M. 1912a. Bemerkungen über mein neues Heteropterensystem. Ofv. Finska Vet.-Soc. Forh. 54(12):54 pp.

Reuter, O. M. 1912b. Zur generischen Teilung der paläarktischen und nearktischen Acanthiaden. Ofv. Finska Vet.-Soc. Forh. 54A (12):24 pp.

Reuter, O. M., and B. Poppius. 1910. Monographia Nabidarum orbis terrestris. Acta Soc. Sci. Fenn. 37:1–62, 1 pl.

Ribaut, H. 1923. Etude sur le genre *Triphleps* (Heteroptera-Anthocoridae). Bull. Soc. Hist. Nat. Toulouse 51:522–538.

Ribeiro, S. T. 1989. Group effects and aposematism in *Jadera haematoloma* (Hemiptera: Rhopalidae). Ann. Entomol. Soc. Am. 83:466–475.

Rieger, C. 1976. Skelett und Muskulatur des Kopfes und Prothorax von *Ochterus marginatus* Latreille. Zoomorphologie 83:109–191.

Rodrigues, P. Duarte, R. J. Pupedis, and C. W. Schaefer. 1982. Taxonomic differences in some sensory structures of the Tingidae (Hemiptera: Heteroptera). J. Kans. Entomol. Soc. 55:117–124.

Rolston, L. H. 1974. Revision of the genus *Euschistus* in Middle America (Hemiptera: Pentatomidae; Pentatomini). Entomol. Am. 48:1–103.

Rolston, L. H. 1981. Ochlerini, a new tribe in Discocephalinae. (Hemiptera: Pentatomidae). J. N.Y. Entomol. Soc. 89:40–42.

Rolston, L. H. 1982a. A brachypterous species of *Alathetus* from Haiti (Hemiptera: Pentatomidae). J. Kans. Entomol. Soc. 55:156–158.

Rolston, L. H. 1982b. A revision of the *Euschistus* Dallas subgenus *Lycipta* Stål (Hemiptera: Pentatomidae). Proc. Entomol. Soc. Wash. 84:281–296.

Rolston, L. H. 1984. Key to males of the nominate subgenus of *Euschistus* in South America, with descriptions of three new species (Hemiptera: Pentatomidae). J. N.Y. Entomol. Soc. 92:352–364.

Rolston, L. H., and R. Kumar. 1975 (1974). Two new genera and two new species of Acanthosomatidae (Hemiptera) from South America, with a key to the genera of the Western Hemisphere. J. N.Y. Entomol. Soc. 82:271–278.

Rolston, L. H., and F. J. D. McDonald. 1979. Keys and diagnoses for the families of Western Hemisphere Pentatomoidea, subfamilies of Pentatomidae and tribes of Pentatominae (Hemiptera). J. N.Y. Entomol. Soc. 87:189–207.

Rolston, L. H., and F. J. D. McDonald. 1981. Conspectus of Pentatomini genera of the Western Hemisphere. Part 2. (Hemiptera: Pentatomidae). J. N.Y. Entomol. Soc. 88:257–282.

Rolston, L. H., and F. J. D. McDonald. 1984. A conspectus of Pentatomini of the Western Hemisphere. Part 3. (Hemiptera: Pentatomidae). J. N.Y. Entomol. Soc. 92:69–86.

Rolston, L. H., F. J. D. McDonald, and D. B. Thomas, Jr. 1980. A conspectus of Pentatomini genera of the Western Hemisphere. Part 1. (Hemiptera: Pentatomidae). J. N.Y. Entomol. Soc. 88:120–132.

Rolston, L. H., R. L. Aalbu, M. J. Murphy, and D. A. Rider. 1993. A catalog of the Tessaratomidae of the world. Papua New Guinea J. Agric. For. Fish. 36:36–108.

Ronderos, R. A. 1960. Polyctenidae americanos. I (Hemiptera-Heteroptera). Actas Trab. 1st Congr. Sudamer. Zool. 1959, La Plata. 3:175–186, 2 pls., 1 map.

Ronderos, R. A. 1962. Nuevos aportes para el conocimiento de los Polyctenidae Americanos (Hemiptera). An. Inst. Nac. Microbiol. 1:67–76.

Rose, H. A. 1965. Two new species of Thaumastocoridae (Hemiptera: Heteroptera) from Australia. Proc. R. Entomol. Soc. Lond. (B) 34:141–144.

Roth, L. M. 1961. A study of odoriferous glands of *Scaptocoris divergens* (Hemiptera: Cydnidae). Ann. Entomol. Soc. Am. 54:900–911.

Rothschild, G. H. L. 1970. Observations on the ecology of the rice ear bug, *Leptocorisa oratorius* (F.) (Hemiptera: Alydidae) in Sarawak (Malaysian Borneo). J. Appl. Ecol. 7:147–167.

Ruckes, H. 1947. Notes and keys on the genus *Brochymena* (Pentatomidae, Heteroptera). Entomol. Am. 26:143-238.

Ruhoff, F. 1968. Bibliography and index to scientific contributions of Carl J. Drake for the years 1914–1967. U.S. Natl. Mus. Bull. 267:81 pp.

Ryckman, R. E. 1984. The Triatominae of North and Central America and the West Indies: a checklist with synonymy (Hemiptera: Reduviidae: Triatominae). Bull. Soc. Vector Ecol. 9:71–844.

Ryckman, R. E. 1986a. The Triatominae of South America: a checklist with synonymy (Hemiptera: Reduviidae: Triatominae). Bull. Soc. Vector Ecol. 11:199–208.

Ryckman, R. E. 1986b. The vertebrate hosts of the Triatominae of North and Central America and the West Indies (Hemiptera: Reduviidae: Triatominae). Bull. Soc. Vector Ecol. 11:221–241.

Ryckman, R. E., and E. F. Archbold. 1981. The Triatominae and Triatominae-borne trypanosomes of Asia, Africa, Australia and the East Indies. Bull. Soc. Vector Ecol. 6:143–166.

Ryckman, R. E., and C. M. Blankenship. 1984. The Triatominae and Triatominae-borne trypanosomes of North and Central America and the West Indies: a bibliography with index. Bull. Soc. Vector Ecol. 9:112–430.

Ryckman, R. E., and M. A. Casidin. 1977. The Polyctenidae of the world, a checklist with bibliography. Calif. Vector News 24(7–8):25–31.

Ryckman, R. E., and R. D. Sjogren. 1980. A catalogue of the Polyctenidae. Bull. Soc. Vector Ecol. 5:1–22.

Ryckman, R. E., and J. L. Zackrison. 1987. A bibliography to Chagas' disease, the Triatominae and Triatominae-borne trypanosomes of South America (Hemiptera: Reduviidae: Triatominae). Bull. Soc. Vector Ecol. 12:1–4464.

Ryckman, R. E., D. G. Bentley, and E. F. Archbold. 1981. The Cimicidae of the Americas and oceanic islands, a checklist and bibliography. Bull. Soc. Vector Ecol. 6:93–142.

Sachtleben, H. 1961. Nachträge zu "Walter Horn und Ilse Kahle: Über entomologische Sammlungen." Beitr. Entomol. 11:481–540.

Sailer, R. I. 1944. The genus *Solubea* (Heteroptera: Pentatomidae). Proc. Entomol. Soc. Wash. 46:105–127.

Salas-Aguilar, J., and L. E. Ehler. 1977. Feeding habits of *Orius tristicolor*. Ann. Entomol. Soc. Am. 70:60–62.

Samy, O. 1969. A revision of the African species of *Oxycarenus* (Hem. Lyg.). Trans. R. Entomol. Soc. Lond. 121:79–165.

Sands, D. P. A. 1978. The biology and ecology of *Leptocorisa* (Hemiptera: Alydidae) in Papua New Guinea. Res. Bull. 18: Dept. Primary Industries, Port Moresby.

Savage, A. A. 1989. Adults of British Aquatic Hemiptera Heteroptera, pp. 1–173. Freshwater Biological Association, Scientific Publication 50.

Say, T. 1831. Descriptions of new species of heteropterous Hemiptera of North America. New Harmony. 39 pp.

Schaefer, C. W. 1964. The morphology and higher classification of the Coreoidea (Hemiptera-Heteroptera): Parts I and II. Ann. Entomol. Soc. Am. 57:670–684.

Schaefer, C. W. 1965. The morphology and higher classification of the Coreoidea (Hemiptera-Heteroptera). Part III. The families Rhopalidae, Alydidae, and Coreidae. Misc. Publ. Entomol. Soc. Am. 5(1):1–76.

Schaefer, C. W. 1966a. The morphology and higher systematics of the Idiostolinae (Hemiptera: Lygaeidae). Ann. Entomol. Soc. Am. 59:602–613.

Schaefer, C. W. 1966b. The nymphs of *Idiostolus insularis* Berg (Hemiptera: Idiostolidae). Occ. Pap. Univ. Conn., Biol. Sci. Ser. 1:13–23.

Schaefer, C. W. 1969. Morphological and phylogenetic notes on the Thaumastocoridae (Hemiptera-Heteroptera). J. Kans. Entomol. Soc. 42:251–256.

Schaefer, C. W. 1972a. A cladistic analysis of the Piesmatinae (Hemiptera: Heteroptera: Piesmatidae). Ann. Entomol. Soc. Am. 65:1258–1261.

Schaefer, C. W. 1972b. Clades and grades in the Alydidae. J. Kans. Entomol. Soc. 45:135–141.

Schaefer, C. W. 1975. Heteropteran trichobothria (Hemiptera: Heteroptera). Int. J. Insect. Morph. Embryol. 4:193–264.

Schaefer, C. W. 1980a. The host plants of the Alydinae, with a note on heterotypic feeding aggregations (Hemiptera: Coreoidea: Alydidae). J. Kans. Entomol. Soc. 53:115–122.

Schaefer, C. W. 1980b. The sound producing structures of some primitive Pentatomoidea (Hemiptera: Heteroptera). J. N.Y. Entomol. Soc. 88:230–235.

Schaefer, C. W. 1981a. The morphology and relationships of the Stenocephalidae and Hyocephalidae (Hemiptera: Heteroptera: Coreoidea). Ann. Entomol. Soc. Am. 74:83–95.

Schaefer, C. W. 1981b. Improved cladistic analysis of the Piesmatidae and consideration of known host plants. Ann. Entomol. Soc. Am. 74:536–539.

Schaefer, C. W. 1981c. Genital capsules, trichobothria and host plants of the Podopinae (Pentatomidae). Ann. Entomol. Soc. America 74:590–601.

Schaefer, C. W. 1982. The genital capsule of the Hydarinae (Hemiptera: Coreidae). Uttar Pradesh J. Zool. 2:1–6.

Schaefer, C. W. 1993. Notes on the morphology and family relationships of Lestoniidae (Hemiptera: Heteroptera). Proc. Entomol. Soc. Wash. 95:453–456.

Schaefer, C. W., and P. D. Ashlock. 1970. A new genus and new species of Saileriolinae (Hemiptera: Urostylidae). Pac. Insects 12:629–639.

Schaefer, C. W., and N. P. Chopra. 1982. Cladistic analysis of the Rhopalidae, with a list of food plants. Ann. Entomol. Soc. Am. 75:224–233.

Schaefer, C. W., and P. L. Mitchell. 1983. Food plants of the Coreoidea (Hemiptera: Heteroptera). Ann. Entomol. Soc. America 76:591–615.

Schaefer, C. W., and R. O'Shea. 1979. Host plants of three coreine tribes (Hemiptera: Heteroptera: Coreidae). Ann. Entomol. Soc. Am. 72: 519–523.

Schaefer, C. W., and R. J. Pupedis. 1981. A stridulatory device in certain Alydinae (Hemiptera: Heteroptera: Alydidae). J. Kans. Entomol. Soc. 54:143–152.

Schaefer, C. W., and R. I. Sailer. 1980. Obituary. Hsiao Tsai-yu. Proc. Entomol. Soc. Wash. 82:714–721.

Schaefer, C. W., and D. B. Wilcox. 1969. Notes on the morphology, taxonomy and distribution of the Idiostolidae (Hemiptera-Heteroptera). Ann. Entomol. Soc. Am. 62:485–502.

Schaefer, C. W., and D. B. Wilcox. 1971. A new species of Thaumastellidae (Hemiptera: Pentatomoidea) from South Africa. J. Entomol. Soc. S. Afr. 34:207–214.

Schaefer, C. W., W. R. Dolling, and S. Tachikawa. 1988. The shieldbug genus *Parastrachia* and its position within the Pentatomoidea (Insecta: Hemiptera). Zool. J. Linnean Soc. 93:283–311.

Schaefer, C. W., J. C. Schaffner, and I. Ahmad. 1989. The *Alydus*-group, with notes on the Alydine genital capsule (Hemiptera: Heteroptera: Alydidae). Ann. Entomol. Soc. Am. 82:500–507.

Schaefer, C. W., L.-Y. Zheng, and S. Tachikawa. 1991. A review of *Parastrachia* (Hemiptera: Cydnidae: Parastraachiinae). Orient. Insects 25:131–144.

Schaffner, J. C. 1964. A taxonomic revision of certain genera of the tribe Alydini (Heteroptera, Coreidae). Ph.D. dissertation, Iowa State University, Ames.

Schaffner, J. C., and P. S. F. Ferreira. 1989. *Froeschnerana mexicana*, a new genus and species of Deraeocorinae from Mexico (Heteroptera: Miridae). J. N.Y. Entomol. Soc. 97:100–104.

Schell, D. V. 1943. The Ochteridae (Hemiptera) of the Western Hemisphere. J. Kans. Entomol. Soc. 16:29–47.

Schilling, P. S. 1829. Hemiptera Heteroptera Silesiae systematice disposuit. Beitr. Entomol. 1:34–92.

Schiødte, J. C. 1869. Nogle nya hovedsaetninger af Rhynchoternes morphologi og systematik. Nat. Tidskr., ser. 3, 4:237–266.

Schiødte, J. M. C. 1870. On some new fundamental principles of the morphology and classification of Rhynchota. Ann. Mag. Nat. Hist., ser. 4, 6:225–249.

Schlee, D. 1969. Morphologie und Symbiose; ihre Beweiskraft für die Verwandtschaftsbeziehungen der Coleorrhyncha. Stuttg. Beitr. Naturk. 210:1–27.

Schmitz, G. 1968. Monographie des espèces africaines du genre

Schouteden, H. 1903. Rhynchota Aethiopica. I. Scutellerinae et Graphosomatinae. Faun. Entomol. Afr. Trop. 1(1):1–132. Ann. Mus. Congo Zool., ser. 3.

Schouteden, H. 1904–1906. Heteroptera Fam. Pentatomidae Subfam. Scutellerinae. *In:* M. P. Wytsman (ed.), Genera Insectorum, fasc. 24, pp. 1–100. Brussels.

Schouteden, H. 1905a. Faun. Entomol. Afr. Trop. vol. 1, fasc. 2: Rhynchota Aethiopica II, Arminae et Tessaratominae. Ann. Mus. Congo, Zool., ser. 3.

Schouteden, H. 1905b. Heteroptera Fam. Pentatomidae Subfam. Graphosomatinae. *In:* M. P. Wytsman (ed.), Genera Insectorum, fasc. 30, pp. 1–46. Brussels.

Schouteden, H. 1907. Heteroptera Fam. Pentatomidae Subfam. Asopinae (Amyoteinae). *In:* M. P. Wytsman (ed.), Genera Insectorum, fasc. 52, pp. 1–82. Brussels.

Schouteden, H. 1913. Heteroptera Fam. Pentatomidae Subfam. Dinidorinae. *In:* M. P. Wytsman (ed.), Genera Insectorum, fasc. 153, pp. 1–19. Brussels.

Schouteden, H. 1931. Catalogues raisonnés de la faune entomologique de Congo Belge. Hémiptères-Réduviidés. Ann. Mus. Congo Belge, Zool., ser. 2, sec. 2, vol. 1, pp. 91–161.

Schouteden, H. 1932. Catalogues raisonnés de la faune entomologique de Congo Belge. Hémiptères-Réduviidés. Ann. Mus. Congo Belge, Zool., ser. 2, sec. 2, vol. 1, pp. 162–218, pl. IV.

Schuh, [R.] T. 1967. The shore bugs (Hemiptera, Saldidae) of the Great Lakes region. Contr. Am. Entomol. Inst. 2(2):35 pp.

Schuh, R. T. 1974. The Orthotylinae and Phylinae (Hemiptera: Miridae) of South Africa with a phylogenetic analysis of the antmimetic tribes of the two subfamilies for the world. Entomol. Am. 47:1–332.

Schuh, R. T. 1975a. The structure, distribution, and taxonomic importance of trichobothria in the Miridae (Hemiptera). Am. Mus. Novit. 2585:26 pp.

Schuh, R. T. 1975b. Wing asymmetry in the thaumastocorid *Discocoris drakei* (Hemiptera). Rev. Peruana Entomol. 18:12–13.

Schuh, R. T. 1976. Pretarsal structure in the Miridae (Hemiptera) with a cladistic analysis of relationships within the family. Am. Mus. Novit. 2601:39 pp.

Schuh, R. T. 1979. [Review of] *Evolutionary Trends in Heteroptera. Part II. Mouthpart-structures and Feeding Strategies*, by R. H. Cobben. Syst. Zool. 28:653–656.

Schuh, R. T. 1984. Revision of the Phylinae (Hemiptera: Miridae) of the Indo-Pacific. Bull. Am. Mus. Nat. Hist. 177:1–462.

Schuh, R. T. 1986a. *Schizopteromiris*, a new genus and four new species of coleopteroid cylapine Miridae from the Australian region (Heteroptera). Ann. Soc. Entomol. Fr., n.s., 22:241–246.

Schuh, R. T. 1986b. The influence of cladistics on heteropteran classification. Ann. Rev. Entomol. 31:67–93.

Schuh, R. T. 1989. Review of *Daleapidea* Knight (Heteroptera: Miridae: Orthotylinae: Orthotylini). J. N.Y. Entomol. Soc. 97:159–166.

Schuh, R. T. 1991. Phylogenetic, host, and biogeographic analyses of the Pilophorini (Heteroptera: Miridae: Phylinae). Cladistics 7:157–189.

Schuh, R. T., and L. H. Herman. 1988. Biography and bibliography. Petr Wolfgang Wygodzinsky (1916–1987). J. N.Y. Entomol. Soc. 96:227–244.

Schuh, R. T., and J. D. Lattin. 1980. *Myrmecophyes oregonensis*, a new species of Halticini (Hemiptera, Miridae) from the western United States. Am. Mus. Novit. 2697:11 pp.

Schuh, R. T., and J. T. Polhemus. 1980a. *Saldolepta kistnerorum* new genus and new species from Ecuador (Hemiptera, Leptopodomorpha), the sister group of *Leptosalda chiapensis*. Am. Mus. Novit. 2698:5 pp.

Schuh, R. T., and J. T. Polhemus. 1980b. Analysis of taxonomic congruence among morphological, ecological and biogeographic data sets for the Leptopodomorpha (Hemiptera). Syst. Zool. 29:1–26.

Schuh, R. T., and M. D. Schwartz. 1984. *Carvalhoma* (Hemiptera: Miridae): revised subfamily placement. J. N.Y. Entomol. Soc. 92:48–52.

Schuh, R. T., and M. D. Schwartz. 1985. Revision of the plant bug genus *Rhinacloa* Reuter with a phylogenetic analysis (Hemiptera: Miridae). Bull. Am. Mus. Nat. Hist. 179:379–470.

Schuh, R. T., and M. D. Schwartz. 1988. Revision of the New World Pilophorini (Heteroptera: Miridae: Phylinae). Bull. Am. Mus. Nat. Hist. 182:101–201.

Schuh, R. T., and G. M. Stonedahl. 1986. Historical biogeography in the Indo-Pacific: a cladistic approach. Cladistics 2:337–355.

Schuh, R. T., and P. Štys. 1991. Phylogenetic analysis of cimicomorphan family relationships (Heteroptera). J. N.Y. Entomol. Soc. 99:298–350.

Schuh, R. T., B. Galil, and J. T. Polhemus. 1987. Catalog and bibliography of Leptopodomorpha (Heteroptera). Bull. Am. Mus. Nat. Hist. 185:243–406.

Schwartz, M. D. 1984. A revision of the black grass bug genus *Irbisia* Reuter (Heteroptera: Miridae). J. N.Y. Entomol. Soc. 92:193–306.

Schwartz, M. D. 1987. Phylogenetic revision of the Stenodemini with a review of the Mirinae (Heteroptera: Miridae). Ph.D. dissertation, City University of New York. 383 pp.

Schwartz, M. D., and R. G. Foottit. 1992. *Lygus* bugs on the prairies. Biology, systematics, and distribution. Agric. Canada, Research Branch, Tech. Bull. 1992-4E. 44 pp., 1 table.

Schwartz, M. D., and R. T. Schuh. 1990. The world's largest isometopine, *Gigantometopus rossi*, new genus and new species (Heteroptera: Miridae). J. N.Y. Entomol. Soc. 98:9–13.

Schwarz, E. A., O. Heidemann, and N. Banks. 1914. Philip Reese Uhler [Biography and bibliography]. Proc. Entomol. Soc. Wash. 16:1–7, 1 pl.

Sclater, P. L. 1858. On the general geographical distribution of the members of the class Aves. J. Linnaean Soc., Zool. 2:130–145.

Scott, D. R. 1977. An annotated list of host plants of *Lygus hesperus* Knight. Bull. Entomol. Soc. Am. 23: 19–22.

Scott, D. R. 1981. Supplement to the bibliography of *Lygus* Hahn. Bull. Entomol. Soc. Am. 27: 275–279.

Scudder, G. G. E. 1957a. The systematic position of *Dicranocephalus* Hahn, 1826 and its allies (Hemiptera: Heteroptera). Proc. R. Entomol. Soc. Lond. (A) 32:147–158.

Scudder, G. G. E. 1957b. The higher classification of the Rhyparochrominae (Hem., Lygaeidae). Entomol. Mon. Mag. 93:152–156.

Scudder, G. G. E. 1959. The female genitalia of the Heteroptera: morphology and bearing on classification. Trans. R. Entomol. Soc. Lond. 111:405–467.

Scudder, G. G. E. 1962a. Results of the Royal Society Expedition to Southern Chile, 1958–59: Lygaeidae (Hemiptera), with the description of a new subfamily. Can. Entomol. 94:1064–1075.

Scudder, G. G. E. 1962b. New Heterogastrinae (Hemiptera) with a key to the genera of the world. Opusc. Entomol. 27:117–127.

Scudder, G. G. E. 1962c. The Ischnorhynchinae of the world

(Hemiptera: Lygaeidae). Trans. R. Entomol. Soc. Lond. 114: 163–194.

Scudder, G. G. E., J. P. E. C. Darlington, and S. B. Hill. 1967. A new species of Lygaeidae (Hemiptera) from the Tamana caves, Trinidad. Ann. Speleol. 22:465–469.

Seidenstücker, G. 1960. Heteropteren aus Iran 1956. III. *Thaumastella aradoides* Horv., eine Lygaeidae ohne Ovipositor. Stuttgart. Beitr. Naturk. 38:1–4.

Seidenstücker, G. 1964. Zur Systematik von *Bledionotus*, *Bethylimorphus* und *Thaumastella* Horvath (Heteroptera, Lygaeidae). Reichenbachia 3(25):269–279.

Seidenstücker, G. 1965a. Beitrag zu *Gampsocoris* (Heteroptera, Berytidae). Reichenbachia 5(31):273–282.

Seidenstücker, G. 1965b. *Stagonomus devius* n. sp., eine neue Schildwanze aus der Türkei (Heteroptera, Pentatomidae). Reichenbachia 5(3):9–19.

Semenov-Tian-Shanski, A. 1910. Vasily Evgrafovich Jakovlev. Horae Soc. Entomol. Rossicae 39:1–57, including portrait.

Sepúlveda, R. 1955. Biología del *Mecistorhinus tripterus* F. (Hem., Pentatomidae) y su posible influencia en la transmisión de la moniliasis del cacao. Cacao en Colombia 4:15–42.

Sharp, D. 1890. On the structure of the terminal segment in some male Hemiptera. Trans. Entomol. Soc. Lond. 1890:399–427, pls. 12–14.

Shaw, J. G. 1933. A study of the genus *Brachymetra* (Hemiptera, Gerridae). Univ. Kans. Sci. Bull. 21:221–233.

Signoret, V. 1863. Révision des Hémiptères du Chili. Ann. Soc. Entomol. Fr. 4:541–588.

Signoret, V. 1866. Notice nécrologique sur J.-B. Amyot. Ann. Soc. Entomol. Fr. (4)6:603–606.

Signoret, V. 1881–1884. Révision du groupe des Cydnides de la famille des Pentatomides. Ann. Soc. Entomol. Fr. [in 14 parts with independent pagination].

Sillen-Tullberg, B., C. Wiklund, and T. Jarvi. 1982. Aposematic coloration in adults and larvae of *Lygaeus equestris* and its bearing on Müllerian mimicry: an experimental study on predation on living bugs by the great tit *Parus major*. Oikos 39:131–136.

Silvestri, F. 1911. Sulla posizione sistematica del genere *Termitaphis* Wasm. con descrizione di due specie nuove. Portice Boll. Lab. Zool. 5:231–236.

Singh-Pruthi, H. 1925. The morphology of the male genitalia in Rhynchota. Trans. Entomol. Soc. Lond. 1925:127–267, pls. 1–32.

Sites, R. W. 1990. Morphological variations in the hemelytra of *Cryphocricos hungerfordi* Usinger (Heteroptera: Naucoridae). Proc. Entomol. Soc. Wash. 92:111–114.

Sites, R. W., and J. E. McPherson. 1982. Life history and laboratory rearing of *Sehirus cinctus cinctus* (Hemiptera: Cydnidae), with descriptions of immature stages. Ann. Entomol. Soc. Am. 75:210–215.

Slater, A. 1985. A taxonomic revision of the Lygaeinae of Australia (Heteroptera: Lygaeidae). Univ. Kans. Sci. Bull. 52:301–481.

Slater, A. 1992. A genus level revision of Western Hemisphere Lygaeinae (Heteroptera: Lygaeidae) with keys to species. Univ. Kans. Sci. Bull. 55:1–56.

Slater, J. A. 1950. An investigation of the female genitalia as taxonomic characters in the Miridae (Hemiptera). Iowa State Coll. J. Sci. 25:1–81.

Slater, J. A. 1955. A revision of the subfamily Pachygronthinae of the world (Hemiptera: Lygaeidae). Philippine J. Sci. 84:1–160.

Slater, J. A. 1957. Nomenclatorial considerations in the family Lygaeidae (Hemiptera: Heteroptera). Bull. Brooklyn Entomol. Soc. 52:35–38.

Slater, J. A. 1964a. Chapter 2. Hemiptera (Heteroptera): Lygaeidae. South African Animal Life 10:15–228.

Slater, J. A. 1964b. A Catalogue of the Lygaeidae of the World. University of Connecticut, Storrs. 2 vols.

Slater, J. A. 1968. A contribution to the systematics of Oriental and Australian Blissinae. Pac. Insects 10:275–294.

Slater, J. A. 1970. *Saxicoris*, a new genus of Psamminae from South Africa (Hemiptera: Lygaeidae). J. Entomol. Soc. S. Afr. 33:261–265.

Slater, J.A. 1971. The biology and immature stages of South African Heterogastrinae, with the description of two new species (Hemiptera: Lygaeidae). Ann. Natal Mus. 20:443–465.

Slater, J. A. 1973. A contribution to the biology and taxonomy of Australian Thaumastocoridae with the description of a new species (Hemiptera: Heteroptera). J. Aust. Entomol. Soc. 12:151–156.

Slater, J. A. 1974. Class Insecta. Order Hemiptera. Suborder Heteroptera. *In:* W. G. H. Coaton (ed.), Status of the Taxonomy of the Hexapoda of Southern Africa. Entomol. Mem. 38:66–74.

Slater, J. A. 1975. On the biology and zoogeography of Australian Lygaeidae (Hemiptera: Heteroptera) with special reference to the southwest fauna. J. Aust. Entomol. Soc. 14:47–64.

Slater, J. A. 1976. Monocots and chinch bugs: a study of host plant relationships in the lygaeid subfamily Blissinae (Hemiptera: Lygaeidae). Biotropica 8:143–165.

Slater, J. A. 1977. The incidence and evolutionary significance of wing polymorphism in lygaeid bugs with particular reference to those of South Africa. Biotropica 9:217–229.

Slater, J. A. 1979. The systematics, phylogeny, and zoogeography of the Blissinae of the world (Hemiptera, Lygaeidae). Bull. Am. Mus. Nat. Hist. 165:1–180.

Slater, J. A. 1982. Hemiptera. *In:* S. Parker (ed.), Synopsis and Classification of Living Organisms, vol. 2, pp. 417–447. McGraw-Hill, New York.

Slater, J. A. 1983. The *Ozophora* of Panama, with descriptions of thirteen new species (Hemiptera, Lygaeidae). Am. Mus. Novit. 2765:29 pp.

Slater, J. A. 1984 (1983). On the biology of cave inhabiting Antillocorini with the description of a new species from New Guinea (Hemiptera: Lygaeidae). J. N.Y. Entomol. Soc. 91:424–430.

Slater, J. A. 1988. Zoogeography of West Indian Lygaeidae (Hemiptera). *In:* J. K. Liebherr (ed.), Zoogeography of Caribbean Insects, pp. 38–60. Cornell University Press, Ithaca, N.Y.

Slater, J. A., and I. Ahmad. 1964. The mole bugs of the genus *Spalacocoris* (Hemiptera: Lygaeidae). Pac. Insects 6:730–740.

Slater, J. A., and R. M. Baranowski. 1978. How to Know the True Bugs. Pictured Key Nature Series. William C. Brown Publishers, Dubuque, Iowa. 256 pp.

Slater, J. A., and R. M. Baranowski. 1990. Lygaeidae of Florida (Hemiptera: Heteroptera). Arthropods of Florida and Neighboring Land Areas, vol. 14. Florida Dep. Agr. and Consumer Services. xv + 211 pp.

Slater, J. A., and H. Brailovsky. 1983. The systematic status of the family Thaumastocoridae with the description of a new species of *Discocoris* from Venezuela (Hemiptera: Heteroptera). Proc. Entomol. Soc. Wash. 85:560–563.

Slater, J. A., and H. Brailovsky. 1986. The first occurrence of the subfamily Artheneinae in the Western Hemisphere with the description of a new tribe (Hemiptera: Lygaeidae). J. N.Y. Entomol. Soc. 94:409–415.

Slater, J. A., and G. F. Gross. 1977. A remarkable new genus of coleopteroid Miridae from southern Australia (Hemiptera: Heteroptera). J. Aust. Entomol. Soc. 16:135–140.

Slater, J. A., and J. T. Polhemus. 1990. Obituary. Peter D. Ashlock. 1929–1989. J. N.Y. Entomol. Soc. 98:113–122.

Slater, J. A., and T. Schuh. 1969. New species of Isometopinae from South Africa (Hemiptera: Miridae). J. Entomol. Soc. S. Afr. 32:351–366.

Slater, J. A., and R. T. Schuh. 1990. A remarkably large new species of *Discocoris* from Colombia (Heteroptera: Thaumastocoridae). J. N.Y. Entomol. Soc. 98:402–405.

Slater, J. A., and M. H. Sweet. 1961. A contribution to the higher classification of the Megalonotinae (Hemiptera: Lygaeidae). Ann. Entomol. Soc. Am. 54:203–209.

Slater, J. A., and M. H. Sweet. 1965. The systematic position of the Psamminae. Proc. Entomol. Soc. Wash. 67:255–262.

Slater, J. A., and M. H. Sweet. 1970. The systematics and ecology of new genera and species of primitive Stygnocorini from South Africa, Madagascar and Tasmania (Hemiptera: Lygaeidae). Ann. Natal Mus. 20:257–292.

Slater, J. A., and M. H. Sweet. 1977. The genus *Plinthisus* Stephens in the Australian region (Hemiptera: Lygaeidae). Entomol. Scand. 8:109–154.

Slater, J. A., and T. E. Woodward. 1982. Lilliputocorini, a new tribe with six new species of *Lilliputocoris*, and a cladistic analysis of the Rhyparochrominae (Hemiptera, Lygaeidae). Am. Mus. Novit. 2754:1–23.

Slater, J. A., and L. Y. Zheng. 1985. Revision of the lygaeid genera *Porta* and *Primierus* (Hemiptera: Heteroptera), with the description of a new genus of Ozophorini from Papua New Guinea (Hemiptera). Syst. Entomol. 10:453–469.

Smith, C. L., and J. T. Polhemus. 1978. The Veliidae (Heteroptera) of America north of Mexico—keys and check list. Proc. Entomol. Soc. Wash. 80(1):56–68.

Smith, M. R. 1967. A new genus and twelve new species of Isometopinae (Hemiptera-Isometopidae) from Ghana. Bull. Entomol. Soc. Nigeria 1:27–42.

Smith, R. L. 1976. Male brooding behavior of the water bug *Abedus herberti*. Ann. Entomol. Soc. Am. 69:740–747.

Snoddy, E. L., W. J. Humphreys, and M. S. Blum. 1976. Observations on the behavior and morphology of the spider predator, *Stenolemus lanipes* (Hemiptera: Reduviidae). J. Georgia Entomol. Soc. 11:55–58.

Solbreck, C. 1986. Wing and flight muscle polymorphism in a lygaeid bug, *Horvathiolus gibbicollis*: determinants and life history consequences. Ecol. Entomol. 11:435–444.

Southwood, T. R. E. 1955. The morphology of the salivary glands of terrestrial Heteroptera (Geocorisae) and its bearing on classification. Tijdschr. Entomol. 98:77–84.

Southwood, T. R. E. 1956. The structure of the eggs of the terrestrial Heteroptera and its relationship to the classification of the group. Trans. R. Entomol. Soc. Lond. 108:163–221.

Southwood, T. R. E. 1961. A hormonal theory of the mechanism of wing polymorphism in Heteroptera. Proc. R. Entomol. Soc. Lond. (A) 36:63–66.

Southwood, T. R. E. 1962. Migration of terrestrial arthropods in relation to habitat. Biol. Rev. 37:171–214.

Southwood, T. R. E., and D. J. Hine. 1950. Further notes on the biology of *Sehirus bicolor* (L.) (Hem., Cydnidae). Entomol. Mon. Mag. 86:299–201.

Southwood, T. R. E., and D. Leston. 1959. Land and Water Bugs of the British Isles. Frederick Warne and Co., London.

Southwood, T. R. E., and G. G. E. Scudder. 1956. The bionomics and immature stages of the thistle lace bugs (*Tingis ampliata* H.-S. and *T. cardui* L.; Hem., Tingidae). Trans. Soc. British Entomol. 12:93–112.

Spangberg, J. 1879. Nekrolog [of Carl Stål]. Entomol. Z. Stettin 40:97–105.

Spangler, P. J. 1986. Two new species of water striders of the genus *Oiovelia* from the tepui Cerro de la Neblina, Venezuela (Hemiptera: Veliidae). Proc. Entomol. Soc. Wash. 88:438–450.

Speiser, P. 1904. Die Hemipterengattung *Polyctenes* Gigl. und ihre Stellung im System. Zool. Jahrb., suppl., 7:373–380, pls. 1–5.

Spence, J. R. 1983. Pattern and process in coexistence of water striders (Heteroptera: Gerridae). J. Anim. Ecol. 52:497–511.

Spence, J. R., and N. M. Andersen. 1994. Biology of water striders: interactions between systematics and ecology. Ann. Rev. Entomol. 39:101–128.

Spence, J. R., and G. G. E. Scudder. 1980. Habitats, life cycles, and guild structure among water striders (Heteroptera: Gerridae) of the Fraser Plateau of British Columbia. Can. Entomol. 112:779–792.

Spinola, M. 1837. Essai sur les genres d'insectes appartenants à l'ordre des Hémiptères, Lin. ou Rhyngotes, Fb. et à la section des Hétéroptères, Dufour. Yves Gravier, Genoa. 383 pp., 4 tables.

Spinola, M. 1850. Tavola sinottica dei generi spettanti alla classe degli Insetti Arthordignati Hemiptera L. Latr. Rhyngota F., Rhynchota Burm. Mem. Matem. Fis. Soc. Ital. Modena 25:1:43–100.

Spooner, C. S. 1938. The phylogeny of the Hemiptera based on a study of the head capsule. Illinois Biol. Monog. 16(3):1–102.

Sprague, E. B. 1956. The biology and morphology of *Hydrometra martini* Kirkaldy. Univ. Kans. Sci. Bull. 38:579–693.

Staddon, B. W. 1979. The scent glands of Heteroptera. In: Advances in Insect Physiology, vol. 14, pp. 351–418. Academic Press, London.

Stål, C. 1859a. Till kannedomen om Reduvini. Oefv. K. Vet.-Acad. Foerh. 16:175–204.

Stål, C. 1859b. Nova methodus Reduvina (Burm.) disponendi. Berl. Entomol. Z. 3:328.

Stål, C. 1859c. Hemiptera species novas descripsit. Konglika svenska Fregattens Eugenies Resa omkring Jorden. III. Zoologi, Insecter. 1859:219–298.

Stål, C. 1862. Hemiptera Mexicana enumeravit specieque novas descripsit. Stett. Entomol. Ztg. 23:437–462.

Stål, C. 1864–1865. Hemiptera Africana. Norstedtiana, Stockholm. (Vol. 1, 256 pp.; vol. 2, 181 pp.)

Stål, C. 1865–1866. Hemiptera Africana. Norstedtiana, Stockholm. (Vol. 3, 200 pp.; vol. 4, 275 pp., corrigenda.)

Stål, C. 1866a. Analecta hemipterologica. Berlin. Entomol. Z. 10:151–172.

Stål, C. 1866b. Bidrag till Reduviidernas kaennedom. Oef. K. Vet.-Akad. Foerh. 23:235–302.

Stål, C. 1867. Bidrag till hemipternas systematik. Oef. K. Vet.-Akad. Foerh. 24:191–560.

Stål, C. 1870. Enumeratio Hemipterorum. 1. Kongl. Svenska Vet.-Akad. Forh. 24:491–560.

Stål, C. 1872. Enumeratio Hemipterorum. 2. Kongl. Svenska Vet.-Akad. Handl. 10(4):1–159.

Stål, C. 1873. Enumeratio Hemipterorum. 3. Kongl. Svenska Vet.-Akad. Handl. 11(2):1–163.

Stål, C. 1874. Enumeratio Hemipterorum. 4. Kongl. Svenska Vet.-Akad. Handl. 12(1):1–186.

Stål, C. 1876. Enumeratio Hemipterorum. 5. Kongl. Svenska Vet.-Akad. Handl. 14(4):1–162.

Stehlik, J. L. 1965. Mission zoologique de l'I.R.S.A.C. en Afrique orientale (P. Basilewsky–N. Lelup, 1957)—Pyrrhocoridae (Het.). Acta Mus. Moraviae 50:211–252.

Stehlik, J. L. 1966. A new genus of Pyrrhocoridae from the Malagasy region (Heteroptera). Acta Mus. Moraviae 51:329–340.

Steinhaus, E. A., M. M. Batey, and C. L. Boerke. 1956. Bacterial symbiotes from the caeca of certain Heteroptera. Hilgardia 24:495–518.

Stern, V. M. 1976. Ecological studies of *Lygus* bugs in developing a pest management program for cotton pests in the San Joaquin valley, California. *In:* D. R. Scott and L. E. O'Keeffe, Lygus Bug: Host Plant Interactions. University of Idaho Press, Moscow.

Stichel, W. 1926. Die Gattung *Microtomus* Illiger (Hem., Het., Reduv.). Dtsch. Entomol. Z. 1926:179–190, pl. 1.

Stichel, W. (ed.). 1955. Illustrierte Bestimmungstabellen der Wanzen. II. Europa (Hemiptera-Heteroptera Europae), pts. 4, 5, pp. 108–163. Hermsdorf, Berlin.

Stichel, W. 1956–1962. Verzeichnis der Paläarktischen Hemiptera-Heteroptera, pts. 1–4. Hermsdorf, Berlin. 362 pp.

Stock, M. W., and J. D. Lattin. 1976. Biology of intertidal *Saldula palustris* (Douglas) on the Oregon coast (Heteroptera: Saldidae). J. Kans. Entomol. Soc. 49:313–326.

Stonedahl, G. M. 1983. New records for Palearctic *Phytocoris* in western North America (Hemiptera: Miridae). Proc. Entomol. Soc. Wash. 85:463–471.

Stonedahl, G. M. 1986. *Stylopomiris,* a new genus and three new species of Eccritotarsini (Heteroptera: Miridae: Bryocorinae) from Viet Nam and Malaya. J. N.Y. Entomol. Soc. 94:226–234.

Stonedahl, G. M. 1988a. Review of the genus *Phytocoris* in western North America with the description of 80 new species. Bull. Am. Mus. Nat. Hist. 188:1–257.

Stonedahl, G. M. 1988b. Revisions of *Dioclerus, Harpedona, Mertila, Myiocapsus, Prodromus,* and *Thaumastomiris* (Heteroptera: Miridae, Bryocorinae: Eccritotarsini). Bull. Am. Mus. Nat. Hist. 187:1–99.

Stonedahl, G. M. 1990. Revision and cladistic analysis of the Holarctic genus *Atractotomus* Fieber (Heteroptera: Miridae: Phylinae). Bull. Am. Mus. Nat. Hist. 198:88 pp.

Stonedahl, G. M. 1991. The Oriental species of *Helopeltis* (Heteroptera: Miridae): a review of economic literature and guide to identification. Bull. Entomol. Res. 81:465–490.

Stonedahl, G. M., and W. R. Dolling. 1991. Heteroptera identification: a reference guide, with special emphasis on economic groups. J. Nat. Hist. 25:1027–1066.

Stonedahl, G. M., and R. T. Schuh. 1986. *Squamocoris* Knight and *Ramentomiris,* new genus (Heteroptera: Miridae: Orthotylinae). A cladistic analysis and description of seven new species from Mexico and the western United States. Am. Mus. Novit. 2852:26 pp.

Stonedahl, G. M., and M. D. Schwartz. 1986. Revision of the plant bug genus *Pseudopsallus* Van Duzee (Heteroptera: Miridae). Am. Mus. Novit. 2842:58 pp.

Stonedahl, G. M., and M. D. Schwartz. 1988. New species of *Oaxacacoris* Schwartz & Stonedahl and *Pseudospallus* Van Duzee, and a new genus, *Presidiomiris,* from Texas (Heteroptera: Miridae: Orthotylini). Am. Mus. Novit. 2928:18 pp.

Stoner, A., A. M. Metcalfe, and R. E. Weeks. 1975. Plant feeding by Reduviidae, a predaceous family (Hemiptera). J. Kans. Entomol. Soc. 48:185–188.

Stork, N. 1981. The structure and function of the adhesive organs on the antennae of male *Harpocera thoracica* (Fallén) (Miridae; Hemiptera). J. Nat. Hist. 15:639–644.

Strawinski, K. 1925. Historja naturalna Korowca sosnowego *Aradus cinnamomeus* Panz. (Hemiptera: Heteroptera). Mocznik. Nauk. Rolniczych i Lesnych. 14:644–693.

Streams, F. A., and S. Newfield. 1972. Spatial and temporal overlap among breeding populations of New England *Notonecta.* Occ. Pap. Univ. Conn., Biol. Sci. Ser. 2:139–157.

Stride, G. O. 1956. On the mimetic association between certain species of *Phonoctonus* (Hemiptera: Reduviidae) and the Pyrrhocoridae. J. Entomol. Soc. S. Afr. 19:12–27.

Stusak, J. M. 1968. A new genus of Neotropical stilt-bugs (Hemiptera: Berytidae). J. N.Y. Entomol. Soc. 76:2–8.

Stusak, J. M. 1989. New and little known Oriental stilt bugs (Heteroptera: Berytidae). Acta Entomol. Bohemoslov. 86:111–120.

Štys, P. 1958. *Ceratocombus (Xylonannus) kunsti* n. sp.—a new species of Dipsocoridae from Czechoslovakia (Heteroptera). Acta Soc. Entomol. Czechoslov. 55:372–379.

Štys, P. 1959. The 5th stage nymph of *Ceratocombus (Ceratocombus) coleoptratus* (Zetterstedt, 11819) and notes on the morphology and systematics of Dipsocoridae (Heteroptera). Acta Entomol. Mus. Natl. Pragae 33(556):377–388.

Štys, P. 1961a. The stridulatory mechanism in *Centrocoris spiniger* (F.) and some other Coreidae (Heteroptera). Acta Entomol. Mus. Natl. Pragae 34(592):427–431.

Štys, P. 1961b. Morphology of the abdomen and female ectodermal genitalia of the trichophorous Heteroptera and bearing on their classification. Trans. 11th Congr. Entomol., Vienna 1:37–43.

Štys, P. 1963. Notes on the taxonomy, distribution and evolution of the Chauliopinae (Lygaeidae: Heteroptera). Acta Univ. Carol. Biol. 1963(2):209–216.

Štys, P. 1964a. Thaumastellidae—a new family of pentatomoid Hemiptera. Acta Soc. Entomol. Cechoslov. 61:236–253.

Štys, P. 1964b. The morphology and relationship of the family Hyocephalidae (Heteroptera). Acta Zool. Acad. Sci. Hung. 10:229–262.

Štys, P. 1965. General outline of the phylogeny of Coreoidea (Heteroptera). Proc. 12th Int. Congr. Entomol., p. 74. London.

Štys, P. 1966a. Morphology of the wings, abdomen and genitalia of *Phaenacantha australiae* Kirk. (Heteroptera, Colobathristidae) and notes on the phylogeny of the family. Acta Entomol. Bohemoslav. 63:266–280.

Štys, P. 1966b. Revision of the genus *Dayakiella* Horv. and notes on its systemical position (Heteroptera: Colobathristidae). Acta Entomol. Bohemoslav. 63:27–39.

Štys, P. 1967a. *Lipokophila chinai* gen. n., sp. n.—a new genus of Plokiophilidae (Heteroptera) from Brasil. Acta Entomol. Bohemoslav. 64:248–158.

Štys, P. 1967b. Medocostidae—a new family of cimicomorphan Heteroptera based on a new genus and two new species from tropical Africa. I. Descriptive part. Acta Entomol. Bohemoslav. 64:439–465.

Štys, P. 1967c. Monograph of Malcinae, with reconsideration of morphology and phylogeny of related groups (Heteroptera, Malcidae). Acta Entomol. Mus. Natl. Pragae 37:351–516.

Štys, P. 1969. Revision of the fossil and pseudofossil Enicocephalidae (Heteroptera). Acta Entomol. Bohemoslav. 66:352–365.

Štys, P. 1970a. On the morphology and classification of the family Dipsocoridae s. lat., with particular reference to the genus *Hypsipteryx* Drake (Heteroptera). Acta Entomol. Bohemoslav. 67:21–46.

Štys, P. 1970b. A review of the Palaearctic Enicocephalidae (Heteroptera). Acta Entomol. Bohemoslav. 67:223–240.

Štys, P. 1970c. Three new aberrant species of *Systelloderes* Blanch. from the Old World, and notes on the tribal classification of Enicocephalidae (Heteroptera, Enicocephalidae). Acta Univ. Carol., Biol. 1968:435–453.

Štys, P. 1971. Distribution and habitats of Joppeicidae (Heteroptera). Acta Faun. Entomol. Mus. Natl. Pragae 14(170):199–207.

Štys, P. 1974. *Semangananus mirus* gen. n., sp. n. from Celebes—a bug with accessory genitalia (Heteroptera, Schizopteridae). Acta Entomol. Bohemoslav. 71:382–397.

Štys, P. 1975. Suprageneric nomenclature of Anthocoridae (Heteroptera). Acta Univ. Carol. Biol. 1973:159–162.

Štys, P. 1976. *Velohebria antennalis* gen. n., sp. n.—a primitive terrestrial Microveliine from New Guinea, and a revised classification of the family Veliidae (Heteroptera). Acta Entomol. Bohemoslav. 73:388–403.

Štys, P. 1977. First records of Dipsocoridae and Ceratocombidae from Madagascar (Heteroptera). Acta Entomol. Bohemoslav. 74:295–315.

Štys, P. 1978. An annotated list of the genera of Enicocephalidae (Heteroptera). Acta Entomol. Mus. Natl. Pragae 39:241–252.

Štys, P. 1980. *Australostolus monteithi* gen. n., sp. n.—first record of an Australian aenictopecheine bug (Heteroptera, Enicocephalidae). Acta Entomol. Bohemoslav. 77:303–321.

Štys, P. 1981a. A new relict subfamily, genus and species of Enicocephalidae from New Caledonia (Heteroptera). Acta Entomol. Bohemoslav. 78:412–429.

Štys, P. 1981b. Unusual sex ratios in swarming and light-attracted Enicocephalidae (Heteroptera). Acta Entomol. Bohemoslav. 78: 430–432.

Štys, P. 1982a. A new Oriental genus of Ceratocombidae and higher classification of the family (Heteroptera). Acta Entomol. Bohemoslav. 79:354–376.

Štys, P. 1982b. New genus, two new species, females and larvae of Monteithostolinae from New Caledonia (Heteroptera, Enicocephalidae). Acta Univ. Carol. Biol. 1980:491–515.

Štys, P. 1983a. A new family of Heteroptera with dipsocoromorphan affinities from Papua New Guinea. Acta Entomol. Bohemoslav. 80:256–292.

Štys, P. 1983b. A new coleopteriform genus and species of Ceratocombidae from Zaire (Heteroptera, Dipsocoromorpha). Vest. Cs. Spolec. Zool. 47:221–230.

Štys, P. 1985a (1984). Soucasny stav beta-taxonomie radu Heteroptera. Prace Slov. Entomol. Spol. SAV, Bratislava 4:205–235.

Štys, P. 1985b. Phallopiratinae—a new subfamily of plesiomorphic Enicocephalidae based on a new genus and four new species from the Oriental region (Heteroptera). Acta Univ. Carol. Biol. 1981:269–310.

Štys, P. 1988. A new aenictopehceine bug from Tasmania (Heteroptera, Enicocephalidae). Vest. Cs. Spolec. Zool. 52:302–315.

Štys, P. 1989. Phylogenetic systematics of the most primitive true bugs (Heteroptera: Enicocephalomorpha, Dipsocoromorpha). Prace Slov. Entomol. Spol. SAV, Bratislava 8:69–85.

Štys, P. 1991. The first species of Plokophilidae from Madagascar (Heteroptera, Cimicomorpha). Acta Entomol. Bohemoslov. 88:425–430.

Štys, P., and M. Daniel. 1957. *Lyctocoris campestris* (F.) (Heteroptera, Anthocoridae) as a human facultative ectoparasite. Casopsis Ceskoslov. Spol. Entomol. 54:88–97.

Štys, P., and J. Davidova. 1979. Taxonomy of *Thyreocoris* (Heteroptera, Thyreocoridae). Annot. Zool. Bot. 134:1–40.

Štys, P., and J. Davidova-Vilimova. 1989. Unusual numbers of instars in Heteroptera: a review. Acta Entomol. Bohemoslav. 86:1–32.

Štys, P., and A. Jansson. 1988. Check-list of recent family-group names of Nepomorpha (Heteroptera) of the world. Acta Entomol. Fenn. 50:1–44.

Štys, P., and I. M. Kerzhner. 1975. The rank and nomenclature of higher taxa in recent Heteroptera. Acta Entomol. Bohemoslav. 72:64–79.

Štys, P., and P. Riha. 1977. An annotated catalogue of the fossil Alydidae (Heteroptera). Acta Univ. Carolinae, Biol. 1974:173–188.

Sweet, M. H. 1964. The biology and ecology of the Rhyparochrominae of New England (Het.: Lygaeidae), pts. 1 and 2. Entomol. Am. 43:1–124, 44:1–201.

Sweet, M. H. 1967. The tribal classification of the Rhyparochrominae (Heteroptera: Lygaeidae). Ann. Entomol. Soc. Am. 60:208–226.

Sweet, M. H. 1979. On the original feeding habits of the Hemiptera (Insecta). Ann. Entomol. Soc. Am. 72:575–579.

Sweet, M. H. 1981. The external morphology of the pre-genital abdomen and its evolutionary significance in the order Hemiptera (Insecta). Rostria, suppl., 33:41–51.

Sweet, M. H., and C. W. Schaefer. 1985. Systematics and ecology of *Chauliops fallax* Scott. Ann. Entomol. Soc. Am. 78:526–536.

Tachikawa, S., and C. W. Schaefer. 1985. Biology of *Parastrachia japonensis* (Hemiptera: Pentatomoidea: ?-idae). Ann. Entomol. Soc. America 78:387–397.

Takahashi, R. 1923. Observations on the Ochteridae. Bull. Brooklyn Entomol. Soc. 18:67–68.

Tallamy, D. W., and R. F. Denno. 1981a. Alternative life history patterns in risky environments: an example from lacebugs. *In:* R. F. Denno and H. Dingle (eds.), Insect Life History Patterns: Habitat and Geographic Variation, pp. 129–147. Springer-Verlag, New York.

Tallamy, D. W., and R. F. Denno. 1981b. Maternal care in *Gargaphia solani* (Hemiptera: Tingidae). Anim. Behav. 29:771–778.

Tamanini, L. 1947. Contributo ad una revisione del genere *Velia* Latr. e descizione di alcune specie nuove (Hemiptera-Heteroptera, Veliidae). Mem. Soc. Entomol. Ital. 26:17–74.

Tamanini, L. 1955. V° contributo allo studio del genere *Velia* Latr., valore specifico delle descrite da Fabricius e posizione sistematica delle specie Europee e circummediterranee (Hem.-Heter., Veliidae). Mem. Soc. Entomol. Ital. 33:201–207.

Tanaka, T. 1926. Homologies of the wing veins of Hemiptera. Annot. Zool. Japonenses 11:33–54.

Taylor, O. R., Jr. 1968. Coexistence and competitive interactions in fall and winter populations of six sympatric *Notonecta* (Hemiptera, Notonectidae) in New England. Occ. Pap. Univ. Conn., Biol. Sci. Ser. 1:109–139.

Teodoro, G. 1924. Sopra un particolare organo esistente nelle elitre degli eterotteri. Redia 15:87–95.

Thomas, D. B. 1992. Taxonomic synopsis of the asopine Pentatomidae (Heteroptera) of the Western Hemisphere. The Thomas Say Foundation. Vol. 16. Entomological Society of America, Lanham, Md. 156 pp.

Thorpe, W. H., and D. J. Crisp. 1947. Studies on plastron respiration. III. The orientation responses of *Aphelocheirus* in relation to plastron respiration; together with an account of specialized pressure receptors in aquatic insects. J. Exp. Biol. 24:310–328.

Thouvenin, M. 1965. Etude préliminaire de uradénies chez certains Hétéroptères pentatomomorphes. Ann. Soc. Entomol. Fr., n.s., 1:973–988.

Todd, E. L. 1955. A taxonomic revision of the family Gelastocoridae (Hemiptera). Univ. Kans. Sci. Bull. 37:277–475.

Torre-Bueno, J. R. de la. 1903. Notes on the stridulation and habits of *Ranatra fusca,* Pal. B. Can. Entomol. 35:235–237.

Torre-Bueno, J. R. de la. 1905. The tonal apparatus of *Ranatra quadridentata,* Stål. Can. Entomol. 37:85–87.

Torre-Bueno, J. R. de la. 1906. Life history of *Ranatra quadridentata.* Can. Entomol. 38:242–252.

Torre-Bueno, J. R. de la. 1937. A Glossary of Entomology. Brooklyn Entomological Society, Brooklyn, N.Y. ix + 336 pp., 9 pls.

Torre-Bueno, J. R. de la. 1939, 1941. A synopsis of the Hemiptera-Heteroptera of America north of Mexico, pts. 1 and 2. Entomol. Am. 19:141–310; 21:41–122.

Truxal, F. S. 1952. The comparative morphology of the male genitalia of the Notonectidae (Hemiptera). J. Kans. Entomol. Soc. 25:30–38.

Truxal, F. S. 1953. A revision of the genus *Buenoa* (Hemiptera, Notonectidae). Univ. Kans. Sci. Bull. 35:1351–1523.

Tullgren, A. 1918. Zur Morphologie und Systematik der Hemipteren. Entomol. Tidskr. 39:113–132.

Tyshchenko, V. P. 1961. On the relations of some spiders of the family Thomisidae to mimicking insects and their models. Vestnik Leningrad Univ., no. 3 (ser. biol. no. 1):133–139. [In Russian].

Ueshima, N. 1968. New species and records of Cimicidae with keys (Hemiptera). Pan-Pac. Entomol. 44:264–279.

Ueshima, N. 1972. New World Polyctenidae (Hemiptera), with special reference to Venezuelan species. Brigham Young Univ. Sci. Bull. 17:13–21.

Ueshima, N. 1979. Hemiptera. II. Heteroptera. Animal Cytogenetics, vol. 3, Insecta 6. Gebrüder Bornträger, Berlin. 117 pp.

Uhler, P. R. 1860. Hemiptera of the North Pacific exploring expedition under Comr.'s Rodgers and Ringgold. Proc. Acad. Nat. Sci. Phil. 12:221–231.

Uhler, P. R. 1863. Hemipterological contributions, no. 1. Proc. Am. Entomol. Soc. 2:155–162.

Uhler, P. R. 1876. List of Hemiptera of the region west of the Mississippi River, including those collected during the Hayden explorations of 1873. Bull. U.S. Geol. Geog. Surv. Terr. 1:267–361.

Usinger, R. L. 1936. Studies in the American Aradidae with descriptions of new species. Ann. Entomol. Soc. Am. 29:490–516.

Usinger, R. L. 1941. Key to the subfamilies of Naucoridae with a generic synopsis of the new subfamily Ambrysinae. Ann. Entomol. Soc. Am. 34:5–16.

Usinger, R. L. 1942a. Revision of the Termitaphididae (Hemiptera). Pan-Pac. Entomol. 18:155–159.

Usinger, R. L. 1942b. The genus *Nysius* and its allies in the Hawaiian Islands (Hemiptera, Lygaeidae, Orsillini). Bull. Bishop Mus. 173:1–167.

Usinger, R. L. 1943. A revised classification of the Reduvioidea with a new subfamily from South America (Hemiptera). Ann. Entomol. Soc. Am. 36:602–617.

Usinger, R. L. 1945. Classification of the Enicocephalidae (Hemiptera, Reduvioidea). Ann. Entomol. Soc. Am. 38:321–342.

Usinger, R. L. 1946. Hemiptera-Heteroptera of Guam. Insects of Guam. II. Bull. Bishop Mus. 38:58–91.

Usinger, R. L. 1950. The origin and distribution of the apterous Aradidae. Proc. 8th Int. Congr. Entomol., pp. 174–179. Stockholm.

Usinger, R. L. 1952. A new genus of Chryxinae from Brazil and Argentina (Hemiptera: Reduviidae). Pan-Pac. Entomol. 28:55–56.

Usinger, R. L. 1954a. A new genus of Aradidae from the Belgian Congo, with notes on stridulatory mechanisms in the family. Ann. Mus. Congo Tervuren, Misc. Zool. 1:540–543.

Usinger, R. L. 1954b. Revision of the genus *Chelonocoris* Miller (Hemiptera, Aradidae). Zool. Meded. 22:259–272.

Usinger, R. L. 1956. Aquatic hemiptera. *In:* R. L. Usinger (ed.), Aquatic Insects of California, with Keys to North American Genera and California Species, pp. 182–228. University of California Press, Berkeley.

Usinger, R. L. 1966. Monograph of Cimicidae (Hemiptera-Heteroptera). The Thomas Say Foundation. Vol. 7. Entomological Society of America, Lanham, Md. 585 pp.

Usinger, R. L. 1972. Robert Leslie Usinger: Autobiography of an Entomologist, ed. E. G. Linsley and J. L. Gressitt. Pacific Coast Entomological Society, San Francisco. 330 pp.

Usinger, R. L., and R. Matsuda. 1959. Classification of the Aradidae (Hemiptera-Heteroptera). Br. Mus. (Nat. Hist.). 410 pp.

Usinger, R. L., and P. Wygodzinsky. 1960. Heteroptera: Enicocephalidae. *In:* Insects of Micronesia, vol. 7, no. 5, pp. 219–230. Bernice P. Bishop Museum, Honolulu.

Usinger, R. L., and P. Wygodzinsky. 1964. Description of a new species of *Mandanocoris* Miller, with notes on the systematic position of the genus (Reduviidae, Hemiptera, Insect). Am. Mus. Novit. 2204:13 pp.

van Doesburg, P. H., Jr. 1966. Heteroptera of Suriname. I. Largidae and Pyrrhocoridae. Studies on the Fauna of Suriname and Other Guyanas, no. 9. M. Nijhoff, The Hague. 60 pp.

van Doesburg, P. H., Jr. 1968. A revision of the New World Species of *Dysdercus* Guerin Meneville (Heteroptera, Pyrrhocoridae). E. J. Brill, Leiden. 215 pp.

van Doesburg, P. H., Jr. 1970. A new genus and species in Velocipedidae (Heteroptera). Zool. Meded. 44:247–250.

van Doesburg, P. H., Jr. 1977. A new species of *Thaumamannia* from Surinam (Heteroptera, Tingidae, Vianaidinae). Zool. Meded. 52:185–189.

van Doesburg, P. H., Jr. 1980. Notes on Velocipedidae (Heteroptera). Zool. Meded. 55:297–299.

van Doesburg, P. H., Jr. 1984. A new species of *Potamocoris* Hungerford, 1941, from Suriname (Heteroptera: Naucoridae). Zool. Meded. 59(2):19–26.

Van Duzee, E. P. 1916. Check list of the Hemiptera of America north of Mexico. New York Entomological Society. 111 pp.

Van Duzee, E. P. 1917. Catalogue of the Hemiptera of America north of Mexico excepting the Aphididae, Coccidae and Aleurodidae. Univ. Calif. Tech. Bull., Entomol. 2:xiv + 902 pp.

Vasarhelyi, T. 1982. A study of the relation of *Mezira tremulae* Germ. and two allied species (Heteroptera: Aradidae). Acta Zool. Acad. Sci. Hung. 28:389–402.

Vasarhelyi, T. 1986. The pretarsus in Aradidae (Heteroptera). Acta Zool. Hung. 32:377–383.

Vasarhelyi, T. 1987. On the relationships of the eight aradid subfamilies (Heteroptera). Acta Zool. Hung. 33:263–267.

Vepsäläinen, K. 1971a. The roles of photoperiodism and genetic switch alary polymorphism in *Gerris* (Gerridae, Heteroptera), a preliminary report. Acta Entomol. Fenn. 28:101–102.

Vepsäläinen, K. 1971b. The role of gradually changing day length in determination of wing length, alary dimorphism and diapause in a *Gerris odontogaster* (Zett.) population (Gerridae, Heteroptera) in south Finland. Acta Acad. Sci. Fenn. A, IV, Biol., 183:1–25.

Vepsäläinen, K. 1974a. Determination of wing length and diapause in water striders (*Gerris* Fabr., Heteroptera). Hereditas 77:163–176.

Vepsäläinen, K. 1974b. The life cycles and wing length of Finnish *Gerris* Fabr. species (Heteroptera, Gerridae). Acta Zool. Fenn. 141:1–73.

Vepsäläinen, K. 1978. Wing dimorphism and diapause in *Gerris*. Determination and adaptive significance. *In:* H. Dingle, (ed.), Evolution of Insect Migration and Diapause, pp. 218–253. Springer-Verlag, New York.

Verhoeff, C. 1893. Vergleichende Untersuchungen über die Abdominal-Segmente der Weibchen Hemiptera-Heteroptera und Homoptera. Entomol. Nachr. 19:369–378.

Viana, M. J., and D. J. Carpintero. 1981. Una nueva especie de "Discocoris" Kormilev, 1955 (Hemiptera, Xylastodoridae). Com. Mus. Arg. Cienc. Nat. Bernardino Rivadavia, Entomol. 1:65–74.

Vidano, C., and A. Arzone. 1976. Sulla collezione Spinola conservta nel Castello di Tassarolo, pp. 253–260. Atti XI Congresso Nazionale Italiano di Entomoogia, Portici-Sorrento.

Villiers, A. 1948. Faune de l'Empire Français. IX. Hémiptères Réduviidés de l'Afrique Noire. Office de la Recherche Scientifique Coloniale, Editions du Muséum, Paris. 489 pp.

Villiers, A. 1958. Faune de Madagascar. VII. Insectes. Hémiptères Enicocéphalidae. Publ. Inst. Rech. Sci. Tananarive-Tsimbazaza. 78 pp.

Villiers, A. 1964. Exploration du Parc National de las Garamba. 43. Reduviidae (Hemiptera Heteroptera). Institut des Parcs Nationaux du Congo et du Rwanda, Bruxelles. 132 pp.

Villiers, A. 1968. Faune de Madagascar. 28. Insectes. Hémiptères Reduviidae (1re partie). ORSTOM, CNRS, Paris. 198 pp.

Villiers, A. 1969a. Révision des Hémiptères Henicocephalidae africains et Malgaches. Ann. Mus. R. Afrique Centr., Zool. 176:232 pp.

Villiers, A. 1969b. Révision des Réduviidés africains. IV. Saicinae. Bull. I. F. A. N. 31:1186–1247.

Villiers, A. 1979. Faune de Madagascar. 49a. Insectes. Hémiptères Reduviidae (2re partie). ORSTOM, CNRS, Paris. 202 pp.

Villiers, A. 1982. Hémiptères Reduviidae africains. Localisation et descriptions. III. Harpactorinae, Rhinocorini. (1). Rev. Fr. Entomol., n.s., 4:126–136.

Vinokurov, N. N. 1979. Heteroptera of Yakutia. Acad. Sci. USSR, Inst. Zool. 328 pp. [In Russian; trans. Amerind Publishing, New Delhi, 1988].

Vinokurov, N. N., V. B. Golub, I. M. Kerzhner, E. V. Kanyukova, G. P. Tshernova. 1988. Homoptera and Hemiptera. *In:* P. A. Lehr (ed.), Keys to the Insects of the Far East of the USSR, vol. 2, pp. 727–930. Nauka, Leningrad. [In Russian].

Voelker, J. 1966. Wasserwanzen als obligatorische Schneckenfresser im Nildelta (*Limnogeton fieberi* Mayr). Z. Tropenmed. Parasitol. 17:155–165.

Wagner, E. 1952. 41. Blindwanzen oder Miriden. Die Tierwelt Deutschlands und der angrenzenden Meersteile. Gustav Fischer, Jena. 218 pp.

Wagner, E. 1955. Bemerkungen zum System der Miridae (Hem. Het.). Dtsch. Entomol. Z., n.f., 2:230–242.

Wagner, E. 1966. 54. Wanzen oder Heteropteren. I. Pentatomomorpha. Die Tierwelt Deutschlands und der angrenzenden Meeresteile. Gustav Fischer. Jena. 235 pp.

Wagner, E. 1967. 55. Wanzen oder Heteropteren: II. Cimicomorpha. Die Tierwelt Deutschlands un der angrenzenden Meeresteile. Jena, Gustav Fischer. 179 pp.

Wagner, E. 1971–1978. Die Miridae Hahn, 1831, des Mittelmeerraumes und der Makaronsesischen Inseln (Hemiptera, Heteroptera), pts. 1–3, addendum. Entomol. Abhandl. (Pt. 1, 1971, 37(suppl.):1–484; pt. 2, 1973, 39(suppl.):1–421; pt. 3, 1975, 40(suppl.):1–483, 42(suppl.):1–96, 1978.)

Wagner, E., and S. Zimmermann. 1955. Beitrag zur Systematik der Gattung *Gerris* F. (Hemiptera-Heteroptera, Gerridae). Zool. Anz. 155:177–190.

Waloff, N. 1983. Absence of wing polymorphism in the arboreal, phytophagous species of some taxa of temperate Hemiptera: an hypothesis. Ecol. Entomol. 8:229–232.

Wasmann, E. 1902. Species novae insectorum termitophilarum ex America meridionali Tijdschr. Entomol. 45:75–107.

Weber, H. 1930. Biologie der Hemipteren. Julius Springer, Berlin. 543 pp.

Weber, H. H. 1976. Dr. h. c. Eduard Wagner—80 Jahre. Mitt. Dtsch. Entomol. Ges. 35:1–51.

Wefelscheid, H. 1912. Über die Biologie und Anatomie von *Plea minutissima* Leach. Zool. Jahr., Abt. Syst. 32:389–474, pls. 14–15.

Wells, M. J. 1954. The thoracic glands of Hemiptera Heteroptera. Q. J. Microsc. Sci. 95:231–244.

Westwood, J. O. 1874. Thesaurus Entomologicus Oxoniensis. Clarendon Press, Oxford. xxiv + 205 pp., 40 pls.

Wheeler, A. G., Jr. 1974. Studies on the arthropod fauna of alfalfa. VI. Plant bugs (Miridae). Can. Entomol. 106:1267–1275.

Wheeler, A. G., Jr. 1976. *Lygus* bugs as facultative predators. *In:* D. R. Scott and L. E. O'Keeffe. Lygus Bug: Host Plant Interactions. University of Idaho Press, Moscow.

Wheeler, A. G., Jr. 1980. The mirid rectal organ: purging the literature. Florida Entomol. 63:481–485.

Wheeler, A. G., Jr. 1991. Plant bugs of *Quercus ilicifolia:* myriads of mirids (Heteroptera) in pitch pine–scrub oak barrens. J. N.Y. Entomol. Soc. 99:405–440.

Wheeler, A. G., Jr. 1992. *Chinaola quercicola* rediscovered in several specialized plant communities in the southeastern United States (Heteroptera: Microphysidae). Proc. Entomol. Soc. Wash. 92:249–252.

Wheeler, A. G., Jr., and J. E. Fetter. 1987. *Chilacis typhae* (Heteroptera: Lygaeidae) and the subfamily Artheneinae new to North America. Proc. Entomol. Soc. Wash. 89:244–289.

Wheeler, A. G., Jr., and T. J. Henry. 1978. Isometopinae (Hemiptera: Miridae) in Pennsylvania: biology and descriptions of fifth instars, with observations of predation on obscure scale. Ann. Entomol. Soc. Am. 71:607–614.

Wheeler, A. G., Jr., and T. J. Henry. 1981. *Jalysus spinosus* and *J. wickhami:* taxonomic clarification, review of host plants and distribution, and keys to adults and 5th instars. Ann. Entomol. Soc. America 74:606–615.

Wheeler, A. G., Jr., and T. J. Henry. 1992. A Synthesis of the Holarctic Miridae (Heteroptera): Distribution, Biology, and Origin, with Emphasis on North America. The Thomas Say Foundation. Entomological Society of America, Lanham, Md. 282 pp.

Wheeler, A. G., Jr., and J. P. McCaffrey. 1984. *Ranzovius contubernalis:* seasonal history, habits, and description of fifth instar, with speculation on the origin of spider commensalism in the genus *Ranzovius* (Hemiptera: Miridae). Proc. Entomol. Soc. Wash. 84:68–81.

Wheeler, A. G., Jr., and C. W. Schaefer. 1982. Review of stilt bug (Hemiptera: Berytidae) host plants. Ann. Entomol. Soc. Am. 75:498–506.

Wheeler, Q. D., and A. G. Wheeler, Jr. 1994. Mycophagous Miridae? Associations of Cylapinae (Heteroptera) with pyrenomycete fungi (Euascomycetes: Xylariaceae). J. N.Y. Entomol. Soc. 102:114–117.

Wheeler, W. C., R. T. Schuh, and R. Bang. 1993. Cladistic relationships among higher groups of Heteroptera: congruence between morphological and molecular data sets. Entomol. Scand. 24:121–137.

Wiese, K. 1972. Das mechanorezeptorische Beuteortungssystem von *Notonecta*. I. Funktion des tarsalen Scolopidialorgans. J. Comp. Physiol. 78:83–102.

Wigglesworth, V. B. 1933. The physiology of the cuticle and of ecdysis in *Rhodnius prolixus* (Triatomidae, Hempiptera), with special reference to the function of the oenocytes and of the dermal glands. Q. J. Microsc. Sci. 76:269–319.

Wigglesworth, V. B. 1948. The epicuticle in an insect, *Rhodnius prolixus* (Hemiptera). Proc. R. Soc., B, 134:163–181.

Wigglesworth, V. B. 1984. Insect Physiology, 8th ed. Chapman Hall, London. 191 pp.

Wilcox, R. S. 1972. Communication by surface waves. Mating

behavior of a water strider (Gerridae). J. Comp. Physiol. 80:255–266.

Wilcox, R. S. 1975. Sound-production mechanisms of *Buenoa macrotibialis* Hungerford (Hemiptera: Notonectidae). Int. J. Insect Morphol. Embryol. 4:169–182.

Wilcox, R. S. 1979. Sex discrimination in *Gerris remigis:* role of a surface wave signal. Science 206:1325–1327.

Wilcox, R. S. 1980. Ripple communication. Oceanus 23:61–68.

Wilcox, R. S., and J. R. Spence. 1986. The mating system of two hybridizing species of water striders (Gerridae). I. Ripple signal functions. Behav. Ecol. Sociobiol. 19:79–85.

Wolcott, G. W. 1936. Insectae Borinquenses. J. Dep. Agric. Porto Rico 20(1):1–627.

Woodruff, L. C. 1956. Herbert Barker Hungerford. Univ. Kans. Sci. Bull. 38:1–xiv + frontispiece.

Woodruff, L. C. 1963. Herbert Barker Hungerford. J. Kans. Entomol. Soc. 36:197–199.

Woodward, T. E. 1950. Ovariole and testis follicle numbers in the Heteroptera. Entomol. Mon. Mag. 86:82–84.

Woodward, T. E. 1956. The Heteroptera of New Zealand. Part II: The Enicocephalidae. With a supplement to Part I (Cydnidae and Pentatomidae). Trans. R. Soc. New Zealand 84:391–430.

Woodward, T. E. 1968a. A new subfamily of Lygaeidae (Hemiptera-Heteroptera) from Australia. Proc. R. Entomol. Soc. Lond. (B) 37(9–10):125–132.

Woodward, T. E. 1968b. The Australian Idiostolidae (Hemiptera: Heteroptera). Trans. R. Entomol. Soc. Lond. 120:253–261.

Wooley, T. A. 1951. The circulatory system of the box elder bug, *Leptocoris trivittatus* (Say). Am. Midl. Nat. 46:634–639.

Wootton, R. J., and C. R. Betts. 1986. Homology and function in the wings of Heteroptera. Syst. Entomol. 11:389–400.

Wroblewski, A. 1974. Prof. Dr. Tadeusz Jaczewski 1899–1974. Polskie Pismo Entomol. 44(4):689–704 + pl. [In Polish].

Wygodzinsky, P. 1943. Contribuição ao conhecimento do gênero *Salyavata* (Salyavatinae, Reduviidae, Hemiptera). Bol. Mus. Nac., Rio de Janeiro, n.s., Zool. 6:27 pp.

Wygodzinsky, P. 1944a. Contribuição ao conhecimento do gênero "Elasmodema" Stål, 1860 (Elasmodemidae, Reduvioidea, Hemiptera). Rev. Brasil. Biol. 4:193–213.

Wygodzinsky, P. 1944b. Notas sôbre a biologia e o desenvolvimento de *Macrocephalus notatus* Westwood (Phymatidae, Reduvioidea, Hemiptera). Rev. Entomol. 15:139–143.

Wygodzinsky, P. 1946. Sôbre un novo gênero e uma nova species de Chryxinae e considerações sôbre a subfamília (Reduviidae, Hemiptera). Rev. Brasil. Biol. 6:173–180.

Wygodzinsky, P. 1947a. Contribuição al conhecimento do gênero *Heniartes* Spinola, 1837 (Apiomerinae, Reudviidae, Hemiptera). Arq. Mus. Nac., Rio de Janeiro 41:3–65.

Wygodzinksy, P. 1947b. Sur le *Trichotonannus setulosus* Reuter, 1891, avec une théorie sur l'origine des harpagones des Hétéroptères mâles (Hemiptera-Heteroptera, Cryptostemmatidae). Rev. Fr. Entomol. 14:118–125.

Wygodzinsky, P. 1948a. On two new genera of "Schizopterinae" (Cryptostemmatidae) from the Neotropical region (Hemiptera). Rev. Brasil. Biol. 8:143–155.

Wygodzinsky, P. 1948b. Studies on some apterous Aradidae from Brazil (Hemiptera). Bol. Mus. Nac., Rio de Janeiro, Zool., 86:1–23, pls. I–XIV.

Wygodzinsky, P. 1950. Schizopterinae from Angola (Cryptostemmatidae, Hemiptera). Publ. Cult. Co. Diam. Angola 7:9–48.

Wygodzinsky, P. 1953. Cryptostemmatinae from Angola (Crytostemmatidae, Hemiptera). Publ. Cult. Co. Diam. Angola 16:27–48.

Wygodzinsky, P. 1966. A monograph of the Emesinae (Reduviidae, Hemiptera). Bull. Am. Mus. Nat. Hist. 133:1–614.

Wygodzinsky, P., and S. Lodhi. 1989. Atlas of antennal trichbothria in the Pachynomidae and Reduviidae (Heteroptera). J. N.Y. Entomol. Soc. 97:371–393.

Wygodzinsky, P., and J. Maldonado Capriles. 1972. Description of the first genus of physoderine assassin bugs (Reduviidae, Hemiptera) from the New World. Am. Mus. Novit. 2505:7 pp.

Wygodzinsky, P., and K. Schmidt. 1991. Revision of the New World Enicocephalomorpha (Heteroptera). Bull. Am. Mus. Nat. Hist. 200:265 pp.

Wygodzinsky, P., and P. Štys. 1970. A new genus of aenictopecheine bugs from the Holarctic (Enicocephalidae, Hemiptera). Am. Mus. Novit. 2411:17 pp.

Wygodzinsky, P., and P. Štys. 1982. Two new primitive genera and species of Enicocephalidae from Singapore (Heteroptera). Acta Entomol. Bohemoslav. 79:127–142.

Wygodzinsky, P., and R. L. Usinger. 1963. Classification of the Holoptinae and description of the first representive from the New World (Hemiptera: Reduviidae). Proc. R. Entomol. Soc. Lond. (B) 32:47–52.

Yang, We-I. 1935. Notes on the Chinese Tessaratominae with description of an exotic species. Bull. Fan. Inst. Biol., Peking 6:103–144.

Yang, We-I. 1938a. A new method for the classification of urostylid insects. Bull. Fan. Inst. Biol., Peking 8:35–48.

Yang, We-I. 1938b. Two new Chinese urostylid insects. Bull. Fan. Inst. Biol., Peking 8:229–236.

Yang, We-I. 1939. A revision of Chinese urostylid insects (Heteroptera). Bull. Fan Inst. Biol., Peking 9:5–66.

Yonke, T. 1972. A new genus and two new species of Neotropical Chariesterini (Hemiptera: Coreidae). Proc. Entomol. Soc. Wash. 74:283–287.

Yonke, T. R. 1991. Order Hemiptera. *In:* F. W. Stehr (ed.), Immature Insects. vol. 2, pp. 22–65. Kendall/Hunt Publishing, Dubuque, Iowa.

Young, D. K. 1984. Field records and observations of insects associated with cantharadin. Great Lakes Entomol. 17:195–199.

Young, O. P. 1986. Host plants of the tarnished plant bug, *Lygus lineolaris* (Heteroptera: Miridae). Ann. Entomol. Soc. Am. 79:747–762.

Zera, A. J., and K. C. Tiebel. 1991. Photoperiodic induction of wing morphs in the waterstrider *Limnoporus canaliculatus* (Gerridae: Hemiptera). Ann. Entomol. Soc. Am. 84:508–516.

Zetterstedt, J. W. 1828. Fauna Insecta Lapponica. Hammone, xx + 563 pp.

Zimmerman, E. C. 1948. Insects of Hawaii, vol. 3. Heteroptera. University of Hawaii Press, Honolulu. 255 pp.

Zimmermann, M. 1984. Population structure, life cycle and habitat of the pondweed bug, *Mesovelia furcata* (Hemiptera, Mesoveliidae). Rev. Suisse Zool. 91:1017–1035.

Zimsen, E. 1964. The type material of I. C. Fabricius. Munksgaard, Copenhagen. 656 pp.

Zrzavý, J. 1990a. Antennal trichobothria in Heteroptera: a phyogenetic approach. Acta Entomol. Bohemoslav. 87:321–325.

Zrzavý, J. 1990b. Evolution of the aposematic colour pattern in some Coreoidea s. lat. (Heteroptera): a point of view. Acta Entomol. Bohemoslav. 87:470–474.

Zrzavý, J. 1990c. Evolution of Antennal sclerites in Heteroptera (Insecta). Acta Univ. Carol. Biol. 34:189–227.

GLOSSARY

accessory fecundation canal a slender, usually weakly sclerotized, tube running along dorsal wall of common oviduct from gynatrial sac to point of entrance into common oviduct; divided into basal thickening and fecundation pump.
accessory gland vermiform gland (q.v.) or accessory salivary gland (q.v.).
accessory male genitalia in some Schizopteridae, complex dorsoabdominal pregenital structures assumed to perform actual insemination after receiving sperm from primary genitalia.
accessory parempodium in Leptopodomorpha, e.g., small, secondary parempodium; *see also* pseudopulvillus.
accessory salivary gland tubular or vesicular gland associated by a duct with principal salivary gland.
accessory scent glands small glands associated with primary reservoir of metathoracic scent gland.
aedeagus that portion of phallus distal to phallobase, including proximal phallotheca (phallosoma) and distal endosoma.
aeropyle fine pores connected to air spaces in outer and inner meshworks of chorion.
airstraps in Belostomatidae, a pair of straplike appendages derived from abdominal segment 8 used like respiratory siphon in Nepidae to obtain atmospheric air.
anal segment proctiger, q.v.
anal tube proctiger, q.v.
androconia in males of Scutelleridae and some other Pentatomomorpha, unicellular glands grouped in patches on abdominal venter, with hollow bristlelike androconium set in an alveolus.
androtraumatic insemination in *Phallopirates* (Heteroptera: Enicocephalidae), presumed mode of insemination, in which male can pass sperm to female only after breaking off tip of his own copulatory organ.
annulus (pl., **annuli**) a ring encircling an article or segment, as in antennae of some Reduviidae; proctiger, q.v.
anteapical claw *see* preapical claw.
anteclypeus inferior (anterior) portion of clypeus, whenever there is a visible transverse line of demarcation.

Adapted, with permission of the New York Entomological Society, from S. W. Nichols, *The Torre-Bueno Glossary of Entomology* (New York: New York Entomological Society, 1989), xvii + 840 pp.

antennal fossa a groove in which antenna is located or concealed, as in some Phymatinae (Reduviidae).
antenniferous tubercle a protuberance of head which bears antenna.
anterior gonapophyses first valvulae, q.v.
anterior pronotal lobe anterior portion of pronotum, bearing calli.
anterior ramus first ramus, q.v.
anterior valvulae first valvulae, q.v.
apical bulb spermathecal bulb, q.v.
apterous completely lacking wings.
arolium (pl., **arolia**) one or two unpaired bristle- or bladderlike, medial pretarsal structures, originating dorsad of unguitractor and between but isolated from bases of claws (sometimes incorrectly called empodium, parempodium, or rarely pulvillus); *see* empodium, pulvillus.
articulatory apparatus system of plates and apodemes for suspension of phallus and attachment of its motor muscles, drawn out along phallosoma into ligamentary processes, comprising (1) basal plates attached to suspensory apodemes and ponticulus transversalis and (2) dorsal connectives, ending in capitate processes.
auricle(s) variously shaped structure on metapleuron of adult Cimicomorpha and Pentatomomorpha, assisting in spreading metathoracic scent-gland products from ostiole onto evaporatorium.
auxilia basipulvillus, q.v.
basal apparatus articulatory apparatus, q.v.
basal foramen entrance to phallic cavity surrounded by basal plates and ponticulus transversalis, closed or not by a septum.
basal plates two major plates of articulatory apparatus; sometimes applied to entire articulatory apparatus.
basiconjunctiva distal membranous part of phallosoma reaching to, but not including, ejaculatory reservoir.
basipulvillus basal portion of pulvillus in Pentatomomorpha; *see* distipulvillus.
Berlese's organ mesospermalege (q.v.) or spermalege (q.v.), as a whole.
bothrium (pl., **bothria**) pit or tubercle from which a trichobothrium arises.
brachypterous with shortened or abbreviated wings, usually incapable of flight.
Brindley's glands in some adult Reduviidae and Pachynomidae, paired glands located in anteriormost portion of abdomen, with openings situated dorsolaterally just posterior to thoracicoabdominal junction.
buccula (pl., **bucculae**) a flange of gena, on each side of basal portion of labium.
bug a term often loosely used for any insects, but strictly applied to members of suborder Heteroptera; true bug, q.v.
bursa copulatrix (pl., **bursae copulatrices**) variously formed structure serving as a vagina, as in Miridae.
callar area middle part of pronotum behind collar and containing calli, and corresponding in size to prothoracic body cavity.
callus (pl., **calli**) paired or fused impression or elevation in anterior part of pronotum behind collar.
capitate processes mushroomlike or bladelike ends of dorsal connectives (apodemes) of basal plates, on which are inserted protractor muscles of phallus.
capsid member of family Miridae.
capsula seminalis (pl., **capsulae seminales**) spermathecal bulb, q.v.
cardinate coxae hinged and elongate hind coxae in pagiopodous taxa.

cephalic glands   maxillary glands, q.v.
cephalic neck   constricted posterior part of head, for most part inserted into prothorax.
Chagas' disease   disease of humans and other mammals in South and Central America, Mexico, and Texas, caused by flagellate protozoan *Trypanosoma cruzi* (Trypanosomatidae) and transmitted by assassin bugs (Reduviidae), especially *Triatoma* and *Rhodnius*
claspers   parameres, q.v.
claval commissure   junction of hemelytra along clavus on midline of body posterior to apex of scutellum and anterior to membrane, developed in most Panheteroptera.
claval furrow   claval suture, q.v.
claval suture   suture of forewing separating clavus from corium.
clavopruina   in Corixidae, a narrow, white frosted area along anterolateral margin of clavus.
clavus (pl., **clavi**)   usually parallel-sided and sharply pointed anal area of hemelytron.
claw hairs   setiform microtrichia on outer surface of a claw, e.g., in some Miridae.
claw plate   unguitractor plate, q.v.
clypeus   that part of head below frons, to which labrum is attached anteriorly; usually weakly to strongly protruding.
coiled duct   in Gerromorpha, accessory fecundation canal, q.v.
coleopteroid   beetlelike in form, often referring to structure of forewings.
collar   rounded or flattened anterior margin of prothorax.
collum (pl., **colla**)   collar, q.v.
common oviduct   proximal portion of female genital ducts, between lateral oviducts (whether ectodermal or endodermal) and vagina.
conceptaculum seminis (pl., **conceptacula seminis**)   in many Cimicoidea, mesodermal organs of sperm storage, being a differentiation of mesodermal oviducts.
conjunctiva (pl., **conjunctivae**)   intersegmental membrane of abdomen; proximal portion of endosoma of phallus in many Heteroptera.
conjunctival appendages   lobes or processes arising from conjunctiva in expanded aedeagus of many Heteroptera.
connexivum   lateral margin of abdomen, formed by dorsal and ventral laterotergites or laterosternites.
copulatory organ   phallus, q.v.
corial glands   in Plokiophilidae, numerous, large, unicellular glands with low conical openings on dorsal surface of corium.
coriopruina   in Corixidae, a white frosted area between anterior apex of corium and clavopruina, q.v.
corium (pl., **coria**)   in Panheteroptera, proximal coriaceous or otherwise differentiated part of forewing exclusive of clavus and distinct from membrane, often being subdivided into anterior (lateral) exocorium and posterior (mesal) endocorium; *see* cuneus and embolium.
costal fracture   in many Heteroptera, a short, usually transverse line of weakness or break in costal margin of forewing separating sometimes well-differentiated cuneus from rest of corium.
cotton stainer   species of Heteroptera that cause discoloration of cotton fibers by piercing unripe bolls for their sap, e.g., *Dysdercus* spp. (Pyrrhocoridae).
cricoid sclerite   in many Pentatomomorpha, a parietal differentiation of endosoma delimiting conjunctiva from vesica.
Cryptocerata   Nepomorpha
ctenidium (pl., **ctenidia**)   in Polyctenidae, comblike rows of flattened spines.
cubital furrow   simple or forked furrow or plica on hind wing posterior to Cu.

cultrate   shaped like a pruning knife.
cuneal incisure   costal fracture, q.v.
cuneus (pl., **cunei**)   in some Heteroptera, usually triangular posterolateral area of corium demarcated by costal fracture.
diadenian type   omphalian or diastomian type of metathoracic scent gland, with gland cells concentrated within paired glandular components of system and with scent reservoir(s) differentiated; *see* diastomian type, omphalian type.
diastomian type   in adult Heteroptera, metathoracic scent-gland apparatus opening by paired, widely spaced orifices associated with metacoxal cavities on metapleuron; *see also* omphalian type.
distipulvillus   distal membranous or setiform portion of pulvillus in Pentatomomorpha; *see* basipulvillus.
dorsal abdominal scent gland   in nymphal Heteroptera, 1–4 paired or unpaired ectodermal abdominal glands with paired or unpaired orifices situated intersegmentally or intrasegmentally, structure and function sometimes persisting into adult stage.
dorsal arolium   arolium, q.v.
dorsal laterotergite(s)   lateral plate of an abdominal tergum, often subdivided into outer and inner laterotergites.
dorsal paratergite   dorsal laterotergite, q.v.
ductus ejaculatorius (pl., **ducti ejaculatorii**)   median ectodermal efferent duct proximal to phallus, merging into ductus seminis.
ductus receptaculi   spermatheca, q.v.
ductus seminis   median ectodermal duct in phallus, from foramen ductus to secondary gonopore, frequently differentiated into ductus seminis proximalis and ductus seminis distalis.
ductus spermathecae   canal through which sperm enter spermatheca from vagina or bursa copulatrix, q.v.
ectospermalege (pl., **ectospermalegia**)   in some Cimicoidea, external pouchlike ectodermal part of spermalege, q.v.
egg cap   a lid, joined to body of egg along a line of weakness, that is forced off by hatching embryo.
ejaculatory reservoir   in Pentatomomorpha, Dipsocoromorpha, and Nepidae, complex differentiation of proximal end of ductus seminis in endosoma.
embolar groove   trough-shaped groove of forewing running parallel with costal margin anterior to medial fracture and often delimiting embolium; *see* medial fracture.
embolium   in forewing of some Heteroptera, broadened submarginal part of corium proximal to costal fracture; exocorium, q.v.
empodium (pl., **empodia**)   distal extension of unguitractor plate, or, often applied to any unpaired structure arising between claws; *see* arolium, parempodia, pulvillus.
endocorium   posterior (mesal in repose) part of corium between exocorium and clavus.
endosoma (pl., **endosomata**)   distal segment of phallus, free of ligamentary processes and surrounding ductus seminis distalis from ejaculatory reservoir (when present) to secondary gonopore.
epipharyngeal sense organ   anterior (x organ) and posterior (y organ) groups of sensilla located in epipharynx, apparently with sensory function related to feeding.
Euheteroptera   that taxon including the Dipsocoromorpha, Gerromorpha, Nepomorpha, Leptopodomorpha, Cimicomorpha, and Pentatomomorpha.
evaporatorium   evaporatory area, q.v.
evaporatory area   in most Cimicomorpha and Pentatomomorpha, area of specialized cuticle on metathoracic pleuron associated with, and usually surrounding, orifice and auricle of metathoracic scent glands, possibly functioning in controlled dissemination and evaporation of scent-gland products.

eversible gland in adult Saldidae, a gland located in intersegmental membrane of dorsal abdominal laterotergites 7 and 8.

exocorium that part of corium lying between R or R+M and costal margin; *see* cuneus, embolium.

expansion skating in Gerroidea, dispersing of fluid (probably saliva) onto water surface, which lowers surface tension causing insect to move much more rapidly than is otherwise possible.

extragenital insemination traumatic insemination, q.v.

false spiracles in Nepidae, hydrostatic organs, q.v.

fecundation canal accessory fecundation canal, q.v.

fecundation pump in some Gerromorpha, a widened area of accessory fecundation canal provided with a pair of platelike flanges.

first gonocoxae first valvifer; *see* valvifers.

first gonocoxopodites first valvifer; *see* valvifers.

first ramus connecting leaf of first valvulae.

first valvulae the outer blades of the ovipositor.

flagellum (pl., **flagella**) gonoporal process, q.v.; the 2 terminal segments of the antennae.

forceps (pl., **forcipes**) parameres, q.v.

fossula spongiosa in many Cimicomorpha, apically on one or more pairs of tibiae, a vesicular hemolymph-filled structure beset with adhesive setae.

frenum (pl., **frena**) lateral groove in upper margin of scutellum into which fits or catches channeled locking device on lower edge of clavus.

genital atrium vagina, q.v.

genital capsule pygophore, q.v.

genital chamber vagina, q.v.

genital segment(s) in males abdominal segment 9, in females abdominal segments 8 and 9; pygophore, q.v.

glochis spur or short vein in hind wing arising distally from Cu.

gonangulum in female, sclerite uniting first valvifer and laterotergite nine; attached internally to first valvula or at base of first ramus; obscured in external view by first valvifer.

gonapophyses (sing., **gonapophysis**) valvulae, q.v.

gonapophysis eight first valvula; *see* first valvulae.

gonapophysis nine second valvula; *see* second valvulae.

gonocoxite eight first valvifer; *see* valvifers.

gonocoxite nine second valvifer; *see* valvifers.

gonocoxites valvifers, q.v.

gonocoxopodites valifers, q.v.

gonoforcipes (sing., **gonoforceps**) parameres, q.v.

gonopods parameres, q.v.

gonoporal process elongate, sometimes coiled, distal portion of ductus seminis.

gonopore secondary gonopore, q.v.

gonostyli (sing., **gonostylus**) parameres, q.v.

guide in some male Enicocephalidae, reduced remnants of external genitalia arising from posteroventral margin of pygophore.

gustatory organ epipharyngeal sense organ, q.v.

Gymnocerata a grade group of Heteroptera with freely movable, conspicuous antennae.

gynatrial complex in Gerromorpha, a term referring to that portion of internal ectodermalia composed of gynatrial sac, spermathecal tube, and accessory fecundation canal; also applied to complex of homologous structures in other Heteroptera.

gynatrial glands ringed glands, q.v.

gynatrial sac vaginal pouch (q.v.) or bursa copulatrix (q.v.)

gynatrium vagina, q.v.

hamus (pl., **hami**) spur or short vein, sometimes pointed, projecting into middle cell of hind wing and representing M.

harpagones (sing., **harpago**) parameres, q.v.

helicoid process in some Pentatomomorpha, cricoid sclerite, q.v.

hemelytron (pl., **hemelytra**) forewing of Heteroptera, especially in Panheteroptera, with distinctly thickened proximal portion and membranous distal portion.

hemocoelic insemination traumatic insemination, q.v.

hemoglobin respiratory pigment found in hemolymph of Anisopinae (Notonectidae).

hood in Tinginae (Tingidae), elevated anterior part of prothorax, often covering head.

humeral angle posterolateral angle of pronotum.

humerus (pl. **humeri**) humeral angle, q.v.

hydranapheuxis in Gerromorpha and Leptopodomorpha, process of deforming meniscus of water surface to allow ascension to adjacent substrate.

hydrostatic organs in Nepidae, 3 pairs of ovoid structures on connexiva of abdominal sterna 3–5 near spiracles but not connected to tracheal system, which function in spatial orientation.

hygropetric pertaining to life on a thin film of water on a rock surface, as in some Gerromorpha.

hypandrium (pl., **hypandria**) process on ventroposterior margin of pygophoral rim (abdominal segment 9).

hypocostal lamina ventrally deflected proximal part of costal margin of forewing.

hypocostal ridge hypocostal lamina, q.v.

hypocular suture in Corixidae, short sulcus on either side of head capsule posteroventral to eyes.

hypopygium (pl., **hypopygia**) pygophore, q.v.

intercalary sclerites in Gerromorpha, Nepomorpha, and a few other groups of Heteroptera, 2 minute sclerotized plates dorsally between segments 3 and 4 of labium.

juga (sing., **jugum**) mandibular plates, q.v.

lacerate-flush feeding in phytophagous Heteroptera, process of lacerating and macerating cells with stylets and then flushing out material with saliva and imbibing it; *see* sawing-clipping feeding.

laciniate ovipositor ovipositor with elongate, often laterally compressed blades (valvulae).

lamina an expanded or platelike region, as body margins of Termitaphididae.

larval organ in many nymphal Saldidae, an apparently sensory structure, in form of depression, located on abdominal sternum 3 just mesad of spiracle.

lateral oviducts paired canals leading from ovaries to common oviduct, most frequently mesodermal, but in certain Heteroptera proximally mesodermal and distally ectodermal.

laterosternites lateral subdivisions of sterna of pregenital abdominal segments (e.g., in some aquatic bugs).

laterotergites dorsal and ventral laterotergites, q.v.

lima (pl., **limae**) stridulitrum, q.v.

lorum (pl., **lora**) maxillary plate, q.v.

m-chromosomes supernumerary autosomes, occurring most commonly in Lygaeoidea and in some Nepomorpha.

macropterous with both fore- and hind wings fully developed and functional.

macrotrichia a relatively large or elongate seta, as on abdomen of some Anthocoridae.

male hooks parameres, q.v.

mandibular plate that portion of head laterad of (posterior to) clypeus and dorsad of maxillary plate.

mating swarm a conspicuous cloud of insects, usually males, dancing or hovering over a marker or in lee of an obstruction, serving to attract solitary members of other sex, e.g., Enicocephalomorpha.

maxillary glands small paired glands opening near bases of maxillae.

maxillary plate that portion of head ventral to mandibular plate.

medial fracture longitudinal furrow delimiting exocorium (or embolium) from endocorium; *see* embolar groove.

mediotergite unpaired plate of an abdominal tergum with delimited, paired laterotergites.

membrane membranous apical portion of hemelytron in Panheteroptera and some other Heteroptera.

mesoscutellum scutellum, q.v.

mesospermalege (pl., **mesospermalegia**) in some Cimicoidea, subintegumental mesodermal portion of spermalege into which spermatozoa are injected; *see* ectospermalege.

metathoracic scent gland in Heteroptera, universally occurring adult system of paired or unpaired scent glands with single or paired opening on metasternum with external outflow channels (ostiolar canals) that transmit glandular products to ostiole located on metepisterna.

micropyle opening in chorion of egg through which sperm pass during process of fertilization; in most Pentatomomorpha manifested externally by elevated tubular or capitate processes.

microtrichia in Gerromorpha, that portion of body hair layer composed of fine spicules.

mutic without spines.

natatorial, natatory fitted for swimming, being generally applied to swimming legs in aquatic bugs (Nepomorpha).

neck cephalic neck, q.v.

Neoheteroptera that taxon including the Gerromorpha, Nepomorpha, Leptopodomorpha, Cimicomorpha, and Pentatomomorpha.

nodal furrow in Corixidae, costal fracture, q.v.

node costal fracture, q.v.

nymph immature form; larva.

ocular setae in many groups of Heteroptera, usually a pair of bristles located in disc of compound eye of early instars, often lost later in development.

odoriferous gland metathoracic scent gland, (q.v.) or dorsal abdominal scent gland (q.v.).

omphalian type in adult Heteroptera, metathoracic scent-gland apparatus usually with a single (rarely double) opening on metasternum (rarely on abdominal sternum 1), with paired or unpaired internal structures; *see* diastomian type.

omphalium prominent metasternal opening of omphalian type (q.v.) of metathoracic scent gland.

operculum (pl., **opercula**) egg cap (q.v.), as for example in most Cimicomorpha.

organ of Berlese spermalege, q.v.

ostiolar canal external outflow pathway of metathoracic scent gland, usually leading from metathoracic venter to metepisternum.

ostiolar groove ostiolar canal, q.v.

ostiolar peritreme in many Cimicomorpha and Pentatomomorpha, a calloused area of variable shape, surrounding the ostiole, and itself often surrounded by the evaporatorium.

ostiole external opening of metathoracic scent gland, often referring to opening on metepisternum.

ovipositor organ by which eggs are deposited, formed in Heteroptera by paired first and second valvulae.

Pagiopoda that unnatural assemblage of Heteroptera in which posterior coxae are usually elongate and articulation is a hinge joint; *see* cardinate coxae, Trochalopoda.

pagiopodous pertaining to Pagiopoda.

pala (pl., **palae**) in Corixidae, tarsus of foreleg modified into a seta-fringed scoop for particle feeding, and in males for attachment to females during mating or sexual display.

palm, palma in Corixidae, that portion of pala, usually pilose, lying between upper and lower row of palmar setae, sometimes furnished with stridulatory pegs.

palmar hairs in Corixidae, usually a row of long setae on lower margin of pala and a row of short setae along upper margin of palm.

Panheteroptera that taxon including Nepomorpha, Leptopodomorpha, Cimicomorpha, and Pentatomomorpha.

paraclypeal lobe mandibular plate, q.v.

paragenital sinus in Cimicoidea, external pocket or channel leading to external aperture of ectospermalege.

paragenital system in many Cimicoidea, various structural differentiations in females correlated with traumatic insemination; *see* spermalege.

parameres paired male genital structures independent of phallus, arising postembryologically from lateral parts of 2 buds (primary phallic lobes), median parts of which give rise to phallus.

parandrium (pl., **parandria**) one of a pair of expansions of external wall of pygophore in lateroventral position, provided with setae but not muscles.

paranota (sing., **paranotum**) in certain Tingidae, flattened or lamellate sides of pronotum.

parasternites laterosternites, q.v.

paratergites laterotergites, q.v.

parempodia (sing., **parempodium**) paired setiform or lamellate processes arising distally from unguitractor plate, between claw bases; *see* arolium, empodium, pulvillus.

pars intermedialis spermathecal pump, q.v.

pedicel the second antennal segment, sometimes subdivided, as in many Pentatomoidea.

peg plates in Gerromorpha and Ochteridae, minute circular depressions bordered by a shallow rim and filled with subconical pegs, generally found on head and body, and in some species also on certain leg segments.

Pendergrast's organ specialized organ found on abdominal venter of some female Acanthosomatidae.

penis (pl., **penes**) phallus, q.v.

penisfilum (pl., **penisfila**) in Saldidae, reel system, q.v.

periadenian type omphalian or diastomian type of metathoracic scent glands, with gland cells uniformly distributed in paired or unpaired components of system without differentiation of scent reservoir; *see* diastomian type, omphalian type.

peritreme ostiolar peritreme, q.v.

phallandrium in *Phallopirates* (Enicocephalidae), conspicuous bulbous copulatory organ composed largely of novel components including genital plates.

phallobase articulatory apparatus, q.v.

phallosoma proximal portion of phallus supported by or incorporating ligamentary process and surrounding ductus seminis proximalis to ejaculatory reservoir (if present), often referred to as phallotheca when sclerotized.

phallotheca sclerotized proximal part of phallosoma (q.v.), especially in Cimicomorpha and Pentatomomorpha.

phallus (pl., **phalli**) intromittent organ, including phallobase, aedeagus, and its various processes; *see* endosoma, phallosoma, phallotheca, vesica.

plastron in Aphelocheiridae and Cryphocricini (Naucoridae), a physical gill formed by a dense mat of microtrichia on ventral body surface.

plate-shaped ovipositor ovipositor with shortened valvulae which may be fused, reduced, and dorsoventrally compressed; *see* laciniate ovipositor.

plectrum ordinarily movable portion of stridulatory mechanism; *see* stridulitrum.

pleustonic of or pertaining to air-water interface, e.g., Gerridae are pleustonic.

ponticulus transversalis large dorsal rodlike transverse superior connection between basal plates in male phallus; *see* articulatory apparatus, basal plates.

pore-bearing plate in Hyocephalidae, an ovoid pore-bearing structure on each side of abdominal sternum 3.

posterior gonapophyses second valvulae, q.v.

posterior pronotal lobe posterior expansion of pronotum overlaying part or more rarely all of mesonotum.

posterior ramus second ramus.

posterior valvulae second valvulae, q.v.

posterior wall in female Miridae, a sclerotized, platelike structure lying between the rami of the second valvulae.

postnodal pruina in Corixidae, a white, frosted area along lateral border of corium posterior to nodal furrow (costal fracture).

preapical claw in Gerroidea, a condition in which pretarsus is inserted proximal to apex of last tarsal segment.

primary gonopore distal end of ductus ejaculatorius before entering phallus at level of basal foramen to merge into ductus seminis; *see* secondary gonopore.

principal salivary gland major salivary gland of paired salivary system, with 2 or more lobes, always associated with an accessory salivary gland.

processus gonopori flagellum, q.v.

proctiger reduced abdominal segment 10, bearing anus, possibly surrounding invaginated abdominal segment 11.

pronotal carina primarily in Tingidae, main or median carina or keel on pronotum.

prosternal furrow in most Reduviidae, a cross-striated longitudinal groove in prosternum, by means of which stridulation is caused by rubbing apex of rostrum in it by up-and-down movements of the head.

pseudarolium (pl., **pseudarolia**) in Miridae, pulvillus, q.v.

pseudomicropyle in eggs of Cimicomorpha, hollow chorionic micropyle-like processes used for gas exchange.

pseudoperculum an egg cap without a distinct sealing bar and in which eclosion is not result of fluid pressure.

pseudoplacental viviparity viviparity in which eggs contain little or no yolk and embryo presumably receives nourishment from a pseudoplacenta, e.g., in Polyctenidae.

pseudopulvilli in Miridae, paired pretarsal structures arising laterally from unguitractor plate, distinct from parempodia and often superficially resembling pulvilli; *see also* accessory parempodium.

pseudospermathecae in Pachynomidae, Reduviidae, and Tingidae, which lack functional spermatheca, 1 or 2 saclike or tubular diverticula arising from vagina or common oviduct, functioning as sperm-storage organs.

pseudospiracle in Nepidae, hydrostatic organ, q.v.

pulvillus (pl., **pulvilli**) in Miridae, some Anthocoridae, nearly all Pentatomomorpha, bladderlike pretarsal structures arising from ventral or mesal surfaces of claws; *see* arolium, basipulvillus, distipulvillus, empodium, parempodia, pseudopulvillus.

pygofer pygophore, q.v.

pygophore abdominal segment 9 of male, enclosing the phallus.

ramus (pl., **rami**) connecting leaf (or arm) of ovipositor valvulae.

raptorial adapted for seizing prey, e.g., forelegs of many predaceous Heteroptera.

receptaculum seminis spermatheca, q.v.

rectal organ proctiger, q.v.

reel system in Saldidae, differentiation of ductus seminis at junction of ductus seminis proximalis and distalis into a coiled tube.

remigium anterior part of wing chiefly involved in flight; the wing anterior to the claval suture.

respiratory siphon in Nepidae, paired caudal structures derived from abdominal tergum 8, forming a channel in nymphs and a long tube in adults, which connect with eighth abdominal spiracles, and which serve to replenish subhemelytral air store in these aquatic insects.

Ribaga's organ ectospermalege or spermalege, q.v.

ringed glands in some Heteroptera, paired or unpaired glands, dorsally or ventrally on vagina or on vaginal pouch, or bursa copulatrix, sometimes ringed by annular sclerotizations known as ring sclerites.

ring sclerite in some Lygaeidae, cricoid sclerite (q.v.); paired or unpaired annular sclerotization encircling ringed glands of vagina, vaginal pouch, or bursa copulatrix.

rostrum combined labium and maxillary and mandibular stylets.

rotatory coxae nearly globose hind coxae with a ball-and-socket articulation; *see* Trochalopoda.

salivary sheath lipoprotein sheath left in plant tissue, formed from hardened salivary secretions, encasing stylets as they penetrate plant tissue.

sawing-clipping feeding method of feeding in which stylets are moved back and forth in a straight line; *see* lacerate-flush feeding.

scape the basal segment of the antennae.

scent glands dorsal abdominal scent gland in nymphs (sometimes persisting to adulthood) and several types of scent glands in adults (*see* Brindley's glands, metathoracic scent gland), producing pheromones, allomones, venoms, and other substances, with often notorious and unpleasant smell for humans.

scent pore ostiole, q.v.

scent reservoir paired or unpaired reservoir of metathoracic scent glands; *see* diadenian type, diastomian type, omphalian type.

sclerotized ring ring sclerite, q.v.

scrobe a groove, as in foretibia of Phymatinae (Reduviidae) for reception of tarsus.

scutellum (pl., **scutella**) triangular part of mesothorax, generally placed between bases of hemelytra, but in some Pentatomoidea partly or completely overlapping them.

sealing bar in eggs of Cimicomorpha, a bar joining cap to rest of chorion, consisting of a very thin layer of resistant endochorion and a thick amber layer.

secondary gonopore opening of ductus seminis at or near apex of phallus.

secondary hypocostal ridge in adult Heteroptera, a secondary modification of hypocostal lamina.

second gonocoxae second valvifers, q.v.

second gonocoxopodites second valvifers, q.v.

second ramus connecting leaf of second valifer.

second valvifers valifers arising from abdominal segment 9.

second valvulae median blades of the ovipositor.

seminal duct ductus seminis, q.v.

seminal reservoir ejaculatory reservoir, q.v.

semiring sclerite in Colobathristidae, cricoid sclerite, q.v.

sieve pores peg plates, q.v.

spermalege (pl., **spermalegia**) in some Cimicoidea, an organ on pregenital abdominal segments receiving sperm during traumatic insemination and lacking a direct communication with genital apparatus itself; usually consisting of an external integumental pouch (ectospermalege) and an internal mesodermal part (mesospermalege).

spermatheca (pl., **spermathecae**) median, dorsal, unpaired, sclerotized diverticulum of vagina serving as sperm-storage receptacle; receptaculum seminis (q.v.); vermiform gland (q.v.);

spermathecal tube (q.v.); *see also* conceptaculum seminis, pseudospermathecae.
spermathecal bulb generally bulb-shaped terminal portion of spermatheca serving actual sperm-storage function.
spermathecal duct ductus spermathecae, q.v.
spermathecal gland vermiform gland, q.v.
spermathecal pump part of spermatheca between ductus spermathecae and spermathecal bulb, frequently differentiated—in true spermatheca—into a muscular pump with flanges.
spermathecal tube in infraorder Gerromorpha and some other Heteroptera, an elongate, looped spermatheca with glandular cells in its walls; *see* spermatheca.
spermatic duct in many Cimicoidea, duct arising from fusion of vasa deferentia.
spermatic furrow in many Cimicoidea, groove of left paramere in which runs interlocked phallus.
spermodes in Cimicidae, intraepithelial network of canals in walls of pedicels and paired oviducts through which spermatozoa pass from conceptacula seminis to ovarioles.
sperm reservoir ejaculatory reservoir, q.v.
spiracular line a line drawn through the spiracles on the abdominal venter of trichophoran Pentatomomorpha, used to refer to position of abdominal trichobothria relative to spiracles.
spongy fossa fossula spongiosa, q.v.
stapes basal plates, q.v.
staphylinoid condition in which hemelytra are reduced and truncate.
static sense organs *see* hydrostatic organs.
strainer in Hyocephalidae, pore-bearing plate, q.v.
stridulitrum ordinarily stationary portion of stridulatory mechanism; *see* plectrum.
strigil, strigile, strigilis stridulitrum, q.v.
styli (sing., **stylus**) parameres, q.v.
subgenital plate in most female Heteroptera abdominal sternum 7, in Enicocephalidae sternum 8.
submacropterous condition of wings in which corium and clavus of forewings are fully developed, with membrane being slightly to greatly reduced, hind wings generally being functional; *see* brachypterous.
supradistal plate in some male Enicocephalidae, dorsal cover of genitalia.
suspensorial apodemes internal muscle attachments to which are affixed basal plates of articulatory apparatus.
suspensory arms suspensorial apodemes, q.v.
suspensory processes suspensorial apodemes, q.v.
swarming in Enicocephalidae, aggregating in a mating swarm, q.v.
swimming fan in some Veliidae, e.g., *Rhagovelia* (Veliidae), fan-like structure usually formed from a modified ventral arolium and which aids in swimming on flowing water.
synthlipsis minimum interocular distance.
tegmen (pl., **tegmina**), **tegminal** a forewing not differentiated into proximal coriaceous and distal membranous part, as for example in Enicocephalomorpha; *see* hemelytron.
third valvulae a sheathlike structure of the ovipositor, fused with second valifers, absent in all Pentatomomorpha.
traumatic insemination in many Cimicoidea and some Nabidae, puncturing of body wall or wall of inner genitalia by phallus during mating and deposition of sperm outside usual reproductive tract.
trichobothrium (pl., **trichobothria**) specialized, slender, hairlike, sensory setae arising from and including tubercles or pits (bothria) on many body regions and appendages in Heteroptera; *see* bothrium.

trichomes modified setae present on certain myrmecophilous insects which give off secretions that ants (Hymenoptera: Formicidae) imbibe, e.g., base of abdomen in Holoptilinae (Reduviidae).
Trichophora those members of Heteroptera with trichobothria on pregenital abdominal sterna, i.e., Pentatomomorpha less Aradoidea.
Trochalopoda that grouping of Heteroptera in which posterior coxae are nearly globose and articulation is a ball-and-socket joint; *see* Pagiopoda.
true bug a heteropteran.
trypanosomiasis a disease caused by infection with *Trypanosoma* (Trypanosomatidae), transmitted by Triatominae (Reduviidae); *see* Chagas' disease.
tylus (pl., **tyli**) distal part of clypeus; anteclypeal region.
tymbal in some Heteroptera, a sound-producing membrane on abdominal segment 1 or segments 1 and 2.
tympanal organ organ sensitive to vibrations, on mesothorax of some Nepomorpha.
unguitractor plate sclerite lying between bases of claws, with which bases of claws articulate distally, to which retractor tendon is attached proximally, and from which parempodia arise distally.
vagina (pl., **vaginae**) ectodermal genital duct distal to common oviduct.
vaginal pouch variably formed pouch that may bear ringed glands, forming part of gynatrial complex.
valvifers (**1st** and **2nd**) in female Heteroptera, 4 plates or blades, 2 from abdominal segment 8, 2 from abdominal segment 9, articulating on corresponding paratergites and bearing first and second valvulae, respectively.
valvulae the blades (in laciniate type) of the ovipositor, which in 2 pairs, form the egg-laying apparatus, and which proximally attach to the body wall via one (or 2) pairs of corresponding rami.
ventral arolium arolium, q.v.
ventral glands glands located in metathorax of some Reduviidae, distinct from metathoracic scent glands.
ventral laterotergites ventrally situated laterotergites, distinct from dorsal laterotergites (q.v.) but generally fused with sternum and usually bearing spiracles.
ventral lobe in Gerromorpha, buccula, q.v.
ventral paratergites ventral laterotergites, q.v.
ventral plates in some male Enicocephalidae, fused genital plates.
ventral spine in Pentatomoidea, a spinelike projection anteriorly or third true abdominal sternum, directed toward head and lying at times between coxae.
vermiform gland in Cimicomorpha, an organ homologous with spermatheca but without sperm-storing function.
vesica (pl., **vesicae**) portion of endosoma of phallus, often sclerotized, surrounding ductus seminus distalis.
vesical process process arising from the vesica.
water bug member of infraorder Nepomorpha.
wing-coupling mechanism microtrichia-bearing structure at posteroventral margin of clavus of forewing, grasping leading edge of hind wing during expansion and flexion of wings and during flight.
wing-locking mechanism in Euheteroptera, modification of costal margin of forewing and mesothorax to retain wing firmly in postion in repose (*see* frenum); *see* wing-coupling mechanism.
x organ epipharyngeal sense organ, q.v.
xyphus, xiphus a spinous triangular process of prosternum or mesosternum, or both.
y organ epipharyngeal sense organ, q.v.

INDEX

abdomen, 49; expansion of, 159, 201; membranous, 164
abdominal gland, 198
abdominal grasping apparatus, 134, 135, 137, 140, 141, 143
abdominal hair pile, 108
abdominal organ, pregenital ventral, 52
abdominal scent glands: functional, in adults, 126, 267, 282; nonfunctional, in nymphs, 111
abdominal sense organs: in Aphelocheiridae, 126; in Naucoridae, 124
abdominal spine, 231
abdominal spiracles, 74, 78
abdominal sutures, fusion of, 97, 182, 269
abdominal tymbal, 160
Abedus, 113; *indentatus*, 112
Abies procera, 22
Abulites, 216
Acacia, 30, 184, 267, 277, 281; *cunninghami*, 169; *maidenii*, 169
Acalifa diversifolia, 233
Acalypta, 184
Acanthaspis, 158; *petax*, 160
Acanthia, 138
Acanthoberytus wygodzinskyi, 247
Acanthocephalini, 277
Acanthocerini, 277
Acanthocorini, 277
Acanthocrompus, 257
Acantholomidea, 238
Acanthomia horrida, 33
Acanthosoma haemorroidale, 217
Acanthosomatidae, 39, 58, 215, 216
Acanthosomatinae, 216
accessory gland, 55, 151
accessory parempodia, 48, 136, 168, 175
accessory salivary gland, 59, 60
Acer negundo, 36, 282
Acestra malayana, 273
Achaearanea tepidariorum, 160

Acinocoris calidus, 268; *stehliki*, 268
Acompocoris pygmaeus, 196
Acromyrmex lundi, 184
acus, 191, 194
adaptationist arguments, 4
Adelphocoris, 179; *lineolatus*, 34
Adenocoris, 212
adhesive pads, of hind coxae, 81–83, 135, 141, 142
Adrisa, 206
aedeagus, 49, 50
Aelia germari, 35; *rostrata*, 35
Aenictopecheidae, 68
Aenictopecheinae, 68
Aenictopecheini, 69
Aenictopechys necopinatus, 69
Aepophilidae, 136
Aepophilus, 137; *bonnairei*, 21, 136, 137
Aeptini, 232
aeropyles, 63, 146
Aesepus, 216
aestivation, 274
Aethus indicus, 33
Afrocimex, 201
Afrocimicinae, 201
Afropiesma, 266
Afrovelia phoretica, 101, 102
Agamedes, 216
Agathyrna ceramica, 277
Agave, 179
aggregation, 22, 238
aggregation pheromone, 279
aggregations, dry season, 274
aggressive behavior, 51
agonistic behavior, 51
Agonoscelis rutila, 35; *versicolor*, 35
Agrammatinae, 182
Agraptocorixa hyalinipennis, 122
Agriopocorinae, 275
Agriopocoris, 275
airstore: abdominal, 114, 128, 130; subhemelytral, 121

airstraps, 113, 114
alary muscles, of heart, 61, 63
Alathetus haitiensis, 232, 233
Alberproseniini, 159
alcohol preservation, 18
alfalfa, 34
alfalfa seed, 32
algae, as food, 122
Alienates, 70, 71; *elongatus*, 72
Alienatinae, 71
alimentary canal, 59, 60, 271, 273
alkaloids, 27, 263
Alloeorhynchus, 188, 189; *mabokei*, 187
allspice, 34
Almeida, 197
Almeidini, 197
Alydidae, 29, 32, 58, 271, 272
Alydinae, 273
Alydus, 273, 274
Amaranthaceae, 267
Amblypelta cocophaga, 33
Ambrysini, 125
Ambrysus magniceps, 126
Amemboa, 105
American Museum of Natural History, 15
Ammianus alberti, 181
Amnestinae, 222
Amnestus, 222; *spinifrons*, 221; *subferrugineus*, 224
Amorbini, 277
Amphaces, 216
Amphibicorisae, 1
Amphibolus venator, 36
Amulius, 157
Amyot, Charles Jean-Baptiste, 1, 6
anal furrow, 44
anal lobe, 44
anal veins, 43, 44
Anasa, 33, 279; *alfaroi*, 277; *guayaquila*, 277; *tristis*, 33
androconium, 58
Androctenes, 202
Andromeda, 36
Aneuraptera cimiciformis, 210
Aneurinae, 210
Aneurus, 208, 210, 213; *laevis*, 211
Angilia, 101
Angilovelia, 51, 101
Anhomoeini, 277
Anisopinae, 128
Anisops, 51, 128, 129; *megalops*, 129
Anisoscelidini, 277
Anolis, 27
Anommatocoris, 183; *coleoptratus*, 180, 184
Anoplocerus elevatus, 278
Anoplocnemis, 277; *curvipes*, 33
ant mimetic, 29
ant mimicry, models for, 29
anteclypeus, 41
antennae, 41, 149;
antennae, concealed below head, 107, 108; flattened, 278; geniculate, 214; modified, 105, 106, 178; pseudosegmented, 156
antennal cleaner, 47, 117
antennal segmentation, 107
Antestiopsis lineaticollis, 35
Anthocoridae, 32, 195, 197

Index 323

Anthocorini, 197
Anthocoris, 197; *nemorum*, 196
Antillocorini, 40, 260
Antiteuchus, 233; *tripterus limbativentris*, 233
ants, attraction of, 160
anus, 49
aorta, 61
Aphelocheiridae, 55
Aphelocheirus, 127; *aestivalis*, 127; *lahu*, 109; *malayensis*, 127
Aphelonotinae, 39, 149
Aphelonotus, 54, 149, 150; *africanus*, 150; *fuscus*, 150
aphids, predators of, 32
Aphylidae, 39, 218
Aphylum bergrothi, 218; *syntheticum*, 218
Apiomerini, 157
Apiomerus, 157
Apodidae, 201
Apoplymus, 248
aposematic coloration, 27, 177, 257, 263, 269, 279, 283
apples, 34
Apterola, 257
aptery, 23
Aquarius, 105; *lacustris*, 63; *paludum*, 104
aquatic bugs. *See* Nepomorpha
aquatic habitats, 21, 107
Araceae, 179
Arachnocorini, 188
Arachnocoris, 21, 186, 188, 192
Arachnophila cubana, 192
Aradacanthia, 210
Aradellini, 154
Aradidae, 39, 208, 209
Aradinae, 210
Aradiolus, 210
Aradus, 208, 210, 211; *cinnamomeus*, 211, 213
Arbela carayoni, 187
areolate sculpturing, 183
Arhaphe, 269; *carolina*, 269
Arilus, 150, 157
Armstrongocoris, 156
arolia, 46, 146. *See also* dorsal arolium; parempodia; ventral arolium
Artabanus lativentris, 211
Artemisia, 36
Artheneinae, 255
Artheneini, 255
arthropod predators, 263
Asclepias curassavica, 22
Asclepios, 105
ash-gray leaf bugs. *See* Piesmatidae
Ashlock, Peter D., 6
Asopinae, 35, 230
aspirator, 17
Aspisocoris termitophilus, 213
Aspongopus, 206
assassin bugs. *See* Reduviidae
Asterocoris australis, 211
Astragalus, 36
Atractotomus, 178
aubergine, 36
Auchenorrhyncha, 5

Augocoris, 239
Auricillocorini, 178
Australian National Insect Collection, 16
Australian Region, 38, 39
Australmeida, 197
Australostolus, 70; *monteithi*, 44, 69
Austronepa, 116
Austronepini, 116
Austropamphantus woodwardi, 255
Austrovelia, 88
autapomorphy, 4
Azalea, 36

Baccharis, 184
Bacillometra, 95
backswimmers. *See* Notonectidae
Baclozygum, 167; *brevipilosum*, 169; *depressum*, 167–169
Bactrodes, 154, 156
Bactrodinae, 154
Bactrophyamixia, 274
Bagrada cruciferarum, 35
balloon vine, 282
bananas, 33, 184, 266
Banksia, 169
Baptista, 101
Barber, Harry G., 6
Barce, 54
bark, as habitat, 163
bark bugs. *See* Aradidae
Barreratalpini, 277
basal articulatory apparatus, 49
basal foramen, 49
basal plates, 49
basiflagellum, 41
basipulvillus, 205
Bassian Subregion, 39
bat parasites. *See* Polyctenidae
Bathycles amarali, 263
Bathycoelia thalassina, 35
bats, as hosts, 201, 203
Beamerella balius, 171; *personatus*, 174
beans, 33–36, 266
beating sheet, 17
bed bugs. *See* Cimicidae
beetles, myrmecomorphic, 29
Belonochilus numenius, 263
Belostoma, 113
Belostomatidae, 45, 111, 112, 114
Belostomatinae, 113
Berg, Carlos, 6
Bergevin, Ernest de, 6
Bergroth, Ernst Evald, 6
Berlese funnel, 17
Bertilia, 201
Berytidae, 32, 246, 248
Berytinae, 248
Berytinus, 248; *hirticornis*, 247; *minor*, 247
Berytus, 248
Betula, 263
Beyeria, 267
bibliographies, 14
big-eyed bugs. *See* Geocorinae
Bilia, 197, 198
Bilianella, 197, 198
Biliola, 197, 198

biological control, 257
Biosystematics Research Centre, Agriculture Canada, 15
Biprorlulus bibax, 35
bird nests, as habitats, 195
birds, as predators, 28
Bishop Museum, 15
bivoltine life cycle, 25, 179, 264
black pepper, 34
Blaptostethini, 197
Blaptostethoides, 197
Blaptostethus, 197
Blatchley, W. S., 6
Blaudusinae, 216
Blaudusini, 216
Bledionotinae, 255
Bledionotini, 255
Bledionotus, 29; *systellonotoides*, 29, 255
Blindwanzen. *See* Miridae
Blissinae, 45, 255, 263
Blissus insularis, 33; *leucopterus*, 33; *leucopterus hirtus*, 33
Bliven, B. P., 7
Bloeteomedes, 161
blood feeding, 21, 159, 263
blue jays, 283
Boisea, 57, 282, 283; *trivittata*, 36, 281, 282
Bolboderini, 159
Boraginaceae, 224
Boreostolus, 68–70; *americanus*, 70; *sikhotalinensis*, 70
Bothriomirini, 176
Botocudo cavernicola, 261
box elder, as host, 282
box elder bug, 36, 282
Boxiopsis madagascariensis, 34
Bozius respersus, 237
Brachelytron angelicus, 232
Brachylybas, 277
Brachymetra, 104
Brachyplatys, 237
brachyptery, 23, 264
Brachysteles, 197
brain. *See* supraesophageal ganglion
Breddin, Gustav, 7
bright coloration, 27, 240
Brindley's gland, 57, 148, 151, 159, 181
broad-headed bugs. *See* Alydidae
broad-shouldered water striders. *See* Veliidae
Brochymena, 233
Bromeliaceae, as habitat, 101
bronze orange bug, 36
brooding behavior, 114
Bryocorinae, 34, 173
Bryocorini, 48, 173
Bryocoris, 173
bucculae, 41
buccular bridge, 41
Buchananiella, 197
Bucimex, 201
Buddleia, 263
Buenoa, 51, 53, 128, 129
Burmeister, Hermann Carl Conrad, 7
burrower bugs. *See* Cydnidae

bursa copulatrix, 50, 51
Burtinus, 274
Butler, Edward A., 7

cacao, 34, 35
Cacodminae, 201
Cacodmus bambusicola, 200
caducous wings, 23
Caenusia, 253
Cajanus cajan, 33
California Academy of Sciences, 15
Caliotis colarata, 198
Calisiinae, 210
Calisiopsis, 210
Calisius, 210
calli, 43
Callichilella grandis, 177
Callidea, 240
calling signals, 106
Callitris preissii, 228
Calocoris norvegicus, 34
Calotropis procera, 22
Camarochilus, 149, 150; *americanus*, 150
camouflage, 160, 235
campaniform sensillum, 46, 89, 96
Camptopus, 273; *lateralis*, 32, 273, 274
Campyloneura virgula, 179
Canopidae, 40, 219
Canopus, 219, 220; *burmeisteri*, 220; *caesus*, 219, 220; *impressus*, 220; *orbicularis*, 220
canopy fogging, 17
Cantacaderinae, 182
Cantacaderini, 182
Cantacader quadricornis, 183
Cantao ocellatus, 240
cantharadin, attraction to, 180
capsids. *See* Miridae
Caravalhoma malcolmae, 174
Carayonia, 151
Carayon's glands. *See* ventral glands
Carcinocorini, 154
Carcinocoris, 45
cardenolides, 27, 28
Cardiastethus, 197
cardinate coxae, 45
Cardiospermum corindum, 282; *halicacabum*, 282
Carpocoris, 35
Carrabas, 224
Carthasini, 188
Carthasis, 187
Carvalhoma, 179; *malcolmae*, 169
Carventinae, 210
Caryophyllaceae, 267
cashew, 34
Casuarina, 277
Catadispon, 216
catalogs, 14
Cataractocorini, 125
Catorhintha, 279; *siblica*, 277
Caulotops, 179
Cavaticovelia, 88, 90
Cavelerius sweeti, 34
Cavernicolini, 159
caves, as habitats, 90, 160, 224, 264

Cecyrina, 230
Ceiba, 35
central duct, 43
central ganglion, 61, 62
central nervous system, 61
Centrocneminae, 154
cephalonotum, 131
Ceraleptus obtusus, 278
Ceratocapsini, 178
Ceratocapsus, 178
Ceratocombidae, 75, 77
Ceratocombinae, 77
Ceratocombini, 77
Ceratocombus, 77; *corticalis*, 77; *mareki*, 76
Ceratocoris, 236, 237
Cercotmetus, 116
cereal crops, 232
Cethera, 156
Cetherinae, 39, 156
Cetherini, 156
Chaetometra, 95, 97
Chagas' disease, 36, 159
character polarity, 4
Chariesterini, 277
Charmatometra, 104
Charmatometrinae, 104
Chartoscirta, 140; *cocksii*, 139
Chauliopinae, 266
Chauliops, 34, 266
Cheirochela, 126
Cheirochelinae, 125
Cheirochelini, 125
chelae, on forelegs, 159
Chelinidea, 33
Chelinideini, 277
Chelonocoris ferrugineus, 211; *javensis*, 211
Chenopodiaceae, 267
Chepuvelia, 93, 94; *usingeri*, 94
Chilacis typhae, 264
Chiloxanthinae, 140
Chiloxanthus, 140
Chimarrhometra, 105, 106
China, William Edward, 7
Chinamyersia, 212
Chinamyersiinae, 39, 212
Chinamyersiini, 212
Chinaola, 163; *quercicola*, 163
chinch bug, 33, 255
Chironomidae, 39
Chiton, 215
Chlamydatus, 178
chlorazol black, 19
chorion, 62
Chorosomini, 282
chromosome numbers, 63, 243, 245, 256
chromosomes, 63
Chrysocoris, 240
Chryxinae, 156
Chryxus, 156
cibarial dilator muscles, 43
cibarial pump, 60
Cimex, 201; *hemipterus*, 32, 201; *lectularius*, 32, 200, 201
Cimicidae, 32, 199, 200
Cimicinae, 201

Cimicoidea, 49, 51
Cimicomorpha, 5, 43, 46, 48, 57, 146
cinchona, 34
Ciorulla, 163
circulatory system, 61, 63
Cistaceae, 267
citrus, 35, 36
cladistics, 4, 38
claspers. *See* parameres
claval commissure, 44, 243
claval suture, 44
Clavigralla elongata, 33; *gibbosa*, 33; *tomentosicollis*, 33
clavus, 43
claws, 46, 48; cleft, 176, 177; spicules on, 176; with subapical tooth, 176–178; toothed, 156; unequal, 192
cleaning, ultrasonic, 18
Cleontes, 157
Clerada, 263; *apicicornis*, 260
Cleradini, 260
cliff swallows, 33
Cligenes subcavicola, 264
Clivinemini, 177
Cloresmini, 277
Closterocoris, 173, 177
clypeus, 41; swollen, 237
Cnethocymatia, 121
Cobben, René H., 7
Cocles, 72
coconuts, 33
Codophila, 35
coffee, 35, 36
Coleoptera larvae, as prey, 233
Coleopterocoris, 123
Coleopterodes, 184
coleopteroid adults, 184
coleoptery, 23
Coleorrhyncha, 5
collar, 41
Collartidini, 156
collecting equipment, 17
collections, 15
collum. *See* collar
Colobathristes chalcocephalus, 249
Colobathristidae, 33, 40, 249
Colobathristinae, 250
color, changing, 30
colorado potato beetle, 35
coloration: abdominal, 116; aposematic, 27, 177, 257, 263, 269, 279, 283; bright, 27, 240; cryptic, 283; deflective, 30, 279; metallic, in nymphs, 269; myrmecomorphic, 29; protective, 30
Colpurini, 277
commensals, 21
competition, 22
compound eyes, 41; absent, 202, 214; reduced, 201
concealment, 31
conjunctiva, 49, 50
connexivum, 49
continental drift, 38
Copium, 184
Coptosoma, 237, 238; *inclusa*, 237; *lyncea*, 22, 238; *scutellatum*, 237

Index 325

copulation method, 257
copulatory position, 126, 134
copulatory tube, 191–193, 195, 198, 199
Coquillettia, 29; *insignis*, 27, 28
coral, as habitat, 97
Corallocoris, 141; *nauruensis*, 142
coral treaders. *See* Hermatobatidae
Corcovadocola, 179
Corecoris, 33
Coreidae, 33, 45, 46, 58, 274, 275
Coreinae, 277
Coreini, 277
corial glands, 56, 59, 190, 192, 193
Coridius, 206; *janus*, 33, 226; *viduatus*, 226
Coridromius, 45
Corimelaena lateralis, 223; *pulicaria*, 223
Corimelaeninae, 222
corium, 44
Corixa dentipes, 122; *jakowleffi*, 122
Corixidae, 45, 120
Corixinae, 121
Corizomorphini, 282
Corizus hyoscyami, 282
corn, 33
corpora allata, 61
corpora cardiaca, 61
Corythaica cyathicollis, 37; *monacha*, 37
Corythuca, 182; *ciliata*, 36; *gossypii*, 36
Cosmocoris, 240
costa, 43, 44
costal fracture, 43, 69, 139, 193
cotton, 33–36, 240
cotton stainers. *See* Pyrrhocoridae
coupling mechanisms, wing-to-body, 44
courtship, 51
courtship signals, 106
coxae, 45
coxal combs, 220, 224, 243
creeping water bugs. *See* Naucoridae
Creontiades pallidus, 34
critical point drying, 18
crop damage, 32
crop plants, 34
Cruciferae, 35
crusader bug, 33
Cryphocricinae, 125
Cryphocricini, 125
Cryphocricos, 24, 125; *hungerfordi*, 24; *latus*, 47, 108, 109
Cryptacrus, 240
Cryptocerata, 1
Cryptophysoderes fairchildi, 158
Cryptorhamphinae, 256
Cryptorhamphus, 256
Cryptostemma, 70, 78; *incurvatum*, 79
Cryptovelia, 88, 90; *terrestris*, 88, 89
ctenidia, 202, 203
Ctypomiris, 179
cubitus, 43, 44
Cucumis, 33
Cucurbitaceae, 33, 226
cuneus, 43, 162, 171
Curicta, 116
Curictini, 116
cursorial legs, 45
Cyclopelta obscura, 33

Cydamus, 274
Cydnidae, 33, 220, 222
Cydninae, 45, 222
Cydnus aterrimus, 221
Cylapinae, 39, 176
Cylapini, 176
Cylapocoris, 179
Cylapus, 175, 179
Cylindrostethinae, 104
Cylindrostethus, 104, 105; *palmaris*, 104
Cyllarini, 277
Cylopelta, 225
Cymatia, 121
Cymatiainae, 121
Cyminae, 30, 256, 263
Cymini, 256
Cymophyes, 258
Cymus, 22; *angustatus*, 254
Cyperaceae, 255
Cyphopelta, 173, 177
Cyrtocorinae, 231
Cyrtocoris trigonus, 233
Cyrtomenus crassus, 223
Cyrtopeltis ebaeus, 175; *tenuis*, 34
Cyrtopeltocoris, 173

Daclera punctata, 273
Daladerini, 277
Dallas, William Sweetland, 7
damsel bugs. *See* Nabidae
Darwinivelia, 88, 90
Dasycnemini, 154
Dasynini, 277
Dayakiella, 249, 250; *brevicornis*, 250; *sumatrensis*, 250
Dayakiellinae, 250
deciduous wings, 23
defensive behavior, 51
defensive reaction, 114
deflective coloration, 30, 279
De Geer, Carl, 7
Deliastini, 156
deotocerebrum, 61
Departamento de Entomología, Museo de La Plata, 16
Department of Biology, Nankai University, Tianjin, China, 16
Deraeocorinae, 177, 179
Deraeocorini, 177
Deraeocoris, 177
dermal glands, 58
Deroplax, 239
deutocerebrum, 62
Diactor, 46
diapause, 264
Diaprepocorinae, 121
Diaprepocoris, 121
Diaspidiini, 157
Diaspidius, 157
Diconocoris capusi, 182
Dicranocephalus, 283; *agilis*, 284; *albipes*, 284; *pallidus*, 284; *setulosus*, 284
Dictyonota strichnocera, 59
Dicyphina, 48, 173
Dicyphini, 173
Dicyphus, 173, 178

Diemeniini, 232
Dieuches, 262
Dilompini, 255
Dilompus, 255
Dinidor, 225, 226
Dinidoridae, 33, 225
Dinidorinae, 225
Dinidorini, 225
Dinomachus, 264
Diocoris, 46
Diolcus irroratus, 240
Diphlebini, 177
Diplonychus, 109, 113; *rusticum*, 113
Diplura macrura, 192
Dipsocoridae, 78
Dipsocoris, 78
Dipsocoromorpha, 46, 48, 74
Discocephalinae, 231
Discocoris, 167, 169; *drakei*, 166, 168
Discogastrini, 277
disc-shaped organs, on female abdomen, 227
dispersal, 22, 263
dispersalist hypotheses, 38
Distant, William Lucas, 7
Distantiella theobroma, 34
distasteful species, 27
distiflagellum, 41
distipulvillus, 205
distribution, 38
Ditomotarsinae, 216
Ditomotarsini, 217
Doesbergiana, 225
Dolichiella, 190
Dolichocephalometra, 95, 97
Dolycoris, 35
dorsal arolium, 48, 91, 93, 96, 98, 109, 135
dorsal vessel, 61, 63
Douglas, John William, 8
Drake, Carl J., 8
Druckknopfsystem, 45
Drymini, 260, 263
Dryophilocoris, 178
ductus ejaculatorius, 49
ductus seminis, 49
Dufour, Leon, 1, 8
Dufouriellini, 197
Dulinius unicolor, 35
dust bugs, 31
dwellings of humans, as habitats, 160
Dysdercus, 23, 28, 35, 36, 40, 58, 270, 271; *bimaculatus*, 270; *blotei*, 271; *discolor*, 271; *fasciatus*, 58, 270, 271; *koenigii*, 271; *obscuratus*, 28

Eccritotarsina, 40, 176, 179
Eccritotarsini, 176
Eccritotarsus, 179
economic importance, 32
Ectinoderini, 157
Ectinoderus, 157
Ectomocoris, 157
ectoparasites, 201, 203
ectospermalege, 63, 198
Ectrichodia gigas, 160
Ectrichodiinae, 156, 160

Edessa, 48, 54
Edessinae, 232
egg burster, 62
egg cap. *See* operculum
egg guarding, 160, 213, 233, 240
eggs, 62, 64; deposition of, 160
ejaculatory reservoir, 49
Ekblom's Organ, 186, 188, 189
Elaeocarpus obovatus, 169
Elasmodema, 154; *erichsoni*, 152, 155
Elasmodeminae, 58, 154
Elasmolomus sordidus, 34
Elasmostethus cruciatus, 217
Elasmucha dorsalis, 217; *putoni*, 217
electric light bugs. *See* Belostomatidae
Ellenia, 173
Elvisurini, 239
Emballonuridae, 204
Embiidina, 193
Embiophila, 192; (*Acladina*), 192; *africana*, 193; *myersi*, 192
Embiophilinae, 192
embolium, 44
embryonic development: in Lyctocoridae, 195; in Plokiophilidae, 193
Emesinae, 156
Emesini, 156
Empicoris, 150
Enalosalda, 138, 140
endocrine glands, 61
endosoma, 49, 50
Engistus, 257
Enicocephalidae, 70, 71
Enicocephalinae, 71
Enicocephalini, 72
Enicocephalomorpha, 39, 43, 45, 46, 48, 67
Enicocephalus, 72
Enithares, 128, 129; *maai*, 129
Entomological Laboratory, Kyushu University, 16
Entomovelia, 99; *doveri*, 100
Eobates, 104
Eocanthocona furcellata, 35
Eoctenes, 202; *spasmae*, 203
Eotrechinae, 105
Eotrechus, 105, 106
epipharyngeal sense organ, 41
Erianotus, 145; *lanosus*, 144
Esaki, Teiso, 8
Esakia, 105
esophagus, 59, 60
Ethiopian Region, 38, 39
Eucalyptus, 218, 281; *camaldulensis*, 218; *globosus*, 169; *trachyphloia*, 169
Eugenia, 229
Eumenotes, 225, 226; *obscura*, 226
Eumenotini, 225
Euphenini, 156
Eupheno, 156
Euphorbiaceae, 233, 283
Eupychodera corrugata, 240
Eurydema, 35
Eurygaster, 36, 239, 240; *integriceps*, 36
Eurygastrinae, 239
Eurygerris, 102, 105

Eurylochus, 158
Eurymetra, 106; *natalensis*, 104
Euryophthalmus, 269
Euschistus, 35, 233
Eusolenophora, 190
Eusthenini, 242
Euthetus, 272, 273
Eutorpe edulis, 169
Euvelia, 99
evaporatory area (evaporatorium), 43, 56, 57, 78, 83, 146, 205
eversible glands, 58, 137, 138, 140
exocorium, 161
exocrine glands, 55
expansion skating, 102
Eysarcoris ventralis, 35

Fabaceae, 226
Fabricius, Johann Christian, 1, 8
Falconia, 173
Fallén, Carl Friedrich, 8
faunistic studies, 15
fecundation canal, 89, 91, 100, 104
fecundation pump, 100, 104
feeding, 20
feeding cone, 20
feeding habits, in Joppeicidae, 165
female genitalia, 50
femora, 45
fertilization, 51, 188, 195; in vitellarium, 190
Feshina, 75; *schmitzi*, 77
Ficus, 217, 264
Fieber, Franz Xavier, 8
figs, 32, 263
filter chamber, 60
first ramus, 50
first thoracic ganglion, 62
first valvifer, 50
first valvula, 50
flake cuticle, 57
flat bugs. *See* Aradidae
flat-podded golden rain tree, 282
flight, loss of, 29
flight cycle, 263
flight intercept traps, 17
flightlessness, 23
floral glands, 58
flower bugs. *See* Anthocoridae
flowers, feeding on, 179
flowing water, as habitat, 106
foam masses, as habitat, 101
food canal, 41, 43
forefemora, 45; enlarged, 45, 117, 148, 159, 263
foregut, 60
foretarsi, 77
foretibiae, spinose, 243
foretibial sense organ, 116
forewing, 43, 44; folding, 219, 220, 228
Fort Morgan virus, 33
fossils, 67, 140, 143
fossorial legs, 45, 224
fossula spongiosa, 47, 48, 148, 151, 159, 186, 190, 195, 199
frenum, 44

Froeschnerana mexicanus, 170
frontal ganglion, 62
frontoclypeal sense organ, 130
fruit feeding, 270
Fucus, 137
Fulviini, 176
Fulvius, 176, 179; *quadristillatus*, 156
fungal mycelia, 213
fungi, as hosts, 21, 179, 213, 220
Fusarium, 33

Galeatus, 182
Galgulus, 117; *ovalis*, 223
gall formation, 184
Gamostolini, 69
Gamostolus, 68–70; *subantarcticus*, 69
Gampsocoris culicinus, 247; *panormimus*, 247
Gargaphia torresi, 37
Garsauria, 222
Garsauriinae, 222
gastric caeca, 59, 60, 205, 271, 273
Gastrodes, 253
Gelastocoridae, 45, 118
Gelastocoris, 30, 117, 118; *oculatus*, 47, 62, 108, 109, 117, 118; *peruensis*, 117
genital capsule, 49; rotated, in male Helotrophidae, 130
genitalia: asymmetrical, 49; as characters, 3; dissection and preparation of, 19; female, 49; male, 49
Geocorinae, 34, 257, 263
Geocoris, 34
Geocorisae, 1, 4
Gerridae, 25, 45, 49, 102, 103
Gerrinae, 105
Gerrini, 105
Gerris, 105, 106; *incurvatus*, 86
Gerromorpha, 45, 46, 48, 87
Ghandi bug, 32
giant water bugs. *See* Belostomatidae
Gigantemetopus rossi, 177
Gigantometra, 102, 104
glandular cuticular structures, 183
glandular pilosity, of host plants, 248
glandular setae, 264, 266; in nymphs, 247, 248. *See also* viscid setae
Gleditsia triacanthos, 184
Glyptocombus, 82; *fluminensis*, 211
Gmelin, Johann Friedrich, 8
Gnostocoris, 212
Godefridus, 161
golden rain tree, 282
gonangulum, 225, 277
gonapophyses. *See* valvulae
Gondwanaland, 39, 40
Gondwana patterns, 39
Gonianotini, 260
Gonocerini, 277
gonocoxae. *See* valvifers
gonocoxopodites. *See* valvifers
gonoplacs. *See* third valvulae
gonopore, secondary, 49
gonostyli. *See* valvulae
Gonystus, 256; *nasutus*, 256
Goondnomdanepa, 116

Goondnomdanepini, 116
Gorpini, 188
Gorpis, 188
gourds, 33
Gramineae. *See* Poaceae
grapes, 34
grasses, 34, 177, 179, 184, 250, 274
gregarious nymphs, 184
grooming comb, 46
grooved setae, 98, 101
ground-living species, 263
Guadua, 274
Guapinannus, 75
guava, 34
gula, 41
gummosis, 33
gut. *See* alimentary canal
Gymnocerata, 1
gynatrial complex, 50
gynatrial sac, 89, 100

habitat partitioning, 129
habitats: permanent, 24, 264; temporary, 24
Hadronema, 27, 180; *uhleri*, 180
Haematosiphoninae, 201
Haematosiphon inodorus, 33
Hahn, Carl Wilhelm, 8
hair pile, 84
Hakea, 217
Hallodapini, 178
Hallodapus, 179; *albofasciatus*, 52
Halobates, 105, 106; *sericeus*, 86
Halobatinae, 105
Halobatini, 105
Halosalda, 140
Halovelia, 99, 101, 102
Haloveliinae, 99
Halovelioides, 99, 102
Halticini, 45, 177
Halticotoma, 179
Halticus, 45, 177
Halyini, 232
Hammacerinae, 156
Handlirsch, Anton, 8
harlequin cabbage bug, 35
harlequin lobe, 63
Harmostini, 282
Harpactorinae, 58, 156, 157
Harpactorini, 157
harpagones. *See* parameres
Harpocera thoracica, 178
Harrisocoris, 157
head, 41
hearing, 53
heart, 61
heartseed vine, 282
Hebridae, 90, 92
Hebrinae, 92
Hebrometra, 92
Hebrovelia, 99, 101
Hebroveliini, 99
Hebrus, 92; *pusillus*, 91, 92; *ruficeps*, 92; *sobrinus*, 85; (*Subhebrus*), 90
Heidemann, Otto, 8
Heissia, 210
Hekista, 173

Helopeltis, 34, 176; *westwoodi*, 171
Helotrephes, 131; *admorsus*, 132; *bouvieri*, 132
Helotrephidae, 45, 131
Helotrephinae, 131
hematophagy. *See* blood feeding
hemelytra, 5, 43, 44; padlike, 199
Hemiptera, classification of, 1
hemoglobin, 129
Henestarinae, 257
Henestaris, 257
Heniartes, 157
Henicocorinae, 257
Henicocoris monteithi, 257
Henschiella, 72
Heraeus triguttatus, 29
Hermatobates, 97; *breddini*, 97; *weddi*, 98
Hermatobatidae, 97
Herrich-Schaeffer, Gottlieb A. W., 8
Hesperocorixa interrupta, 122
Hesperoctenes, 202–204; *fumarius*, 203
Hesperocteninae, 202
Hesperolabops, 179
Heteroblissus anomilis, 51
Heterocleptes, 95, 96; *hoberlandti*, 96
Heterocleptinae, 95
Heterocorixa, 121
Heterocorixinae, 121
Heterogaster, 257
Heterogastrinae, 257, 264
Heteroptera, 1, 45; classification of, 3, 5; as prey, 188
Heteropterodea, 5
Heteroscelis, 230; *lepida*, 231
Heterosceloides lepida, 231
Heterotermes convexinotatus, 215
Heterotrephes admorsus, 132
heterozygotes, 25
Hibiscus, 35
hind femora, 45
hind gut, 59, 60
hind legs, modified, in *Rheumatobates*, 106
hind tibiae, expanded, 278
hind wings, 44
Hipposideridae, 204
Hirundinidae, 201
historical biogeography, 38
Hoffmanocoris, 157
Holarctic, 39
Holoptilinae, 39, 58, 154
Holoptilini, 154
Homalocoris, 156
Homoeocerini, 277
homology, 4
Homoptera, 1, 5
homozygotes, lethal, 25
Hoplitocoris, 72
Horcias scutellatus, 174
Horvath, Geza, 9
Horvathinia, 113
Horvathiniinae, 113
Horvathiolus gibbicollis, 24
host choice, in Coreidae, 278
host plants, with glandular hairs, 179
Hotea, 239
hot springs, as habitats, 132

house sparrows, 33
Hsiao, Tsai-Yu, 9
human blood, as food, 195
human food: Belostomatidae as, 114; Corixidae as, 122
humans, as hosts, 201
humeral angles. *See* pronotum
Hungerford, Herbert Barker, 9
Husseyella, 101
Hyaliodini, 177
Hyalochloria, 179
Hyalonysius pallidomaculatus, 263
Hyalopeplini, 177
Hyalymenus, 273
Hydara, 275
Hydarini, 277
Hydrocorisae, 1
Hydrocyrius, 113
Hydrometra, 52, 95–97; *martini*, 96; *stagnorum*, 96
Hydrometridae, 95
Hydrometrinae, 95
hydrostatic organs. *See* static sense organs
Hydrotrephes, 108, 109
hygropetric species, 106
Hymenocoris brunneocephalis, 72
Hyocephalidae, 279
Hyocephalus aprugnus, 280
hypocerebral ganglion, 61
hypocostal lamina (hypocostal ridge), 43
Hypoctenes, 202
Hypselosoma, 82; *hirashimai*, 59, 81; *matsumurae*, 81
Hypselosomatinae, 82
Hypsipterygidae, 80
Hypsipteryx, 80; *ecpaglus*, 80; *machadoi*, 79, 80; *ugandensis*, 79, 80
Hyrcaninae, 92
Hyrcanus, 90, 92; *capitatus*, 91

Idiocarus, 125
Idiocorinae, 131
Idiocoris, 131; *lithophilus*, 132
Idiostolidae, 39, 251
Idiostolus insularis, 251, 252
Iella, 190
ileorectal valve. *See* pyloric valve
ileum. *See* pylorus
immature fruits, feeding on, 233
Indo-Pacific distribution patterns, 40
infraorders, phylogenetic relationships of, 5
inquilines, 21, 184; in Embiidina webs, 192; in spider webs, 192
insemination, 51; androtraumatic, 73; traumatic, 188, 192, 194, 195, 201, 202; traumatic intravaginal, 188
Institute of Zoology, Academia Sinica, Beijing, 16
Instituto de Biología, Universidad Autónomo de México, 16
internal anatomy, 3
intertidal dwarf bugs. *See* Omaniidae
intertidal zone, as habitat, 21, 90, 137, 140, 141
intracellular symbionts, 61
intrapediceloid, 148

intrinsic musculature, of heart, 61
intromittent organ. *See* aedeagus
Ioscytus, 140
Irbisia, 34, 177, 179, 180
Irochrotus, 58
Ischnodemus, 253, 255
Ischnorhynchinae, 257
Isoderminae, 39, 212
Isodermus, 212, 213; *planus*, 209
Isometoparia, 177
Isometopinae, 177, 179
Isometopini, 177
Issidomimini, 77
Issidomimus, 77
Ithamar, 283; *hawaiiensis*, 281

Jaczewski, Tadeusz L., 9
Jadera, 57, 282, 283; *haematoloma*, 282
Jakovlev, Vasiliy E., 9
Jalysus, 248; *spinosus*, 32, 248; *wickhami*, 32, 248
Japetus, 269
Jeannel, René, 9
Joppeicidae, 40, 164
Joppeicus, 164, 165; *paradoxus*, 164, 165
juga. *See* mandibular plates
juvenile hormone, 61

kairomones, 57
Keltonia, 178
killing bottles, 17
Kiritshenko, Alexandr Nikolayevich, 9
Kirkaldy, George Willis, 9
kissing bugs. *See* Triatominae
Kleidocerys, 53, 257, 263
Knight, Harry Hazelton, 10
Kodormus, 152, 155
Koelreuteria elegans, 282; *paniculata*, 282
Kumaressa, 212; *scutellata*, 209
Kvamula, 75, 77; *coccinelloides*, 77

Labiatae, 264
labium, 41, 121; swollen, 237
Labops, 34, 179
laboratory animals, 263, 271; bugs as, 159
labrum, 41
Laccocorinae, 125
Laccocoris, 52; *hoogstraali*, 109
Laccophorella, 216
Laccophorellini, 216
Laccotrephes, 109, 114, 116
lace bugs. *See* Tingidae
lacerate-flush feeding, 20
Lagenaria, 33
Lamiaceae, 224
Lampracanthia, 140
Lanopini, 216
Lantana, 37, 184
large milkweed bug, 263
Largidae, 58, 268
Largus rufipennis, 268
Larix, 213
larval organ, 135, 138, 140
Lasiella, 190
Lasiochilidae, 190

Lasiochilus, 190, 191; *fusculus*, 191; *pallidulus*, 191
Lasiocolpus, 190
lateral gland, 55
laterosternites, 49
laterotergites, 77; dorsal, 49, 190; inner, 49; ventral, 49
Lathrovelia, 101
Latimbini, 277
Latreille, Pierre-André, 1, 10
Latrocimex, 201
Latrocimicinae, 201
lawn grasses, 33
leaf-curl disease, 35
leaf-footed bugs. *See* Coreidae
legs, 45, 47
legumes, 274, 278
Leguminosae, 278
Leistarchini, 156
Lentia, 156; *corcovadensis*, 156
Leotichiini, 144
Leotichius, 134, 145; *shiva*, 135
Lepidoptera: as predators of eggs, 32; larvae as prey, 233
Lepionysiini, 257
Leptocimex duplicatus, 200
Leptocoris, 282, 283; *hexophthalmus*, 36
Leptocorisa, 32, 273; *acuta*, 22, 32, 273, 274; *chinensis*, 32, 273; *oratorius*, 32
Leptocorisinae, 273
Leptocorisini, 273
Leptodema, 157
Leptoglossus, 40, 58, 278, 279; *ashmeadi*, 278; *clypealis*, 33; *corculus*, 33, 278; *fulvicornis*, 278; *gonagra*, 33, 278; *occidentalis*, 33; *phyllopus*, 33, 278
Leptonannus, 77
Leptophya capitata, 183
Leptopodidae, 143
Leptopodinae, 144
Leptopodini, 145
Leptopodomorpha, 46, 48, 134, 136
Leptoterna, 34
Leptopus, 145; *marmoratus*, 144
Leptosalda, 145; *chiapensis*, 143, 144
Leptosaldinae, 143
Leptoscelidini, 277
Leston, Dennis, 10
Lestonia grossi, 228; *haustorifera*, 227, 228
Lestoniidae, 39, 40, 227
Lestonocorini, 232
Lethaeini, 261
Lethierry, Lucien, 10
Lethocerinae, 113
Lethocerus, 113, 114; *griseus*, 113; *maximus*, 107; *niloticus*, 113
Leucophoroptera, 178, 180
Leucophoropterini, 178
Libyaspis Group, 237, 238
life histories, in Miridae, 179
light traps, 18
Ligyrocoris diffusus, 52
Lilliputocorini, 262
Limacodidae, 35
Limnobates, 97
Limnobatodes, 87, 95, 96; *paradoxus*, 95

Limnobatodinae, 95
Limnocorinae, 125
Limnocoris, 125
Limnogeton, 45, 110–114
Limnogonus, 105
Limnonabis, 190
Limnoporus, 105; *canaliculatus*, 25
Lindberg, Hakan, 10
Linnaeus, Carl, 10
Linshcosteus, 159
Liorhyssus hyalinus, 36, 281
Lipogomphus, 90, 92
Lipokophila, 48, 147, 192; *chinai*, 192, 193; *eberhardi*, 56, 193
Lippomanus, 197
Litadeini, 182
litchi, 36
literature sources, 14
littoral zone, as habitat, 141
Lizarda, 158
lizards, as predators, 28
Llaimocorini, 212
Llaimocoris penai, 212
Lomagostus jeanneli, 69
Lophoscutus, 154
Lopidea, 27, 28, 178; *instabile*, 27; *nigridia*, 27
Lopodytes, 157
lora. *See* maxillary plates
Loricula, 162, 163; *pilosella*, 163; *pselaphiformis*, 163, 164
low-frequency vibrations, 160
Lucerocoris brunneus, 256
Luffa, 33
Lund University Museum of Zoology and Entomology, 16
Lupinus caudatus, 27
Lyctocoridae, 194
Lyctocoris, 195; *beneficus*, 195; *campestris*, 194, 195; *menieri*, 194; *nidicola*, 195; *tuberosus*, 194
Lygaeidae, 28, 33, 39, 40, 58, 251, 253
Lygaeinae, 257, 263
Lygaeospilus, 257
Lygaeus, 27; *equestris*, 28, 282
Lygocoris, 177
Lygus, 34, 42, 57, 177, 179; *elisus*, 34; *hesperus*, 34; *lineolaris*, 34; *pratensis*, 34; *rugulipennis*, 34; spp., as predators, 34
Lyrnessus, 274

m chromosome. *See* supernumerary chromosomes
Macaranga roxburghi, 240
Maccevethini, 282
Macchiademus diplopterus, 34
Machadonannus ocellatus, 81
Macrocephalinae, 46
Macrocephalini, 154
Macrocephalus notatus, 160
Macrocytus brunneus, 223
Macropes femoralis, 254
macroptery, 23, 264
Macrosalda, 140
macrotrichia, 67, 84
Macrovelia, 93, 94; *hornii*, 85, 94

Macroveliidae, 93
Madagascar, 39
Madeovelia, 88
Madeoveliinae, 39, 88
Maevius indecorus, 279, 280
Magnocellus transvaalensis, 169
Mahea, 216
main salivary gland, 59
Malcidae, 34, 40, 264, 266
Malcinae, 266
Malcus, 264, 266
Malcus flavidipes, 265, 266; *furcatus*, 265; *japonicus*, 265, 266
male genitalia, 50; asymmetrical, 78, 82, 169, 172
Mallochiola, 163
Malpighian tubules, 59, 60, 267
Malvaceae, 240; as hosts, 28
Malvales, 270
Manangocorinae, 157
Manangocoris horridus, 157
mandibles. *See* mandibular stylets
mandibular levers, 41
mandibular plates, 41; produced, 166, 167, 237
mandibular stylets, 41, 43; coiled, 208, 212, 214
Manocoreoini, 277
Mantodea, as predators, 28
Maoristolinae, 69
Maoristolus, 70
marginal riffles, as habitats, 125
marine bugs. *See* Aepophilidae
marine habitats, 105
marsh treaders. *See* Hydrometridae
Martarega mexicana, 128
Martiniola, 145
maternal care, 184, 217, 224, 233, 236, 240
mating position, in Joppeicidae, 165; in Saldidae, 139
mating ritual, 263
Matsuda, Ryuichi, 10
maxillae. *See* maxillary stylets
maxillary glands, 60
maxillary levers, 41
maxillary plates, 41
maxillary stylets, 41, 43; coiled, 208, 212, 214
Mcateella, 266, 267
mechanoreceptors, 53
Mecideini, 232
Mecistorhinus tripterus, 35
Mecistoscelini, 177
Mecocnemini, 277
media, 43, 44
medial furrow, 43
Medicago, 34
mediotergites, 49
Medocostes, 184–186; *lestoni*, 184–186
Medocostidae, 40, 184
Megaderma spasmae, 203
Megadermatidae, 204
Megalonotini, 262
Megalotomus, 274; *quinquespinosus*, 272
Megarididae, 40, 228
Megaris, 229; *hemisphaerica*, 228; *laevicollis*, 228; *majusculus*, 229; *puertoricensis*, 229; *rotunda*, 228; *semiamicta*, 229
Megenicocephalinae, 73
Megenicocephalus, 70, 73
Megymeninae, 225
Megymenini, 225
Megymenum, 225, 226; *gracilicorne*, 59; *insulare*, 226
Melanerythrus, 257
membrane, 44; deciduous, 212, 213
Mendanocoris, 158
meniscus, ascension of, 102
Meranoplus mucronatus, 238
Merocorini, 278
Merocoris, 278
Meropachydinae, 278
Meropachydini, 278
Merragata, 90, 92; *brunnea*, 91
Mertila malayensis, 174
mesoscutellum. *See* scutellum
mesoscutum, 43
Mesosepis papua, 155
mesospermalege, 188
mesosternal carina, 215, 217
mesosternum, 43
Mesovelia, 88, 90; *amoena*, 90; *furcata*, 64, 90; *mulsanti*, 89, 90
Mesoveliidae, 88
Mesoveliinae, 88
Mesoveloidea, 88, 90; *williamsi*, 89
Metacanthinae, 248
metamorphosis, 61
Metapterini, 156
metasternum, 43; produced, 232
Metatropiphorus, 186
metepisternum, 43
Metrargini, 257
Metrobates, 105; *trux*, 86
Metrocorini, 105
Metrocoris, 105
Mezira, 48; *reducta*, 213; *tremulae*, 211
Mezirinae, 212
Micrelytra, 274; *fossularum*, 273
Micrelytrinae, 274
Microchiroptera, 204
Micronecta, 121
Micronectinae, 121
Microphysa, 162
Microphysidae, 161
microptery, 23
micropylar canals, 63
micropylar processes, 63, 64, 205
micropyles, 63, 278
Microtomus, 156
microtrichia, 84
Microvelia, 25, 37, 92, 99, 101, 102; *diluta*, 102; *douglasi*, 25; *longipes*, 106; *pulchella*, 37, 100; *reticulata*, 100
Microveliinae, 99
Microveliini, 99
Mictini, 277
Mictis profana, 33
midgut, 59, 60; discontinuous, 60
Miespa, 266
migration, 270
Miller, Norman Cecil Egerton, 10
millet, 35
millipedes, as prey, 160
mimicry, 27, 270; aggressive, 27, 28; of ants, 29; Batesian, 27, 28, 263; of beetles, 30; Mertensian, 27; Müllerian, 27, 28, 263; in Rhopalidae, 282; of substrate, 30; Wasmannian, 27; of wasps, 29, 177
Mimocoris rugicollis, 29
minute pirate bugs. *See* Anthocoridae
Miraradus, 210
Miridae, 27, 34, 45, 46, 49, 169, 172
Mirinae, 177
Mirini, 177
Mirperus jaculus, 32
mites: as prey, 195; predators of, 32
Mixotrephes hoberlandti, 132
Mniovelia, 88, 90; *kuscheli*, 88, 89
Molossidae, 204
Monalocoris, 173; *americanus*, 175
Monaloniina, 34, 48, 176
Monalonion, 34, 176, 179
Monocotoledoneae, 255
monographs, 15
Mononyx, 117
monophagy, 179
monophyletic groups, 38
monophyly, 5
Monosteira unicostata, 36
Montandon, Arnold Lucien, 10
Montandoniola maraquesi, 32
Monteithocoris hirsutus, 251
Monteithostolini, 73
Monteithostolus genitalis, 69, 70, 73
morphology, 42
Morus bombycis, 266
mosquito larvae, as prey, 130
mosquitoes, 36, 37
mosses, 184
mounting: card, 18; pin, 18; point, 18; slide, 18
mouthparts, 43; in Aradidae, 211; in Termitaphididae, 211
Muatianvuaia, 77; *barrosmachadoi*, 76
multivoltine life cycle, 179
Murgantia histrionica, 35
Murphyanella, 70; *aliquantula*, 70
Murphyanellinae, 70
Musaceae, 179
Musée Royal de l'Afrique Centrale, 16
Museo Argentino de Ciencias Naturales "Bernardino Rivadavia," 16
Muséum National d'Histoire Naturelle, Paris, 16
museums, 15
Musgraveia sulciventris, 36, 241
Mutillidae, as models, 269
mycetomes, 61
mycophagy, 213, 215, 237
Myiomma, 177
Myiommaria, 177
Myodocha serripes, 254
Myodochini, 262
myrmecomorphic body shape, 29
myrmecomorphic coloration, 29
myrmecomorphic nymphs, 274

myrmecomorphy, 28, 179, 250, 255, 263, 269, 273, 274; sexual, 29
Myrmecophyes oregonensis, 171
Myrmedobia, 163; *exilis*, 163, 164
Myrmica, 217; *ruginodis*, 217
Myrmus miriformis, 281
Myrocheini, 232
Myrtaceae, 229

Nabicula propinqua, 188
Nabidae, 40, 186
Nabinae, 187
Nabini, 188
Nabis, 47, 54, 188, 189; *americoferus*, 187; *ferus*, 187; *limbatus*, 64
Naboidea, 48
Naniella, 173
Nasutitermes exitiosus, 160
Natalicolinae, 242
natatorial legs, 45
National Collection of Insects, Plant Protection Research Institute, Pretoria, 16
National Museum of Natural History, Rio de Janeiro, 16
National Museum of Natural History, Smithsonian Institution, 15
National Museum, Prague, 16
National Science Museum (Natural History), Tokyo, 16
Natural History Museum, Budapest, 16
Natural History Museum, London, 16
Naturhistorisches Museum, Vienna, 16
Naucoridae, 23, 45, 124
Naucorinae, 125
Naucoris cimicoides, 126
Neacoryphus, 263; *bicrucis*, 27, 263
Neadenocoris, 212
Nearctic Region, 38
neck, 41
Neella, 179
negro bugs. *See* Cydnidae
Neides, 248; *muticus*, 248
nematoceran larvae, as prey, 122
Nematopodini, 277
Neoalardus typicus, 100
Neocentrocnemis signoreti, 155
Neochauliops, 266
Neocratoplatys salvazai, 237
Neogerris parvulus, 104
Neogorpis, 188
Neolocoptiris, 154
Neoncylocotis, 72
Neopamera bilobata, 29
Neoplea, 130; *striola*, 130
neoteny, 24
Neotimasius, 92
Neotrephes, 131; *usingeri*, 132
Neotrephinae, 131
Neotropical Region, 38
Nepa, 115, 116; *cinerea*, 115, 116
nephrocytes, 61
Nepidae, 24, 45, 55, 116
Nepinae, 116
Nepini, 116
Nepomorpha, 41, 46, 48, 107, 110

Nereivelia, 88, 90
Nerthra, 30, 117, 118; *amplicollis*, 117; *annulipes*, 117; *hungerfordi*, 117
nervous system, 61, 62
Neseis, 258
Nesenicocephalus, 71
nests, as habitats, 160
nets, 17
Neuroctenus pseudonymus, 213
neurosecretory system. *See* endocrine glands
Nezara, 233; *viridula*, 35, 64
Nichomachini, 178
Nichomachus, 178
Nidicola, 199
Niesthreini, 282
Ninini, 256
Noliphini, 273
Nothochromini, 255
Nothochromus maoricus, 255
Nothofagus, 251
Notonecta, 53, 128, 129; *glauca*, 129; *undulata*, 44
Notonectidae, 45, 127
Notonectinae, 128
Noualhieridia, 216
Nycteridae, 204
nymphal organ, 137
nymphal thoracic glands, 60
Nymphocorinae, 70
Nymphocoris, 70; *hilli*, 70; *maoricus*, 70
nymphs: first-instar, 46; myrmecomorphic, 29; of Reduviidae, 160
Nysiini, 257
Nysius, 258, 263; *ericae*, 34; *raphanus*, 34; *thymi*, 254; *vinitor*, 34

Oaxacacoris guadalajara, 174
occipital apodemes, 100, 101, 105
ocean, as habitat, 106
oceanic distributions, 40
Oceanides, 258
ocelli, 41
Ocellovelia, 98, 101; *germari*, 100
Ocelloveliinae, 101
Ochteridae, 45
Ochterus, 109, 118; *barberi*, 119; *caffer*, 108; *marginatus*, 118, 119; *seychellensis*, 119
ocular seta, 56
Odoniellina, 34, 48, 176
Odontoscelis, 58
Odontotarsinae, 239
Odontotarsus, 36
odors, 160
Oebalus pugnax, 35
Oeciacus, 201; *vicarius*, 33
Oedancala dorsalis, 254
Oiovelia, 101; *spumicola*, 101
oligophagy, 179
olives, 32
Omania, 141; *coleoptrata*, 135, 142
Omaniidae, 141
Ommatides, 82
ommatidia, 41
omphalus, 199

Oncacontias, 216
Oncomerinae, 242
Oncomerini, 242
Oncopeltus, 22, 27, 28, 263; *cingulifer*, 22; *fasciatus*, 22, 24, 28, 43, 64, 254, 263; *sandarachatus*, 22
Oncylocotis, 71–73; *curculio*, 48; *tasmanicus*, 73
Ontiscini, 256
Onychotrechus, 105, 106; *rhexenor*, 104
Onymocoris, 167; *hackeri*, 169; *izzardi*, 166
operculum, 62, 64
optic lobes, 61, 62
Opuntia, 179
Oravelia, 93, 94; *pege*, 85, 94
Orchidaceae, 179
Orectoderus, 29; *obliquus*, 170
Oriental Region, 38
Oriini, 46, 197
Orius, 32, 197–199
Orsillinae, 40, 257, 263
Orsillini, 257
Orsillus depressus, 254
Orthophrys, 138, 140; *pygmaeum*, 140
Orthotylinae, 40, 177
Orthotylini, 178
Oshanin, Vasiliy F., 10
osmoregulation, 184
ostia, of heart, 61
ostracods, as food, 130
ovaries, 165
ovariole numbers, 3, 51
overwintering, 184
ovipositor, 49, 50; laciniate, 49; platelike, 50
ovoviviparous species, 263
Oxycareninae, 58, 258
Oxycarenus, 34, 258, 264; *hyalinipennis*, 34, 254
Oxythyreus cylindricornis, 151
Ozophora, 251; *singularis*, 253
Ozophorini, 40, 262

Pachycoleus, 78
Pachycorinae, 239
Pachycoris torridus, 238, 240
Pachygrontha, 258
Pachygronthinae, 30, 258
Pachygronthini, 258, 263
Pachynomidae, 39, 41, 148, 149
Pachynominae, 149
Pachynomus, 148, 149; *picipes*, 149
Pagasa, 188; *luticeps*, 186
pagiopodous coxae, 45
pala, 119, 121
Palaucorina, 176
Palaucoris unguidentatus, 175
Palearctic Region, 38
Paleotropical, 40
palm bugs. *See* Xylastodorinae
palms, 35, 169
Pameridea, 176
Pamphantini, 29, 255
Pandanaceae, 179
Pandanus, 158
Pantochlora, 232

Index   331

Parabryocoropsis, 171
Paracalisiopsis, 210
Paracalisius, 210
Paracimex, 201
Paracopium, 184
Paradacerla, 29
Paradindymus madagascariensis, 271
paragenital glands. See ventral abdominal glands
Paragonatas divergens, 54
Paralosalda, 140
parameres, 49; articulated, 218; as copulatory organs, 195, 199, 202
parandria, 138, 139, 208
Paranisops, 128, 129
paranota, 183
Paraphrynovelia, 93; *brincki*, 93
Paraphrynoveliidae, 40, 92
paraphyletic group, Lygaeidae as, 252
Parapiesma, 266
Paraplea, 130
Pararachnocoris, 186
parasitoids, 57
parastigmal pits, 148, 149, 186–189
Parastrachia, 220, 222, 224; *japonensis*, 224
Parastrachiinae, 222
Paratrephes, 131
Paravelia, 85, 101; *rescens*, 48, 52
parempodia, 46, 48, 98, 109, 136, 137, 141, 168, 175, 178; asymmetrical, 176; fleshy, 171, 177; increased numbers of, 107; reduced, 135, 192
parental care. See maternal care
Paropsis, 218
pars stridens. See stridulitrum
Parshley, Howard Madison, 11
parthenogenetic species, 90, 179
Paskia, 131
Patalochirus, 158
Patapius, 145; *spinosus*, 144, 145; *thaiensis*, 144
peaches, 34
peanuts, 34
pears, 34, 36
pedicel, 41, 148, 149
peg plates, 84
Peirates, 157
Peiratinae, 40, 157
Pelocoris femoratus, 125; *shoshone*, 124
Pelogonus, 118
Pendergrast's organs, 215, 217
Pentacora, 140; *grossi*, 139; *signoreti*, 138–140
Pentatomidae, 35, 58, 229, 230
Pentatominae, 232
Pentatomini, 232
Pentatomoidea, 41, 43
Pentatomomorpha, 5, 43, 46, 48, 57, 60, 205
Pephricus paradoxus, 276
Peregrinator biannulipes, 36
Peridontopyge spinosissima, 160
Perillus, 35; *bioculatus*, 35, 233
peritreme, 43, 57; produced, 247, 248
Peritropis, 172

Perittopinae, 101
Perittopus, 100, 101
Perkinsiella saccharicida, 34
Peruda, 51
pet food, Corixidae as, 122
Petascelidini, 277
Phaenacantha, 33, 250; *australiae*, 250; *saccharicida*, 250; *saileri*, 250
phallandrium, 73
Phallopirates, 73
Phallopiratinae, 73
phallotheca (phallosoma), 49, 50
Pharnaceum aurantium, 244
Phaseolus, 32
Phasmosomini, 262
Phasmosomus, 253; *araxis*, 253
Phatnomini, 182
pheromones, 57
Phimophorinae, 158
Phimophorus, 158
Phloea, 235
Phloeidae, 40, 41, 234
phloem vessels, 20
Phloeophana, 234, 235; *longirostris*, 235, 236
Phonoctonous fasciatus, 63
Phonoctonus, 28, 36, 270
phoretic male, 102
Phorticini, 188
Phorticus, 188
photoperiod, 25
Phrynovelia, 88, 90
Phthia picta, 33
Phthirocorinae, 73
Phthirocorini, 73
Phthirocoris subantarcticus, 73
Phylinae, 178
Phylini, 178
Phyllocephalinae, 60, 232
Phyllomorphini, 277
phylogenetic relationships: of Cimicomorpha, 146; of Gerromorpha, 84; of infraorders, 5; of Leptopodomorpha, 134; of Nepomorpha, 110
Phymata, 47, 52, 58, 154; *crassipes*, 160; *erosa*, 155; *pennsylvanica*, 30, 155; *pennsylvanica americana*, 160
Phymatinae, 45, 58, 154, 160
phymatine complex, 154
Phymatini, 154
physical gill, 53, 137
physiology, 263
Physoderes, 158
Physoderinae, 40, 158
Physopelta famelica, 268
Phytelephas, 169
Phytocoris, 22, 177, 179; *calli*, 174; *neglectus*, 22; *nobilis*, 22; *populi*, 170
phytophagy, 20
Piesma, 266, 267; *capitatum*, 35, 267; *cinereum*, 35, 267; *maculatum*, 267; *quadratum*, 35, 53, 267
Piesmatidae, 35, 266
Piesmatinae, 266
Piezoderus hybneri, 35
Piezosternini, 242

Piezosternum, 242; *calidium*, 242; *subulatum*, 242
Pilophorini, 178
Pilophorus, 46, 56, 180; *kockensis*, 171
pimentos, 33
Pinus, 33, 179, 213
Piper, 264
Pirates hybridus, 155
Pisilus, 157
pistachio, 33
pit, at base of claval commissure, 128
pitfall traps, 18
pit organs, 52, 101
plant bugs. See Miridae
plant pathogens, 35
plant vascular system, feeding in, 233, 278
plastron, 137
plastron respiration, 125–127, 136
Platanus, 36, 263
Plataspidae, 40, 58, 236
plates, 52
Platygerris, 104
Platylygus, 177, 179
Platytatini, 242
Plea, 130; *atomaria*, 130; *minutissima*, 130
plectrum, 51
Pleidae, 45
Plinthisini, 39, 262
Plinthisus flindersi, 253
Plochiocoris, 190
Ploiariini, 156
Plokiophila, 192; *cubana*, 192
Plokiophilidae, 46, 59, 190, 192
Plokiophilinae, 39, 192
Plokiophiloides, 192; *asolen*, 193
Poaceae, 255
pod-sucking bugs, 32
Podisus, 35, 57; *maculiventris*, 233
Podopinae, 232
Poeantius, 28
Poecilocoris, 240
Poecilometis eximus, 229
Poisson, Raymond A., 11
pollen, 179, 184, 199
Polychisme ferruginosus, 255
Polychismini, 255
Polyctenes, 202; *molossus*, 203
Polyctenidae, 46, 202
Polycteninae, 202
polyphagy, 179
Polytoxus, 158
Pompilidae, as models, 274
pond skaters. See Gerridae
Poppius, Robert Bertil, 11
population density, 25
pore-bearing organ (pore-bearing plate), 279–281
postclypeus, 41
postcubitus, 44
postocular lobe, 67
Potamobates, 105
Potamocoris, 123; *nieseri*, 123; *parvus*, 123
Potamometra, 105
potatoes, 34
poultry bug, 33
predation, 21

332 Index

predators, 217; learning by, 29; visual, 159
predatory habits, in Asopinae, 231
prenatal care, 213
prepedicellite, 41, 150, 188
pressure receptors, 125
pretarsus, 46, 48, 109, 135, 175; in Miridae, 178
prey location, 106
prickly pear cactus, 33
Primicimex, 201
Primicimicinae, 201
Primierus quadrispinosus, 261
Prionogastrini, 242
Prionotylini, 277
Procamptini, 277, 278
Procamptus segrex, 278
processus corial. *See* stub
processus gonopori. *See* acus
Procryphocricos perplexus, 24
proctiger, 49
procurrent nerve, 61
Pronotacantha, 56
pronotum: anterior lobe, 41; humeral angles, 43; posterior lobe, 43
Propicimex, 201
Prostemma, 47, 52, 54, 188, 189; *guttula*, 187
Prostemmatinae, 41, 45, 188
Prostemmatini, 188
prosternum, 43
Prosympiestinae, 212
Prosympiestini, 212
Prosympiestus, 212
Protea, 264
Proteaceae, 264
protective resemblance, 263
Protenor, 274
prothoracotropic hormone, 61
prothorax, 41
protocerebrum, 61, 62
Psacasta, 58
Psallopinae, 178
Psallops, 178; *oculatus*, 171
Psamminae, 258
Psammium, 258
pseudarolia. *See* pulvilli
Pseudatomoscelis, 178
Pseudocetherini, 156
Pseudocnemodus canadensis, 51
pseudocompetition, 263
Pseudomonas, 61
Pseudophloeinae, 278
Pseudopsallus, 178–180; *lattini*, 176
pseudopulvilli, 171, 175, 176, 178. *See also* accessory parempodia
Pseudosaldula, 140; *chilenis*, 138
pseudospermatheca, 50, 148–151, 155, 159, 181, 183
Pseudotheraptus wayi, 33
Pseudovelia, 101, 102
Psocoptera, 5
psocopteran webs, as habitats, 160
Psotilnus, 283
pterygopolymorphism, 188
Ptilocerus ochraceus, 160
Ptilomera, 105, 106

Ptilomerinae, 39, 105
pulvilli, 46, 48, 167, 168, 171, 175, 176, 197, 205; absent, in Aradidae, 210
pumpkins, 33
Punctius, 148, 149; *alutaceus*, 149
Puton, Jean-Baptiste Auguste, 11
Pycnoderes, 175, 179
pygmy backswimmers. *See* Pleidae
pygopher. *See* genital capsule
pygophore. *See* genital capsule
Pylorgus, 257
pyloric valve, 60
pylorus, 59, 60
Pyrrhocoridae, 28, 35, 40, 58, 270
Pyrrhocoris apterus, 58, 270, 271

Quercus, 35
quiet water, as habitat, 106
Quilnus, 210

radial sector, 44
radius, 43, 44
Ranatra, 47, 51, 108, 115, 116; *compressicollis*, 115; *drakei*, 115
Ranatrinae, 116
Ranatrini, 116
Ranzovius, 21, 180
rape seed, 34
raptorial legs, 45, 67, 107
Rasahus, 157; *sulcicollis*, 155
rectal gland (rectal pad), 60
rectum, 60; eversible, 179
recurrent nerve, 61
Reduviidae, 28, 31, 35, 49, 150, 151
Reduviinae, 158
Reduvioidea, 48
Reduvius, 148, 158; *personatus*, 31, 158
reel system, of aedeagus, in Saldidae, 137
relict distributions, 40
remigium, 44
respiration, 114
respiratory siphon, 116
Resthenini, 177
Restionaceae, 256
reticulate body surface, 266
retractor tendon, 46
Reuter, Odo Moranal, 2, 3, 11
Reuteroscopus, 178
Rhagadotarsinae, 105
Rhagadotarsus, 105, 106; *kraepelini*, 104
Rhagovelia, 48, 52, 54, 101, 102; *distincta*, 86
Rhagoveliinae, 101
Rhamphocoris, 188
Rhaphidosoma, 157
Rhaphidosomini, 157
Rheumatobates, 102, 105, 106; *meinerti*, 104
Rhinacloa, 178; *forticornis*, 179
Rhinocoris, 64
Rhinocylapus, 179
Rhinolophidae, 204
Rhodainiella, 157
Rhodniini, 159
Rhodnius, 58; *prolixus*, 57, 159
Rhododendron, 36, 263

Rhopalidae, 36, 58, 281, 282
Rhopalimorpha, 216
Rhopalinae, 282
Rhopalini, 282
Rhopalomorpha, 217
Rhynchota, 1
Rhyngota. *See* Rhynchota
Rhynocoris, 28, 157
Rhyparochrominae, 29, 45, 258, 259
Rhyparochromini, 262
Rhytocoris, 53
rice, 32, 35, 232
rice planthoppers, 37
riffle bugs. *See* Veliidae
ring gland, 184
riparian habitats, 116, 118
ripple communication, 106
Riptortus, 32, 273
rodent nests, as habitats, 195
rostral groove, 91
rotatory coxae, 45
rowing, 45
Roystonea regia, 169
Ruckes, Herbert, 11
Ruckesona, 246; *vitrella*, 246
Rupisalda, 140
rushes, 263
Rutaceae, 36
rutherglen bug, 34

Sagocorini, 125
Sagriva, 226
Sahlbergella singularis, 34
Saica, 158
Saicinae, 158
Saicini, 158
Saileriola, 246; *sandakanensis*, 246
Saileriolinae, 245
Salda, 138; *littoralis*, 140
Saldidae, 39, 45, 46, 58, 140
Saldinae, 140
Saldini, 140
Saldoida, 140
Saldoidini, 140
Saldolepta, 145; *kistnerorum*, 143, 144
Saldula, 140; *dentulata*, 139; *fucicola*, 64, 139; *laticollis*, 139–141; *orthochila*, 141; *pallipes*, 135; *sibiricola*, 139
Saldundula, 140
Saldunculini, 140
salivary canal, 41, 43
salivary glands, 59; function, 59; structure, 59
salivary pump, 59
saltatorial legs, 45
Salyavata, 158; *variegata*, 160
Salyavatinae, 158
Sandaliorrhyncha, 120
Sangarius, 217; *paradoxus*, 217
sap feeding, 60, 263
Sapindus saponaria var. *drummondii*, 282
Saturniomirini, 177
savoy virus, 35
saxicolous, 141
Saxicoris, 258; *verrucosus*, 258
Say, Thomas, 11

scale insects, 179
scanning electron microscopy, 19
scape, 41
Scaptocorinae, 224
Scaptocoris, 45, 46; *castaneus*, 33, 224; *divergens*, 33, 57, 221
scavengers, in spider webs, 180
Scelionidae, 57, 233
scent-gland channel, 56
scent-gland fluids: composition, 57; fungistatic role, 57; shooting, 236
scent-gland opening, 281; in coxal cavities, 282; obsolete, 281
scent glands: adult, 57; function, 57; metathoracic, 55; nymphal, 56; structure, 56, 57
scentless plant bugs. See Rhopalidae
Schaffneria, 180
Schiodte, 3
Schizoptera, 82
Schizopteridae, 80, 82
Schizopterinae, 82
Schoepfia jasminodora, 224
Schouteden, Henri, 11
Sciocorini, 232
Sciocoris microphthalmus, 231
scolopidia, 53
Scolopini, 58, 198
scolopophorous organs (scolopidial organs), 53, 107
Scoloposcelis flavicornis, 198
Scotinophara, 35, 232; *lurida*, 35
Scotomedes, 161, 184; *alienus*, 162; *borneensis*, 162
Scott, John, 11
Scutellera, 240
Scutelleridae, 36, 58, 238, 239
Scutellerinae, 239
Scutellerini, 239
scutellum, 43, 232; enlarged, 218, 219, 228, 231, 232, 236, 238, 240
Seabranannnus immitator, 81
sealing bar, 62
second valifer, 50
second valvula, 50
secondary fecundation canal, 84
secretory cells, subhypodermal, 183
sedges, 263
seed bugs. See Lygaeidae
seeds, feeding on, 28, 29, 233, 244, 263, 270, 281
Sehirinae, 224
Sehirus, 224; *bicolor*, 224; *cinctus*, 223, 224; *cinctus albonotatus*, 221
Seidenstücker, Gustav, 12
Semangananus mirus, 81
semiaquatic bugs. See Gerromorpha
seminal conceptacles, 195, 199
Semium, 173
Senecio, 263; *smallii*, 27
sensory patches, on male abdomen, 231
Sepinini, 242
sequences, rDNA, 5
Serbana, 30; *borneensis*, 232
Serbaninae, 232
Serenthiinae, 182

Serinethinae, 282
Serjania brachycarpa, 282
Serville, Jean-Guillaume Audinet, 1, 12
Severiniella, 237
sex chromosomes, 63
sexual dimorphism, 163; of wings, 213
shell. See chorion
shield bugs. See Scutelleridae
shore bugs. See Saldidae
sieve plate. See pore-bearing organ
Sigara, 121; *sahlbergi*, 122
Signoret, Victor, 12
Sinea diadema, 28
Sinotagini, 277
Sirthenea, 157
Slaterobius, 29
Slaterocoris, 178
small grains, 34
small water striders. See Veliidae
snail predator, 114
Snow Entomological Museum, University of Kansas, 16
soapberry bug, 282
soapberry tree, 282
socket glands, 58
Socratea montana, 169
Solanaceae, 37, 266
Solanum, 34
Solenopsis, 29
Solidago, 30
soun bug, 36
sound production, 51; in *Phymata*, 160
soybeans, 35
Spalacocoris philippinensis, 256
Spartocerini, 277
Spathophorini, 278
species recognition, 106
Speovelia, 88, 90
sperm, 63, 140; storage of, 146
spermatheca, 50, 205, 229, 231; absent, in Idiostolidae, 251; invaginated, 232
spermathecal pump, 100
spermathecal tube, 89, 91
spermatic pocket, 195, 199
spermatophore, 63
spermatozoa, 239
Sphaeridopinae, 158
Sphaeridops, 158
Sphaerocorini, 239
Sphaerocoris annulus, 240
Sphagnum, 78, 92, 164
Sphedanolestes, 157
spider predators, 160
spiders, myrmecomorphic, 29
spider webs, as habitats, 160
Spilostethus, 27, 28, 34; *pandurus*, 34
spines, nymphal, 181, 265
Spinola, Maxmillian, 12
spinose body, 278
spiracle cover, 108
spiracles, 43, 50, 108, 189; abdominal, 49
spiracular rosettes, 126, 127
Spiraea, 263
spongy fossa. See fossula spongiosa
Squamocoris, 179
squash bug, 33

Stagonomus amoenus, 231
staining, 19
Stål, Carl, 2, 12
staphylinoidy, 23, 24, 163, 243
static sense organs, 55, 108, 111, 113–116
Stemmocrypta antennata, 83
Stemmocryptidae, 82
Stenaptula, 257
Stenocephalidae, 58, 283
Stenocoris, 273; *tipuloides*, 272
Stenocorixa protrusa, 121
Stenocorixinae, 121
Stenodema, 30
Stenodemini, 177
Stenolemus, 152, 160; *lanipes*, 160
Stenonabis, 188
Stenophyella, 258
Stenopirates, 72, 73
Stenopodainae, 158
Stephanitis, 36, 179, 182; *pyri*, 36, 183; *rhododendri*, 35
sternal glands. See ventral glands
Sternorrhyncha, 5
sternum 7 cleft, in female, 268
Stethoconus, 35, 179; *frappai*, 35; *japonicus*, 35
Sthenaridea australis, 54, 175
sticky setae. See viscid setae
Stilbocoris, 61, 188, 263
stilt bugs. See Berytidae
stink bugs. See Pentatomidae
Stollia fabricii, 59
stomach, 60
stomodeal nervous system (stomogastric nervous system), 61
stones, as habitat, 141
Stongylovelia, 99
strainer. See pore-bearing organ
strawberries, 34
stretch receptors, 106
stridulation, 129, 224, 263; defensive, 160; by nymphs, 160
stridulatory structures: abdominal sternum–hind leg, 51, 52, 211; in Aradidae, 213; base of labium–femoral apex, 53; connexival margin–hind femur, 51, 101, 125; femoral ridge–coxal peg, 53; forecoxa–forecoxal cavity, 51; forewing edge–hind femur, 51, 52, 132, 140, 179; head–forefemur, 51; hypocostal lamina, 53; margin of pygophore, 53; maxillary plate, 119; maxillary plate–forefemur, 53, 121; metapleuron–middle femur, 51; metathoracic wing–abdomen, 53, 145, 224, 232, 244; propleuron–forefemur, 51; prosternal groove, 53, 151, 159, 166; tibial comb–labial prong, 51, 128, 129; underside of clavus, 53
stridulatory sulcus, 52
stridulitrum, 51
Stridulivelia, 51, 101, 102
strigil, 121, 122
Strombosoma, 224
Strongylovelia, 99
stub, 162, 163, 171, 186
Stygnocorini, 39, 262

stylets, 41
stylet-sheath feeding, 20
styloids. See third valvulae
Stylopomiris malayensis, 176
subcosta, 44
subesophageal ganglion, 62
subgenital plate, 50, 67, 74
submacroptery, 23
subrectal gland, 58, 156
sugar beets, 35
sugar cane, 33, 34, 250
sugar cane leafhopper, predator of, 34
supernumerary chromosomes, 64
supraesophageal ganglion, 61
Surinamellini, 177
swallows, as hosts, 201
swarming, 70, 73
sweating, 183
Swedish Museum of Natural History, 16
sweeping nets, 18
sweet potato, 33
swifts, as hosts, 201
swimming, inverted, 129, 130
swimming fan, 99, 101
sycamore lacebug, 36
Sycanus collaris, 36
symbiont transmission, 61
symbionts, gut, 60. See also gastric caeca; intracellular symbionts; mycetomes
sympathetic nervous system. See stomodeal nervous system
Sympeplus, 258
Symphylax musiphthora, 250
syncytial bodies, 194
Systelloderes, 67, 73
Systelloderini, 73

Tachygerrini, 105
Tachygerris, 105
Tahitocoris cheesmanae, 233
Tanycricini, 125
Targarema stali, 253
Targaremini, 262
tarsi, 46, 47; cleft, 99; reduced, 224; swollen, 173, 176
Tarsotrechus, 105
Taylorilygus pallidulus, 34
tea, 34
Tectocoris, 58, 239; *diophthalmus*, 239, 240
Tegeini, 157
tegumentary glands, 58
Teleonemia scrupulosa, 37
Telmatometra, 105
Teloleuca, 140
temporary habitats, 264
Tenagobia, 121
Tenagogonus, 105
Tengella radiata, 192
Tenodera ardifolia sinensis, 28
Teracriini, 258
Termatophylini, 177
Termitaphididae, 214
Termitaphis, 214
Termitaradus, 214, 215; *guianae*, 48, 56, 214; *jamaicensis*, 215; *panamensis*, 211

termites: inquilines of, 213, 215; as prey, 156
territorial defense, 106; male, 278
Tessaratoma javanica, 242; *papillosa*, 36, 57, 242
Tessaratomidae, 36, 241
Tessaratominae, 242
Tessaratomini, 242
testes, 137, 196
testis follicle numbers, 3
Tetraphleps, 197; *latipennis*, 196
Tetraripis, 101, 102
Tetroda histeroides, 232
Thaicoris, 266
Thaicorinae, 266
Thalmini, 225
Thasus, 279; *acutangulus*, 275, 279
Thaumamannia, 183; *vanderdrifti*, 183
Thaumastaneis montandoni, 269
Thaumastella, 243; *aradoides*, 243, 244; *elizabethae*, 243, 244; *namaquensis*, 243, 244
Thaumastellidae, 243
Thaumastocoridae, 165, 167
Thaumastocorinae, 49, 167
Thaumastocoris, 167; *australicus*, 169; *hackeri*, 167
theca. See phallotheca
Themnocoris kinkalanus, 155
Themonocorini, 154
Theraneis, 269
Theseus modestus, 231
third valvulae, 49
thoracic ganglion, first, 61
thorax, modification of, 106
thornlike outgrowths. See grooved setae
Thyanta calceata, 233
Thyreocoridae, 222
Thyreocorinae, 224
Thyreocoris, 224; *scarabaeoides*, 223, 224
Thysanoptera, predators of, 32
tibiae, 45; annulate, 202; spinose, 220
tibial appendix, 49, 167, 168
tibial comb, 47, 168
tibial extensor pendant sclerite, 46
tibial flexor sclerite, 46
Timahocoris, 70; *paululus*, 70
Timasius, 92; *ventralis*, 91
Tingidae, 36, 180, 182; as prey, 179
Tinginae, 182
Tingini, 182
Tingis ampliata, 183; *cardui*, 183
Tinna wagneri, 36
toad bugs. See Gelastocoridae
toads, as predators, 283
Tollius, 274
tomato stilt bug, 32
tomatoes, 32–35
Tonkuivelia, 101
Tornocrusus, 69; *penai*, 69
Torre-Bueno, José R., 12
torrential streams, as habitats, 126
torrents, as habitats, 106
Trachelium, 274
Transpacific, 40
tree holes, as habitats, 101

Trephotomas, 131
Trephotomasinae, 131
Trepobates, 105; *taylori*, 104
Trepobatinae, 105
Tretocorini, 212
Tretocoris, 212
Triatoma, 35, 159; *nigromaculata*, 152; *rubrofasciata*, 44, 155, 159
Triatominae, 40, 159, 160
Triatomini, 159
Tribelocephala, 159
Tribelocephalinae, 159
Tricentrus, 250
trichobothria: abdominal, 54, 148, 150, 161, 186, 188, 205, 227, 243, 244, 250, 252, 253, 265; antennal, 54, 148–150, 155, 159; in Aphelonotinae, 149; cephalic, 54, 84, 95, 102, 134; femoral, 49, 54, 171, 178; in Miridae, 171, 178; in Nabidae, 54, 186; in Pachynomidae, 148, 150; in Pachynominae, 149; in Pentatomomorpha, 205; in Prostemmatinae, 188; in Reduviidae, 150, 159; scutellar, 54, 188; in Thaumastellidae, 243; in Velocipedidae, 161
Trichocentrus, 51
Trichocorixa reticulata, 120
trichome, abdominal, 160
Trichophora, 205, 263
Trichotelocera, 74
Trichotonanninae, 77
Trichotonannus, 77; *dundo*, 76, 79; *oidipos*, 77, 79
Tridemula pilosa, 155
Trigonotylus, 34
Trisecus armatus, 251; *pictus*, 251, 252
tritocerebrum, 61
trochalopodous coxae, 45
trochanters, 45
Trochopus, 101
Tropidotylus, 237
Trypanosoma cruzi, 159
tubercular sense organs. See corial glands
Tullgren, A., 2, 4
tur pod bug, 33
Tylocryptus egenus, 277
tylus. See clypeus
tymbals, 53
tympanal organs, 53
Typha, 263, 264
Tytthus, 178; *mundulus*, 34

Udeocorini, 39, 262
Uhler, Phillip Reese, 12
unguitractor plate, 109, 135
unique-headed bugs. See Enicocephalomorpha
University Zoological Museum, Helsinki, 16
univoltine life cycle, 25, 264
unques. See claws
unquitractor plate, 46
uradenia, 198, 199
uradenies. See ventral abdominal glands
Urentius hystricellus, 36
Urnacephala californica, 72

Urolabida, 245
Urostylidae, 245
Urostylinae, 246
Usinger, Robert L., 12

Valleriola, 135, 143, 145
valvifers: first, 49; second, 49
valvulae: first, 49; second, 49; third, 49
Van Duzee, Edward Payson, 12
Vannius, 172
vascular system, feeding in, 20
vectors, 159
Velia, 101, 102; *caprai*, 100
Veliidae, 25, 37, 45, 49, 98, 99
Veliinae, 101
Veliohebria, 101
Veliohebriini, 99
Veliometra, 95–97; *schuhi*, 96
Velocipeda, 161
Velocipedidae, 161
Veloidea, 101
velvet water bugs. *See* Hebridae
velvety shore bugs. *See* Ochteridae
Ventidius, 105
Ventocoris fischeri, 36
ventral abdominal glands, 58
ventral arolium, 48, 91, 93, 96, 98, 100, 101, 109, 134, 138
ventral glands, 58, 151, 159
vermiform gland, 50, 146, 155, 159, 184–187, 190, 191
Vernonia, 184
vescia, 49, 159
Vesciinae, 40, 159
Veseris, 158
vesica, 50
Vianaida, 183
Vianaidinae, 182

vicariance biogeographic model, 38
Vilga, 278
Villiers, André, 13
Viola, 224
Visayanocorini, 158
viscid setae, 49, 158–160
viviparity, 204
Volesus, 158

Wagner, Eduard, 13
Walambianisops, 128
wasp mimics, 273
water boatmen. *See* Corixidae
water bugs. *See* Nepomorpha
water measurers. *See* Hydrometridae
water scorpions. *See* Nepidae
water striders. *See* Gerridae
water surface, as habitat, 21, 84
water treaders. *See* Mesoveliidae
Waterhouseana, 178
waxy secretions, 184
web lovers. *See* Plokiophilidae
webs: living in, 21; of spiders, as habitats, 188
Wechina, 167
western equine encephalitis, 33
wheat, 34–36
wherrymen. *See* Gerridae
Whiteiella, 190; *rostralis*, 191
wing coupling, 44, 69
wing dimorphism, sexual, 23
wing folding, 227, 228, 237
wing loss, 213
wing muscle histolysis, 270
wing polymorphism, 70; environmental determination, 24; genetic determination, 24, 25; hormomal control, 24; polygenic determination, 25

wing reduction, 23
wings, 43, 44; folding, 236, 237; reduced, 277
Wollastoniella ferruginea, 196; *obesula*, 196; *punctata*, 198
Wollastoniola, 197
Woodward, Thomas Emmanuel, 13
Wygodzinsky, Petr (Pedro) Wolfgang, 13
Wygodzinskyella, 156

Xenobates, 99
Xenoblissus lutzi, 253
Xenocylapus, 179
Xenogenus, 283
Xiphovelia, 99, 102
Xiphoveloidea, 99; *major*, 100
Xosa, 216
Xylastodorinae, 46, 48, 167
Xylastodoris, 167, 169; *luteolus*, 167
Xylocorini, 199
Xylocoris, 199; *afer*, 196; *cacti*, 196; *galactinus*, 198
xyphus, 43

Ypsotingini, 182
Yucca, 179

Zelurus, 158
Zelus, 28, 157
Zetekella minuscula, 48
Zetterstedt, J. W., 13
Zoological Institute, St. Petersburg, 16
Zoological Museum, Copenhagen, 16
Zoological Museum, University of Hamburg, 16
Zootermopsis nevadensis, 213

About the Authors

Randall T. Schuh was born in Corvallis, Oregon, on May 11, 1943. He received his B.S. degree in business administration from Oregon State University, Corvallis, his M.S. degree in entomology from Michigan State University, East Lansing, and his Ph.D. degree in entomology from the University of Connecticut, Storrs, under the direction of James A. Slater. His dissertation dealt with the systematics and phylogeny of Miridae from southern Africa. He is currently George Willett Curator of Entomology at the American Museum of Natural History, New York, where he is responsible for one of the world's largest collections of true bugs; Adjunct Professor, Department of Entomology, Cornell University, Ithaca, New York; and Adjunct Professor, Department of Biology, City College, City University of New York. His research is focused on the Miridae and Leptopodomorpha, as well as on the phylogenetic relationships and historical biogeography of the Heteroptera. He has collected extensively in North America, South Africa, South America, and Malaya.

James A. Slater was born in Belvidere, Illinois, on January 10, 1912. He received his B.A. and M.S. degrees in entomology from the University of Illinois, Urbana, maintaining as well a strong interest in snakes and other squamates. Under the direction of H. H. Knight, he earned his Ph.D. degree in entomology from Iowa State College, Ames, where he worked on the structure and taxonomic value of female genitalia in the Miridae. As Professor of Biology, University of Connecticut, Storrs (1953–1988), and Research Associate, Department of Entomology, American Museum of Natural History, New York, he conducts research on the systematics of the Heteroptera and has advised numerous M.S. and Ph.D. candidates. His primary interest has long been the Lygaeidae. He has collected in South Africa, Australia, and the Caribbean, and has published extensively on those faunas.

WITHDRAWN